CHEVY
MALIBU·CHEVELLE·MONTE CARLO
1970-1980
SHOP MANUAL

By
JIM COMBS

ERIC JORGENSEN
Editor

JEFF ROBINSON
Publisher

CLYMER PUBLICATIONS

World's largest publisher of books
devoted exclusively to automobiles and motorcycles

12860 MUSCATINE STREET · P.O. BOX 20 · ARLETA, CALIFORNIA 91331

Copyright © 1975 Clymer Publications

*All rights reserved. No part of this publication may
be reproduced, stored in a retrieval system, or transmitted,
in any form or by any means, electronic, mechanical,
photocopying, recording or otherwise,
without the written permission of Clymer Publications.*

FIRST EDITION
First Printing August, 1980
Second Printing May, 1981

Printed in U.S.A.

ISBN: 0-89287-319-1

Production Coordinator, Christine Robertson

•

COVER:

Photographed by Mike Brown/Visual Imagery, Los Angeles, Calif.

Assisted by Chris Ryfle.

Automobile courtesy of Chevrolet Division, General Motors Corp.

Chapter One
General Information

Chapter Two
Troubleshooting

Chapter Three
Lubrication, Maintenance and Tune-up

Chapter Four
Engine

Chapter Five
Fuel and Exhaust Systems

Chapter Six
Emission Control Systems

Chapter Seven
Cooling, Heating and Air Conditioning

Chapter Eight
Electrical System

Chapter Nine
Clutch and Transmissions

Chapter Ten
Brakes

Chapter Eleven
Front Suspension and Steering

Chapter Twelve
Rear Suspension and Drive Shaft

Index

CONTENTS

QUICK REFERENCE DATA IX

CHAPTER ONE
GENERAL INFORMATION 1

    Manual organization
    Service hints
    Safety first
    Expendable supplies
    Shop tools
    Emergency tool kit
    Troubleshooting and tune-up equipment

CHAPTER TWO
TROUBLESHOOTING 9

    Starting system
    Charging system
    Engine performance
    Engine oil pressure light
    Fuel system (carburetted)
    Fuel system (fuel injected)
    Fuel pump test (mechanical and electric)
    Emission control systems
    Engine noises
    Electrical accessories
    Cooling system
    Clutch
    Manual transmission/transaxle
    Automatic transmission
    Brakes
    Steering and suspension
    Tire wear analysis
    Wheel balancing

CHAPTER THREE
LUBRICATION, MAINTENANCE, AND TUNE-UP 33

    Gas stop checks
    Periodic maintenance
    Lubrication and lubricants
    Engine
    Chassis
    Engine tune-up
    Spark plug removal
    Tune-up specifications

CHAPTER FOUR
ENGINE 78

    Engine removal
    Engine installation
    General overhaul sequence
    General reassembly sequence
    Manifold assembly (inline engines)
    Intake manifold (V6 and V8 engines)
    Exhaust manifold (V6 and V8 engines)
    Rocker arm cover (all models)
    Valve mechanism
    Hydraulic valve lifters
    Valve stem oil seal and/or valve spring
    Cylinder head
    Oil pan replacement
    Oil pump
    Rear main oil seal
    Main bearings
    Connecting rod bearing
    Connecting rod and piston
    Camshaft
    Crankshaft
    Cylinder block
    Engine specifications

CHAPTER FIVE
FUEL AND EXHAUST SYSTEMS 142

    Fuel system
    Rochester 1MV-1ME
    Rochester 2GV-2GC-2GE
    Rochester M2M
    Rochester 4MV
    Rochester M4MC-M4MCA
    Holley 4150 carburetor
    Fuel pump
    Fuel tank
    Exhaust systems
    Muffler and tailpipe
    Exhaust pipe
    Catalytic converter

CHAPTER SIX
EMISSION CONTROL SYSTEMS 189

    Positive crankcase ventilation
    Controlled combustion system
    Air injection reactor system
    Combined emission control system
    Fuel evaporation control system
    Exhaust gas recirculation
    Carburetor calibration
    Distributor calibration
    Catalytic converter
    Early fuel evaporation system

CHAPTER SEVEN
COOLING, HEATING, AND AIR CONDITIONING 206

    Components
    Maintenance
    Water pump
    Radiator
    Heater
    Air conditioning
    System operation
    Get to know your vehicle's system
    Routine maintenance
    Refrigerant
    Troubleshooting
    Discharging the system

CHAPTER EIGHT
ELECTRICAL SYSTEM 224

    Battery maintenance
    Battery testing
    Charging system
    10-SI Delcotron troubleshooting
    1D Delcotron troubleshooting
    Delcotron removal/installation
    Alternator service
    Starter motor
    Body electrical system
    Fuses/circuit breakers
    Lighting switch
    Dimmer switch
    Stoplight switch
    Back-up switch
    Neutral safety switch
    Distributor
    High energy ignition

CHAPTER NINE
CLUTCH AND TRANSMISSION 257

    Clutch
    Transmission replacement
    Shift linkage adjustment
    Saginaw 3-speed manual transmission
    Saginaw 4-speed manual transmission

CHAPTER TEN
BRAKES 286

    Brake inspection   Wheel cylinder service
    Bleeding   Pad replacement
    Brake linings   Caliper
    Service brake adjustment   Power brake unit
    Parking brake adjustment   Master cylinder

CHAPTER ELEVEN
FRONT SUSPENSION AND STEERING 306

    Wheel alignment   Front wheel hub
    Shock absorbers   Wheel bearings
    Front springs   Steering adjustments
    Ball-joints   Steering gear
    Stabilizer bar   Steering linkage
    Steering knuckle   Power steering pump

CHAPTER TWELVE
REAR SUSPENSION AND DRIVE SHAFT 324

    Rear suspension   Rear axle

INDEX 333

WIRING DIAGRAMS .... END OF BOOK

QUICK REFERENCE DATA

TIMING MARKS

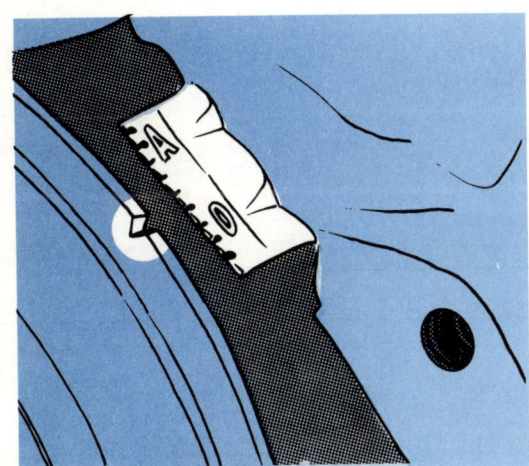

L6 ENGINES

V8 ENGINES

TUNE-UP SPECIFICATIONS*

Cylinder Head Bolt Torque
  L6 — 95 ft.-lb.
  Small block V8 and 200-229 V6 — 65 ft.-lb.
  Mark IV V8 and 231 V6 — 80 ft.-lb.

Valve Clearance
  Hydraulic lifters—adjustment is automatic
  Mechanical lifters—see chart on page XII

Spark Plugs
  Type — See Chapter Three—Tune-Up Up
  Gap
    1967-1974 (all) — 0.035 in.
    1975 (all) — 0.060 in.
    1976-on (except 231 V6) — 0.045 in.
    1978-on 231 V6 — 0.060 in.

Point Gap
  1967-1974 (all) — 0.016 in. (used); 0.019 in. (new)

*Refer to Vehicle Emission Control Information sticker in engine compartment. If sticker is missing, use these specifiations.

(continued)

TUNE-UP SPECIFICATIONS (continued)

Dwell Angle (continued)
L6—1970-1974 (all)   31-34 degrees
V8—1970-1971 (all)   28-32 degrees
V8—1972-1974 (all)   29-31 degrees

Ignition Timing   See Chapter Three, Table 4

Idle Speed   See Chapter Three, Table 4

Compression
L6   130 psi
V8-V6   150-160 psi

Firing order
L6   1-5-3-6-2-4 (No. 1 cylinder — front)
V8   1-8-4-3-6-5-7-2 (No. 1 cylinder — left front)
V6   1-6-5-4-3-2

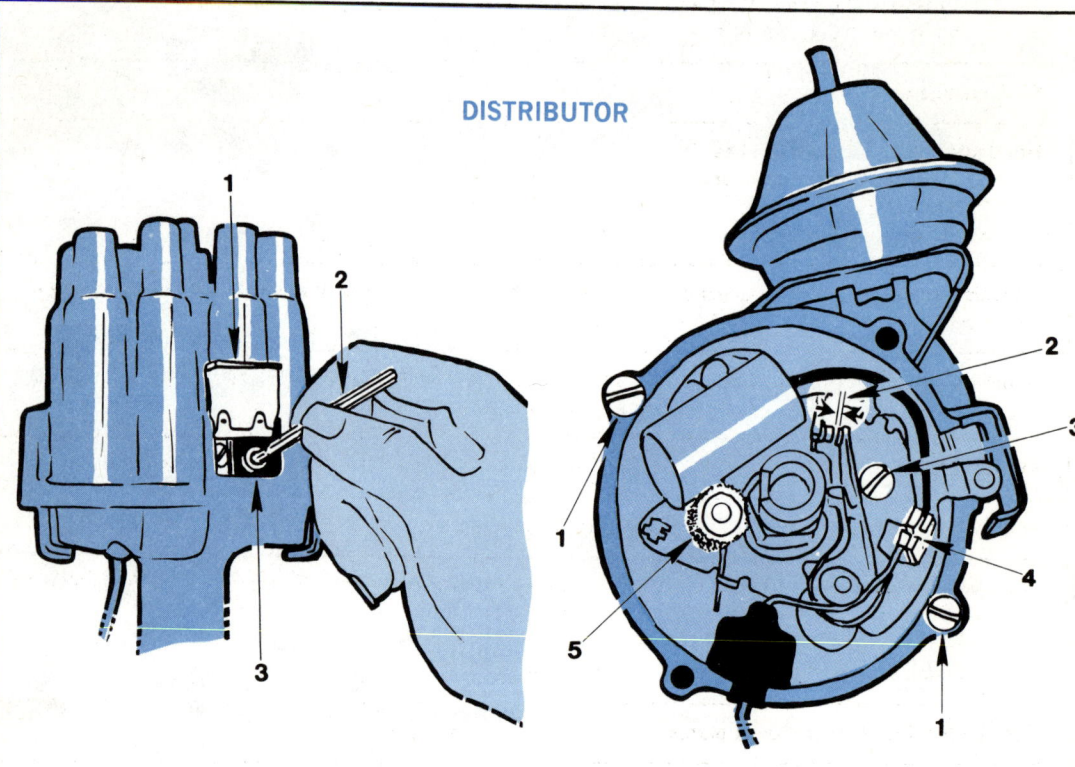

DISTRIBUTOR

1967-1974 V8 ENGINES
1. Window
2. Hexagon wrench
3. Adjusting screw

1967-1974 L6 ENGINES
1. Breaker plate attaching screws
2. Point gap
3. Contact set attaching screws
4. Quick disconnect terminal
5. Cam lubricator

RECOMMENDED LUBRICANTS AND FLUIDS

Use	Recommendation
Engine oil	Product bearing letter designation SE
Power steering	GM power steering fluid or, if not available, DEXRON® or DEXRON® II automatic transmission fluid
Differential—standard	SAE 80 or SAE 90 GL-5 gear lubricant (SAE 80 GL-5 in Canada)
Differential—Positraction	GM part No. 1051022 (lubricant) or equivalent
Steering gear—manual	GM part No. 1051052 (lubricant) or equivalent
Transmission—manual	SAE 80 or SAE 90 GL-5 gear lubricant (SAE 80 GL-5 in Canada)
Brake master cylinder	DOT-3 or Delco Supreme 11 brake fluid
Clutch linkage (manual transmission) 1. Pivot points 2. Cross-shaft grease fitting and push rod to clutch fork joint	Engine oil EP chassis lubricant meeting requirements of GM specification GM 6031-M
Shift linkage (all)	Engine oil
Hood latch and hinge assemblies 1. Pivots and spring anchor 2. Release pawl 3. Hinges	Engine oil Chassis lubricant Engine oil
Chassis lubrication (front suspension, steering linkage, etc.)	Chassis lubricant meeting requirements of GM specification GM 6031-M
Constant velocity universal joint	Lubricant, GM part No. 1050679, or meeting requirements of GM specification GM 6040-M
Automatic transmission	DEXRON® or DEXRON® II automatic transmission fluid
Parking brake cables	Chassis lubricant
Front wheel bearings	Chassis lubricant meeting the requirements of GM specification GM 6031-M
Door, tailgate, seat and trunk hinges	Engine oil
Convertible door-to-lock wedge plates	Stick-type lubricant
Windshield washer	GM Optikleen washer solvent, part No. 1050001 or equivalent
Battery	Clean, odorless drinking water
Engine cooling system	Mixture of water and high quality Ethylene Glycol base antifreeze, meeting requirements of GM specification 1899-M

LAMP USAGE (MONTE CARLO — MALIBU CLASSIC)

Circuit	1970	1971	1972	1973	1974	1975
Headlamp	6012A	6014	6014	6014	6014	6014
Front park/turn	1157NA	1157NA	1157NA	1157NA	1157NA	1157NA
Tail stop/turn	1157	1157	1157	1157	1157	1157
Turn signal ind.	194	194	194	194	168	168
High beam ind.	194	194	194	194	168	168
Instrument lamps	194	194	194	168	168	168
Courtesy lamp	211	211	211	211	211	211
License plate lamp	67	67	194	67	168	168
Radio dial lamp	293/1893	1816	1816	1816	1816	1816
Brake alarm	194	194	194	168	168	168
Back-up lamp	1156	1156	1156	1156	1156	1156
Glove box lamp	1893	1893	1893	1893	1893	1891
Heater cont. panel	1445	1445	1445	1445	1445	1445
A.C. control panel	1445	1445	1445	1445	1445	1445

Circuit	1976	1977	1978	1979	1980
Headlamp, upper[1]	4652	4652	4652	4652	N.A.
Headlamp, lower[2]	4651	4651	4651	4651	
Front park/turn	1157NA	1156NA	1156NA	1156NA	
Tail stop/turn	1157	1157	1157	1157	
Turn signal ind.	168	168	168	168	
High beam ind.	168	168	168	168	
Instrument lamps	168	168	168	168	
Courtesy lamp (dome)	211-2	211-2	211-2	211-2	
License plate lamp	168	194	194	168	
Radio dial lamp	1816	216/1893	216/1893	216/1893	
Brake alarm	168	168	168	168	
Back-up lamp	1156	1156	1156	1156	
Glove box lamp	1893	1893	1893	1893	
Heater cont. panel	1445	194	194	194	
A.C. control panel	1445	194	194	194	

1. Or outboard. 2. Or inboard.

VALVE LASH SPECIFICATIONS

Year	Engine	Intake (Hot)	Exhaust (Hot)
1970	350 cu. in., 370 hp	0.030 in.	0.030 in.
1970	402 cu. in., 375 hp	0.024 in.	0.028 in.
1970	454 cu. in., 450 hp	0.024 in.	0.028 in.

CHEVY
MALIBU·CHEVELLE·MONTE CARLO
1970-1980
SHOP MANUAL

INTRODUCTION

This detailed, comprehensive manual covers all 1970-1980 Chevy Malibu, Chevelle and Monte Carlo models. The expert text gives complete information on maintenance, tune-up, repair, and overhaul. Hundreds of photos and drawings guide you through every step. The book includes all you need to know to keep your car running right.

Where repairs are practical for the owner/mechanic, complete procedures are given. Equally important, difficult jobs are pointed out. Such operations are usually more economically performed by a dealer or independent garage.

A shop manual is a reference. You want to be able to find information fast. As in all Clymer books, this one is designed with this in mind. All chapters are thumb tabbed. Important items are extensively indexed at the rear of the book. Finally, all the most frequently used specifications and capacities are summarized on the *Quick Reference* pages at the front of the book.

Keep the book handy. Carry it in your glove box. It will help you to better understand your car, lower repair and maintenance costs, and generally improve your satisfaction with your vehicle.

CHAPTER ONE

GENERAL INFORMATION

The troubleshooting, tune-up, maintenance, and step-by-step repair procedures in this book are written for the owner and home mechanic. The text is accompanied by useful photos and diagrams to make the job as clear and correct as possible.

Troubleshooting, tune-up, maintenance, and repair are not difficult if you know what tools and equipment to use and what to do. Anyone not afraid to get their hands dirty, of average intelligence, and with some mechanical ability can perform most of the procedures in this book.

In some cases, a repair job may require tools or skills not reasonably expected of the home mechanic. These procedures are noted in each chapter and it is recommended that you take the job to your dealer, a competent mechanic, or machine shop.

MANUAL ORGANIZATION

This chapter provides general information and safety and service hints. Also included are lists of recommended shop and emergency tools as well as a brief description of troubleshooting and tune-up equipment.

Chapter Two provides methods and suggestions for quick and accurate diagnosis and repair of problems. Troubleshooting procedures discuss typical symptoms and logical methods to pinpoint the trouble.

Chapter Three explains all periodic lubrication and routine maintenance necessary to keep your vehicle running well. Chapter Three also includes recommended tune-up procedures, eliminating the need to constantly consult chapters on the various subassemblies.

Subsequent chapters cover specific systems such as the engine, transmission, and electrical systems. Each of these chapters provides disassembly, repair, and assembly procedures in a simple step-by-step format. If a repair requires special skills or tools, or is otherwise impractical for the home mechanic, it is so indicated. In these cases it is usually faster and less expensive to have the repairs made by a dealer or competent repair shop. Necessary specifications concerning a particular system are included at the end of the appropriate chapter.

When special tools are required to perform a procedure included in this manual, the tool is illustrated either in actual use or alone. It may be possible to rent or borrow these tools. The inventive mechanic may also be able to find a suitable substitute in his tool box, or to fabricate one.

The terms NOTE, CAUTION, and WARNING have specific meanings in this manual. A NOTE provides additional or explanatory information. A CAUTION is used to emphasize areas where equipment damage could result if proper precautions are not taken. A WARNING is used to stress those areas where personal injury or death could result from negligence, in addition to possible mechanical damage.

SERVICE HINTS

Observing the following practices will save time, effort, and frustration, as well as prevent possible injury.

Throughout this manual keep in mind two conventions. "Front" refers to the front of the vehicle. The front of any component, such as the transaxle, is that end which faces toward the front of the vehicle. The "left" and "right" sides of the vehicle refer to the orientation of a person sitting in the vehicle facing forward. For example, the steering wheel is on the left side. These rules are simple, but even experienced mechanics occasionally become disoriented.

Most of the service procedures covered are straightforward and can be performed by anyone reasonably handy with tools. It is suggested, however, that you consider your own capabilities carefully before attempting any operation involving major disassembly of the engine.

Some operations, for example, require the use of a press. It would be wiser to have these performed by a shop equipped for such work, rather than to try to do the job yourself with makeshift equipment. Other procedures require precision measurements. Unless you have the skills and equipment required, it would be better to have a qualified repair shop make the measurements for you.

Repairs go much faster and easier if the parts that will be worked on are clean before you begin. There are special cleaners for washing the engine and related parts. Brush or spray on the cleaning solution, let it stand, then rinse it away with a garden hose. Clean all oily or greasy parts with cleaning solvent as you remove them.

WARNING
Never use gasoline as a cleaning agent. It presents an extreme fire hazard. Be sure to work in a well-ventilated area when using cleaning solvent. Keep a fire extinguisher, rated for gasoline fires, handy in any case.

Much of the labor charge for repairs made by dealers is for the removal and disassembly of other parts to reach the defective unit. It is frequently possible to perform the preliminary operations yourself and then take the defective unit in to the dealer for repair, at considerable savings.

Once you have decided to tackle the job yourself, make sure you locate the appropriate section in this manual, and read it entirely. Study the illustrations and text until you have a good idea of what is involved in completing the job satisfactorily. If special tools are required, make arrangements to get them before you start. Also, purchase any known defective parts prior to starting on the procedure. It is frustrating and time-consuming to get partially into a job and then be unable to complete it.

Simple wiring checks can be easily made at home, but knowledge of electronics is almost a necessity for performing tests with complicated electronic testing gear.

During disassembly of parts keep a few general cautions in mind. Force is rarely needed to get things apart. If parts are a tight fit, like a bearing in a case, there is usually a tool designed to separate them. Never use a screwdriver to pry apart parts with machined surfaces such as cylinder head and valve cover. You will mar the surfaces and end up with leaks.

Make diagrams wherever similar-appearing parts are found. You may think you can remember where everything came from — but mistakes are costly. There is also the possibility you may get sidetracked and not return to work for days or even weeks — in which interval, carefully laid out parts may have become disturbed.

Tag all similar internal parts for location, and mark all mating parts for position. Record number and thickness of any shims as they are removed. Small parts such as bolts can be iden-

GENERAL INFORMATION

tified by placing them in plastic sandwich bags that are sealed and labeled with masking tape.

Wiring should be tagged with masking tape and marked as each wire is removed. Again, do not rely on memory alone.

When working under the vehicle, do not trust a hydraulic or mechanical jack to hold the vehicle up by itself. Always use jackstands. See **Figure 1**.

Disconnect battery ground cable before working near electrical connections and before disconnecting wires. Never run the engine with the battery disconnected; the alternator could be seriously damaged.

Protect finished surfaces from physical damage or corrosion. Keep gasoline and brake fluid off painted surfaces.

Frozen or very tight bolts and screws can often be loosened by soaking with penetrating oil like Liquid Wrench or WD-40, then sharply striking the bolt head a few times with a hammer and punch (or screwdriver for screws). Avoid heat unless absolutely necessary, since it may melt, warp, or remove the temper from many parts.

Avoid flames or sparks when working near a charging battery or flammable liquids, such as brake fluid or gasoline.

No parts, except those assembled with a press fit, require unusual force during assembly. If a part is hard to remove or install, find out why before proceeding.

Cover all openings after removing parts to keep dirt, small tools, etc., from falling in.

When assembling two parts, start all fasteners, then tighten evenly.

The clutch plate, wiring connections, brake shoes, drums, pads, and discs should be kept clean and free of grease and oil.

When assembling parts, be sure all shims and washers are replaced exactly as they came out.

Whenever a rotating part butts against a stationary part, look for a shim or washer. Use new gaskets if there is any doubt about the condition of old ones. Generally, you should apply gasket cement to one mating surface only, so the parts may be easily disassembled in the future. A thin coat of oil on gaskets helps them seal effectively.

Heavy grease can be used to hold small parts in place if they tend to fall out during assembly. However, keep grease and oil away from electrical, clutch, and brake components.

High spots may be sanded off a piston with sandpaper, but emery cloth and oil do a much more professional job.

Carburetors are best cleaned by disassembling them and soaking the parts in a commercial carburetor cleaner. Never soak gaskets and rubber parts in these cleaners. Never use wire to clean out jets and air passages; they are easily damaged. Use compressed air to blow out the carburetor, but only if the float has been removed first.

Take your time and do the job right. Do not forget that a newly rebuilt engine must be broken in the same as a new one. Refer to your owner's manual for the proper break-in procedures.

SAFETY FIRST

Professional mechanics can work for years and never sustain a serious injury. If you observe a few rules of common sense and safety, you can enjoy many safe hours servicing your vehicle. You could hurt yourself or damage the vehicle if you ignore these rules.

1. Never use gasoline as a cleaning solvent.
2. Never smoke or use a torch in the vicinity of

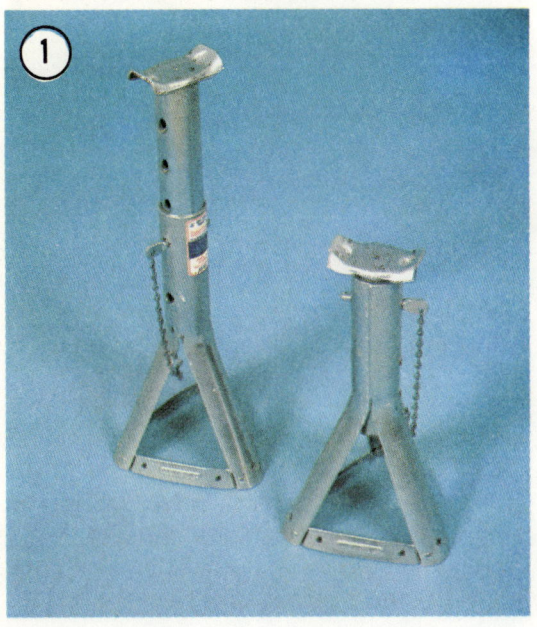

flammable liquids such as cleaning solvent in open containers.

3. Never smoke or use a torch in an area where batteries are being charged. Highly explosive hydrogen gas is formed during the charging process.

4. Use the proper sized wrenches to avoid damage to nuts and injury to yourself.

5. When loosening a tight or stuck nut, be guided by what would happen if the wrench should slip. Protect yourself accordingly.

6. Keep your work area clean and uncluttered.

7. Wear safety goggles during all operations involving drilling, grinding, or use of a cold chisel.

8. Never use worn tools.

9. Keep a fire extinguisher handy and be sure it is rated for gasoline (Class B) and electrical (Class C) fires.

EXPENDABLE SUPPLIES

Certain expendable supplies are necessary. These include grease, oil, gasket cement, wiping rags, cleaning solvent, and distilled water. Also, special locking compounds, silicone lubricants, and engine cleaners may be useful. Cleaning solvent is available at most service stations and distilled water for the battery is available at most supermarkets.

SHOP TOOLS

For proper servicing, you will need an assortment of ordinary hand tools (**Figure 2**).

As a minimum, these include:

a. Combination wrenches
b. Sockets
c. Plastic mallet
d. Small hammer
e. Snap ring pliers
f. Gas pliers
g. Phillips screwdrivers
h. Slot (common) screwdrivers
i. Feeler gauges
j. Spark plug gauge
k. Spark plug wrench
l. Torque wrench

Special tools necessary are shown in the chapters covering the particular repair in which they are used.

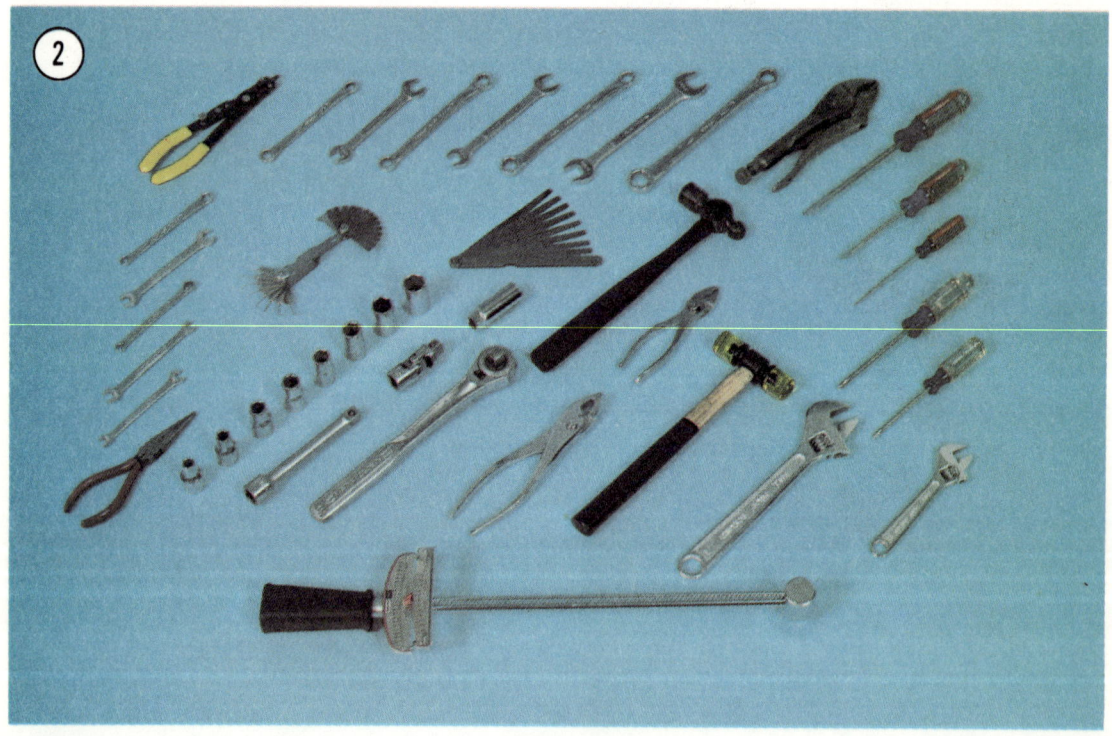

GENERAL INFORMATION

Engine tune-up and troubleshooting procedures require other special tools and equipment. These are described in detail in the following sections.

EMERGENCY TOOL KIT

A small emergency tool kit kept in the trunk is handy for road emergencies which otherwise could leave you stranded. The tools listed below and shown in **Figure 3** will let you handle most roadside repairs.

a. Combination wrenches
b. Crescent (adjustable) wrench
c. Screwdrivers — common and Phillips
d. Pliers — conventional (gas) and needle nose
e. Vise Grips
f. Hammer — plastic and metal
g. Small container of waterless hand cleaner
h. Rags for cleanup
i. Silver waterproof sealing tape (duct tape)
j. Flashlight
k. Emergency road flares — at least four
l. Spare drive belts (cooling fan, alternator, etc.)

TROUBLESHOOTING AND TUNE-UP EQUIPMENT

Voltmeter, Ohmmeter, and Ammeter

For testing the ignition or electrical system, a good voltmeter is required. For automotive use, an instrument covering 0-20 volts is satisfac-

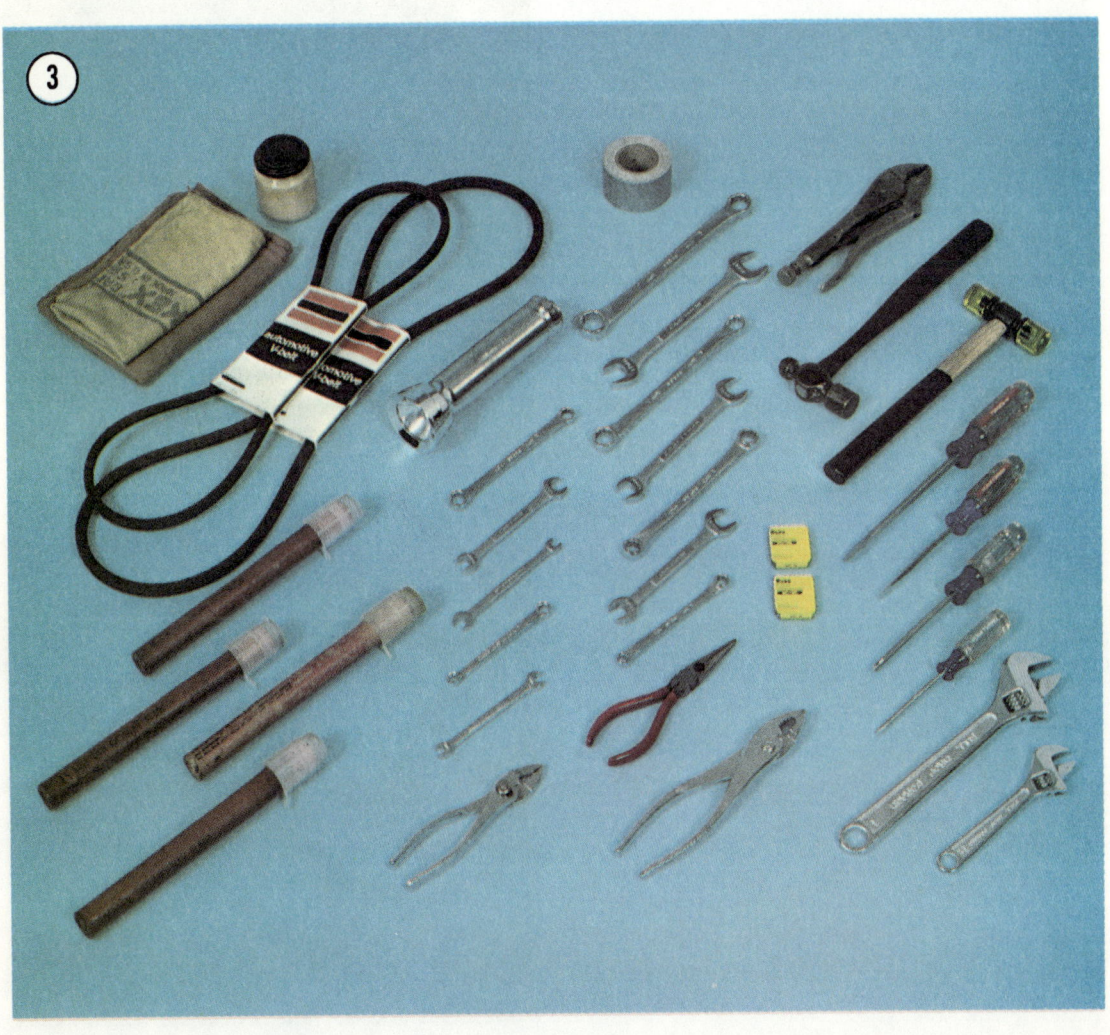

tory. One which also has a 0-2 volt scale is necessary for testing relays, points, or individual contacts where voltage drops are much smaller. Accuracy should be ± ½ volt.

An ohmmeter measures electrical resistance. This instrument is useful for checking continuity (open and short circuits), and testing fuses and lights.

The ammeter measures electrical current. Ammeters for automotive use should cover 0-50 amperes and 0-250 amperes. These are useful for checking battery charging and starting current.

Several inexpensive VOM's (volt-ohm-milliammeter) combine all three instruments into one which fits easily in any tool box. See **Figure 4**. However, the ammeter ranges are usually too small for automotive work.

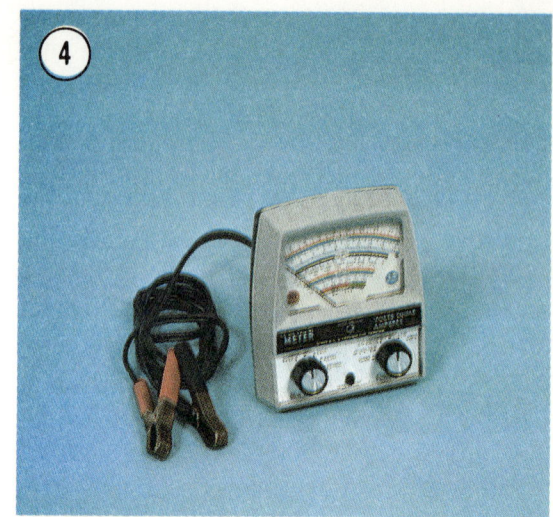

Hydrometer

The hydrometer gives a useful indication of battery condition and charge by measuring the specific gravity of the electrolyte in each cell. See **Figure 5**. Complete details on use and interpretation of readings are provided in the electrical chapter.

Compression Tester

The compression tester measures the compression pressure built up in each cylinder. The results, when properly interpreted, can indicate general cylinder and valve condition. See **Figure 6**.

Most compression testers have long flexible extensions built-in or as accessories. Such an extension is necessary since the spark plug holes are deep inside the metal air cooling covers.

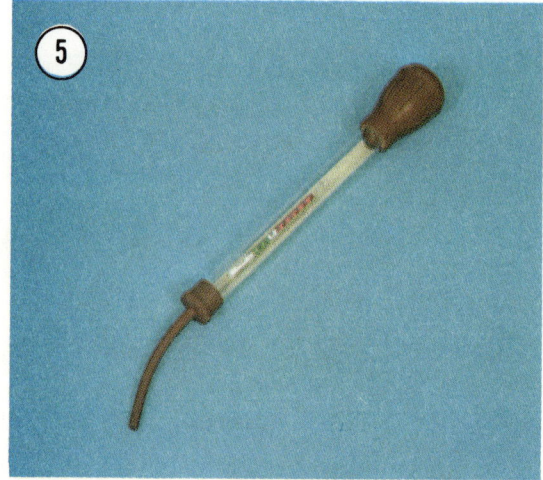

Vacuum Gauge

The vacuum gauge (**Figure 7**) is one of the easiest instruments to use, but one of the most difficult for the inexperienced mechanic to interpret. The results, when interpreted with other findings, can provide valuable clues to possible trouble.

To use the vacuum gauge, connect it to a vacuum hose that goes to the intake manifold. Attach it either directly to the hose or to a T-fitting installed into the hose.

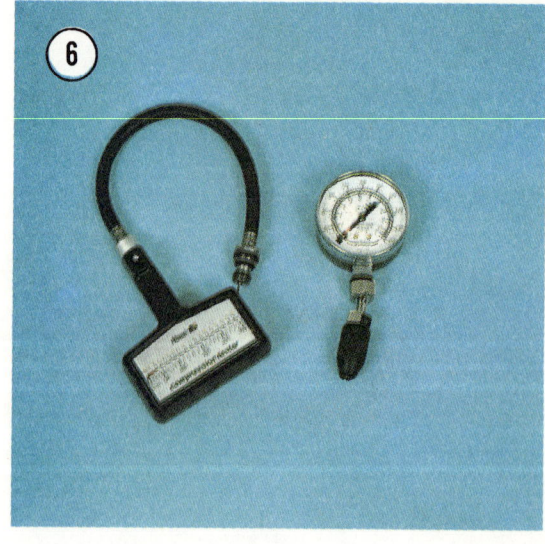

GENERAL INFORMATION

NOTE: *Subtract one inch from the reading for every 1,000 ft. elevation.*

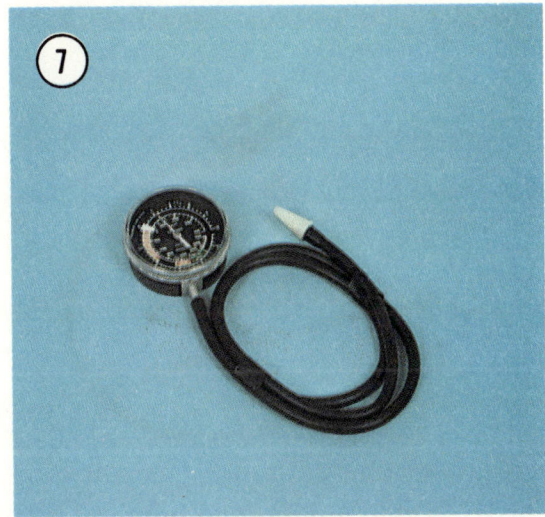

Fuel Pressure Gauge

This instrument is invaluable for evaluating fuel pump performance. Fuel system troubleshooting procedures in this manual use a fuel pressure gauge. Usually a vacuum gauge and fuel pressure gauge are combined.

Dwell Meter (Contact Breaker Point Ignition Only)

A dwell meter measures the distance in degrees of cam rotation that the breaker points remain closed while the engine is running. Since this angle is determined by breaker point gap, dwell angle is an accurate indication of breaker point gap.

Many tachometers intended for tuning and testing incorporate a dwell meter as well. See **Figure 8**. Follow the manufacturer's instructions to measure dwell.

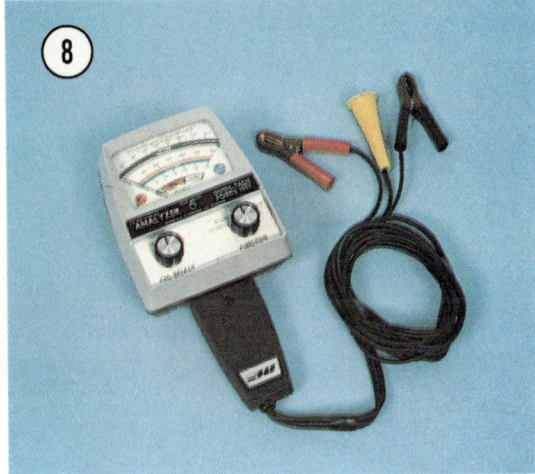

Tachometer

A tachometer is necessary for tuning. See **Figure 8**. Ignition timing and carburetor adjustments must be performed at the specified idle speed. The best instrument for this purpose is one with a low range of 0-1,000 or 0-2,000 rpm for setting idle, and a high range of 0-4,000 or more for setting ignition timing at 3,000 rpm. Extended range (0-6,000 or 0-8,000 rpm) instruments lack accuracy at lower speeds. The instrument should be capable of detecting changes of 25 rpm on the low range.

Strobe Timing Light

This instrument is necessary for tuning, as it permits very accurate ignition timing. The light flashes at precisely the same instant that No. 1 cylinder fires, at which time the timing marks on the engine should align. Refer to Chapter Three for exact location of the timing marks for your engine.

Suitable lights range from inexpensive neon bulb types ($2-3) to powerful xenon strobe lights ($20-40). See **Figure 9**. Neon timing lights are difficult to see and must be used in dimly lit areas. Xenon strobe timing lights can be used

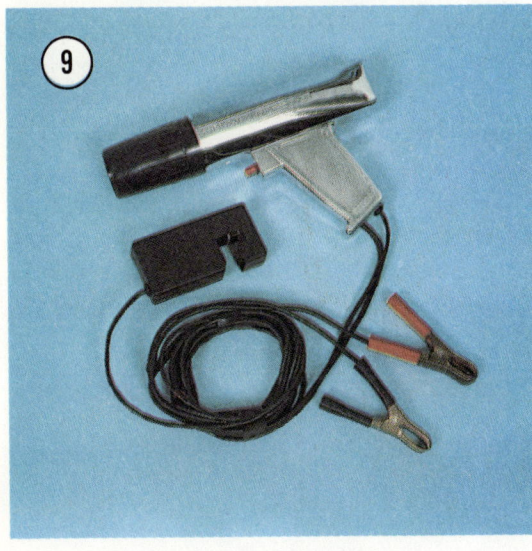

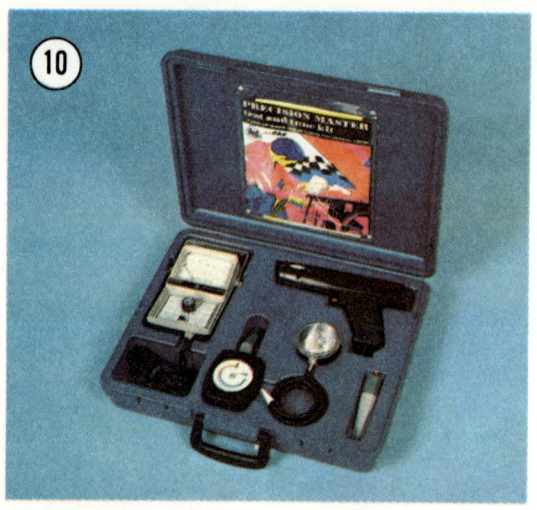

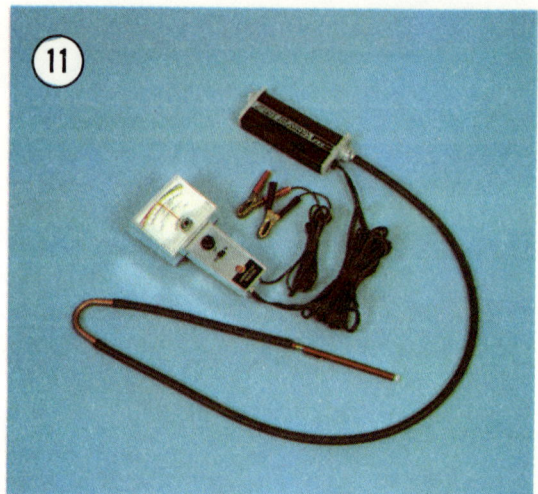

outside in bright sunlight. Both types work on this vehicle; use according to the manufacturer's instructions.

Tune-up Kits

Many manufacturers offer kits that combine several useful instruments. Some come in a convenient carry case and are usually less expensive than purchasing one instrument at a time. **Figure 10** shows one of the kits that is available. The prices vary with the number of instruments included in the kit.

Exhaust Gas Analyzer

Of all instruments described here, this is the least likely to be owned by a home mechanic. This instrument samples the exhaust gases from the tailpipe and measures the thermal conductivity of the exhaust gas. Since different gases conduct heat at varying rates, thermal conductivity of the exhaust is a good indication of gases present.

An exhaust gas analyzer is vital for accurately checking the effectiveness of exhaust emission control adjustments. They are relatively expensive to buy ($70 and up), but must be considered essential for the owner/mechanic

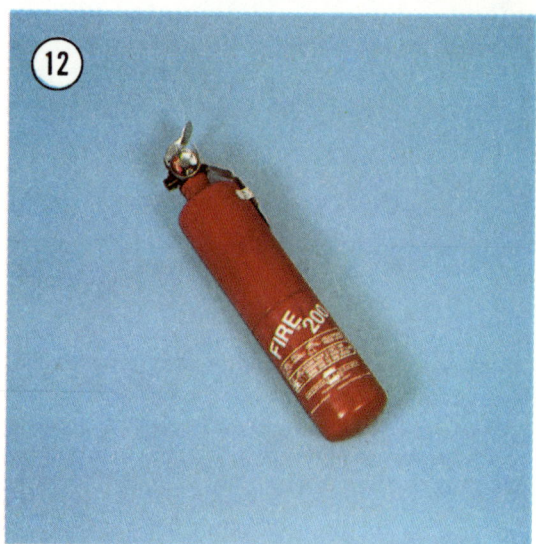

to comply with today's emission laws. See **Figure 11**.

Fire Extinguisher

A fire extinguisher is a necessity when working on a vehicle. It should be rated for both *Class B* (flammable liquids — gasoline, oil, paint, etc.) and *Class C* (electrical — wiring, etc.) type fires. It should always be kept within reach. See **Figure 12**.

CHAPTER TWO

TROUBLESHOOTING

Troubleshooting can be a relatively simple matter if it is done logically. The first step in any troubleshooting procedure must be defining the symptoms as closely as possible. Subsequent steps involve testing and analyzing areas which could cause the symptoms. A haphazard approach may eventually find the trouble, but in terms of wasted time and unnecessary parts replacement, it can be very costly.

The troubleshooting procedures in this chapter analyze typical symptoms and show logical methods of isolation. These are not the only methods. There may be several approaches to a problem, but all methods must have one thing in common — a logical, systematic approach.

STARTING SYSTEM

The starting system consists of the starter motor and the starter solenoid. The ignition key controls the starter solenoid, which mechanically engages the starter with the engine flywheel, and supplies electrical current to turn the starter motor.

Starting system troubles are relatively easy to find. In most cases, the trouble is a loose or dirty electrical connection. **Figures 1 and 2** provide routines for finding the trouble.

CHARGING SYSTEM

The charging system consists of the alternator (or generator on older vehicles), voltage regulator, and battery. A drive belt driven by the engine crankshaft turns the alternator which produces electrical energy to charge the battery. As engine speed varies, the voltage from the alternator varies. A voltage regulator controls the charging current to the battery and maintains the voltage to the vehicle's electrical system at safe levels. A warning light or gauge on the instrument panel signals the driver when charging is not taking place. Refer to **Figure 3** for a typical charging system.

Complete troubleshooting of the charging system requires test equipment and skills which the average home mechanic does not possess. However, there are a few tests which can be done to pinpoint most troubles.

Charging system trouble may stem from a defective alternator (or generator), voltage regulator, battery, or drive belt. It may also be caused by something as simple as incorrect drive belt tension. The following are symptoms of typical problems you may encounter.

1. *Battery dies frequently, even though the warning lamp indicates no discharge* — This can be caused by a drive belt that is slightly too

CHAPTER TWO

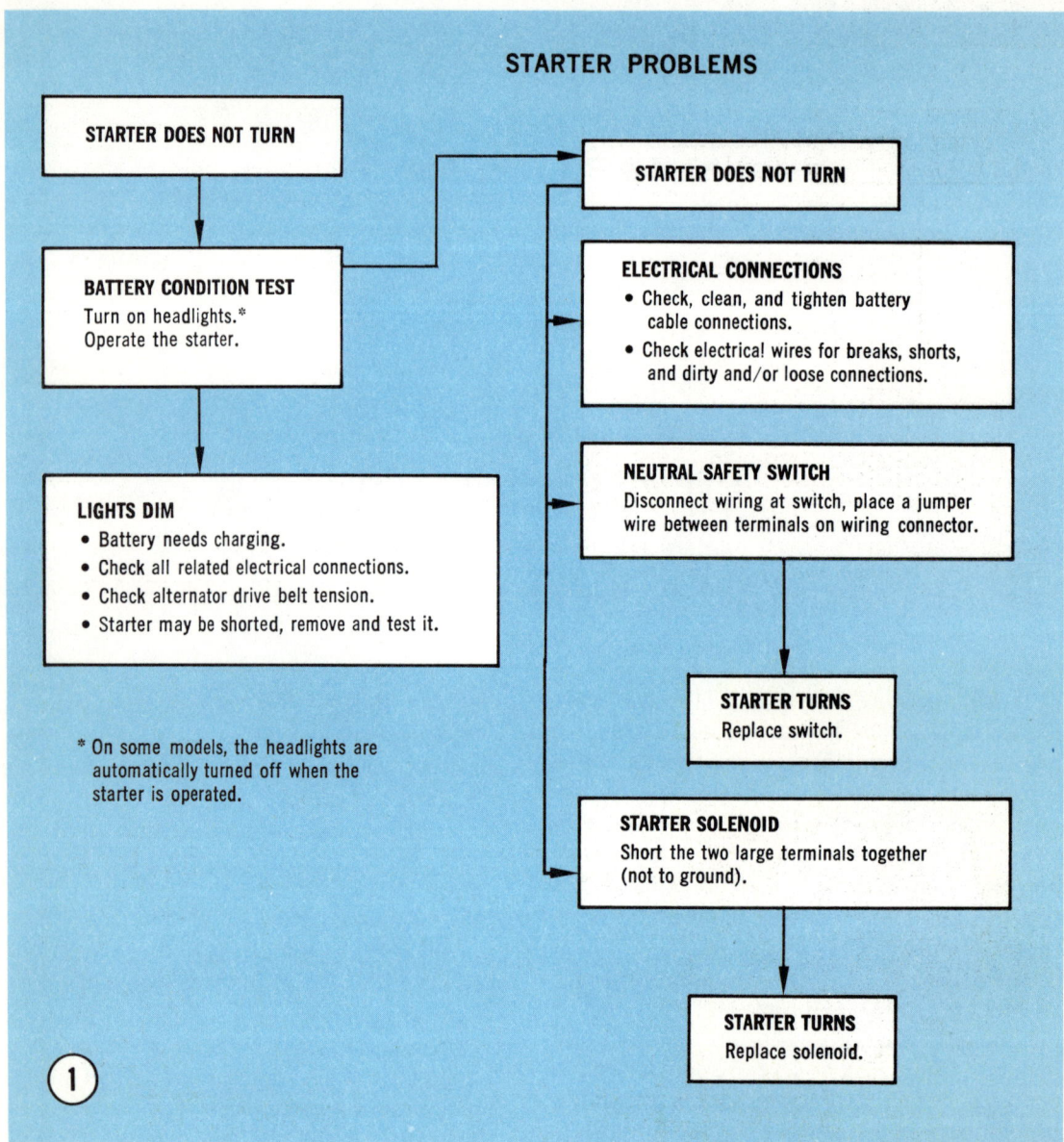

1.

loose. Grasp the alternator (or generator) pulley and try to turn it. If the pulley can be turned without moving the belt, the drive belt is too loose. As a rule, keep the belt tight enough that it can be deflected about ½ in. under moderate thumb pressure between the pulleys (**Figure 4**). The battery may also be at fault; test the battery condition.

2. *Charging system warning lamp does not come on when ignition switch is turned on* — This may indicate a defective ignition switch, battery, voltage regulator, or lamp. First try to start the vehicle. If it doesn't start, check the ignition switch and battery. If the car starts, remove the warning lamp; test it for continuity with an ohmmeter or substitute a new lamp. If the lamp is good, locate the voltage regulator and make sure it is properly grounded (try tightening the mounting screws). Also, the alternator (or generator) brushes may not be making contact. Test the alternator (or generator) and voltage regulator.

3. *Alternator (or generator) warning lamp comes on and stays on* — This usually indicates

TROUBLESHOOTING

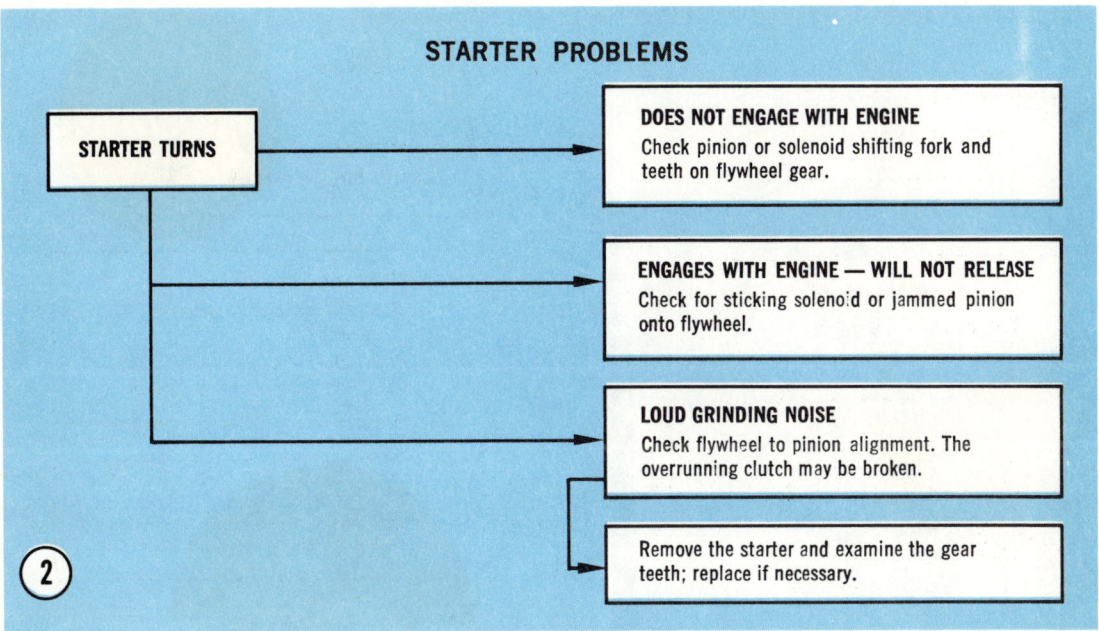

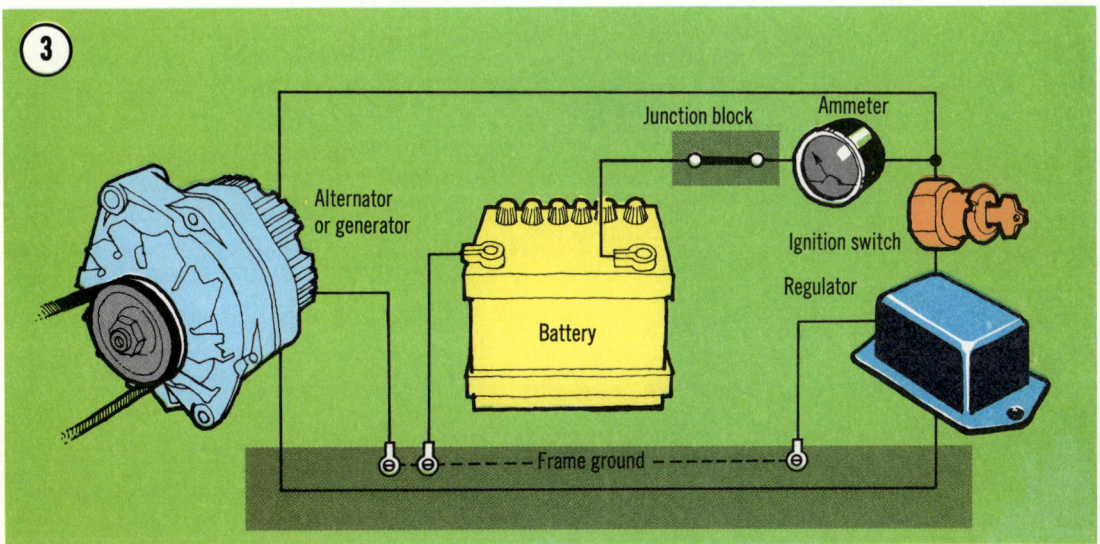

that no charging is taking place. First check drive belt tension (**Figure 4**). Then check battery condition, and check all wiring connections in the charging system. If this does not locate the trouble, check the alternator (or generator) and voltage regulator.

4. *Charging system warning lamp flashes on and off intermittently* — This usually indicates the charging system is working intermittently. Check the drive belt tension (**Figure 4**), and check all electrical connections in the charging system. Check the alternator (or generator). *On generators only*, check the condition of the commutator.

5. *Battery requires frequent additions of water, or lamps require frequent replacement* — The alternator (or generator) is probably overcharging the battery. The voltage regulator is probably at fault.

6. *Excessive noise from the alternator (or generator)* — Check for loose mounting brackets and bolts. The problem may also be

CHAPTER TWO

worn bearings or the need of lubrication in some cases. If an alternator whines, a shorted diode may be indicated.

IGNITION SYSTEM

The ignition system may be either a conventional contact breaker type or an electronic ignition. See electrical chapter to determine which type you have. **Figures 5 and 6** show simplified diagrams of each type.

Most problems involving failure to start, poor performance, or rough running stem from trouble in the ignition system, particularly in contact breaker systems. Many novice troubleshooters get into trouble when they assume that these symptoms point to the fuel system instead of the ignition system.

Ignition system troubles may be roughly divided between those affecting only one cylinder and those affecting all cylinders. If the trouble affects only one cylinder, it can only be in the spark plug, spark plug wire, or portion of the distributor associated with that cylinder. If the trouble affects all cylinders (weak spark or no spark), then the trouble is in the ignition coil, rotor, distributor, or associated wiring.

In order to get maximum spark, the ignition coil must be wired correctly. Make sure that the double wire from the battery is attached to terminal No. 15 on the ignition coil and that the single wire from the distributor is attached to terminal No. 1 on the ignition coil.

The troubleshooting procedures outlined in **Figure 7** (breaker point ignition) or **Figure 8** (electronic ignition) will help you isolate ignition problems fast. Of course, they assume that the battery is in good enough condition to crank the engine over at its normal rate.

ENGINE PERFORMANCE

A number of factors can make the engine difficult or impossible to start, or cause rough running, poor performance and so on. The majority of novice troubleshooters immediately suspect the carburetor or fuel injection system. In the majority of cases, though, the trouble exists in the ignition system.

The troubleshooting procedures outlined in **Figures 9 through 14** will help you solve the majority of engine starting troubles in a systematic manner.

Some tests of the ignition system require running the engine with a spark plug or ignition coil wire disconnected. The safest way to do this is to disconnect the wire with the engine

TROUBLESHOOTING

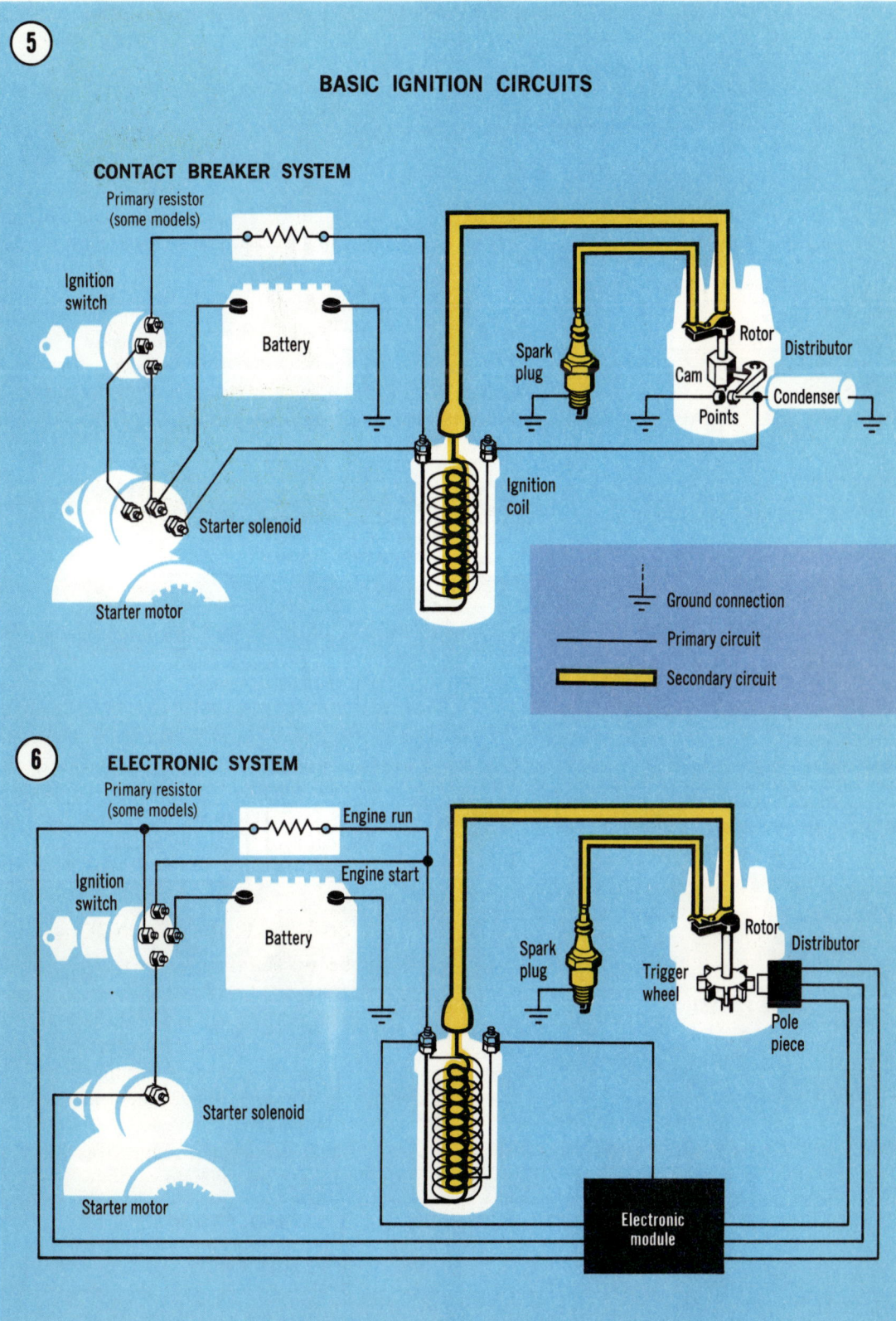

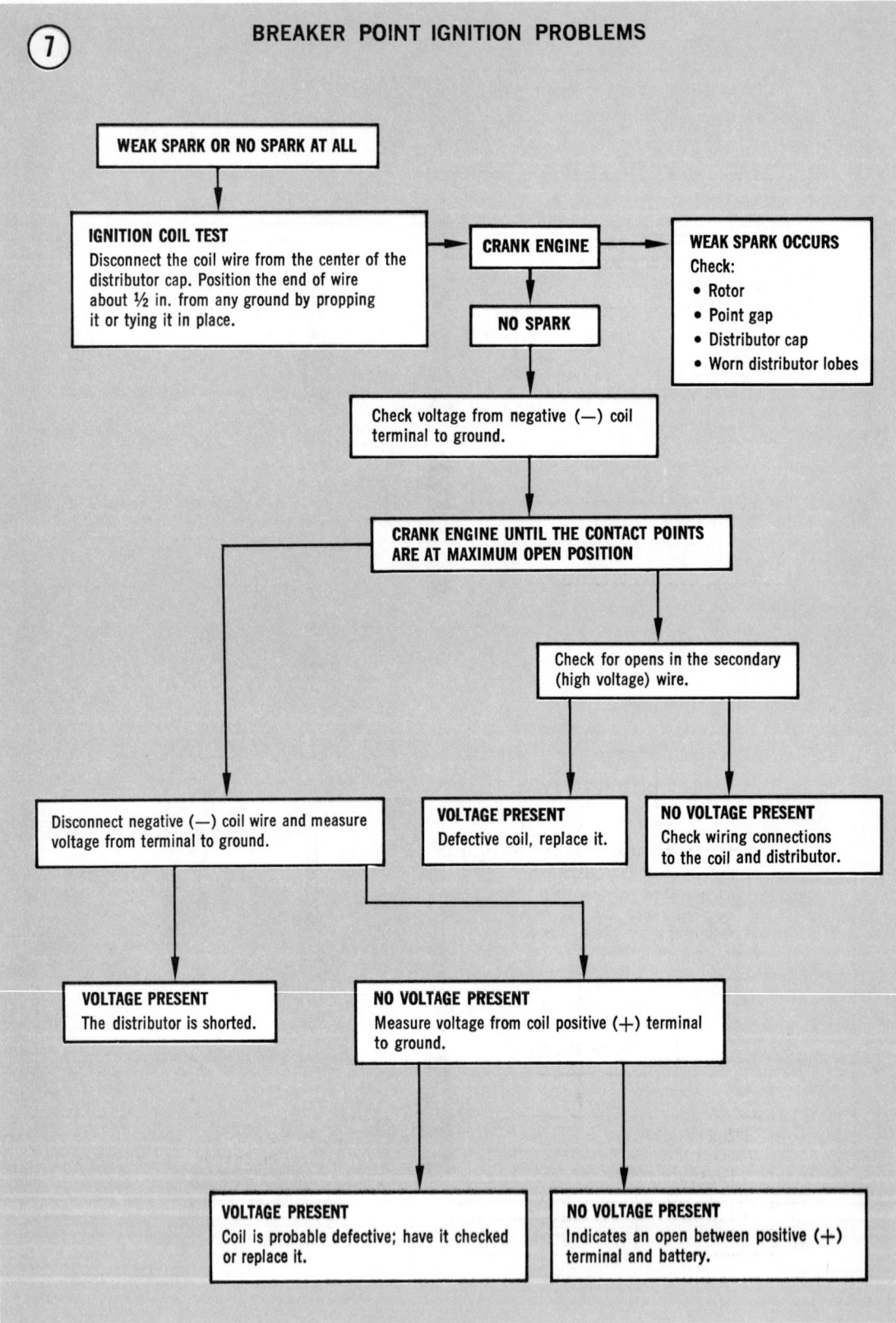

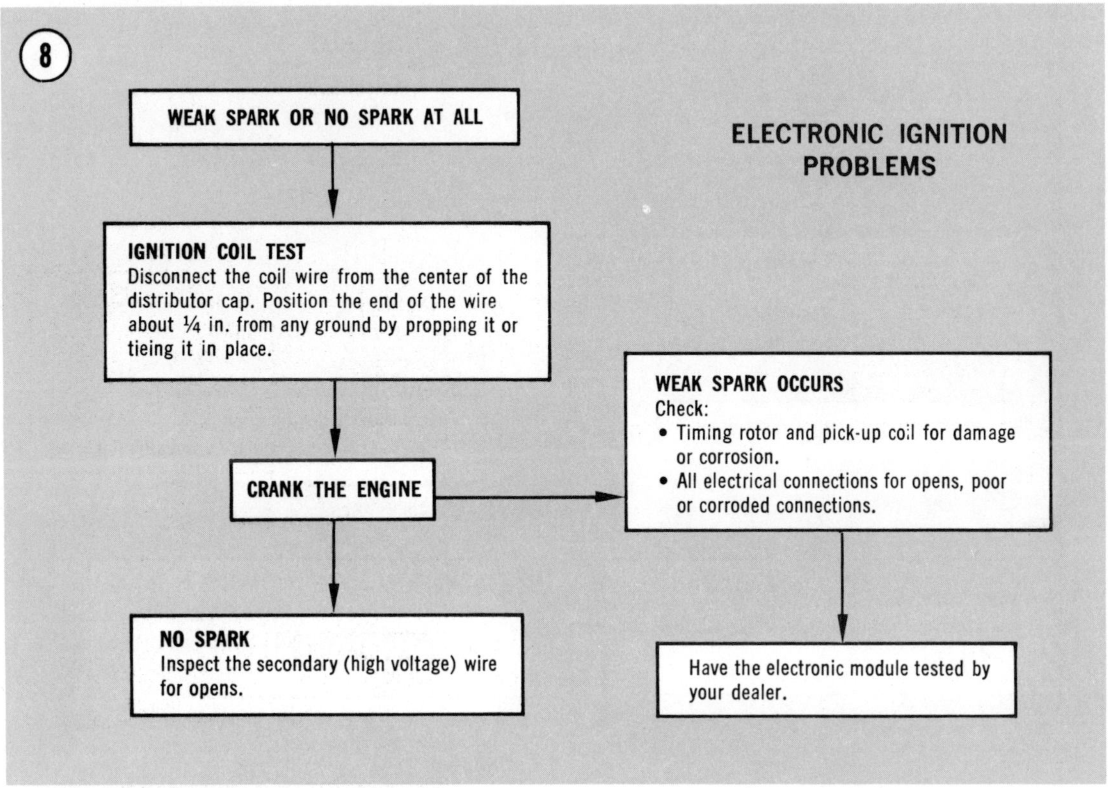

TROUBLESHOOTING

(8)

WEAK SPARK OR NO SPARK AT ALL
↓
IGNITION COIL TEST
Disconnect the coil wire from the center of the distributor cap. Position the end of the wire about ¼ in. from any ground by propping it or tieing it in place.
↓
CRANK THE ENGINE →
↓
NO SPARK
Inspect the secondary (high voltage) wire for opens.

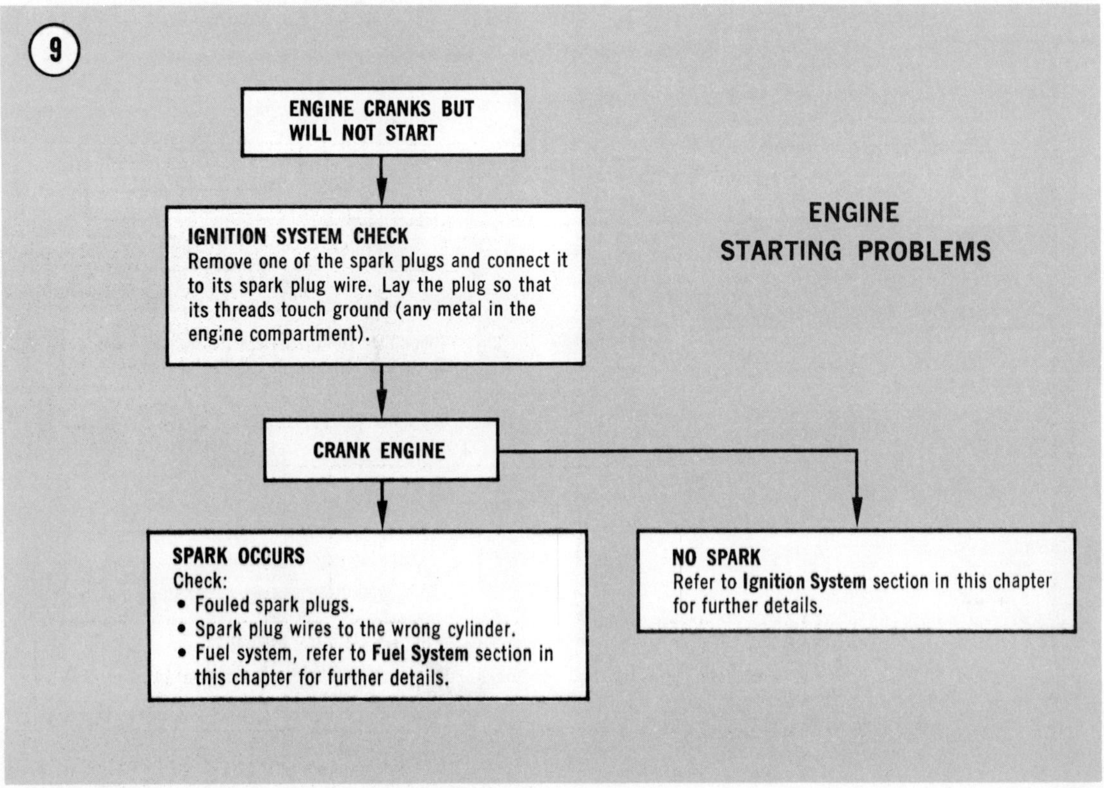

WEAK SPARK OCCURS
Check:
- Timing rotor and pick-up coil for damage or corrosion.
- All electrical connections for opens, poor or corroded connections.
↓
Have the electronic module tested by your dealer.

ELECTRONIC IGNITION PROBLEMS

---

(9)

ENGINE CRANKS BUT WILL NOT START
↓
IGNITION SYSTEM CHECK
Remove one of the spark plugs and connect it to its spark plug wire. Lay the plug so that its threads touch ground (any metal in the engine compartment).
↓
CRANK ENGINE →
↓
SPARK OCCURS
Check:
- Fouled spark plugs.
- Spark plug wires to the wrong cylinder.
- Fuel system, refer to **Fuel System** section in this chapter for further details.

NO SPARK
Refer to **Ignition System** section in this chapter for further details.

ENGINE STARTING PROBLEMS

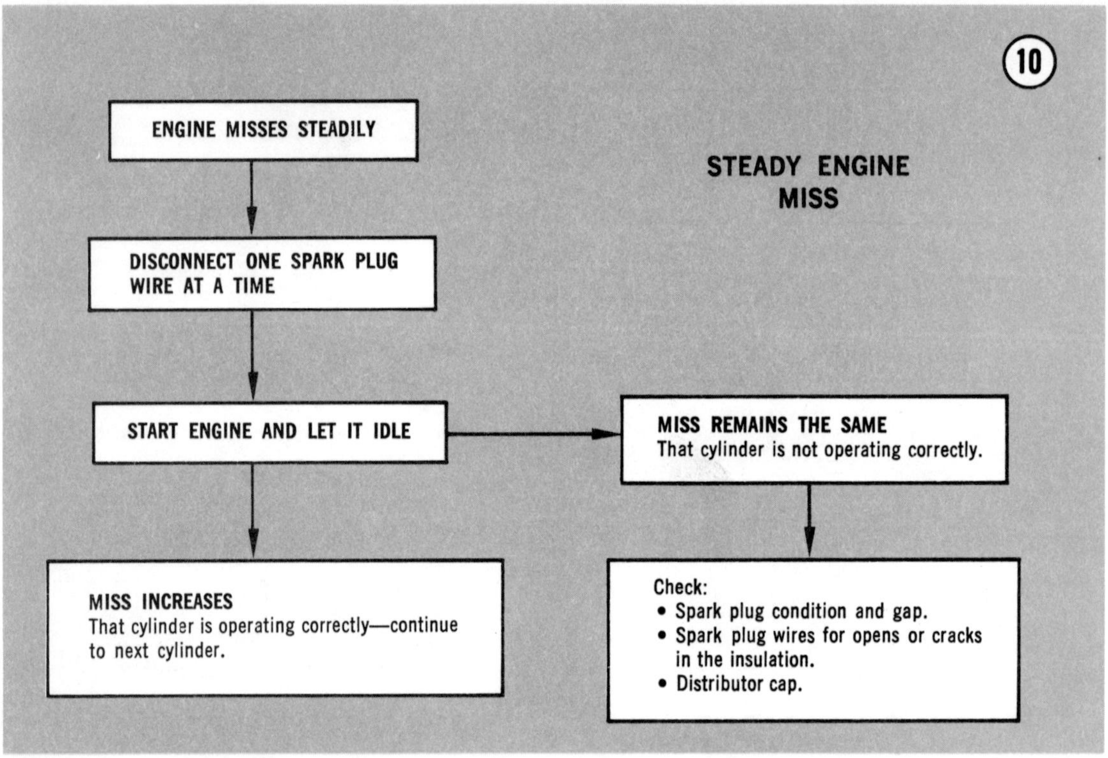

CHAPTER TWO

⑩ STEADY ENGINE MISS

ENGINE MISSES STEADILY
↓
DISCONNECT ONE SPARK PLUG WIRE AT A TIME
↓
START ENGINE AND LET IT IDLE → MISS REMAINS THE SAME
That cylinder is not operating correctly.
↓
MISS INCREASES
That cylinder is operating correctly—continue to next cylinder.

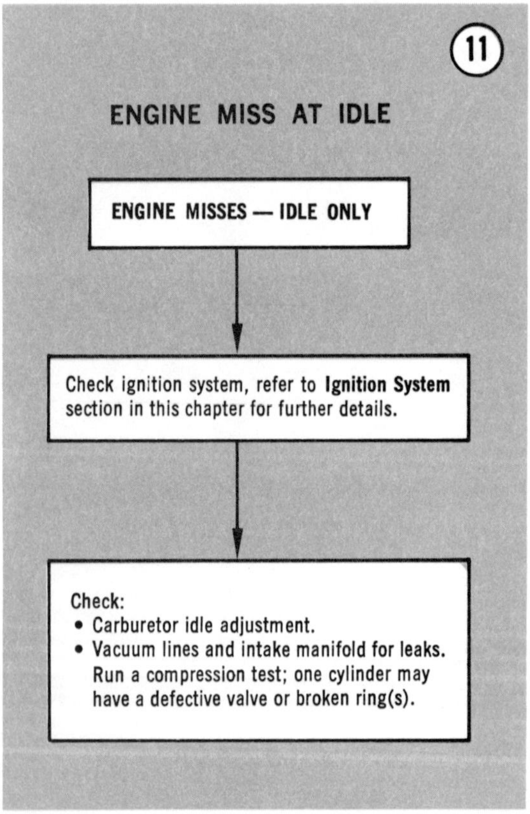

Check:
- Spark plug condition and gap.
- Spark plug wires for opens or cracks in the insulation.
- Distributor cap.

⑪ ENGINE MISS AT IDLE

ENGINE MISSES — IDLE ONLY
↓
Check ignition system, refer to **Ignition System** section in this chapter for further details.
↓
Check:
- Carburetor idle adjustment.
- Vacuum lines and intake manifold for leaks. Run a compression test; one cylinder may have a defective valve or broken ring(s).

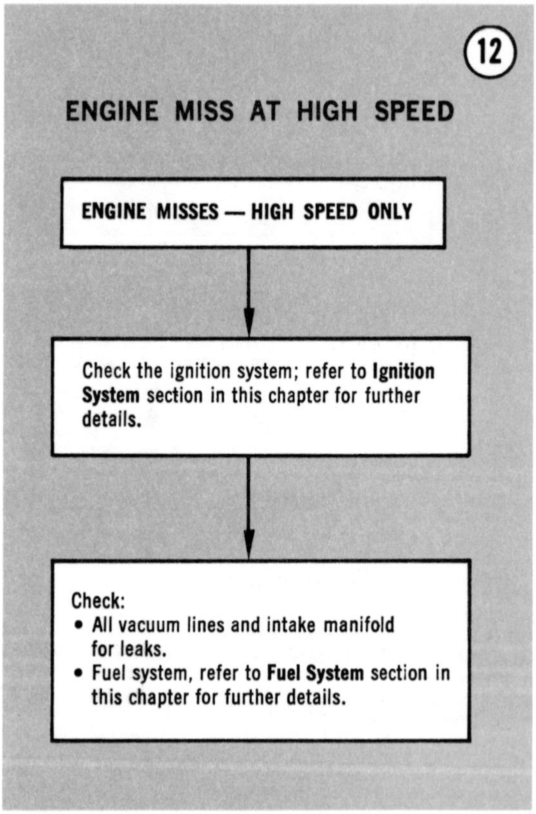

⑫ ENGINE MISS AT HIGH SPEED

ENGINE MISSES — HIGH SPEED ONLY
↓
Check the ignition system; refer to **Ignition System** section in this chapter for further details.
↓
Check:
- All vacuum lines and intake manifold for leaks.
- Fuel system, refer to **Fuel System** section in this chapter for further details.

TROUBLESHOOTING

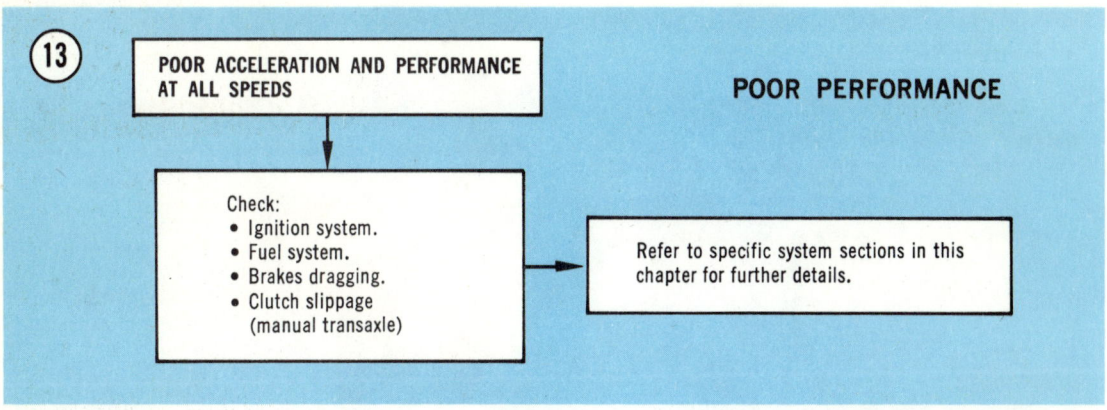

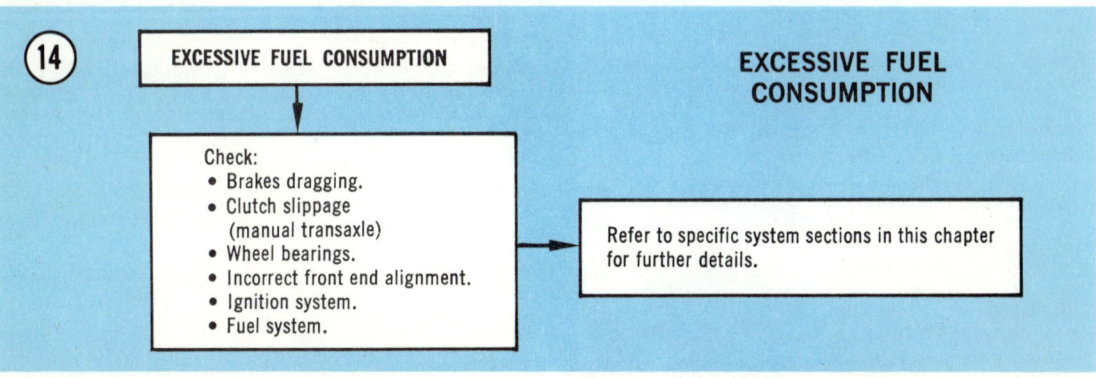

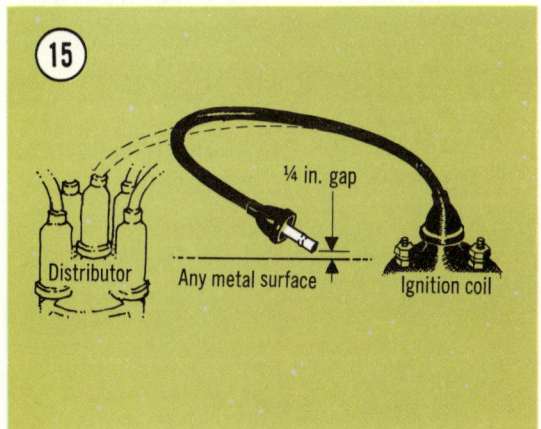

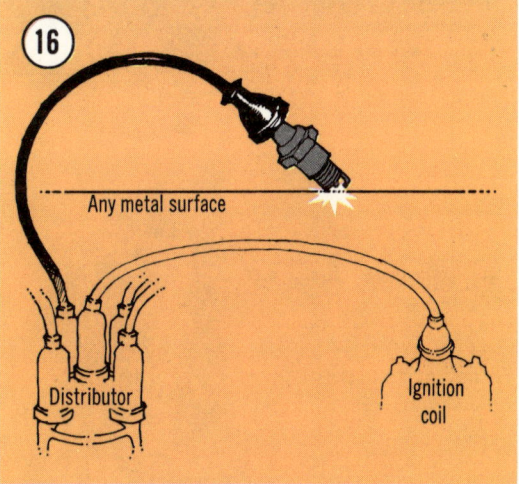

stopped, then prop the end of the wire next to a metal surface as shown in **Figures 15 and 16**.

WARNING
Never disconnect a spark plug or ignition coil wire while the engine is running. The high voltage in an ignition system, particularly the newer high-energy electronic ignition systems could cause serious injury or even death.

Spark plug condition is an important indication of engine performance. Spark plugs in a properly operating engine will have slightly pitted electrodes, and a light tan insulator tip. **Figure 17** shows a normal plug, and a number of others which indicate trouble in their respective cylinders.

NORMAL
- Appearance—Firing tip has deposits of light gray to light tan.
- Can be cleaned, regapped and reused.

CARBON FOULED
- Appearance—Dull, dry black with fluffy carbon deposits on the insulator tip, electrode and exposed shell.
- Caused by—Fuel/air mixture too rich, plug heat range too cold, weak ignition system, dirty air cleaner, faulty automatic choke or excessive idling.
- Can be cleaned, regapped and reused.

OIL FOULED
- Appearance—Wet black deposits on insulator and exposed shell.
- Caused by—Excessive oil entering the combustion chamber through worn rings, pistons, valve guides or bearings.
- Replace with new plugs (use a hotter plug if engine is not repaired).

LEAD FOULED
- Appearance — Yellow insulator deposits (may sometimes be dark gray, black or tan in color) on the insulator tip.
- Caused by—Highly leaded gasoline.
- Replace with new plugs.

LEAD FOULED
- Appearance—Yellow glazed deposits indicating melted lead deposits due to hard acceleration.
- Caused by—Highly leaded gasoline.
- Replace with new plugs.

OIL AND LEAD FOULED
- Appearance—Glazed yellow deposits with a slight brownish tint on the insulator tip and ground electrode.
- Replace with new plugs.

FUEL ADDITIVE RESIDUE
- Appearance — Brown-colored, hardened ash deposits on the insulator tip and ground electrode.
- Caused by—Fuel and/or oil additives.
- Replace with new plugs.

WORN
- Appearance — Severely worn or eroded electrodes.
- Caused by—Normal wear or unusual oil and/or fuel additives.
- Replace with new plugs.

PREIGNITION
- Appearance — Melted ground electrode.
- Caused by—Overadvanced ignition timing, inoperative ignition advance mechanism, too low of a fuel octane rating, lean fuel/air mixture or carbon deposits in combustion chamber.

PREIGNITION
- Appearance—Melted center electrode.
- Caused by—Abnormal combustion due to overadvanced ignition timing or incorrect advance, too low of a fuel octane rating, lean fuel/air mixture, or carbon deposits in combustion chamber.
- Correct engine problem and replace with new plugs.

INCORRECT HEAT RANGE
- Appearance—Melted center electrode and white blistered insulator tip.
- Caused by—Incorrect plug heat range selection.
- Replace with new plugs.

TROUBLESHOOTING

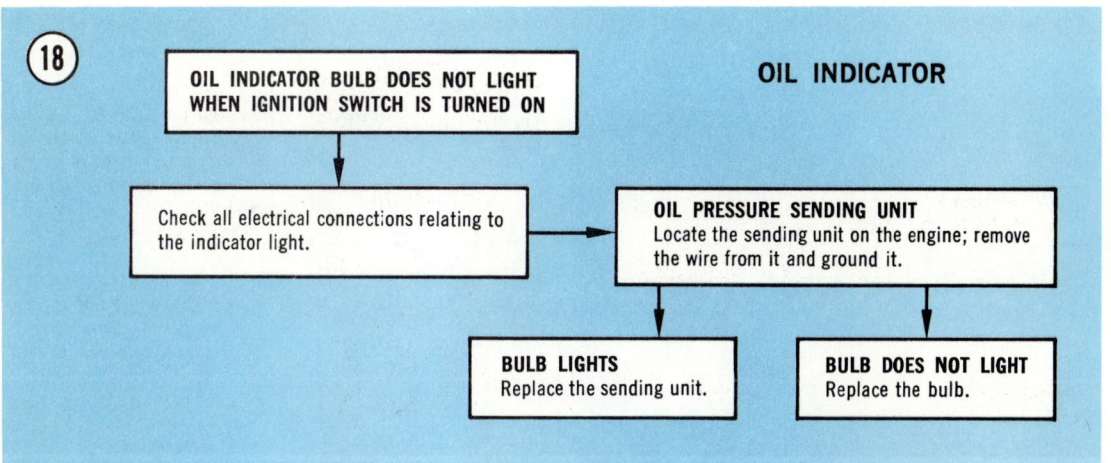

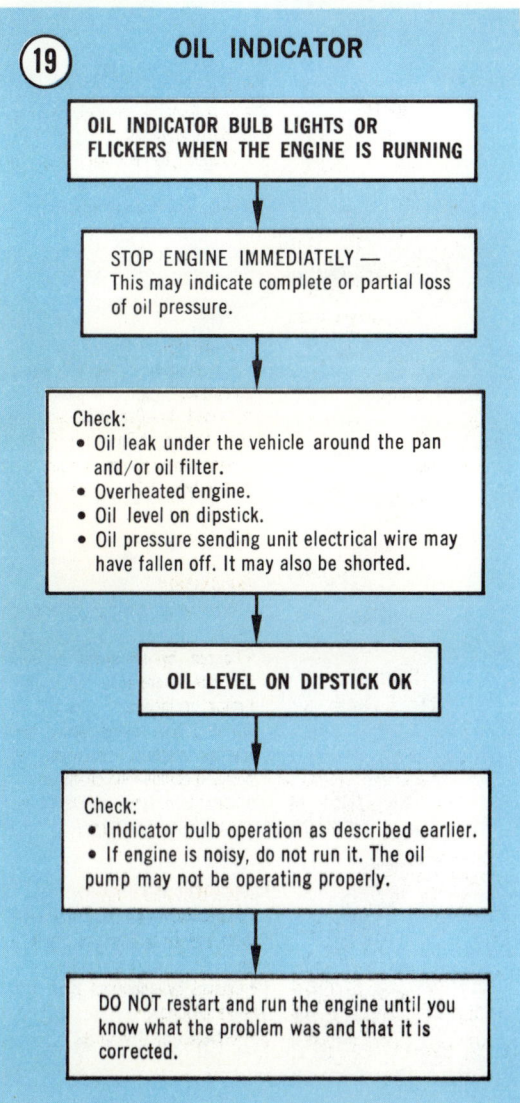

ENGINE OIL PRESSURE LIGHT

Proper oil pressure to the engine is vital. If oil pressure is insufficient, the engine can destroy itself in a comparatively short time.

The oil pressure warning circuit monitors oil pressure constantly. If pressure drops below a predetermined level, the light comes on.

Obviously, it is vital for the warning circuit to be working to signal low oil pressure. Each time you turn on the ignition, but before you start the vehicle, the warning light should come on. If it doesn't, there is trouble in the warning circuit, not the oil pressure system. See **Figure 18** to troubleshoot the warning circuit.

Once the engine is running, the warning light should stay off. If the warning light comes on or acts erratically while the engine is running there is trouble with the engine oil pressure system. *Stop the engine immediately*. Refer to **Figure 19** for possible causes of the problem.

FUEL SYSTEM (CARBURETTED)

Fuel system problems must be isolated to the fuel pump (mechanical or electric), fuel lines, fuel filter, or carburetor(s). These procedures assume the ignition system is working properly and is correctly adjusted.

1. *Engine will not start* — First make sure that fuel is being delivered to the carburetor. Remove the air cleaner, look into the carburetor throat, and operate the accelerator

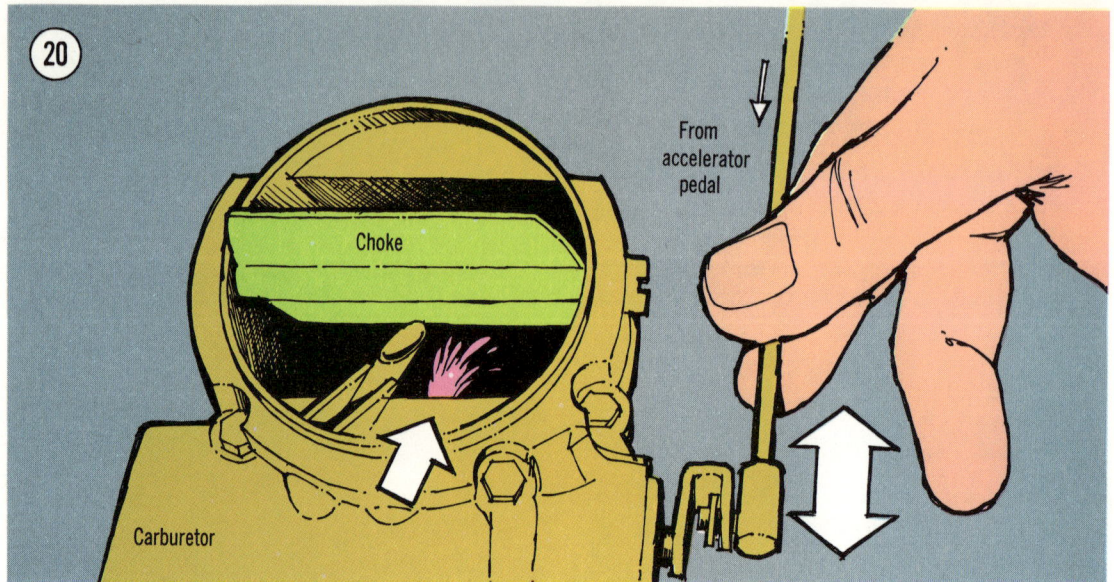

linkage several times. There should be a stream of fuel from the accelerator pump discharge tube each time the accelerator linkage is depressed (**Figure 20**). If not, check fuel pump delivery (described later), float valve, and float adjustment. If the engine will not start, check the automatic choke parts for sticking or damage. If necessary, rebuild or replace the carburetor.

2. *Engine runs at fast idle* — Usually this is caused by a defective automatic choke heater element. Ensure that the heater wire is connected and making good contact. Check the idle speed, idle mixture, and decel valve (if equipped) adjustment.

3. *Rough idle or engine miss with frequent stalling* — Check idle mixture and idle speed adjustments.

Poor idle may also be caused by a defective or dirty electromagnetic cutoff valve. Check that the electromagnetic cutoff valve wire is connected to the valve (on the carburetor) and making good contact. If it is, turn the ignition switch on, disconnect the wire and touch it to the valve terminal. If the valve is working, there should be a slight click heard each time the wire touches. If the valve is defective, turn the small setscrew on the end of the valve fully counterclockwise. This permanently opens the valve,

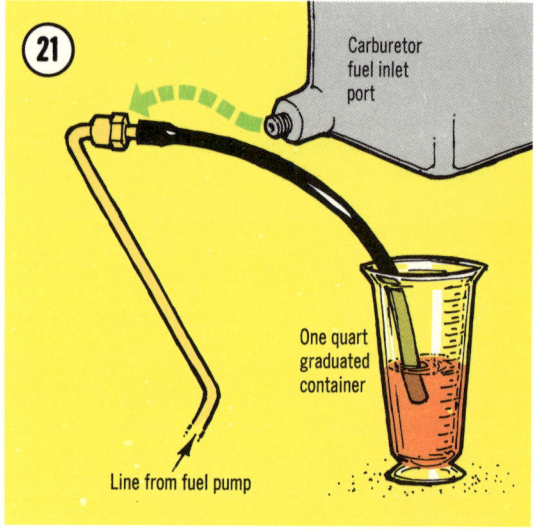

permitting the car to idle properly until the valve can be cleaned or replaced.

NOTE: *The engine may "diesel" in this condition. Replace the valve as soon as possible.*

4. *Engine "diesels" (continues to run) when ignition is switched off* — Check idle mixture (probably too rich), ignition timing, and idle speed (probably too fast). Check the throttle solenoid (if equipped) and electromagnetic cutoff valve for proper operation. Check for overheated engine.

TROUBLESHOOTING

5. *Stumbling when accelerating from idle* — Check the idle speed and mixture adjustments. Check the accelerator pump.

6. *Engine misses at high speed or lacks power* — This indicates possible fuel starvation. Check fuel pump pressure and capacity as described in this chapter. Check float needle valves. Check for a clogged fuel filter or air cleaner.

7. *Black exhaust smoke* — This indicates a badly overrich mixture. Check idle mixture and idle speed adjustment. Check choke setting. Check for excessive fuel pump pressure, leaky floats, or worn needle valves.

8. *Excessive fuel consumption* — Check for overrich mixture. Make sure choke mechanism works properly. Check idle mixture and idle speed. Check for excessive fuel pump pressure, leaky floats, or worn float needle valves.

FUEL SYSTEM (FUEL INJECTED)

Troubleshooting a fuel injection system requires more thought, experience, and know-how than any other part of the vehicle. A logical approach and proper test equipment are essential in order to successfully find and fix these troubles.

It is best to leave fuel injection troubles to your dealer. In order to isolate a problem to the injection system make sure that the fuel pump is operating properly. Check its performance as described later in this section. Also make sure that fuel filter and air cleaner are not clogged.

FUEL PUMP TEST (MECHANICAL AND ELECTRIC)

1. Disconnect the fuel inlet line where it enters the carburetor or fuel injection system.

2. Fit a rubber hose over the fuel line so fuel can be directed into a graduated container with about one quart capacity. See **Figure 21**.

3. To avoid accidental starting of the engine, disconnect the secondary coil wire from the coil.

4. Crank the engine for about 30 seconds.

5. If the fuel pump supplies the specified amount (refer to the fuel chapter later in this book), the trouble may be in the carburetor or fuel injection system. The fuel injection system should be tested by your dealer.

6. If there is no fuel present or the pump cannot supply the specified amount, either the fuel pump is defective or there is an obstruction in the fuel line. Replace the fuel pump and/or inspect the fuel lines for air leaks or obstructions.

7. Also pressure test the fuel pump by installing a T-fitting in the fuel line between the fuel pump and the carburetor. Connect a fuel pressure gauge to the fitting with a short tube (**Figure 22**).

8. Reconnect the primary coil wire, start the engine, and record the pressure. Refer to the fuel chapter later in this book for the correct pressure. If the pressure varies from that specified, the pump should be replaced.

9. Stop the engine. The pressure should drop off very slowly. If it drops off rapidly, the outlet valve in the pump is leaking and the pump should be replaced.

EMISSION CONTROL SYSTEMS

Major emission control systems used on nearly all U.S. models include the following:

a. Positive crankcase ventilation (PCV)
b. Thermostatic air cleaner
c. Air injection reaction (AIR)
d. Fuel evaporation control
e. Exhaust gas recirculation (EGR)

Emission control systems vary considerably from model to model. Individual models contain variations of the five systems described here. In addition, they may include other special systems. Use the index to find specific emission control components in other chapters.

Many of the systems and components are factory set and sealed. Without special expensive test equipment, it is impossible to adjust the systems to meet state and federal requirements.

Troubleshooting can also be difficult without special equipment. The procedures described below will help you find emission control parts which have failed, but repairs may have to be entrusted to a dealer or other properly equipped repair shop.

22

23

EXHAUST EMISSION PROBLEMS

CO CONTENT TOO LOW →

Check:
- Idle speed too low.
- Idle mixture adjustment (too lean).
- Carburetor jets and channels; clean and/or replace.
- Engine condition with a compression and vacuum test.

CO CONTENT TOO HIGH →

Check:
- Idle mixture adjustment (too rich).
- Defective automatic choke.
- Air filter warm air inlet door for sticking.
- Carburetor jets and channels; clean and/or replace.
- Engine condition with a compression and vacuum test.

HYDROCARBON LEVEL TOO HIGH →

Check:
- Throttle valve closes completely.
- Spark plug condition and gap.
- Ignition timing (too advanced) and breaker point gap (if so equipped).
- Intake manifold for leaks.
- Valve clearance (too small).
- Engine condition with a compression and vacuum test.

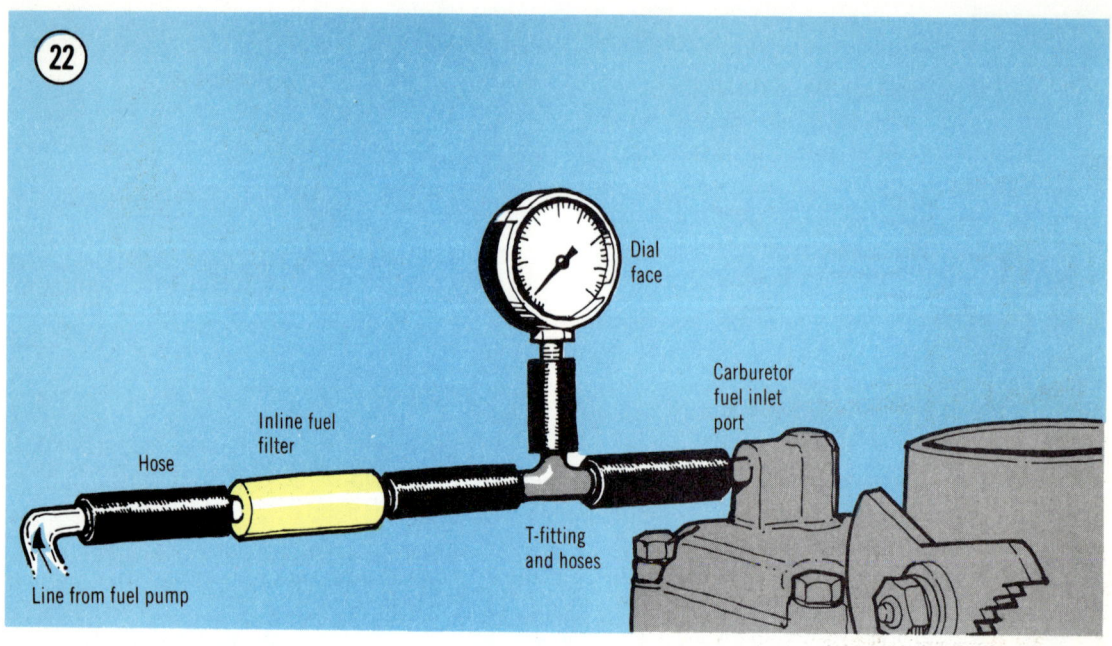

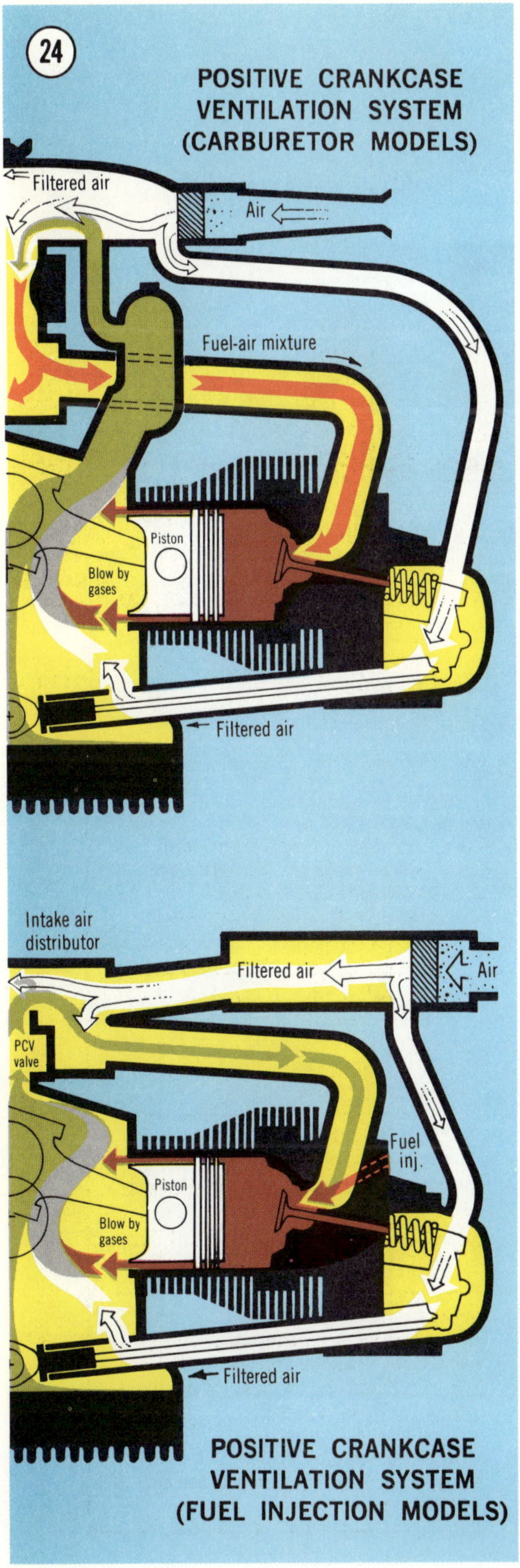

TROUBLESHOOTING

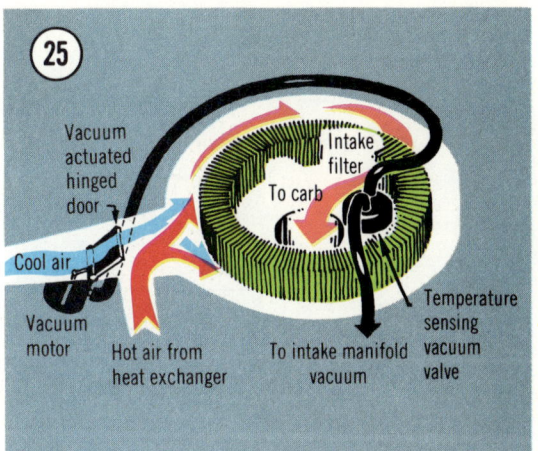

With the proper equipment, you can test the carbon monoxide and hydrocarbon levels. **Figure 23** provides some sources of trouble if the readings are not correct.

Positive Crankcase Ventilation

Fresh air drawn from the air cleaner housing scavenges emissions (e.g., piston blow-by) from the crankcase, then the intake manifold vacuum draws emissions into the intake manifold. They can then be reburned in the normal combustion process. **Figure 24** shows a typical system.

Thermostatic Air Cleaner

The thermostatically controlled air cleaner maintains incoming air to the engine at a predetermined level, usually about 100°F or higher. It mixes cold air with heated air from the exhaust manifold region. The air cleaner includes a temperature sensor, vacuum motor, and a hinged door. See **Figure 25**.

The system is comparatively easy to test. See **Figure 26** for the procedure.

Air Injection Reaction System

The air injection reaction system reduces air pollution by oxidizing hydrocarbons and carbon monoxide as they leave the combustion chamber. See **Figure 27**.

The air injection pump, driven by the engine, compresses filtered air and injects it at the exhaust port of each cylinder. The fresh air mixes with the unburned gases in the exhaust and pro-

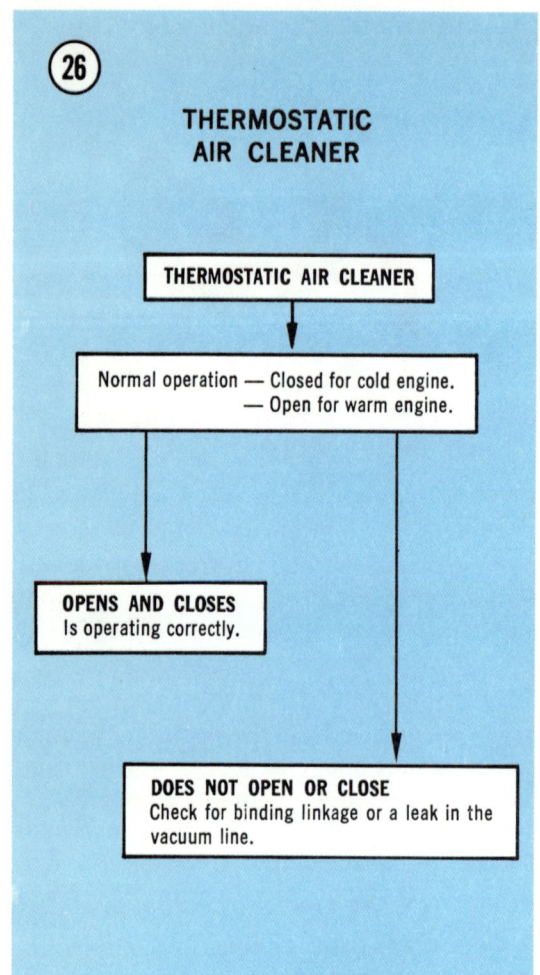

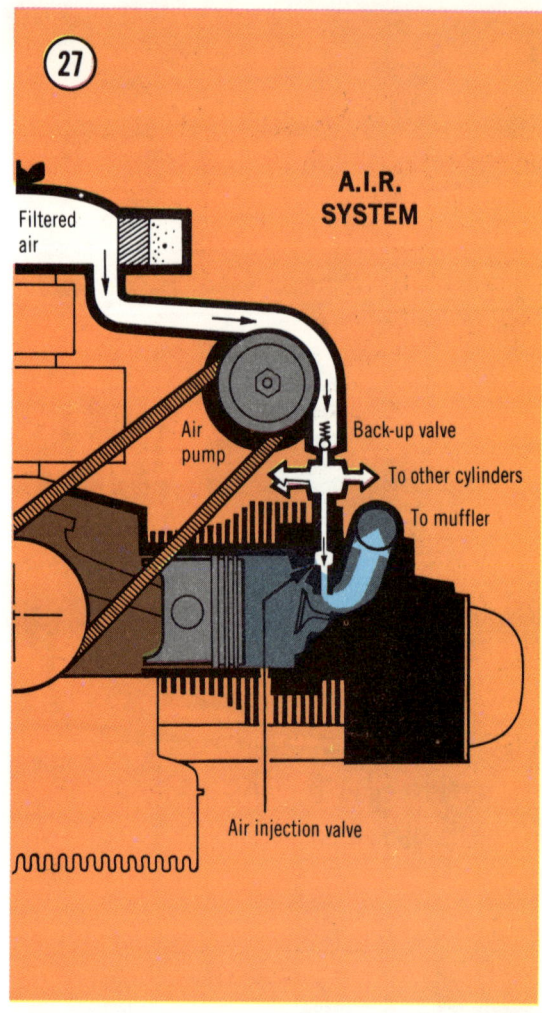

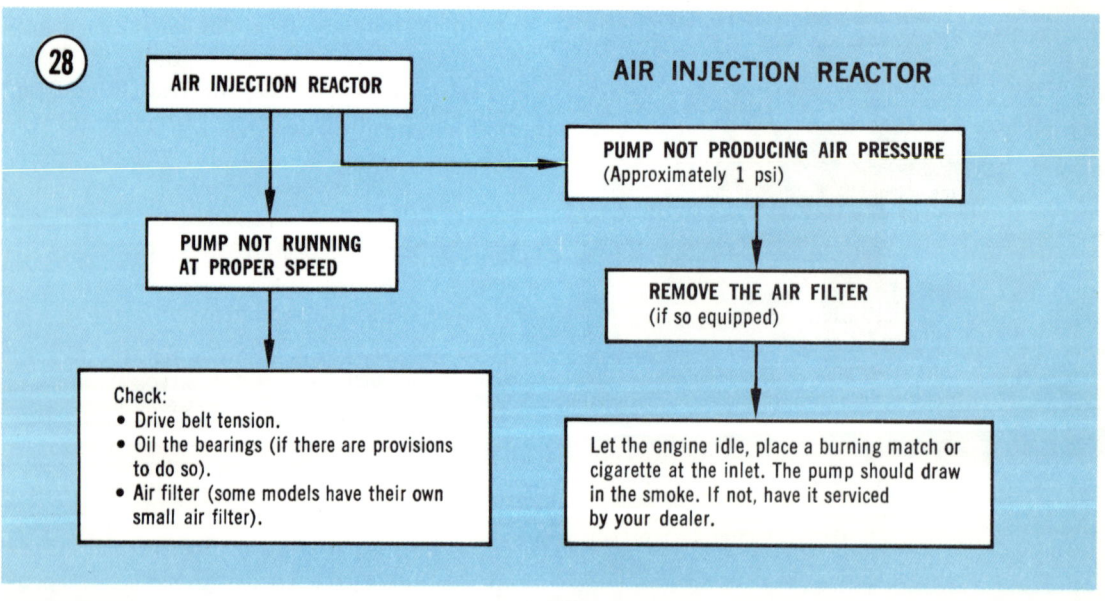

TROUBLESHOOTING

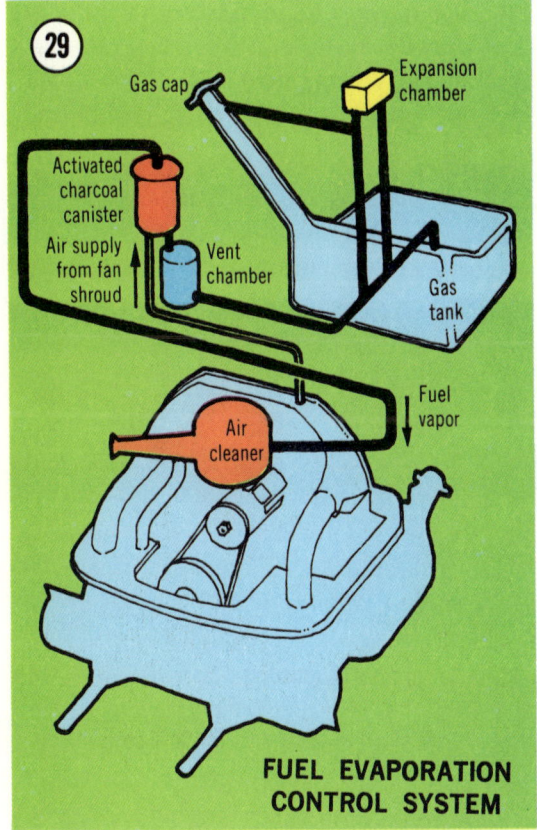

FUEL EVAPORATION CONTROL SYSTEM

motes further burning. A check valve prevents exhaust gases from entering and damaging the air pump if the pump becomes inoperative, e.g., from a drive belt failure.

Figure 28 explains the testing procedure for this system.

Fuel Evaporation Control

Fuel vapor from the fuel tank passes through the liquid/vapor separator to the carbon canister. See **Figure 29**. The carbon absorbs and stores the vapor when the engine is stopped. When the engine runs, manifold vacuum draws the vapor from the canister. Instead of being released into the atmosphere, the fuel vapor takes part in the normal combustion process.

Exhaust Gas Recirculation

The exhaust gas recirculation (EGR) system is used to reduce the emission of nitrogen oxides (NOx). Relatively inert exhaust gases are introduced into the combustion process to slightly reduce peak temperatures. This reduction in temperature reduces the formation of NOx.

Figure 30 provides a simple test of this system.

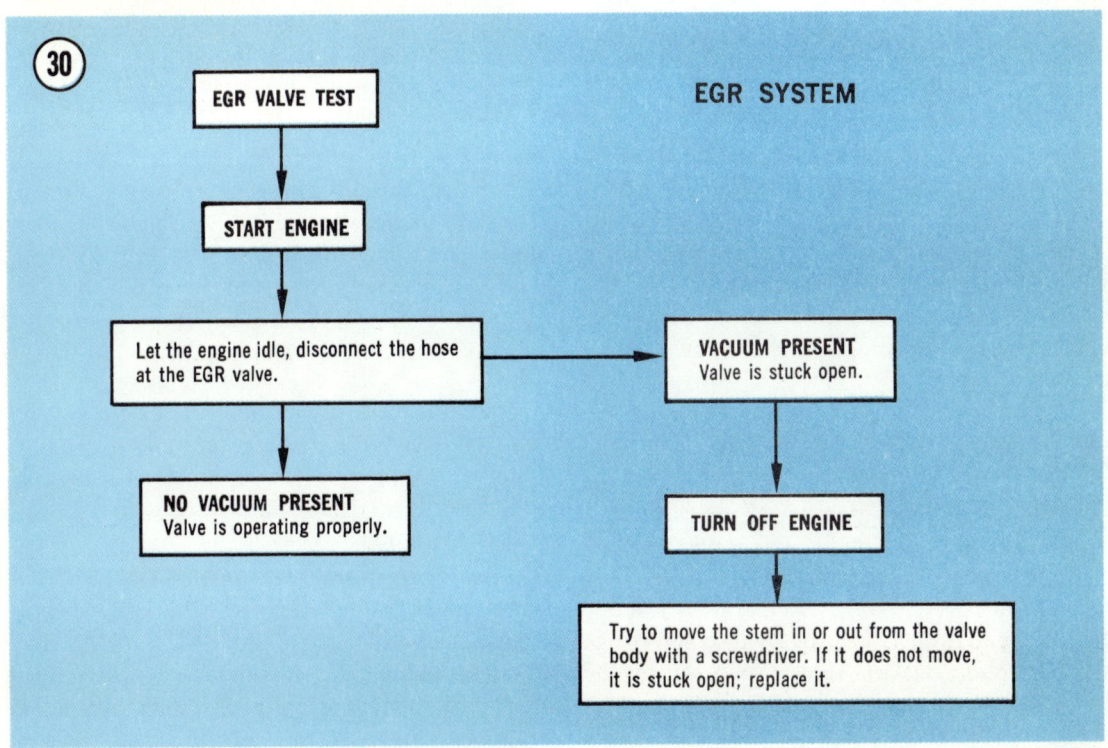

ENGINE NOISES

Often the first evidence of an internal engine trouble is a strange noise. That knocking, clicking, or tapping which you never heard before may be warning you of impending trouble.

While engine noises can indicate problems, they are sometimes difficult to interpret correctly; inexperienced mechanics can be seriously misled by them.

Professional mechanics often use a special stethoscope which looks similar to a doctor's stethoscope for isolating engine noises. You can do nearly as well with a "sounding stick" which can be an ordinary piece of doweling or a section of small hose. By placing one end in contact with the area to which you want to listen and the other end near your ear, you can hear sounds emanating from that area. The first time you do this, you may be horrified at the strange noises coming from even a normal engine. If you can, have an experienced friend or mechanic help you sort the noises out.

Clicking or Tapping Noises

Clicking or tapping noises usually come from the valve train, and indicate excessive valve clearance.

If your vehicle has adjustable valves, the procedure for adjusting the valve clearance is explained in Chapter Three. If your vehicle has hydraulic lifters, the clearance may not be adjustable. The noise may be coming from a collapsed lifter. These may be cleaned or replaced as described in the engine chapter.

A sticking valve may also sound like a valve with excessive clearance. In addition, excessive wear in valve train components can cause similar engine noises.

Knocking Noises

A heavy, dull knocking is usually caused by a worn main bearing. The noise is loudest when the engine is working hard, i.e., accelerating hard at low speed. You may be able to isolate the trouble to a single bearing by disconnecting the spark plugs one at a time. When you reach the spark plug nearest the bearing, the knock will be reduced or disappear.

Worn connecting rod bearings may also produce a knock, but the sound is usually more "metallic." As with a main bearing, the noise is worse when accelerating. It may even increase further just as you go from accelerating to coasting. Disconnecting spark plugs will help isolate this knock as well.

A double knock or clicking usually indicates a worn piston pin. Disconnecting spark plugs will isolate this to a particular piston, however, the noise will *increase* when you reach the affected piston.

A loose flywheel and excessive crankshaft end play also produce knocking noises. While similar to main bearing noises, these are usually intermittent, not constant, and they do not change when spark plugs are disconnected.

Some mechanics confuse piston pin noise with piston slap. The double knock will distinguish the piston pin noise. Piston slap is identified by the fact that it is always louder when the engine is cold.

ELECTRICAL ACCESSORIES

Lights and Switches (Interior and Exterior)

1. *Bulb does not light* — Remove the bulb and check for a broken element. Also check the inside of the socket; make sure the contacts are clean and free of corrosion. If the bulb and socket are OK, check to see if a fuse has blown. The fuse panel (**Figure 31**) is usually located under the instrument panel. Replace the blown fuse. If the fuse blows again, there is a short in that circuit. Check that circuit all the way to the battery. Look for worn wire insulation or burned wires.

If all the above are all right, check the switch controlling the bulb for continuity with an ohmmeter at the switch terminals. Check the switch contact terminals for loose or dirty electrical connections.

2. *Headlights work but will not switch from either high or low beam* — Check the beam selector switch for continuity with an ohmmeter at the switch terminals. Check the switch contact terminals for loose or dirty electrical connections.

TROUBLESHOOTING

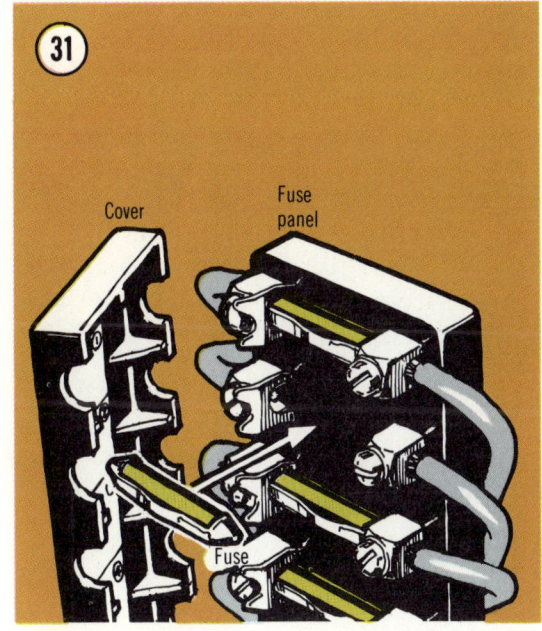

3. *Brake light switch inoperative* — On mechanically operated switches, usually mounted near the brake pedal arm, adjust the switch to achieve correct mechanical operation. Check the switch for continuity with an ohmmeter at the switch terminals. Check the switch contact terminals for loose or dirty electrical connections.

4. *Back-up lights do not operate* — Check light bulb as described earlier. Locate the switch, normally located near the shift lever. Adjust switch to achieve correct mechanical operation. Check the switch for continuity with an ohmmeter at the switch terminals. Bypass the switch with a jumper wire; if the lights work, replace the switch.

Directional Signals

1. *Directional signals do not operate* — If the indicator light on the instrument panel burns steadily instead of flashing, this usually indicates that one of the exterior lights is burned out. Check all lamps that normally flash. If all are all right, the flasher unit may be defective. Replace it with a good one.

2. *Directional signal indicator light on instrument panel does not light up* — Check the light bulbs as described earlier. Check all electrical connections and check the flasher unit.

3. *Directional signals will not self-cancel* — Check the self-cancelling mechanism located inside the steering column.

4. *Directional signals flash slowly* — Check the condition of the battery and the alternator (or generator) drive belt tension (**Figure 4**). Check the flasher unit and all related electrical connections.

Windshield Wipers

1. *Wipers do not operate* — Check for a blown fuse and replace it. Check all related terminals for loose or dirty electrical connections. Check continuity of the control switch with an ohmmeter at the switch terminals. Check the linkage and arms for loose, broken, or binding parts. Straighten out or replace where necessary.

2. *Wiper motor hums but will not operate* — The motor may be shorted out internally; check and/or replace the motor. Also check for broken or binding linkage and arms.

3. *Wiper arms will not return to the stowed position when turned off* — The motor has a special internal switch for this purpose. Have it inspected by your dealer. Do not attempt this yourself.

Interior Heater

1. *Heater fan does not operate* — Check for a blown fuse.. Check the switch for continuity with an ohmmeter at the switch terminals. Check the switch contact terminals for loose or dirty electrical connections.

2. *Heat output is insufficient* — Check that the heater door(s) and cable(s) are operating correctly and are in the open position. Inspect the heat ducts; make sure that they are not crimped or blocked.

3. *Exhaust fumes in passenger compartment* — Open all windows and inspect heat exchangers and heating system immediately.

WARNING
Do not continue to operate the vehicle with deadly carbon monoxide fumes present in the passenger compartment.

COOLING SYSTEM

Engine cooling is provided by an engine driven fan which draws in outside air for the cylinders and cylinder heads. Thermostatically controlled air flaps limit the amount of cold air when engine is cold to provide rapid warm up.

If the engine is running abnormally hot, check fan drive condition and tension, air control ring adjustment and/or air control thermostat.

If overheating is extreme, the engine will have to be removed and the cooling duct system removed and inspected.

CLUTCH

All clutch troubles except adjustments require removal of the engine/transaxle assembly to identify and cure the problem.

1. *Slippage* — This is most noticeable when accelerating in a high gear at relatively low speed. To check slippage, park the vehicle on a level surface with the handbrake set. Shift to 2nd gear and release the clutch as if driving off. If the clutch is good, the engine will slow and stall. If the clutch slips, continued engine speed will give it away.

Slippage results from insufficient clutch pedal free play, oil or grease on the clutch disc, worn pressure plate, or weak springs. Also check for binding in the clutch cable and lever arm which may prevent full engagement.

CAUTION
This is a severe test. Perform this test only when slippage is suspected, not periodically.

2. *Drag or failure to release* — This trouble usually causes difficult shifting and gear clash, especially when downshifting. The cause may be excessive clutch pedal free play, warped or bent pressure plate or clutch disc, excessive clutch cable guide sag, broken or loose linings, lack of lubrication in gland nut bearing or felt ring. Also check condition of main shaft splines.

3. *Chatter or grabbing* — A number of things can cause this trouble. Check tightness of

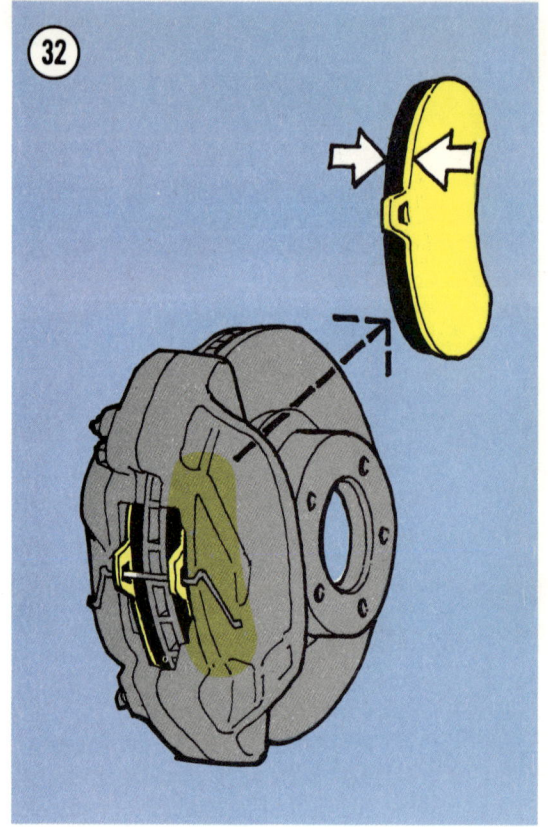

engine mounts and engine-to-transmission mounting bolts. Check for worn or misaligned pressure plate and misaligned release plate, or excessive cable guide sag.

4. *Other noises* — Noise usually indicates a dry or defective release or pilot bearing. Check the bearings and replace if necessary. Also check all parts for misalignment and uneven wear.

MANUAL TRANSAXLE

Transaxle troubles are evident when one or more of the following symptoms appear:

  a. Difficulty changing gears
  b. Gears clash when downshifting
  c. Slipping out of gear
  d. Excessive noise in NEUTRAL
  e. Excessive noise in gear
  f. Oil leaks

Transaxle repairs, except for one oil seal, are **not possible without expensive special tools.**

TROUBLESHOOTING

The main shaft oil seal, however, is easily replaced after removing the engine.

Transaxle troubles are sometimes difficult to distinguish from clutch troubles. Eliminate the clutch as a source of trouble before installing a new or rebuilt transaxle.

AUTOMATIC AND SEMI-AUTOMATIC TRANSAXLE

Most automatic and semi-automatic transaxle repairs require considerable specialized knowledge and tools. It is impractical for the home mechanic to invest in the tools, since they cost more than a properly rebuilt transmission.

Check fluid level and condition frequently to help prevent future problems. If the fluid is orange or black in color or smells like varnish, it is an indication of some type of damage or failure within the transmission. Have the transmission serviced by your dealer or competent automatic transmission service facility.

Refer to transaxle chapter for specific troubleshooting procedures.

BRAKES

Good brakes are vital to the safe operation of the vehicle. Performing the maintenance specified in Chapter Three will minimize problems with the brakes. Most importantly, check and maintain the level of fluid in the master cylinder, and check the thickness of the linings on the disc brake pads (**Figure 32**) or drum brake shoes (**Figure 33**).

If trouble develops, **Figures 34 through 36** will help you locate the problem. Refer to the brake chapter for actual repair procedures.

STEERING AND SUSPENSION

Trouble in the suspension or steering is evident when the following occur:

a. Steering is hard
b. Vehicle pulls to one side
c. Vehicle wanders or front wheels wobble
d. Steering has excessive play
e. Tire wear is abnormal

Unusual steering, pulling, or wandering is usually caused by bent or otherwise misaligned suspension parts. This is difficult to check without proper alignment equipment. Refer to the suspension chapter in this book for repairs that you can perform and those that must be left to a dealer or suspension specialist.

If your trouble seems to be excessive play, check wheel bearing adjustment first. This is the most frequent cause. Then check ball-joints as described below. Finally, check tie rod end ball-joints by shaking each tie rod. Also check steering gear, or rack-and-pinion assembly to see that it is securely bolted down.

TIRE WEAR ANALYSIS

Abnormal tire wear should be analyzed to determine its causes. The most common causes are the following:

a. Incorrect tire pressure
b. Improper driving
c. Overloading
d. Bad road surfaces
e. Incorrect wheel alignment

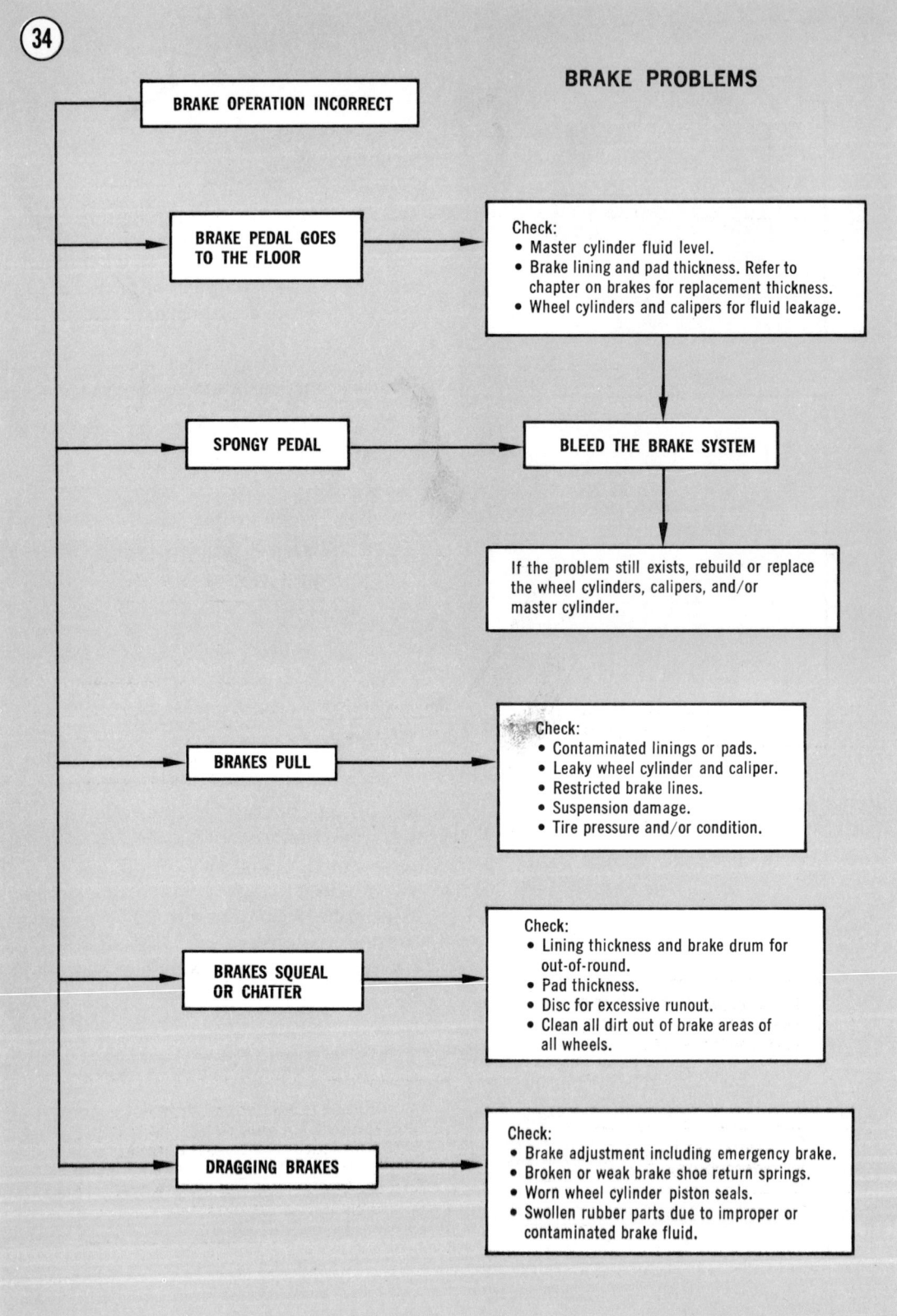

TROUBLESHOOTING 31

㉟ **BRAKE PROBLEMS**

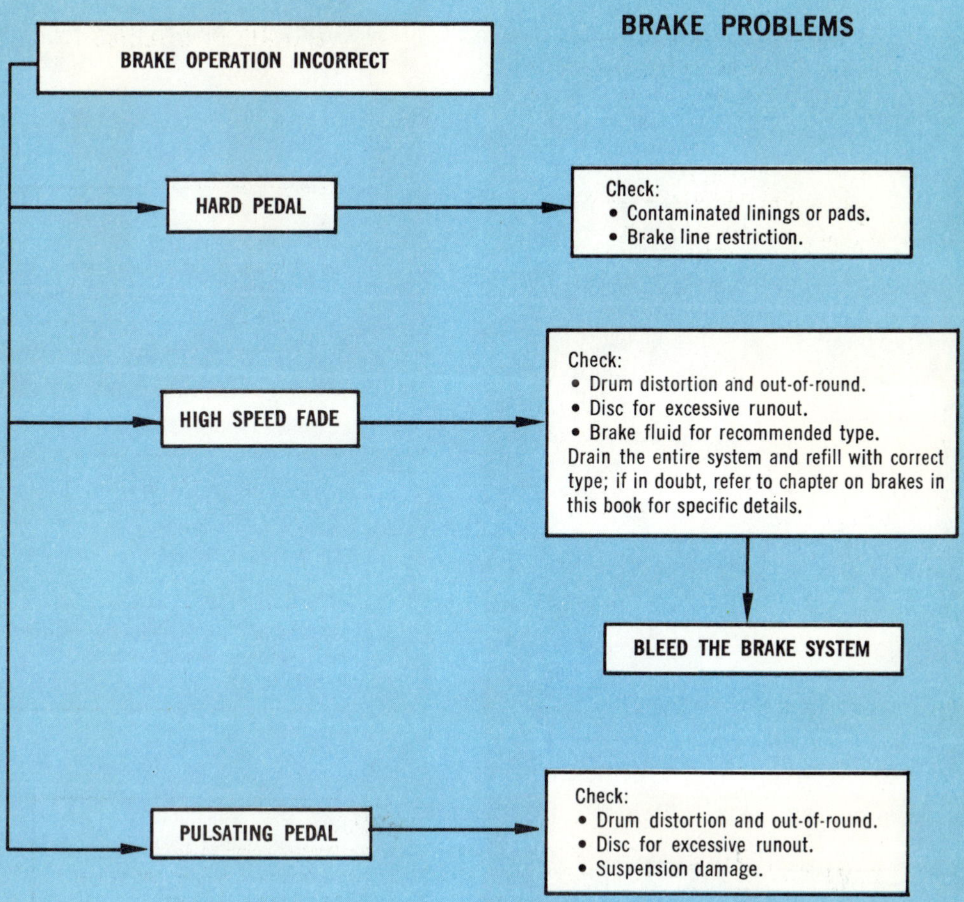

BRAKE OPERATION INCORRECT

→ HARD PEDAL → Check:
- Contaminated linings or pads.
- Brake line restriction.

→ HIGH SPEED FADE → Check:
- Drum distortion and out-of-round.
- Disc for excessive runout.
- Brake fluid for recommended type.

Drain the entire system and refill with correct type; if in doubt, refer to chapter on brakes in this book for specific details.

↓ BLEED THE BRAKE SYSTEM

→ PULSATING PEDAL → Check:
- Drum distortion and out-of-round.
- Disc for excessive runout.
- Suspension damage.

㊱ **BRAKE PROBLEMS**

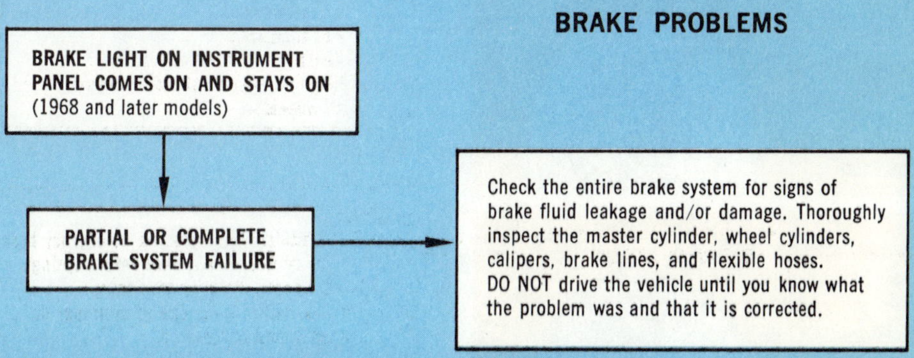

BRAKE LIGHT ON INSTRUMENT PANEL COMES ON AND STAYS ON
(1968 and later models)

↓

PARTIAL OR COMPLETE BRAKE SYSTEM FAILURE → Check the entire brake system for signs of brake fluid leakage and/or damage. Thoroughly inspect the master cylinder, wheel cylinders, calipers, brake lines, and flexible hoses. DO NOT drive the vehicle until you know what the problem was and that it is corrected.

CHAPTER TWO

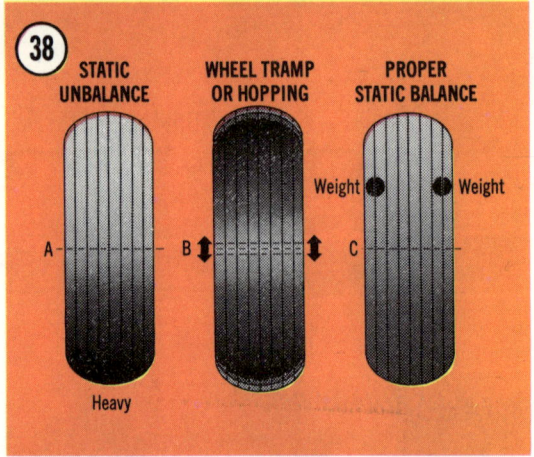

Figure 37 identifies wear patterns and indicates the most probable causes.

WHEEL BALANCING

All four wheels and tires must be in balance along two axes. To be in static balance (**Figure 38**), weight must be evenly distributed around the axis of rotation. (A) shows a statically unbalanced wheel; (B) shows the result — wheel tramp or hopping; (C) shows proper static balance.

To be in dynamic balance (**Figure 39**), the centerline of the weight must coincide with the centerline of the wheel. (A) shows a dynamically unbalanced wheel; (B) shows the result — wheel wobble or shimmy; (C) shows proper dynamic balance.

CHAPTER THREE

LUBRICATION, MAINTENANCE, AND TUNE-UP

A carefully followed program of maintenance will pay dividends in longer engine and vehicle life and fewer repair bills. Such a program is especially important if the vehicle is used off the road or in out-of-the-way places. Breakdowns or failures are much less likely to occur if the vehicle has been well maintained.

Certain maintenance tasks and checks should be made at each gas stop. Others should be performed at certain time or mileage intervals. Still others should be done whenever certain symptoms develop.

GAS STOP CHECKS

Many of these procedures are made as a matter of routine by some service station attendants. You may want to develop the habit of looking over the attendant's shoulder or of doing the tasks yourself. Even though the checks are simple, they are important to the care and trouble-free driving of any automobile. They also give an indication of the need for other maintenance.

Items to be checked at gas stops include:

a. Engine oil level
b. Coolant level
c. Battery electrolyte level
d. Windshield washer fluid level

Engine Oil Level

NOTE: *Oil should be checked as the last step at a fuel stop. This will allow oil in the upper part of the engine to drain back into the crankcase.*

To check engine oil level, remove the dipstick (**Figure 1**, typical), wipe it with a cloth or paper towel, replace the dipstick until it seats firmly, then remove it again and read the level of the oil on the lower end. Reinsert the dipstick after taking the reading.

The level of oil should be maintained somewhere between the ADD or ADD OIL mark and the FULL mark. See **Figure 2**. Oil should be added whenever the level drops below the ADD mark. However, do not overfill the engine. Too much oil can sometimes be as harmful to the engine as too little. See **Table 1** for type of oil recommended.

Coolant Level

On vehicles equipped with coolant recovery systems, check coolant level by observing the liquid level in the recovery system reservoir. See **Figure 3**. Do not remove the radiator cap. Coolant should be at the COLD FULL mark on the reservoir after the engine cools to ambient temperature. Coolant should be at the HOT FULL mark when the engine is at normal.

> WARNING
> *The radiator cap should not be removed when engine is warm, especially if an air conditioner has been in use. The cap allows the cooling system to be pressurized to 15 lb., which permits the engine to operate safely at cooling temperatures between 245°F and 260°F (depending upon year model). Removal of radiator cap allows pressure to drop and the heat in excess of 212°F will be dissipated by conversion of water to steam. Since steam may form in engine water passages, it may blow coolant out of the upper radiator hose and tank top, causing loss of coolant and possible injury to bystanders.*

The fluid level on vehicles without coolant recovery systems should be maintained 3 inches below the bottom of the filler neck when the cooling system is cold.

> NOTE: *Any significant loss of coolant could mean a leak in the cooling system. If the system consistently loses coolant, see Chapter Seven for check procedure.*

Battery Electrolyte Level

> NOTE: *Late models are equipped with a sealed, maintenance-free battery. This battery does not require periodic electrolyte level checks.*

Check battery electrolyte level by removing the vent plug from each cell and observing the level in each vent well. The level in each cell should reach the bottom of the split vent well. See **Figure 4**. If the level is low, add clear, odorless drinking water until the level contacts the bottom of the vent well. Do not overfill, as this will result in loss of electrolyte and shorter battery life.

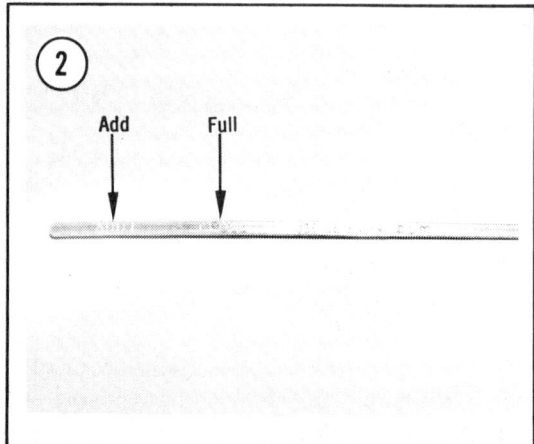

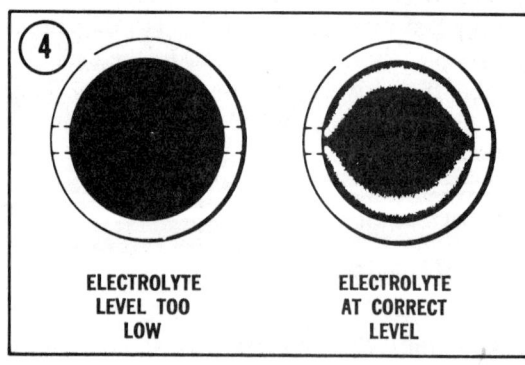

ELECTROLYTE LEVEL TOO LOW | ELECTROLYTE AT CORRECT LEVEL

LUBRICATION, MAINTENANCE, AND TUNE-UP

Table 1 RECOMMENDED LUBRICANTS AND FLUIDS

Use	Recommendation
Engine oil	Product bearing letter designation SE
Power steering	GM power steering fluid or, if not available, DEXRON® or DEXRON® II automatic transmission fluid
Differential—standard	SAE 80 or SAE 90 GL-5 gear lubricant (SAE 80 GL-5 in Canada)
Differential—Positraction	GM part No. 1051022 (lubricant) or equivalent
Steering gear—manual	GM part No. 1051052 (lubricant) or equivalent
Transmission—manual	SAE 80 or SAE 90 GL-5 gear lubricant (SAE 80 GL-5 in Canada)
Brake master cylinder	DOT-3 or Delco Supreme 11 brake fluid
Clutch linkage (manual transmission) 1. Pivot points 2. Cross-shaft grease fitting and push rod to clutch fork joint	Engine oil EP chassis lubricant meeting requirements of GM specification GM 6031-M
Shift linkage (all)	Engine oil
Hood latch and hinge assemblies 1. Pivots and spring anchor 2. Release pawl 3. Hinges	Engine oil Chassis lubricant Engine oil
Chassis lubrication (front suspension, steering linkage, etc.)	Chassis lubricant meeting requirements of GM specification GM 6031-M
Constant velocity universal joint	Lubricant, GM part No. 1050679, or meeting requirements of GM specification GM 6040-M
Automatic transmission	DEXRON® or DEXRON® II automatic transmission fluid
Parking brake cables	Chassis lubricant
Front wheel bearings	Chassis lubricant meeting the requirements of GM specification GM 6031-M
Door, tailgate, seat and trunk hinges	Engine oil
Convertible door-to-lock wedge plates	Stick-type lubricant
Windshield washer	GM Optikleen washer solvent, part No. 1050001 or equivalent
Battery	Clean, odorless drinking water
Engine cooling system	Mixture of water and high quality Ethylene Glycol base antifreeze, meeting requirements of GM specification 1899-M

Windshield Washer Fluid

The windshield washer is an important safety feature and an adequate supply of fluid should be present in the reservoir (**Figure 5**) at all times. A quick glance at the reservoir while performing other under-the-hood service should become a habit. If the reservoir is less than ½ full, it should be refilled. If an additive is used, follow the manufacturer's instructions.

PERIODIC MAINTENANCE

Table 2 provides a complete vehicle maintenance and lubrication schedule, recommended by General Motors. The schedule is intended only as a guide. If your vehicle is subjected to conditions such as heavy dust, continuous short trips, or pulling trailers (or use of leaded fuel in 1974 and later models), more frequent servicing will be required.

The following is a brief explanation of each of the services listed in **Table 2**. Use only those services that apply to your vehicle.

1. *Chassis* — Lubricate all grease fittings in front suspension, steering linkage, and universal joint. Shift linkage and cable guides should also be lubricated.

2. *Fluid levels* — Check levels in brake master cylinder, power steering pump, battery, engine crankcase, rear axle (differential), transmission, and windshield washer. Also check engine coolant for proper level and freeze protection to $-20°F$ or the lowest temperature expected (whichever is lower). Any significant loss of fluid in any of these systems should be investigated as it could mean that trouble is developing and that immediate corrective action is required. Low level in the front reservoir of the brake master cylinder could be an indicator that disc brake pads (on cars so equipped) need replacing.

3. *Engine oil* — Change oil at the intervals indicated if vehicle was operated under normal conditions. Change oil every 2 or 3 months or every 3,000 miles when the vehicle is operated under any of the following conditions:

   a. Heavy dust
   b. Trailer towing
   c. Extensive idling

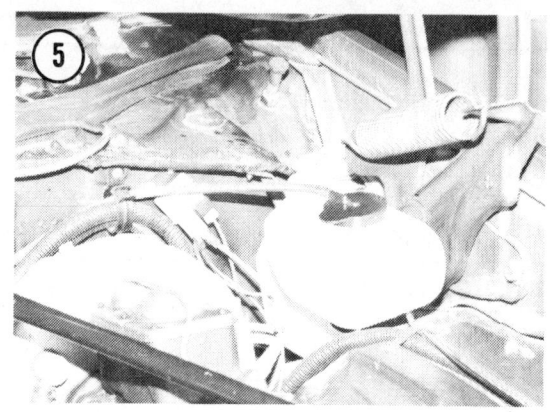

   d. Short trips, especially under freezing temperatures with engine not thoroughly warmed up.

4. *Air conditioning* — Check condition of air conditioner hoses and have them replaced if need is indicated. Check refrigerant charge at sight glass. Flowing refrigerant should be clear, although a few bubbles may be present in mild weather. Foam in the flow indicates that refrigerant charge is low. Absence of flow and no temperature difference between compressor inlet and outlet lines indicates no charge.

5. *Tires* — Steel belted radial tires should be rotated (see **Figure 6**) at first 7,500 miles and every 15,000 miles thereafter. Bias belted tires should be rotated every 6,000 miles. Tires should be inflated to the pressures shown on the tire placard on the rear face of the driver's door (late models) or to the values given in the owner's manual.

6. *Engine oil filter* — Replace at first oil change, then every second oil change thereafter (every oil change on 1967 models). If driving under severe conditions described in Step 3 above, change filter with every oil change.

7. *Rear axle* — Check at intervals shown on **Table 2**. Change lubricant at first 12,000 miles (15,000 miles on 1975 and later models) on Positraction axles. Change lubricant every 12,000 miles (7,500 miles on 1975 and later models) on all types of rear axles or final drives on vehicles used extensively to pull trailers.

8. *Cooling system* — At intervals shown in **Table 2**, test system for leaks (see Chapter Seven) and check or replace pressure cap.

LUBRICATION, MAINTENANCE, AND TUNE-UP

Table 2   MAINTENANCE SCHEDULE

Frequency	Item No.	Service
Section A — Lubrication and General Maintenance		
1975-on—every 6 months or 7,500 miles Others—every 4 months or 6,000 miles	1	Chassis lubrication
	2	Fluid levels
	3	Engine oil
1975-on—every 12 months Others—every 6 months	4	Air conditioner
See explanation in text	5	Tire rotation
All years—at first oil change, then every second oil change	6	Engine oil filter
See explanation in text	7	Rear axle
1975-on—every 12 months or 15,000 miles Others—every 12 months or 12,000 miles	8	Cooling system
1975-on—every 30,000 miles Others—every 24,000 miles	9	Wheel bearings
1975-on—every 30,000 miles Others—every 24,000 miles	10	Automatic transmission
1975-on—every 30,000 miles Others—every 36,000 miles	11	Manual steering gear and clutch cross shaft
Section B — Safety Maintenance		
1975-on—every 4 months or 7,500 miles Others—every 4 months or 6,000 miles	12	Owner safety checks
	13	Tires and wheels
	14	Exhaust system
	15	Drive belts
	16	Suspension and steering
	17	Brakes and power steering
	18	Disc brakes
1975-on—every 12 months or 15,000 miles Others—every 12 months or 12,000 miles	19	Drum and parking brakes
	20	Throttle linkage
	21	Headlights
	22	Underbody
	23	Bumpers

(continued)

Table 2 MAINTENANCE SCHEDULE (continued)

Frequency	Item No.	Services
Section C — Emission Control Maintenance		
1975-1976 and 1977-on Schedule I— at 6 months or 7,500 miles, then at 18-month/22,500-mile intervals 1977-on Schedule II—At 6 months or 7,500 miles, then at 24 month/30,000 mile intervals Others—first at 4 months or 6,000 miles, then at 12-month/12,000-mile intervals	24	Thermostatically controlled air cleaner
	25	Carburetor choke
	26	Timing, dwell, carburetor idle, distributor and coil
	27	Manifold heat valve (if equipped)
	28	Carburetor mounting torque
1975-1976 and 1977-on Schedule I — Every 22,500 miles 1977-on Schedule II — Every 30,000 miles Others — Every 12,000 miles	29	Spark plug replacement
1975-on—every 12 months or 15,000 miles Others—every 12 months or 12,000 miles Note: 1977-on Schedule I, perform Item 33 at 7,500 miles, at 22,500 miles, and every 22,500 miles thereafter. Perform item 36 every 22,500 miles.	30	EGR system (1973-1974)
	31	Carburetor fuel inlet filter
	32	Thermal vacuum switch and hoses (if equipped)
	33	Vacuum advance solenoid and hoses
	34	Transmission control switch (1974 and earlier)
	35	Idle stop solenoid
	36	PCV system
1975-on—every 24 months or 30,000 miles Others—every 24 months or 24,000 miles	37	Engine compression
	38	ECS system
	39	Fuel cap, tank and lines
	40	AIR system (1972-1974)
1975-on—every 30,000 miles Others—every 24,000 miles	41	Mechanical valve lifters (1970 models so equipped)
	42	Air cleaner element
1975-1976 and 1977-on Schedule I—every 22,500 miles 1977-on Schedule II—every 30,000 miles Others—first 24 months or 24,000 miles, then every 12,000 miles or 12 months	43	Spark plug and ignition coil wires
	44	Carburetor vacuum break (1977 — Schedule I only)
	45	Oxygen change sensor (1977 — Schedule II, if so equipped)
	46	VD and DVDS valves (1977 — Schedule II only)
1975-on—first at 6 months or 7,500 miles, then at 18-month/22,500-mile intervals	47	EFE valve

LUBRICATION, MAINTENANCE, AND TUNE-UP

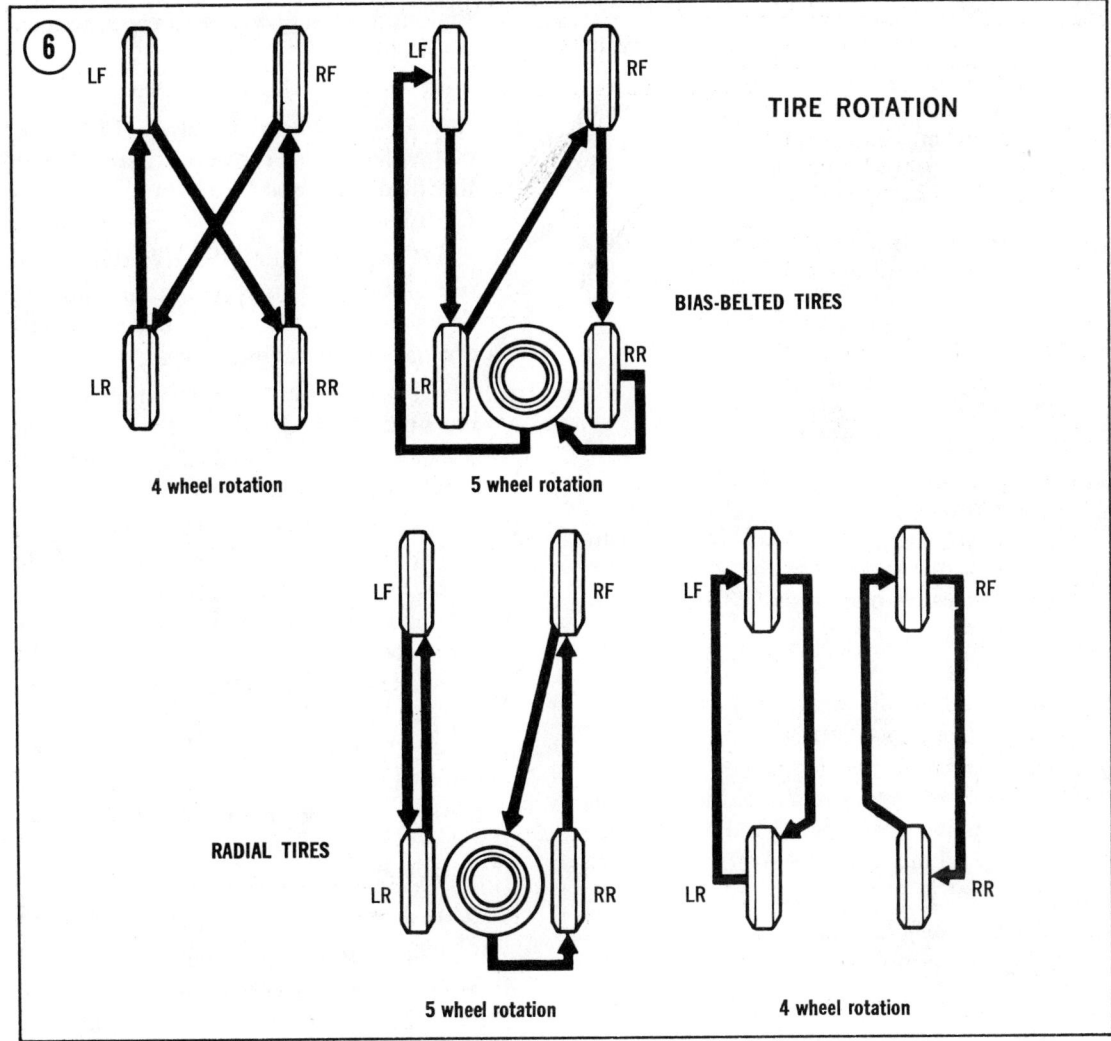

Tighten hose clamps and inspect condition of all cooling and heater hoses. Replace all hoses every 24 months, or earlier if chipped, swollen or otherwise deteriorated. Clean exterior of radiator and air conditioner every 12 months (preferably using compressed air). Drain and flush radiator and replace coolant every 24 months or 24,000 miles (30,000 miles on 1975 and later models).

9. *Wheel bearings* — Clean and repack wheel bearings at intervals shown on chart. See **Table 1** for recommended lubricant.

10. *Automatic transmission fluid* — Under normal driving conditions, change fluid and service sump filter every 24,000 miles (30,000 miles, 1975 and later models). Cut intervals in half if vehicle has been operated under severe conditions, such as trailer pulling, and so forth.

11. *Manual steering gear* — Check around pitman arm and housing for seal leakage and oozing. Leakage (solid grease, not just oily film) should be corrected immediately. Lubricate clutch cross shaft lever.

12. *Owner safety checks* — Items listed below should be checked at the intervals given in **Table 2** and deficiencies should be corrected as soon as possible.

   a. *Steering column lock* (if so equipped) — Key should turn to LOCK position only when transmission selector is in PARK (automatic transmission) or REVERSE (manual transmission).

b. *Parking brake and transmission* PARK *mechanism* — Check holding ability by restraining vehicle with parking brake only on a fairly steep hill. Then check PARK mechanism (automatic transmission) by placing transmsission selector in PARK (P) and releasing all brakes.

c. *Starter safety switch* — Starter should operate only in PARK (P) or NEUTRAL (N) positions (automatic transmissions) or in NEUTRAL with clutch fully depressed (manual transmissions, if equipped with starter safety switch).

d. *Transmission shift indicator* — Verify that automatic shift indicator accurately indicates the shift position selected.

e. *Steering* — Verify that the steering mechanism operates freely and does not have excessive play or make harsh sounds when turning or parking.

f. *Wheel alignment and balance* — Check for abnormal tire wear, then road test the vehicle. The need for alignment may be indicated by a pull to either the right or left on a straight, level road. The need for wheel balancing is usually indicated by excessive vibration of the steering wheel or the front of the vehicle while driving at normal highway speeds.

g. *Brakes* — Observe brake warning light (if so equipped) during brake action. Also check for changes in braking action, such as pulling to one side, unusual sounds, or increased brake pedal travel. See Chapter Ten for additional checks.

h. *Exhaust system* — Be alert to any smell of fumes in the vehicle, or to any change in the sound of the exhaust system that might indicate leakage.

i. *Windshield wipers and washers* — Check operation of wipers and condition and alignment of wiper blades. Check operation of washers (amount and direction of sprayed fluid) and fill fluid reservoir.

j. *Defrosters* — Turn on heater, then move control to DEFROST (DEF) and check amount of air directed to windshield.

k. *Rear view mirrors and sun visors* — Verify that friction joints are adjusted so that mirrors and visors stay in selected position.

l. *Horn* — Verify that horn still works.

m. *Lap and shoulder belts* — Check all components for proper operation. Verify that all anchor bolts are tight. Check belts for fraying.

n. *Head restraints* — Verify that head restraints, if so equipped, adjust up and down properly and that no components are missing, loose or damaged.

o. *Seat back latches* — Verify that latches are holding by pulling forward on seat backs with doors closed (if equipped with automatic seat back latches).

p. *Lights and buzzers* — Verify that all lights and buzzers are working. These include seat belt reminder light and buzzer, ignition key buzzer, interior lights, instrument panel illuminating and warning lights, headlights, license plate lights, side marker lights, parking lights, turn directional signals, back-up lights, and hazard warning flashers.

q. *Glass* — Check for any condition that could obscure vision or be a safety hazard. Correct as required.

r. *Door latches* — Verify positive closing, latching, and locking action.

s. *Hood latches* — Verify that hood closes firmly by lifting up on hood after closing it. Check for missing or damaged parts.

t. *Fluid leaks* — Check under vehicle after it has been parked for a while for evidence of fuel, water, or oil leaks. (Water dripping from air conditioner after use is normal.) Immediately determine and correct cause of any gasoline fumes or liquid to avoid possible fire or explosion.

13. *Tires and wheels* — Check tires for nails, cuts, excessive wear or other damage. Verify that tread wear indicators are not visible. Verify that tires are inflated to pressures given on tire placard (on rear face of driver's door, late models) or in owner's manual.

14. *Exhaust system* — Check entire exhaust system, from exhaust manifold to tailpipe, including catalytic converter, if so equipped. Also

LUBRICATION, MAINTENANCE, AND TUNE-UP

check nearby areas of body and trunk lid for broken, damaged, missing, or mispositioned parts, open seams, holes, loose connections, or any other defect that could permit exhaust fumes to enter the trunk or passenger compartment. The presence of dust or water in the trunk could be an indication of a problem in one of these areas. Whenever a muffler is replaced, all system components to the rear of the muffler (exhaust pipes, resonators, etc.) should also be replaced to maintain the integrity of the system.

15. *Drive belts* — Check all engine drive belts (fans, Delcotron, AIR pump, power steering pump, and air conditioner compressor belts) for cracks, fraying and tension. Adjust or replace, as necessary.

16. *Suspension and steering* — Check for loose, damaged, or missing parts, signs of excessive wear, and lack of lubrication. Repair or correct, as required.

17. *Brakes and power steering* — Check all hoses and lines for proper connections and leaks or deterioration. Immediately replace any questionable parts. If abrasion or other wear is evident on hoses, find and correct cause.

18. *Disc brakes* — Check brake pads and rotor condition while wheels are removed for tire rotation. Condition should be checked more often if driving conditions and habits result in frequent brake application.

19. *Drum brakes and parking brakes* — Remove drums, springs, etc. Verify that cylinders are not leaking. Check parking brake adjustment for drag, and lubricate parking brake cable passages and lever mechanism at chassis lube period. More frequent brake application.

20. *Throttle linkage* — Inspect for damaged or missing parts, interference or binding. Correct deficiencies immediately.

21. *Headlights* — Have headlights checked for proper aim. More frequent checks should be made if illumination area ahead of car seems inadequate.

22. *Underbody* — If car is operated in areas where salt is used on roads for snow removal, or other corrosive materials are used, underside of car should be flushed and inspected yearly, preferably in the spring. Particular attention should be given to areas where mud and other materials have collected.

23. *Bumpers* — Check front and rear bumper systems for loose bolts and alignment.

24. *Thermostatically controlled air cleaner* — Verify that all hoses and ducts are properly connected and that valve is working properly.

25. *Carburetor choke* — Verify that choke mechanism is operating freely and remove any petroleum gum formation on choke shaft or correct any other damage or condition that may be causing binding.

26. *Timing, dwell, carburetor idle, distributor and coil* — Tune up engine at the intervals indicated, using procedures given in this chapter.

27. *Manifold heat valve* — Inspect (if so equipped) and verify free operation. Correct, if binding is present, by application of penetrating oil such as Liquid Wrench. Do not use motor oil.

28. *Carburetor mounting* — Tighten carburetor mounting bolts at the interval shown, then every 24,000 miles, to compensate for gasket compression and/or vibration.

29. *Spark plugs* — Replace at the intervals shown. If misfiring occurs earlier, remove and clean spark plugs if condition permits, otherwise replace.

30. *Exhaust Gas Recirculation System (EGR)* — At the interval shown (and if vehicle is so equipped), remove, inspect and clean EGR valve. Inspect and clean, if required, EGR passages in inlet manifold. If EGR valve is damaged, it must be repaired or replaced.

31. *Carburetor fuel inlet fiter* — Replace at the intervals indicated, or more frequently if filter becomes clogged.

32. *Thermal vacuum switch and hoses* — Check for proper operation and replace switch if not functioning properly. Check hoses for proper connection and damage. Replace if damaged.

33. *Vacuum advance solenoid and hoses* — Check both vacuum and electrical functions and replace if not operating properly. Check all electrical wires and vacuum hoses and replace those found defective or reconnect all improper connections.

CHAPTER THREE

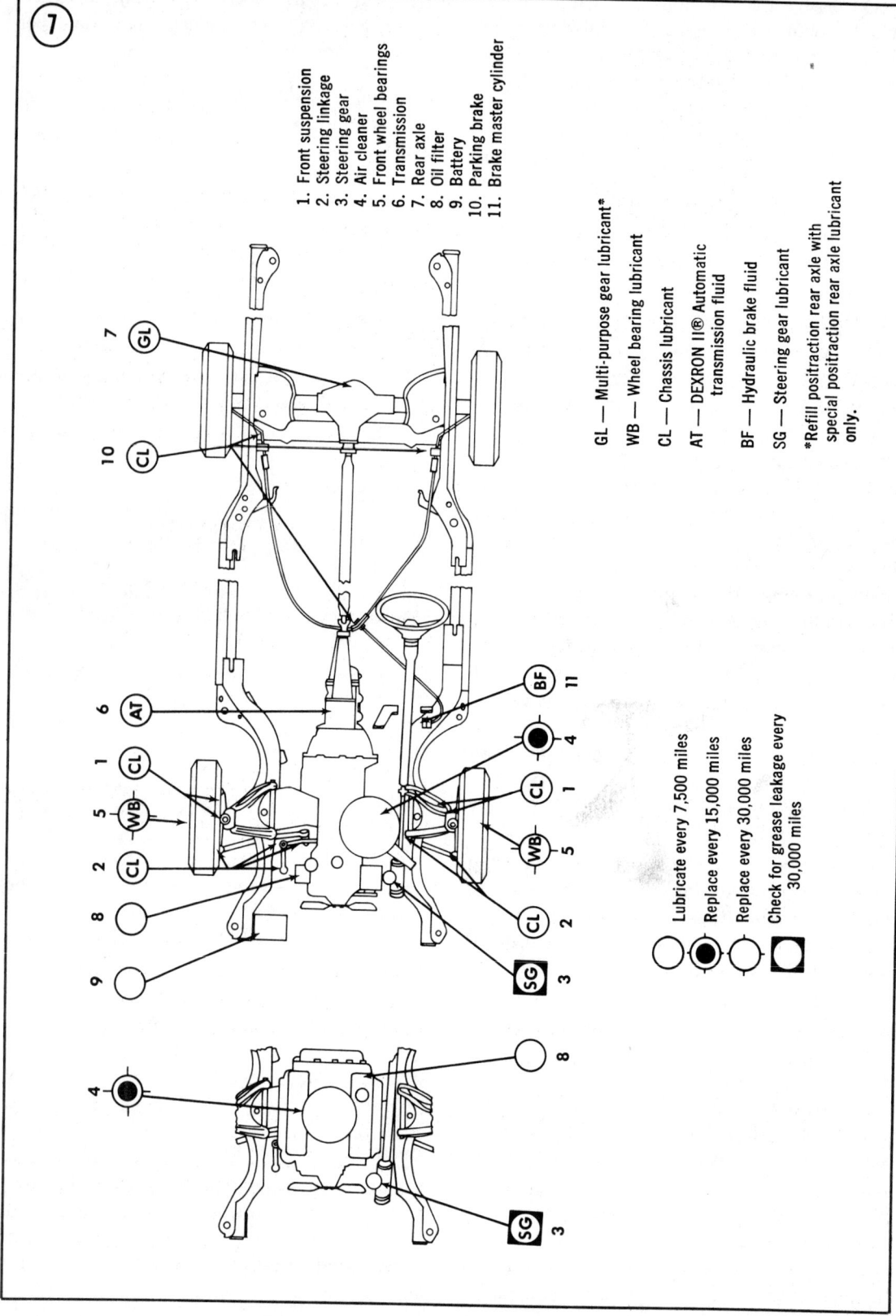

LUBRICATION, MAINTENANCE, AND TUNE-UP

34. *Transmission control switch* — Verify switch operation and check all wires and connections. Replace defective parts.

35. *Idle stop solenoid* — Verify proper operation and replace if defective.

36. *Positive Crankcase Ventilation system* (PCV) — Check operation at the intervals given in **Table 2** and replace every 24,000 miles. Clean and inspect hoses, replacing any damaged ones. Replace PCV filter (if so equipped) at 12 month/12,000 mile intervals.

37. *Engine compression* — Check compression at the intervals listed in **Table 2**, using the procedure given this chapter.

38. *Evaporation Control System* (ECS) — Check all hoses and lines for proper routing, connections and condition. Check vapor canister for cracks or other damage. Replace damaged parts as required. Replace filter in canister lower section.

39. *Fuel cap, lines and tank* — Inspect for damage that could cause leakage. Inspect cap for sealing ability. Replace damaged parts.

40. *Air Injection Reactor (AIR) system* — Check hoses for deterioration and improper connections. Correct or replace, as required.

41. *Mechanical valve lifters* (if so equipped) — Adjust clearance at intervals shown in **Table 2**.

42. *Air cleaner element* — Replace at the intervals shown in **Table 2**, or more frequently if vehicle is being operated in dusty areas.

43. *Spark plug and ignition coil wires* — Inspect and test at intervals shown on **Table 2**. See procedure in this chapter. Replace faulty or deteriorated wires.

44. *Carburetor vacuum break* — Inspect linkage for proper operation. Correct any binding of linkage. Check for deterioration and correct connection, and replace any damaged parts.

45. *Oxygen change sensor* — Replace sensor at the intervals shown (or when sensor warning signal is visible in speedometer face). Reset signal. See Chapter Five for procedures.

46. *Vacuum differential valve (VDV) and Differential Vacuum Delay and Separator Valve* (DVDSV) — Check valve and connecting hoses for correct routing and proper connections, and damage and deterioration. Replace faulty parts as required.

47. *Early Fuel Evaporation (EFE) valve* — Check for proper operation. Any binding condition must be corrected. Check switch for proper operation. Check for and replace damaged hoses.

LUBRICATION AND LUBRICANTS

The selection of lubricants and fluids is as important to automobile care as the actual maintenance program itself. Only those types of lubricants and fluids recommended by the manufacturers should be used. This is especially true if the vehicle is still under warranty, as the use of non-recommended types could cause cancellation of the warranty.

Table 1 lists the various lubricants and fluids recommended by General Motors Corporation. **Figure 7** shows the lubrication points.

ENGINE

Crankcase Capacity

All 1970 and later 6- and 8-cylinder engines use 4 qt. plus 1 qt. at filter change.

Engine Oil and Filters

Engine oil should be selected to meet the demands of the temperatures and driving conditions anticipated. General Motors recommends the use of oil with the letter designation SE for all vehicles regardless of model year and previous oil quality recommendations. This designation is plainly marked on the oil can. See **Figure 8**.

If the temperatures expected during the period the oil will be used are between −30 and +20 degrees F, the use of 5W-20 or 5W-30 viscosity oil is recommended. However, 5W-20 oil is not recommended for sustained high-speed driving.

If the expected temperatures will be between 0 and +60 degrees F, 10W, 5W-30, 10W-30, or 10W-40 viscosity oils are recommended. SAE 30 oil may be used if the temperature is not expected to drop below 40 degrees F.

If the temperature is not expected to drop below +20 degrees F, 20W-20, 10W-30, 10W-40, 20W-40, or 20W-50 viscosity oils are recommended.

If the vehicle is to be operated exclusively in Canada, 5W-30 viscosity oil should be used in all seasons.

The recommended frequency of oil and filter changes are given in **Table 2**. The recommended periods should be cut in half under any of the following conditions:

1. Driving in dusty areas. Operation in a dust storm may require an immediate oil change.
2. Trailer pulling
3. Extensive engine idling
4. Short trip operation in sub-freezing weather

CAUTION
*General Motors specifically warns against the use of non-detergent, low quality oil, or any oil not having the SE designation. The regular use of oil additives is specifically not recommended. Occasional use of additives to correct specific problems is authorized. For instance, the use of Super Engine Oil Supplement (available at Chevrolet dealers) is recommended to reduce varnish and sludge deposits in older engines. As a rule, a Chevrolet dealer should be consulted before using additives. Oil filters should be changed at the intervals recommended in **Table 2**. Use a high quality filter.*

Engine oil does not "wear out." Instead, it becomes diluted by fuel vapor leaking by the pistons and piston rings, and by the condensation of water vapor on the cylinder walls and in the crankcase. The detergents which provide the engine cleaning capability of the oil also tend to carry dirt and other contamination in suspension.

Leakage of fuel or fuel vapors into the crankcase occurs mostly during warm-up periods, when fuel is not always thoroughly vaporized and burned. Water vapor enters the crankcase through normal crankcase ventilation (especially on pre-1970 models) and through exhaust gas blow-by. When the engine is not completely warmed up, the water vapor tends to condense, combine with the condensed fuel and exhaust gases, and form acid compounds. When the temperatures in the crankcase are hot enough to prevent condensation, no harm is done. In extremely cold climates, however, the engine does not, as a rule, warm up sufficiently to prevent acid formation (especially on short runs). The acid can cause serious etching or pitting and thus cause very rapid wear on piston pins, bearings, and other moving parts. Fortunately, modern engines are equipped with a number of automatic devices which minimize the danger of crankcase dilution.

The thermostat, mounted in the cylinder head water outlet, restricts the flow of water to the radiator until a pre-selected temperature is reached. This cuts down the amount of time required for the engine to reach an efficient operating temperature. This, in turn, cuts down on the time that engine temperatures are low enough to allow condensation.

Engines also have a water bypass in the cooling system which allows limited circulation until the thermostat opens. This helps to eliminate hot spots during warm-up, and also helps prolong engine life.

A thermostatic heat control on the exhaust manifold directs hot exhaust gases against the center of the intake manifold during the warm-up period. This greatly aids in the vaporization of fuel.

The automatic choke (if so equipped) reduces the likelihood of unvaporized fuel entering the combustion chambers and leaking into the crankcase.

An efficient crankcase ventilation system helps draw off fuel and other vapors and aids in the evaporation of fuel and water.

LUBRICATION, MAINTENANCE, AND TUNE-UP

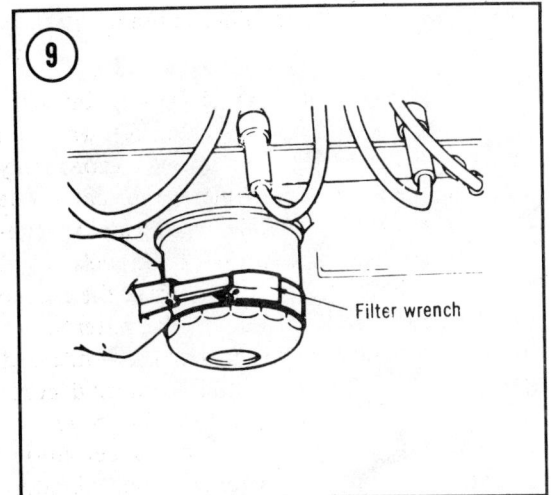

the drain pan into a gallon bleach bottle. Tighten the cap and throw it in your household trash (if not prohibited).

1. Warm engine to operating temperature, then shut it off.
2. Put drain pan under oil drain plug, and remove plug with a wrench.
3. Let oil drain for at least 10 minutes.
4. Uscrew oil filter counterclockwise by hand or use a filter wrench. See **Figure 9** (typical).
5. Wipe the gasket surface of the engine block with a clean, line-free cloth.
6. Coat the neoprene gasket on the new filter with clean oil. See **Figure 10**.
7. Screw the filter onto the engine *by hand* until the filter gasket just touches the base, i.e., until you feel the slightest resistance when turning the filter. Then tighten the filter *by hand* ⅔ turn or more.

CAUTION
Do not overtighten and do not use a filter wrench or the filter will leak.

8. Install oil drain plug, and tighten securely.
9. Remove oil filter cap.
10. Pour in 4 quarts (5 quarts with filter change) of oil recommended in **Table 1**.
11. Start engine and let it idle. The oil pressure light on instrument panel will remain on for a short time (15-30 seconds), then it will go out.

CAUTION
Do not rev engine to make oil light go out. It takes time for the oil to reach all areas of the engine and excessive engine speed could damage dry parts.

12. While the engine is running, make sure that the drain plug and oil filter are not leaking.
13. Turn the engine off and check the oil level with the dipstick. See **Figures 1 and 2**. Add oil if necessary to bring it up to FULL mark, but *do not overfill.*

From the above discussion, the need for regular oil changes can be seen, especially if the vehicle is used exclusively on short trips or in extremely cold temperatures.

To drain the oil and change the filter, you will need:

a. Drain pan
b. Funnel
c. Can opener or pour spout
d. Filter wrench
e. ⁹⁄₁₆-inch wrench (14mm will also work)
f. 4 or 5 quarts of oil (See **Table 2**)
g. Oil filter (AC PF-25; Lee LF25; Fram PH-30; Purolator PER-49).

There are a number of ways of discarding the old oil safely. The easiest way is to pour it from

Air Cleaners

Two basic types of air cleaners were used during the period covered by this book. One type was the semi-permanent polyurethane foam type and the other was the replaceable paper element type.

Service to the polyurethane type used in some early models consists of removal, washing in solvent, soaking in engine oil, squeezing to wring out excess oil, and replacing the cleaner on the vehicle.

Service to the paper element type used in later models consists only of replacement. Elements must not be cleaned with air hose, tapped, washed or oiled.

To remove the filter element, remove the wing nut holding the air cleaner cover in place (**Figure 11**, typical) and lift out the filter element. Installation is the reverse. Securely tighten the wing nut.

Battery Terminal Washers

Some older models have batteries equipped with felt washers on the terminals. The purpose of these oil-soaked washers is to control corrosive action. The washers should be saturated with engine oil every 6,000 miles.

Fuel Filter

A clogged fuel filter can cause "stumbling" or "cutting out" at high speeds and will eventually lead to complete fuel starvation. Filters should be changed at the intervals recommended in **Table 2** or, of course, whenever they become clogged.

The fuel filter is located in the carburetor fuel inlet fitting on all models. See **Figure 12** (typical). Disconnect the fuel line from the fuel inlet and remove the fuel filter nut. Remove old filter and filter spring. Install spring and new filter element in carburetor. Install a new gasket on the filter nut and install and securely tighten nut. Install the fuel line and tighten connector.

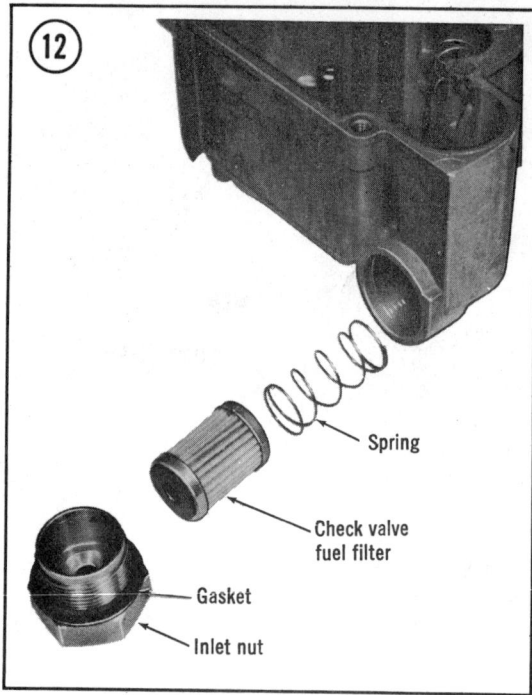

Rear Axle and Manual Transmission Lubrication

CAUTION
Operating conditions at high speeds are very severe and damage to differential and transmission gears may occur if improper lubricants are used. Straight mineral oil gear lubricant must not be used, especially in hypoid rear axles. Instead, use SAE 80 or SAE 90 GL-5 Gear Lubricant. Use SAE 80 GL-5 in vehicles used in Canada or other areas where the temperatures are expected to be consistently below 32°F.

Rear axle and manual transmission lubricants normally do not require changing during the life of the vehicle. However, if the vehi-

LUBRICATION, MAINTENANCE, AND TUNE-UP

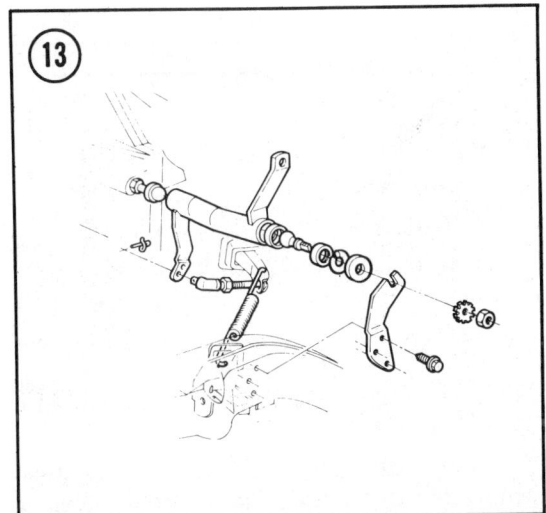

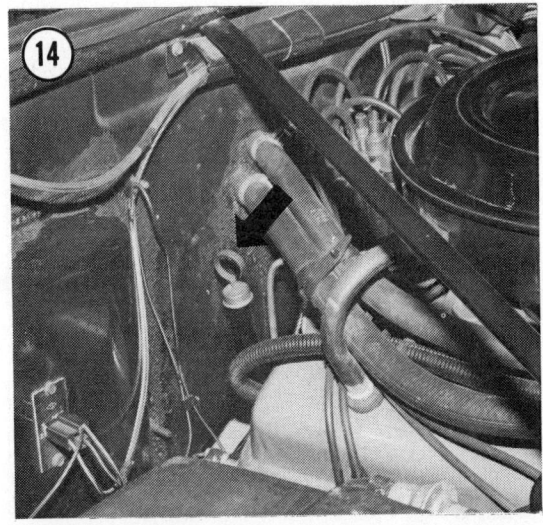

cle is used consistently for trailer pulling, rear axle lubricant should be changed at 12,000 mile intervals (15,000 miles for 1975 and later models).

Positraction rear axles require a special lubricant (General Motors Part No. 1051052 or equivalent). Lubricant levels should be checked at operating temperatures and at the intervals recommended in **Table 2**. Add only lubricant of the same type already in the housing. When at operating temperature, lubricant level should be even with bottom of filler plug hole. Rear axle filler hole is located on the right side of the differential carrier (**Figure 7**). Transmission filler hole is located on the right side of the transmission housing (**Figure 7**).

Transmission Shift Linkage (Manual and Automatic)

Shift linkage and manual transmission floor control lever contacting faces should be lubricated with engine oil every 4 months or 6,000 miles, whichever comes first.

NOTE: *If chassis lubricant has been used previously on linkage, continue using this lubricant.*

Clutch Cross Shaft

Periodic lubrication of the clutch cross shaft is not required. However, at 36,000 miles (or sooner, if required), remove plug, install grease fitting and apply chassis lubricant with a grease gun.

An exploded view of the cross shaft is shown in **Figure 13**.

AUTOMATIC TRANSMISSIONS

Automatic transmission fluids identified with the names DEXRON II or DEXRON are recommended for use in all Chevrolet automatic transmissions. These fluids are available at Chevrolet dealers and also at service stations.

Although several models of automatic transmissions were supplied during the period covered herein, the service procedure is basically the same for all models.

1. Check fluid level at the interval recommended in **Table 2**.
2. Operate the automobile for several miles to bring transmission up to normal operating temperature.
3. Park in a level space and leave engine running, with parking brake set and transmission selector in PARK.
4. Remove transmission dipstick (**Figure 14**), wipe clean, and reinsert until cap fully seats. Remove dipstick and note reading.
5. If fluid is at or below the ADD mark, add sufficient fluid to bring level to FULL mark.

NOTE: *One pint of fluid is required to bring the level from the* ADD *mark to the* FULL *mark in all models.*

CAUTION
Do not overfill the transmission. To do so could cause damage.

Under normal driving conditions the automatic transmission should be drained and refilled at the intervals indicated in **Table 2**. If the car is driven extensively in heavy city traffic in hot weather, or is used to pull a trailer, cut the recommended intervals in half. See **Table 3** for the fluid capacities of automatic transmissions.

After replacing automatic transmission fluid, always recheck the fluid level, using the procedure given above.

To change the transmission fluid, you will need the following:

a. The required amount of DEXRON II Automatic Transmission Fluid (see **Table 3**)
b. Funnel
c. Pour spout
d. Container large enough to hold the old fluid.

> CAUTION
> *Do not attempt to change the transmission fluid if you must work outdoors. Cleanliness is very important when working with an automatic transmission. Blowing dust can settle on internal parts and cause serious damage. If you do not have access to a clean, enclosed garage, let your dealer or local repair shop change the fluid.*

1. Drive the vehicle for at least 5 miles to warm up the transmission fluid.
2. Raise the vehicle and support it on jackstands. Also place a suitable jack under the transmission. Remove the transmission crossmember support, if necessary.
3. Place a suitable drain pan under the transmission and remove the attaching bolts from the front and side of the oil pan.
4. Unscrew the rear oil pan attaching bolts about 4 turns and then carefully pry the oil pan loose and allow fluid to drain.
5. Remove the rear oil pan attaching screws and remove the oil pan and gasket. Discard the gasket.
6. Drain the remaining fluid from the oil pan and clean the pan with solvent. Make sure the pan is thoroughly clean and then dry it with compressed air.
7. Remove the attaching screws and then remove the strainer and gasket from the valve body. Discard the gasket.
8. If the strainer is to be reused, thoroughly clean it in solvent and blow it dry with compressed air.
9. Install the strainer (or new filter), using a new gasket and the 2 attaching screws.
10. Install the oil pan, using a new gasket. Tighten the attaching screws to 12 ft.-lb.

> CAUTION
> *Do not overtighten the transmisssion oil pan attaching screws, as this could damage the gasket and cause transmission fluid leaks. This, in turn, could cause serious transmission damage.*

11. Lower the vehicle and fill the transmission with the proper amount of DEXRON II automatic transmission fluid. Use a clean funnel inserted in the filler tube (inside the engine compartment), and a clean pouring spout installed in the fluid container to help prevent spillage.
12. Place the transmission selector lever in PARK position; set the handbrake, start the engine, and let it idle.

> CAUTION
> *Do not race the engine at this time, as the transmission fluid has not yet had an opportunity to circulate properly and damage could result from lack of lubrication.*

13. Move the transmission selector lever through each position and then return it to PARK. Check the transmission fluid level with

Table 3
AUTOMATIC TRANSMISSION FLUID CAPACITY

Transmission	Capacity
Turbo Hydramatic 400	7½ U.S. pints
Turbo Hydramatic 375	7½ U.S. pints
Turbo Hydramatic 350	5 U.S. pints
Turbo Hydramatic 250	5 U.S. pints
Powerglide	3 U.S. pints

LUBRICATION, MAINTENANCE, AND TUNE-UP

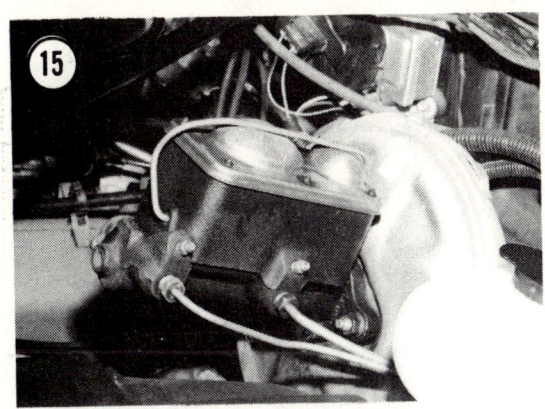

the dipstick (**Figure 14**) and add enough fluid, if necessary, to bring the level to ¼ inch *below* the ADD mark.

> **CAUTION**
> *Do not overfill the transmission, as this could cause foaming of the fluid and inadequate lubrication, resulting in damage. Remember that the transmission fluid you have just added is cold, and that the marks on the transmission dipstick are intended for measuring hot fluid (the fluid expands as it heats).*

14. Test drive the car for several miles to allow the transmission to reach operating temperature and then recheck the level with the transmission in PARK. The level now should be between the ADD and FULL marks on the dipstick. Add fluid as required to bring the level up to, but not above, the FULL mark. The distance between the ADD and FULL marks represents about one pint.

CHASSIS

Chassis Lubricant

General Motors recommends a water resistant EP chassis lubricant meeting the requirements of GM Specification GM 6031-M. This lubricant is designed to be applied by commercial pressure gun equipment. Chassis should be lubricated at the intervals shown in **Table 2**.

Front Wheel Bearings

Front wheel bearings must be lubricated every 30,000 miles (1976 and later models), or 24,000 miles (1970-1974 models), according to model year. The lubricant recommended is a high-melting point, water resistant, front wheel bearing lubricant meeting the requirements of GM Specification GM 6031-M or equivalent. If the vehicle is equipped with disc brakes, use wheel bearing lubricant, GM Part No. 1051344, or equivalent. This, according to General Motors, is a premium, high-melting point lubricant.

> **CAUTION**
> *General Motors warns against the use of "long fiber" or "viscous" type lubricants on front wheel bearings, and also against the mixing of lubricant types. To avoid bearing damage, use the recommended lubricants and thoroughly clean bearings and hubs of old lubricants before repacking.*

To lubricate bearings, remove front wheels and hub assemblies (see Chapter Ten for disc brake removal and replacement procedure). Remove, clean, and inspect bearing assemblies and replace with new ones if necessary. Pack bearings with lubricant, clean hub assembly, and reinstall bearings. Reinstall wheel and hub assemblies and adjust bearings, using the procedure given in Chapter Ten.

> **CAUTION**
> *Do not pack hub between inner and outer bearings or the dust cups as excess lubricant may destroy seals and work into the brake drums and linings, causing brake deterioration or failure.*

> **CAUTION**
> *The proper adjustment of wheel bearings is important to both safety and proper operation. The procedure given in Chapter Ten should be read, understood, and closely followed before attempting adjustment.*

Brake Master Cylinder

At the intervals shown in **Table 2**, check the fluid level in the master cylinder reservoir. Clean dirt and grime from edge of cover so that it will not fall into the reservoir. Pry the wire retainers back with a screwdriver (**Figure 15**) and lift the cover off. The level should be ¼ inch below the lowest edge of the reservoir (see **Figure 16**). Top up if necessary with hydraulic brake fluid clearly marked DOT-3 or GM Hydraulic Brake Fluid Supreme No. 11.

Parking Brake

Apply chassis lubricant to parking brake cable, cable guides, and all operating linkages and levers at the intervals recommended in **Table 2**.

Manual Steering Gear

> NOTE: *The steering gear is filled at the factory. Seasonal change is not required and no lubrication is required for the life of the steering gear.*

Every 36,000 miles (30,000 on 1975 and later models), the steering gear should be inspected for seal leakage. (Worry about actual solid grease; an oily film can be ignored.) If the steering gear is overhauled or a seal is replaced, refill housing with GM Part No. 1051052 Steering Gear Lubricant.

> CAUTION
> *Do not use chassis lubricant to fill the steering gear housing. Also, do not overfill the housing.*

Power Steering System

Check power steering fluid at each oil change period. Add GM Power Steering Fluid (DEXRON II or DEXRON Automatic Transmission Fluid) as required to bring level into proper range on filler cap dipstick, depending upon fluid temperature (see **Figure 17**). If at operating temperature (1972-on models), fluid should be between HOT and COLD marks. If at room temperature, fluid should be between ADD and COLD marks. On earlier models, fluid should be brought up to FULL mark. The fluid does not require periodic changing.

Hood Latches and Hinges

At each oil change period lubricate hood latch and hood hinge assemblies as follows:
1. Wipe latch and hinge areas clear of accumulated dirt or contamination.
2. Apply Lubriplate or equivalent to latch pilot bolts and locking plate.
3. Apply engine oil to all pivot points in hood release mechanism and primary and secondary latch mechanisms.
4. Lubricate hinges with engine oil.
5. Make a functional check to verify that all parts of the hinge and latch assemblies are functioning properly.

ENGINE TUNE-UP

The tune-up consists of a series of inspections, adjustments and part replacements to compensate for wear and deterioration of engine components. Regular tune-ups are especially important to modern, high powered, high-performance engines. Emission control systems, improved electrical systems, and other advance make these engines especially sensitive to improperly operating or incorrectly adjusted parts.

Since proper engine operation depends upon a number of interrelated system functions, a

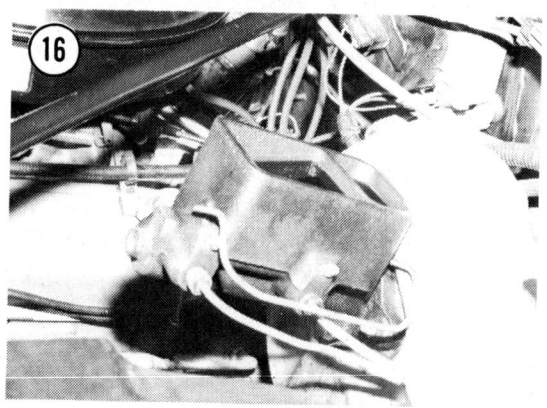

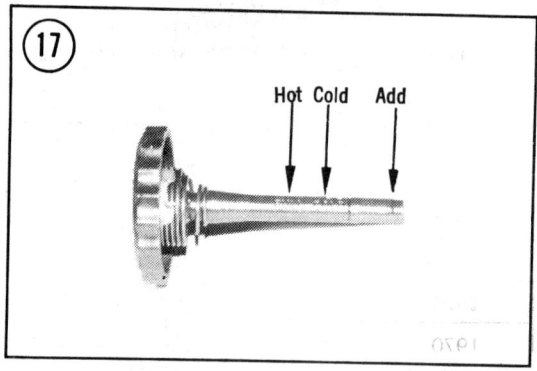

LUBRICATION, MAINTENANCE, AND TUNE-UP

tune-up consisting of only one or 2 corrections will seldom get lasting results. Instead, a thorough, systematic procedure of analysis and correction will pay dividends in improved power, performance, and operating economy.

The procedure presented in this chapter requires a series of visual and mechanical checks and adjustments, followed by an instrument checkout. The instruments required are described in detail in Chapter One. Tune-up specifications are given in **Table 4**.

The tune-up for 1974 and older models consists of the following, in the order listed:

1. Valve lash adjustment (if equipped with mechanical valve lifters)
2. Engine compression check
3. Ignition adjustment and timing, consisting of:
   a. Spark plug replacement
   b. Breaker point replacement and/or adjustment and condenser replacement
   c. Distributor cap and rotor inspection (and replacement, if required)
   d. Spark plug wire inspection (and replacement, if required)
   e. Ignition timing adjustment
4. Carburetor adjustment

NOTE: *All 1975 and later vehicles have breakerless ignition systems.*

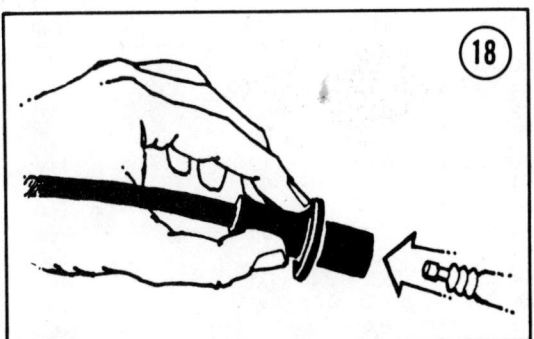

The above steps also apply to 1975 and later cars, except for valve lash adjustment and breaker point replacement and adjustment.

Firing Order

The cylinder firing order for all Chevrolet engines is: 1-5-3-6-2-4 for inline 6-cylinder and 1-8-4-3-6-5-7-2 for all V8 engines. The firing order for V6 engines is 1-6-5-4-3-2.

Valve Clearance Adjustment (Mechanical Lifters)

1. Start engine and warm up to normal operating temperature.
2. Remove rocker arm covers and gaskets.

CAUTION
Do not pry rocker arm cover loose. Bump end of rocker arm cover with palm of hand or rubber mallet to break loose.

3. Use a socket wrench on self-locking arm stud nut and adjust as required to obtain clearance specified in **Table 5** for each valve, measured with leaf-type feeler gauge between rocker arm and end of valve stem.

4. Stop engine, clean gasket sealing surfaces on cylinder heads and rocker arm covers. Install rocker arm covers, using new gaskets, and torque bolts to 50-60 in.-lbs.

SPARK PLUG REMOVAL

1. Remove all foreign matter from around spark plugs, using compressed air if available. A tire pump, vacuum cleaner, or a small paintbrush can also be used.
2. Disconnect spark plug wires by grasping the *boot* portion of the wire and applying only enough twisting force to remove the wire from the plug. See **Figure 18**.

Table 5   VALVE LASH SPECIFICATIONS

Year	Engine	Intake (Hot)	Exhaust (Hot)
1970	350 cu. in., 370 hp	0.030 in.	0.030 in.
1970	402 cu. in., 375 hp	0.024 in.	0.028 in.
1970	454 cu. in., 450 hp	0.024 in.	0.028 in.

Table 4 TUNE-UP SPECIFICATIONS

1. A = Automatic transmission
   M = Manual transmission
2. Curb idle speed with solenoid energized
3. Solenoid de-energized
4. Lean drop idle mixture
5. All except California cars
6. DR = in drive range
   N = in neutral
7. Fed. = All except California cars
   Calif. = California cars only
   All = All cars
   High alt. = Cars equipped for high altitude operation
8. With idle or A.C. solenoid energized
9. Use narrow timing mark on harmonic balancer with handheld timing light. Wide mark is for use with magnetic equipment.
10. 500 (DR) if not equipped with idle solenoid.
11. Without turbocharger
12. With turbocharger

Carburetor (1970 Models)	Engine			Distributor			Spark Plugs	
	Model[1]	CID	Option	Points Setting	Dwell Angle	RPM[2]	Type (AC)	Gap
1 bbl	L6 M	250	—	0.016 used, 0.019 new	31-34°	750	R46N	0.035
1 bbl	L6 A	250	—	0.016 used, 0.019 new	31-34°	600	R46N	0.035
2 bbl	V8 M	307	—	0.016 used, 0.019 new	29-31°	700	R45	0.035
2 bbl	V8 A	307	—	0.016 used, 0.019 new	29-31°	600	R45	0.035
2 bbl	V8 M	350	L-65	0.016 used, 0.019 new	29-31°	750	R44	0.035
2 bbl	V8 A	350	L-65	0.016 used, 0.019 new	29-31°	600	R44	0.035
4 bbl	V8 M	350	L-48	0.016 used, 0.019 new	29-31°	700	R44	0.035
4 bbl	V8 A	350	L-48	0.016 used, 0.019 new	29-31°	600	R44	0.035
4 bbl	V8 M	396	L-34	0.016 used, 0.019 new	29-31°	700	R44T	0.035
4 bbl	V8 A	396	L-34	0.016 used, 0.019 new	29-31°	600	R44T	0.035
4 bbl	V8 M	396	L-78	0.016 used, 0.019 new	29-31°	750	R43T	0.035
4 bbl	V8 A	396	L-78	0.016 used, 0.019 new	29-31°	700	R43T	0.035
2 bbl	V8 M	400	LF-6	0.016 used, 0.019 new	29-31°	700	R44	0.035
2 bbl	V8 A	400	LF-6	0.016 used, 0.019 new	29-31°	600	R44	0.035
4 bbl	V8 M	400	LS-3	0.016 used, 0.019 new	29-31°	700	R44T	0.035
4 bbl	V8 A	400	LS-3	0.016 used, 0.019 new	29-31°	600	R44T	0.035
4 bbl	V8 M	454	L-54	0.016 used, 0.019 new	29-31°	700	R44T	0.035
4 bbl	V8 M	454	L-55	0.016 used, 0.019 new	29-31°	700	R43T	0.035
4 bbl	V8 A	454	L-54	0.016 used, 0.019 new	29-31°	600	R44T	0.035
4 bbl	V8 A	454	L-55	0.016 used, 0.019 new	29-31°	600	R43T	0.035
4 bbl	V8 M	454	L-56	0.016 used, 0.019 new	29-31°	750	R43T	0.035
4 bbl	V8 A	454	L-56	0.016 used, 0.019 new	29-31°	700	R43T	0.035

Carburetor (1970 Models)	Engine			Timing (°BTC)	Idle Speeds			Compression (psi)
	Model[1]	CID	Option		Initial	Final		
1 bbl	L6 M	250	—	0	830	750	400	130
1 bbl	L6 A	250	—	4	630	600	400	130
2 bbl	V8 M	307	—	2	800	700	450	150
2 bbl	V8 A	307	—	8	630	600	450	150
2 bbl	V8 M	350	L-65	0	830	750	450	160
2 bbl	V8 A	350	L-65	4	630	600	450	160
4 bbl	V8 M	350	L-48	0	775	700	—	160
4 bbl	V8 A	350	L-48	4	630	600	—	160
4 bbl	V8 M	396	L-34	0	700	700	—	160
4 bbl	V8 A	396	L-34	4	630	600	—	160
4 bbl	V8 M	396	L-78	4	750	750	—	160
4 bbl	V8 A	396	L-78	4	700	700	—	160
2 bbl	V8 M	400	LF-6	4	800	700	450	160
2 bbl	V8 A	400	LF-6	8	630	600	450	160
4 bbl	V8 M	400	LS-3	4	775	700	—	160
4 bbl	V8 A	400	LS-3	4	630	600	—	160
4 bbl	V8 M	454	L-54	6	700	700	—	160
4 bbl	V8 M	454	L-55	6	700	700	—	160
4 bbl	V8 A	454	L-54	6	630	600	—	160
4 bbl	V8 A	454	L-55	6	630	600	—	160
4 bbl	V8 M	454	L-56	4	750	750	—	150
4 bbl	V8 A	454	L-56	4	700	700	—	150

LUBRICATION, MAINTENANCE, AND TUNE-UP

Table 4 TUNE-UP SPECIFICATIONS (continued)

Carburetor (1971 Models)	Engine Model[1]	CDI	Option	Distributor Points Setting	Dwell Angle	RPM[3]	Spark Plugs Type (AC)	Gap
1 bbl	L6 M	250	—	0.016 used, 0.019 new	31-34°	500	R46TS	0.035
1 bbl	L6 A	250	—	0.016 used, 0.019 new	31-34°	550	R46TS	0.035
2 bbl	V8 M	307	—	0.016 used, 0.019 new	29-31°	600	R45TS	0.035
2 bbl	V8 A	307	—	0.016 used, 0.019 new	29-31°	550	R45TS	0.035
2 bbl	V8 M	350	L-65	0.016 used, 0.019 new	29-31°	600	R45TS	0.035
2 bbl	V8 A	350	L-65	0.016 used, 0.019 new	29-31°	550	R45TS	0.035
4 bbl	V8 M	350	L-48	0.016 used, 0.019 new	29-31°	600	R44TS	0.035
4 bbl	V8 A	350	L-48	0.016 used, 0.019 new	29-31°	550	R45TS	0.035
4 bbl	V8 M	402	LS-3	0.016 used, 0.019 new	29-31°	600	R44TS	0.035
4 bbl	V8 M	454	LS-5	0.016 used, 0.019 new	29-31°	600	R43TS	0.035
4 bbl	V8 A	402	LS-3	0.016 used, 0.019 new	29-31°	600	R44TS	0.035
4 bbl	V8 A	454	LS-5	0.016 used, 0.019 new	29-31°	600	R43TS	0.035
4 bbl	V8 M	454	LS-6	0.016 used, 0.019 new	29-31°	700	R43TS	0.035
4 bbl	V8 A	454	LS-6	0.016 used, 0.019 new	29-31°	700	R43TS	0.035

Carburetor (1971 Models)	Engine Model[1]	CID	Option	Timing (°BTC)	Idle Speeds Initial	Final	CEC	Compression (psi)
1 bbl	L6 M	250	—	4	625	550	850	130
1 bbl	L6 A	250	—	4	530	500	650	130
2 bbl	V8 M	307	—	4	700	600	900	150
2 bbl	V8 A	307	—	8	580	550	650	150
2 bbl	V8 M	350	L-65	2	700	600	900	160
2 bbl	V8 A	350	L-65	6	580	550	650	160
4 bbl	V8 M	350	L-48	4	675	600	900	160
4 bbl	V8 M	402	LS-3	8	580	550	650	160
4 bbl	V8 M	454	LS-5	8	675	600	850	160
4 bbl	V8 A	402	LS-3	8	675	600	850	160
4 bbl	V8 A	454	LS-5	8	630	600	650	160
4 bbl	V8 A	454	LS-5	8	630	600	650	160
4 bbl	V8 M	454	LS-6	8	700	700	900	150
4 bbl	V8 A	454	LS-6	12	700	700	750	150

Carburetor (1972 Models)	Engine Model[1]	CDI	Usage	Distributor Points Setting	Dwell Angle	RPM[2]	Spark Plugs Type (AC)	Gap
1 bbl	L6 M	250	—	0.016 used, 0.019 new	31-34°	700	R46T	0.035
1 bbl	L6 A	250	—	0.016 used, 0.019 new	31-34°	600	R46T	0.035
2 bbl	V8 M	307	—	0.016 used, 0.019 new	29-31°	900	R44T	0.035
2 bbl	V8 A	307	—	0.016 used, 0.019 new	29-31°	600	R44T	0.035
2 bbl	V8 M	350	—	0.016 used, 0.019 new	29-31°	900	R44T	0.035
2 bbl	V8 A	350	—	0.016 used, 0.019 new	29-31°	600	R44T	0.035
4 bbl	V8 M	350	—	0.016 used, 0.019 new	29-31°	900	R44T	0.035
4 bbl	V8 A	350	—	0.016 used, 0.019 new	29-31°	600	R44T	0.035
2 bbl	V8 A	400	—	0.016 used, 0.019 new	29-31°	600	R44T	0.035
4 bbl	V8 M	402	5	0.016 used, 0.019 new	29-31°	750	R44T	0.035
4 bbl	V8 A	402	5	0.016 used, 0.019 new	29-31°	600	R44T	0.035
4 bbl	V8 M	454	5	0.016 used, 0.019 new	29-31°	750	R44T	0.035
4 bbl	V8 A	454	5	0.016 used, 0.019 new	29-31°	600	R44T	0.035

Table 4 TUNE-UP SPECIFICATIONS (continued)

Carburetor (1972 Models)	Engine			Timing (°BTC)	Idle Speeds[6]		Compression (psi)
	Model[1]	CID	Usage		[3]	[4]	
1 bbl	L6 M	250	—	4	500	800/700	130
1 bbl	L6 A	250	—	4	500 (DR)	630/600 (DR)	130
2 bbl	V8 M	307	—	4	500	1,000/900	150
2 bbl	V8 A	307	—	8	500 (DR)	650/600 (DR)	150
2 bbl	V8 M	350	—	6	500	1,050/900	160
2 bbl	V8 A	350	—	6	500 (DR)	650/600 (DR)	160
4 bbl	V8 M	350	—	4	500	1,000/900	160
4 bbl	V8 A	350	—	8	500 (DR)	630/600 (DR)	160
2 bbl	V8 A	400	—	6	500 (DR)	650/600 (DR)	160
4 bbl	V8 M	402	[5]	8	500	750/750	160
4 bbl	V8 A	402	[5]	8	500(DR)	600/600(DR)	160
4 bbl	V8 M	454	[5]	8	500	750/750	160
4 bbl	V8 A	454	[5]	8	500(DR)	600/600(DR)	160

Carburetor (1973 Models)	Engine			Distributor			Spark Plugs	
	Model[1]	CID	Usage	Points Setting	Dwell Angle	RPM[2]	Type (AC)	Gap
1 bbl	L6 M	250	—	0.016 used, 0.019 new	31-34°	700	R46T	0.035
1 bbl	L6 A	250	—	0.016 used, 0.019 new	31-34°	600	R46T	0.035
2 bbl	V8 M	307	—	0.016 used, 0.019 new	29-31°	900	R44T	0.035
2 bbl	V8 A	307	—	0.016 used, 0.019 new	29-31°	600	R44T	0.035
2 bbl	V8 A	350	—	0.016 used, 0.019 new	29-31°	600	R44T	0.035
4 bbl	V8 M	350	—	0.016 used, 0.019 new	29-31°	900	R44T	0.035
4 bbl	V8 A	350	—	0.016 used, 0.019 new	29-31°	600	R44T	0.035
2 bbl	V8 A	400	—	0.016 used, 0.019 new	29-31°	600	R44T	0.035
4 bbl	V8 M	454	—	0.016 used, 0.019 new	29-31°	900	R44T	0.035
4 bbl	V8 A	454	—	0.016 used, 0.019 new	29-31°	600	R44T	0.035

Carburetor (1973 Models)	Engine			Timing (°BTC)	Idle Speeds		Compression (psi)
	Model[1]	CID	Usage		[3]	[4]	
1 bbl	L6 M	250	—	6	500	750/700	130
1 bbl	L6 A	250	—	6	500 (DR)	630/600(DR)	130
2 bbl	V8 M	307	—	4	500	950/900	150
2 bbl	V8 A	307	—	8	500 (DR)	630/600(DR)	150
2 bbl	V8 A	350	—	8	400	630/600(DR)	160
4 bbl	V8 M	350	—	8	500	920/900	160
4 bbl	V8 A	350	—	12	500	620/600(DR)	160
2 bbl	V8 A	400	—	6	500	630/600(DR)	160
4 bbl	V8 M	454	—	10	500	925/900	160
4 bbl	V8 A	454	—	10	500	625/600(DR)	160

LUBRICATION, MAINTENANCE, AND TUNE-UP

Table 4   TUNE-UP SPECIFICATIONS (continued)

Carburetor (1974 Models)	Engine			Distributor			Spark Plugs	
	Model[1]	CID	Usage[7]	Points Setting	Dwell Angle	RPM[2]	Type (AC)	Gap
1 bbl	L6 A	250	Fed.	0.016 used, 0.019 new	31-34°	600	R46T	0.035
1 bbl	L6 M	250	All	0.016 used, 0.019 new	31-34°	850	R46T	0.035
1 bbl	L6 A	250	Calif.	0.016 used, 0.019 new	31-34°	600	R46T	0.035
2 bbl	V8 A	350	Fed.	0.016 used, 0.019 new	29-31°	600	R44T	0.035
2 bbl	V8 A	400	Fed.	0.016 used, 0.019 new	29-31°	600	R44T	0.035
2 bbl	V8 M	350	Fed.	0.016 used, 0.019 new	29-31°	900	R44T	0.035
4 bbl	V8 M	350	Fed.	0.016 used, 0.019 new	29-31°	900	R44T	0.035
4 bbl	V8 A	350	Fed.	0.016 used, 0.019 new	29-31°	600	R44T	0.035
4 bbl	V8 A	400	Fed.	0.016 used, 0.019 new	29-31°	600	R44T	0.035
4 bbl	V8 M	350	Calif.	0.016 used, 0.019 new	29-31°	900	R44T	0.035
4 bbl	V8 M	400	Calif.	0.016 used, 0.019 new	29-31°	900	R44T	0.035
4 bbl	V8 A	350	Calif.	0.016 used, 0.019 new	29-31°	600	R44T	0.035
4 bbl	V8 A	400	Calif.	0.016 used, 0.019 new	29-31°	600	R44T	0.035
4 bbl	V8 M	454	All	0.016 used, 0.019 new	29-31°	800	R44T	0.035
4 bbl	V8 A	454	All	0.016 used, 0.019 new	29-31°	600	R44T	0.035

Carburetor (1974 Models)	Engine			Timing (° BTC)	Idle Speeds		Compression (psi)
	Model[1]	CID	Usage[7]		[3]	[4]	
1 bbl	L6 A	250	Fed.	8	450(DR)	650/600(DR)	130
1 bbl	L6 M	250	All	8	450	950/850	130
1 bbl	L6A	250	Calif.	8	450(DR)	630/600(DR)	130
2 bbl	V8 A	350	Fed.	8	500 (DR)	650/600(DR)	150-160
2 bbl	V8 A	400	Fed.	8	500 (DR)	650/600(DR)	150-160
2 bbl	V8 M	350	Fed.	0	500	1,000/900	150
4 bbl	V8 M	350	Fed.	8	500	950/900	160
4 bbl	V8 A	350	Fed.	8	500 (DR)	650/600(DR)	160
4 bbl	V8 A	400	Fed.	8	500 (DR)	650/600(DR)	160
4 bbl	V8 M	350	Calif.	4	500	950/900	160
4 bbl	V8 M	400	Calif.	4	500	950/900	160
4 bbl	V8 A	350	Calif.	8	500(DR)	630/600(DR)	160
4 bbl	V8 A	400	Calif.	8	500(DR)	630/600(DR)	160
4 bbl	V8 M	454	All	10	500	850/800	160
4 bbl	V8 A	454	All	10	500(DR)	630/600(DR)	160

Carburetor (1975 Models)	Engine			Spark Plugs		Timing (° BTC)
	Model[1]	CID	Usage[7]	Type (AC)	Gap	
1 bbl	L6M	250	Fed.	R46TX	0.060	10
1 bbl	L6A	250	Fed.	R46TX	0.060	10
1 bbl	L6A	250	Calif.	R46TX	0.060	10
2 bbl	V8M	350	Fed.	R44TX	0.060	6
2 bbl	V8A	350	Fed.	R44TX	0.060	6
4 bbl	V8M	350	Fed.	R44TX	0.060	6
4 bbl	V8M	400	Fed.	R44TX	0.060	6
4 bbl	V8A	350	Fed.	R44TX	0.060	8
4 bbl	V8A	400	Fed.	R44TX	0.060	8
4 bbl	V8M	350	Calif.	R44TX	0.060	4
4 bbl	V8A	350	Calif.	R44TX	0.060	6
4 bbl	V8A	400	Calif.	R44TX	0.060	8
4 bbl	V8A	454	Fed.	R44TX	0.060	16

Table 4 TUNE-UP SPECIFICATIONS (continued)

Carburetor (1975 Models)	Engine Model[1]	CID	Usage[7]	Idle Speeds[6] Slow	Fast	Compression (psi)
1 bbl	L6M	250	Fed.	850	1,800	130
1 bbl	L6A	250	Fed.	550(DR)	1,700(N)	130
1 bbl	L6A	250	Calif.	600(DR)	1,700(N)	130
2 bbl	V8M	350	Fed.	800	N.A.	150
2 bbl	V8A	350	Fed.	600	N.A.	150
4 bbl	V8M	350	Fed.	800	1,600	150-160
4 bbl	V8M	400	Fed.	800	1,600	150-160
4 bbl	V8A	350	Fed.	600(DR)	1,600(N)	150-160
4 bbl	V8A	400	Fed.	600(DR)	1,600(N)	150-160
4 bbl	V8M	350	Calif.	800	1,600	150-160
4 bbl	V8A	350	Calif.	600(DR)	1,600(N)	150-160
4 bbl	V8A	400	Calif.	600(DR)	1,600(N)	160
4 bbl	V8A	454	Fed.	600(DR)	1,600(N)	160

Carburetor (1976 Models)	Engine Model[1]	CID	Usage[7]	Spark Plugs Type (AC)	Gap	(°BTDC)	Idle Speed Slow	Fast
1 bbl	L6 M	250	All	R46TS	0.035	6	850	1,800
1 bbl	L6 A	250	Fed.	R46TS	0.035	6	550(DR)	1,700(N)
1 bbl	L6 A	250	Calif.	R46TS	0.035	6	600(DR)	1,700(N)
2 bbl	V8 M	305	All	R45TS	0.045	6	800	N.A.
2 bbl	V8 A	305	Fed.	R45TS	0.045	8	600(DR)	N.A.
2 bbl	V8 A	305	Calif.	R45TS	0.045	TDC	600(DR)	N.A.
2 bbl	V8 A	350	All	R45TS	0.045	6	600(DR)	N.A.
4 bbl	V8 M	350	Fed.	R45TS	0.045	8	800	N.A.
4 bbl	V8 M	350	Calif.	R45TS	0.045	6	800	N.A.
4 bbl	V8 A	350	Fed.	R45TS	0.045	8	600(DR)	N.A.
4 bbl	V8 A	350	Calif.	R45TS	0.045	6	600(DR)	N.A.
4 bbl	V8 A	400	All	R45TS	0.045	8	600(DR)	N.A.

Carburetor (1977 Models)	Engine Model[1]	CID	Usage[7]	Spark Plugs Type (AC)	Gap	(°BTDC)	Idle Speed Slow	Fast
1 bbl without A.C.	L6 M	250	All	R46TS	0.035	6	750	N.A.
1 bbl with A.C.	L6 M	250	All	R46TS	0.035	6	800	N.A.
1 bbl	L6 A	250	Fed.	R46TS	0.035	8	600(DR)	N.A.
1 bbl	L6 A	250	Calif.	R46TS	0.035	6	550(DR)	N.A.
2 bbl	V8 M	305	All	R45TS	0.045	8	700	N.A.
4 bbl	V8 A	350	All	R45TS	0.045	8	500	650
4 bbl	V8 A	350	High alt.	R45TS	0.045	8	600	650

LUBRICATION, MAINTENANCE, AND TUNE-UP

Table 4 TUNE-UP SPECIFICATIONS (continued)

Carburetor (1978 Models)	Engine			Spark Plugs		(°BTDC)	Idle Speed	
	Model[1]	CID	Usage[7]	Type (AC)	Gap		Slow	Fast
2 bbl	V6M	200	Fed.	R45TS	0.045	8	700(N)	1,300
2 bbl	V6A	200	Fed.	R45TS	0.045	8	600(DR)	1,600
2 bbl	V6M	231	Fed.	R46TSX	0.060	15[9]	600/800(N)	—
2 bbl	V6A	231	All	R46TSX	0.060	15[9]	600/760(DR)	—
2 bbl	V8M	305	Fed.	R45TS	0.045	4	600(N)	—
2 bbl	V8A	305	Fed.	R45TS	0.045	4	500(DR)	—
2 bbl	V8A	305	Calif.	R45TS	0.045	6	500/600(DR)	—
2 bbl	V8A	305	HighAlt.	R45TS	0.045	8	600/700(DR)	—
4 bbl	V8A	350	Fed.	R45TS	0.045	6	500/600(DR)	1,600
4 bbl	V8A	350	Calif.	R45TS	0.045	8	500/600(DR)	1,600
4 bbl	V8A	350	HighAlt.	R45TS	0.045	8	500/650(DR)	1,600

Carburetor (1979 Models)	Engine			Spark Plugs		(°BTDC)	Idle Speed	
	Model[1]	CID	Usage[7]	Type (AC)	Gap		Slow	Fast
2 bbl	V6M	200	Fed.	R45TS	0.045	8	700/800(N)	1,300
2 bbl	V6A	200	Fed.	R45TS	0.045	12	600/700(DR)	1,600
2 bbl	V6A	231	Fed.	R45TSX	0.060	15[9]	560/670(DR)[10]	2,200
2 bbl	V6A	231	Calif./HighAlt.	R45TSX	0.060	15[9]	600(DR)	2,200
2 bbl	V8M	267	Fed.	R45TS	0.045	4	600/700(N)	1,600
2 bbl	V8A	267	Fed.	R45TS	0.045	8	500/600(DR)	1,600
2 bbl	V8M	305	Fed./Calif.	R45TS	0.045	4	700(N)	1,300
2/4 bbl	V8A	305	Fed./Calif.	R45TS	0.045	4	500/600(DR)	1,600
2/4 bbl	V8A	305	HighAlt.	R45TS	0.045	8	600/650(DR)	1,750
4 bbl	V8A	350	All	R45TS	0.045	8	600/650(DR)	1,750

Carburetor (1980 Models)	Engine			Spark Plugs		(°BTDC)	Idle Speed	
	Model[1]	CID	Usage[7]	Type (AC)	Gap		Slow	Fast
2 bbl	V6A	229	Fed.	R45TS	0.045	12	600/675(DR)	1,750
2 bbl	V6M	229	Fed.	R45TS	0.045	8	700/800(N)	1,300
2 bbl	V6A	231	Fed.	R45TSX	0.060	15[9]	560/670(DR)	2,200
2 bbl[11]	V6A	231	Calif.	R45TSX	0.060	15[9]	600(DR)	2,200
2 bbl[12]	V6A	231	Calif.	R45TSX	0.060	15[9]	600(DR)	2,200
2 bbl[12]	V6A	231	Fed.	R45TSX	0.060	15[9]	550(DR)	2,200
2 bbl	V8A	267	Fed.	R45TS	0.045	4	500/600(DR)	1,850
4 bbl	V8M	305	Fed.	R45TS	0.045	4	700(N)	1,500
4 bbl	V8A	305	Fed.	R45TS	0.045	4	500/600(DR)	1,850
4 bbl	V8A	305	Calif.	R45TS	0.045	4	550/650(DR)	2,200

CAUTION
Spark plug wires should never be removed by yanking on the wire, as damage could result.

NOTE: *Tag each spark plug wire — a strip of masking tape will do — and mark the proper cylinder number. The cylinder closest to the front of the car is No. 1. Note that on V6 and V8 engines one bank of cylinders is slightly offset from the other. The bank which is farther forward contains cylinders 1, 3, 5, (and 7) and the opposite bank cylinders 2, 4, 6, (and 8) from front to back.*

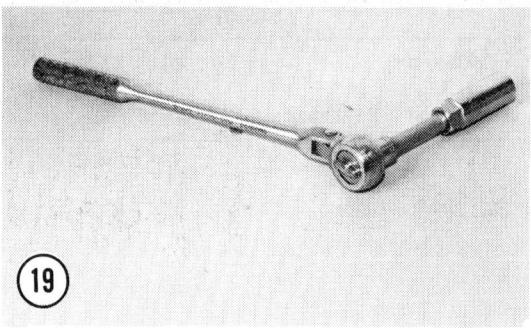

3. Using a spark plug wrench (**Figure 19**), remove the spark plugs in order.

4. Inspect each spark plug as it is removed and compare its appearance with Figure 17, Chapter Two. Electrode appearance is a good indicator of performance in each cylinder and permits early recognition of trouble.

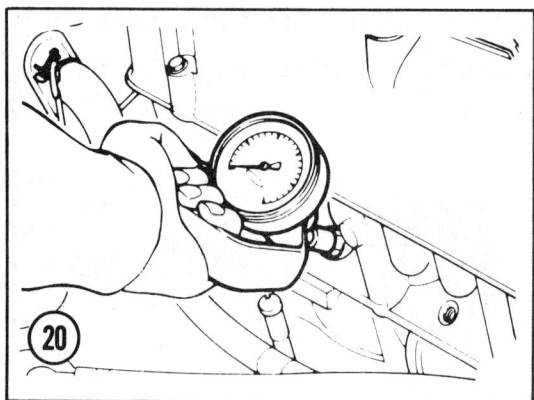

COMPRESSION TEST

An engine with low or uneven compression cannot be properly tuned. In view of this, the engine should be given a compression test before proceeding with the tune-up.

1. Remove air cleaner from carburetor and block throttle and choke in wide open position.
2. Remove the distributor primary lead from the negative post on the coil. Remove the spark plugs.
3. Hook up remote starter per manufacturer's instructions.
4. Firmly insert a compression gauge in each spark plug hole, in order, and crank the engine through at least 4 compression strokes to obtain the highest possible reading. See **Figure 20**. Record the reading for each cylinder.

NOTE: *If there is more than 20 pounds difference between the high and low reading cylinders, the engine cannot be properly tuned until repairs have been made. The remainder of this procedure may be performed to help determine the kind of repairs that are needed.*

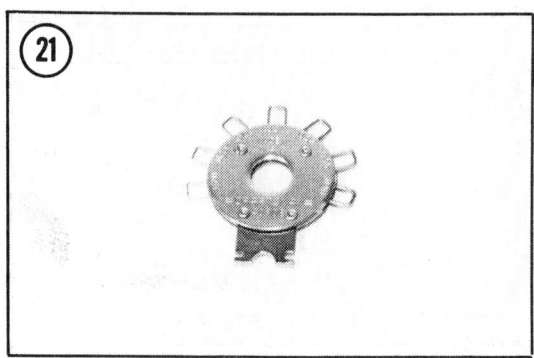

5. If low or uneven compression readings were recorded for one or more cylinders, inject about a tablespoon of motor oil thorugh the spark plug hole of each low-reading cylinder. Crank the engine through several compression strokes and recheck compression. If compression improves, the problem is probably worn rings. If no improvement is noted, in all probability the valves are burned, sticking, or not seating properly. If 2 adjacent cylinders read low and the oil injection does not increase compression, the problem may be a defective head gasket.

LUBRICATION, MAINTENANCE, AND TUNE-UP

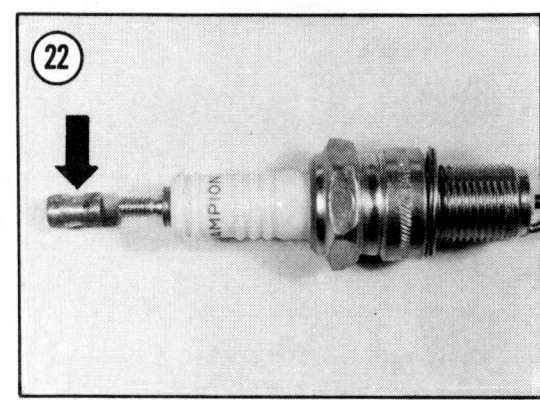

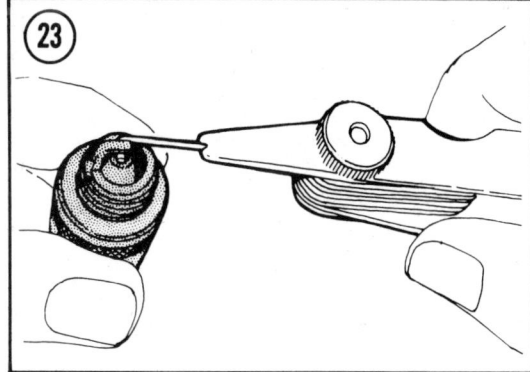

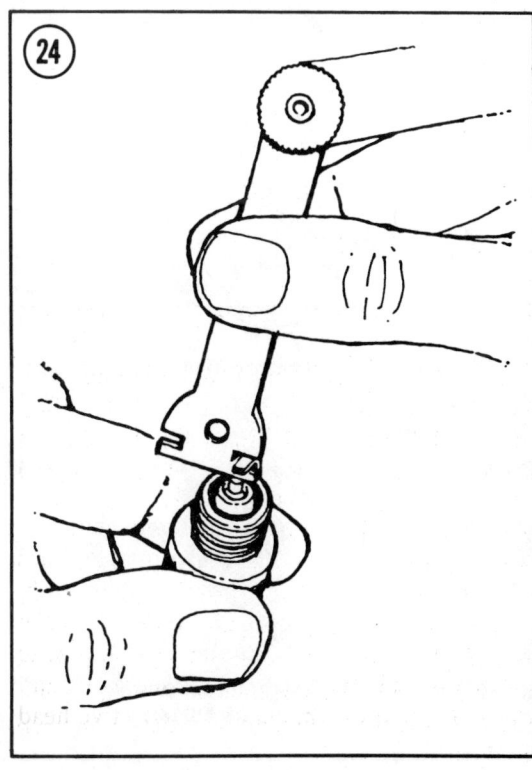

SPARK PLUG REPLACEMENT

Spark plugs should be replaced at the intervals shown in **Table 2**. If misfiring occurs earlier than these intervals, spark plugs in good condition can often be cleaned, regapped and reinstalled with acceptable results. If all new plugs are being installed, skip to Step 3 in the following procedure.

New plugs should be carefully gapped to ensure a reliable, consistent spark. You must use a special spark plug gapping tool with a wire gauge. See **Figure 21**.

1. Inspect plugs and discard and replace those with badly worn electrodes and/or glazed, blistered, or broken porcelain insulators.

2. Clean the serviceable plugs with an abrasive cleaner, such as sandblast. File center electrode flat.

3. Verify that all plugs to be installed are of the same make and of the proper heat range number (**Table 4**).

4. Remove new plugs from box and screw on the small end pieces that are loose in each box. See **Figure 22**.

5. Insert the correct (**Table 4**) gauge wire between the center and side electrode of each spark plug. See **Figure 23**. If the gap is correct, you will feel a slight drag as you pull the wire through. If there is no drag, or the gauge won't pass through, bend the side electrode *with the gapping tool* (see **Figure 24**) to set proper gap.

6. Inspect spark plug hole threads and clean before installing plugs. If required, corrosion can be removed with a 14mm x 1.25 SAE spark plug thread chaser (use grease on the chaser to catch chips).

CAUTION
Use extreme care when using thread chaser to avoid cross-threading. Also crank engine several times to blow out any dislodged material from the engine.

7. Apply a thin film of oil — a drop from engine dipstick will do — to spark plug threads. Install plug in hole and torque to 15 ft.-lb. (all 1971 and later; and 1970 14mm x $^{13}/_{16}$) or 25 ft.-lb. (1970 14mm x $^{5}/_{8}$).

NOTE: *If a torque wrench is not available, tighten spark plugs as tight as*

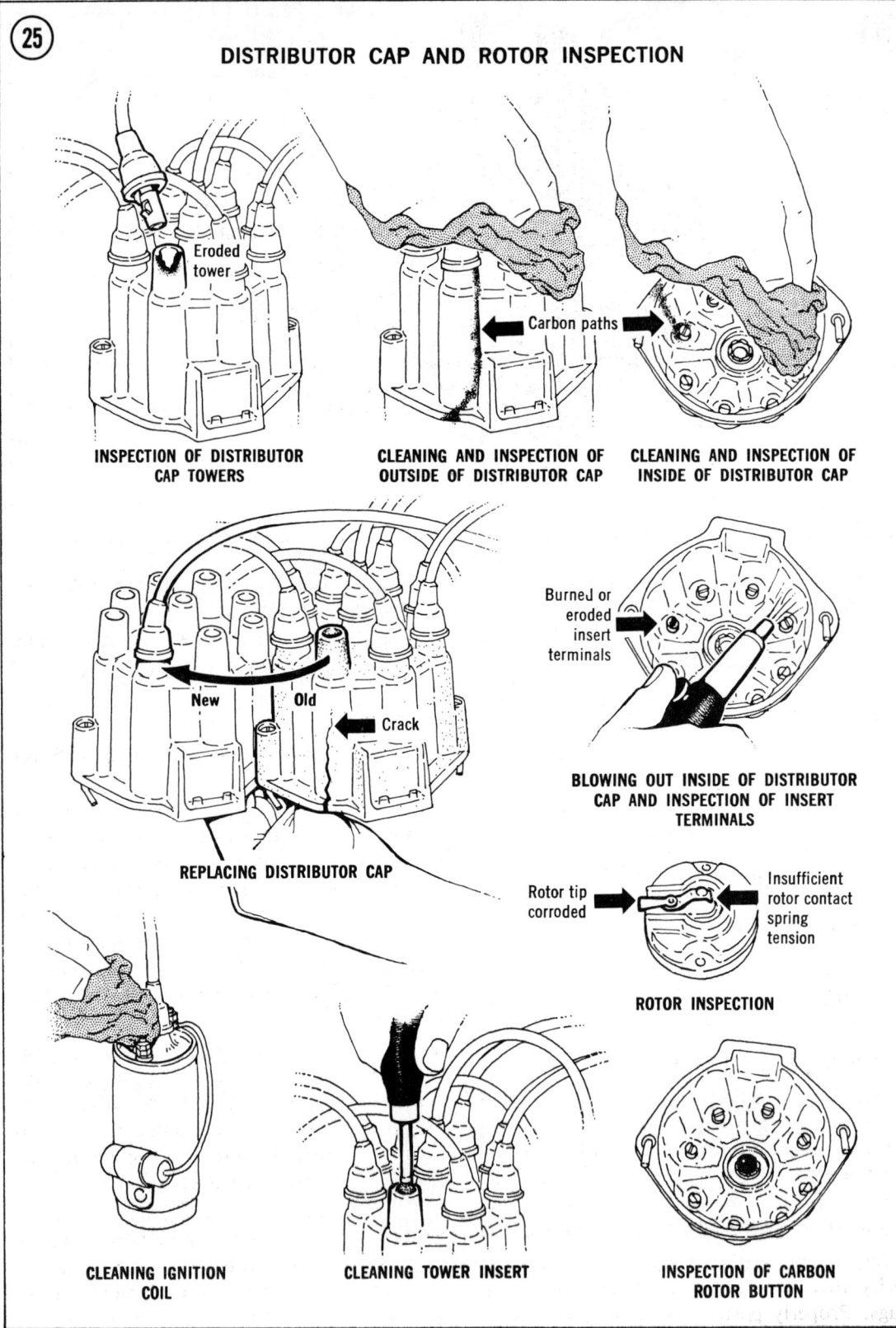

LUBRICATION, MAINTENANCE, AND TUNE-UP

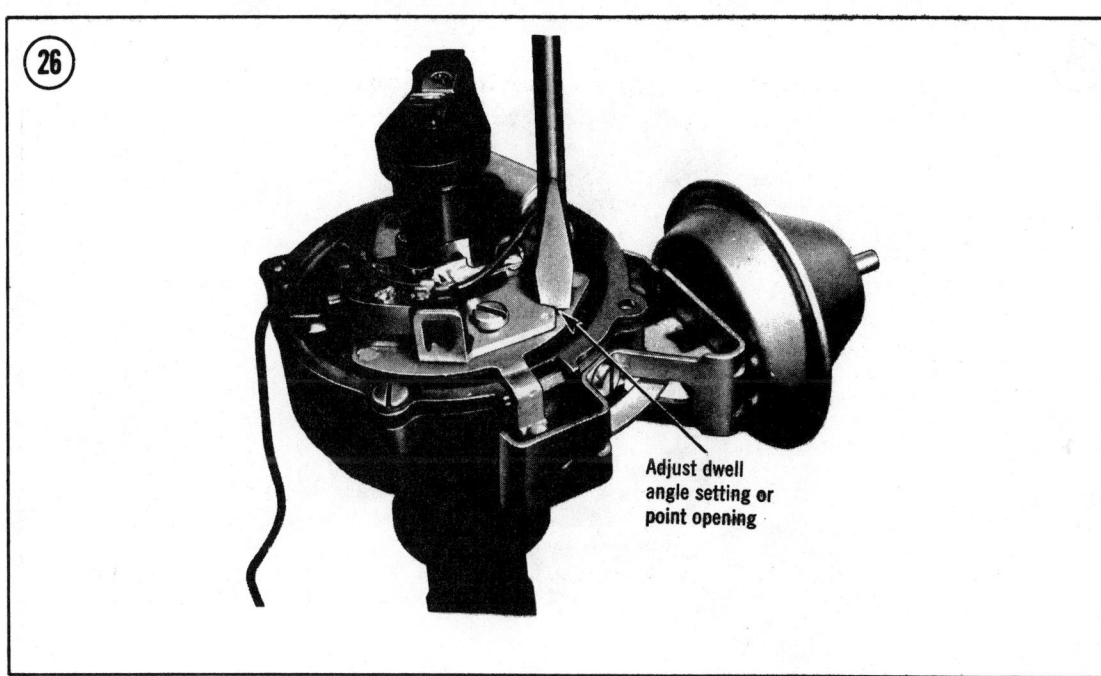

㉖ Adjust dwell angle setting or point opening

possible by hand, then, using wrench, tighten another ½ turn. Do not overtighten. Excessive torque may change gap setting or squash the gasket so badly it cannot seal.

8. Reconnect the spark plug wires.

   NOTE: *If spark plug wires have been in service for a year or longer, or appear cracked, oil soaked, or brittle, check them with an ohmmeter. Any wire with a resistance over 20,000 ohms or "infinity" should be replaced. Use only brand names (Delco, Belden, etc.) replacement resistance cable.*

BREAKER POINT IGNITION SYSTEMS

Ignition System Inspection

1. Remove and clean distributor cap. Inspect for cracks, carbon tracks, and burned, worn or corroded terminals, etc. (**Figure 25**). Replace cap if necessary.
2. Clean rotor and inspect for damage or deterioration (**Figure 25**). Replace if necessary.
3. Inspect spark plug wiring (see note above). Verify that all wires are installed to proper plugs. Properly position wires in supports to avoid contact with engine and to avoid cross-firing.
4. Tighten all ignition system connections and replace any loose, frayed or damaged wires.

Distributor Service

Refer to **Figure 26** (6-cylinder) **or 27** (V8) for this procedure.

1. Check the centrifugal advance system by turning the distributor rotor clockwise as far as it will go. Release the rotor. This spring should return the rotor to its original position. If this does not happen, or if the return action is sluggish, the mechanism must be disassembled, cleaned and inspected for malfunctioning parts. See *Distributor Adjustments and Repairs* section below.
2. Check the vacuum advance system by turning the moveable breaker plate counterclockwise as far as it will go. Release the breaker plate and see if the spring returns it to the original retarded position. Correct any interference or binding condition.
3. Examine breaker point contact surfaces. Points with an even overall gray color and only slight roughness or pitting need not be replaced, but may be dressed with a clean point file as follows:

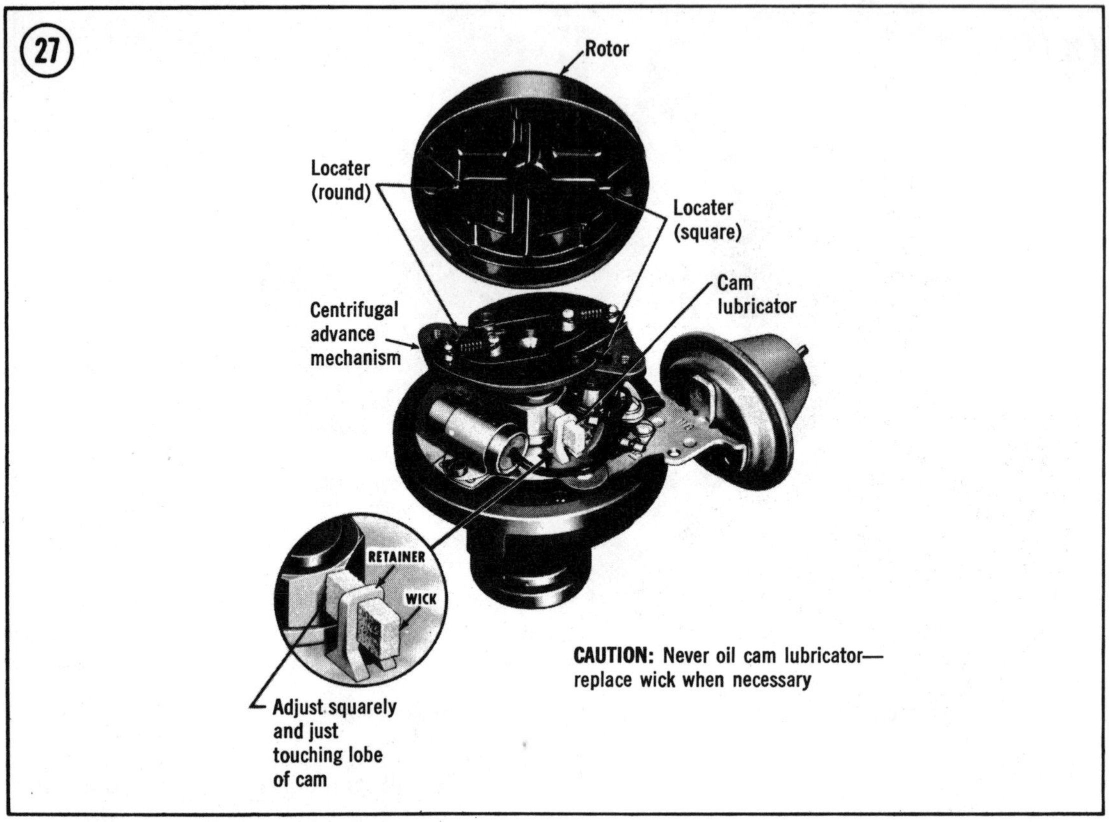

27

Rotor
Locater (round)
Locater (square)
Cam lubricator
Centrifugal advance mechanism
RETAINER
WICK
Adjust squarely and just touching lobe of cam
CAUTION: Never oil cam lubricator—replace wick when necessary

a. File the points, using only a few strokes and a clean, fine-cut, contact point file. Never use sandpaper or emery cloth, and do not attempt to remove all irregularities — just the scale or dirt.
b. Check the alignment of the points and correct as necessary. See **Figure 28**.
c. Clean the distributor cam with cleaning solvent and a clean, lint-free cloth. Rotate cam lubricator wick ½ turn or 180 degrees. Replace wick every 24,000 miles.
d. Using a feeler gauge, set points to 0.016 in. (used points). Breaker arm rubbing block must be on high point of a distributor cam lobe during adjustment.

4. If points are badly pitted or burned, they must be replaced. However, these conditions are usually caused by improper conditions in other parts of the ignition system or by dirt or other contamination in the distributor. The cause must be corrected (see *Distributor Adjustments and Repairs* section below) when installing new points, or the same condition will

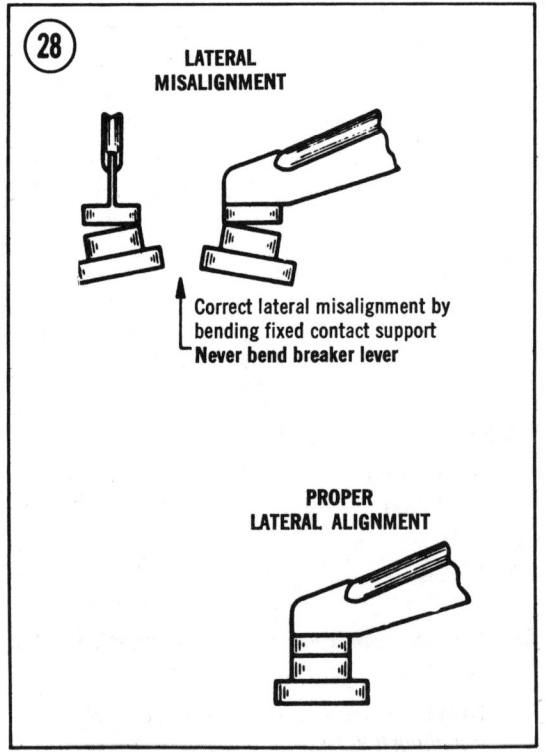

28 LATERAL MISALIGNMENT

Correct lateral misalignment by bending fixed contact support
Never bend breaker lever

PROPER LATERAL ALIGNMENT

LUBRICATION, MAINTENANCE, AND TUNE-UP

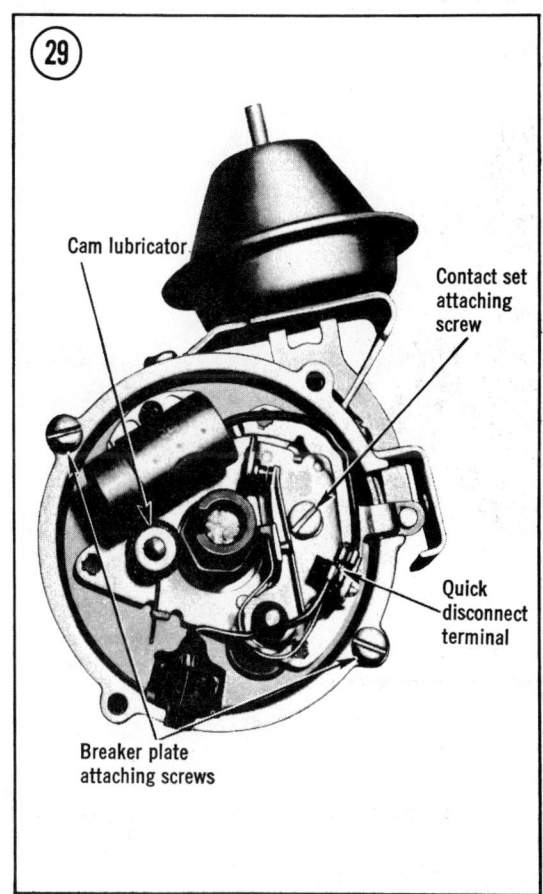

29. Cam lubricator / Contact set attaching screw / Quick disconnect terminal / Breaker plate attaching screws

rapidly develop again. To replace points (for a 6-cylinder engine) proceed as follows:

a. Remove distributor cap by releasing hold-down screws. Move cup out of the way.
b. Remove rotor.
c. Remove condenser and primary lead wires from quick disconnet terminal (**Figure 29**).
d. Remove contact set attaching screw and remove point set.
e. If condenser is to be replaced, remove attaching screw and remove condenser from breaker plate.
f. Using a clean, lint-free cloth, wipe breaker plate free of oil and smudge.
g. Install new point set on brake plate and attach with screw.

NOTE: *Pilot on point set must engage matching hole in breaker plate.*

h. If condenser was removed, install new condenser on breaker plate with attaching screw.
i. Connect primary and condenser lead wires to quick disconnect terminal on point set.
j. Using a spring gauge, measure the tension of the breaker arm spring. Pressure must be between 19 and 23 oz.

NOTE: *Hook spring gauge over the breaker lever (moveable arm of the point set) and read tension just as points separate. Spring tension can be adjusted by carefully bending the breaker lever spring. Decrease pressure by carefully pinching the spring. To increase pressure, remove point set from distributor and bend spring away from lever. Excessive distortion of the spring must be avoided.*

k. Using a feeler gauge, set the new points to 0.019 inch. Rubbing block of points must be resting on the highest point of a cam lobe when this adjustment is made (see *Dwell Angle* procedure below).
l. Rotate cam lubricator 180 degrees at 12,000 mile intervals and replace lubricator every 24,000 miles.
m. Replace rotor and distributor cap. Lock cap to distributor housing with hold-down screws.
n. Start engine and check dwell angle (see below).

5. To replace points on an 8-cylinder engine, proceed as follows:

NOTE: *The contact point set (and condenser on 1974 models) is replaced as a single unit. Breaker lever tension and point gap are preset at the factory, and only dwell angle requires adjustment after replacement.*

a. Remove distributor cap by placing a screwdriver blade in the slotted head of each latch screw, pressing down, and rotating a quarter turn in either direction. Move cap out of way.

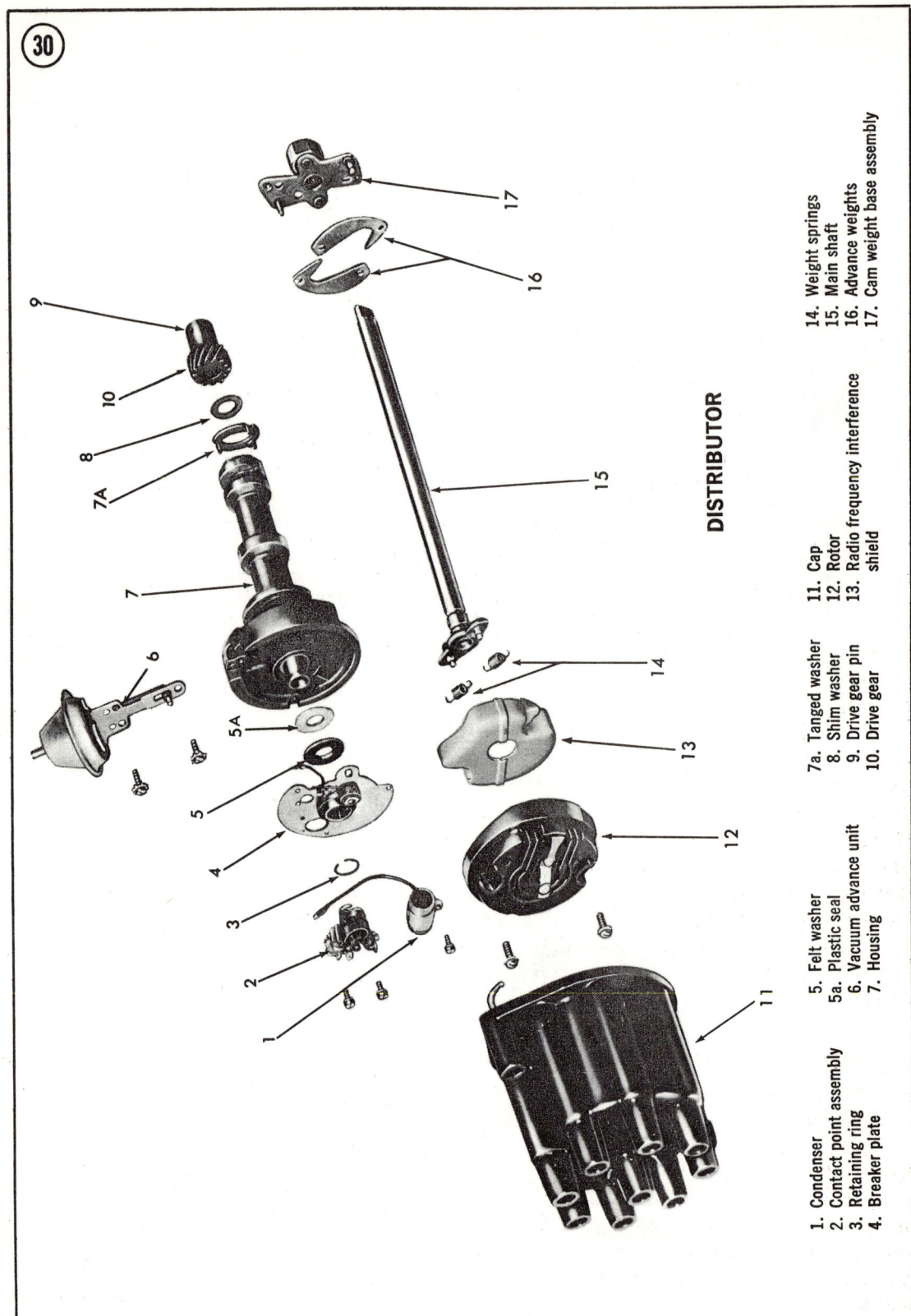

LUBRICATION, MAINTENANCE, AND TUNE-UP

b. Remove rotor attaching screws, then remove rotor (see **Figure 30**).

c. Remove the RFI shield, if so equipped.

d. Loosen the 2 point set assembly attaching screws and remove the set from the breaker plate. Disconnect the primary.

e. Remove the primary and condenser lead wires from the nylon insulated connector in the point set.

f. If condenser is to be replaced (1973 and earlier), remove attaching screw and then remove condenser.

g. Install new condenser, using attaching screw (if old condenser was removed).

h. Install new point set, using attaching screws.

i. Connect primary and condenser lead wires to nylon insulated connector on point set. Make certain lead wires are routed so they will not interfere with point set operation.

j. Replace rotor and install and tighten the attaching screws.

k. Rotate cam lubricator 180 degrees at 12,000 mile intervals. Replace every 24,000 miles.

l. Replace distributor cap and lock into place with latches.

m. Start engine and check dwell angle (see below).

DWELL ANGLE

The preferred method for setting dwell angle is to first set the breaker point gap with a feeler gauge (6-cylinder engines only — points for 8-cylinder engines are factory preset) and then check the setting with a dwell meter. It is very important that breaker points be set to the proper gap. Points set too closely tend to burn and pit rapidly. Points with an excessive gap result in a weak spark at high speeds. New points must be set to a wider gap then used points to compensate for wear of the rubbing block while seating to the distributor cam.

Adjusting Dwell Angle (6-Cylinder Engines)

1. Remove distributor cap and rotor.

2. Verify alignment of points as per **Figure 28**. If necessary, align by bending fixed contact arm. Do not bend the breaker lever (moveable arm). Do not attempt to align used points. Instead, replace them if serious misalignment is present.

3. Crank or turn engine until rubbing block on breaker lever is resting on high point of a distributor cam lobe (this provides maximum point gap).

4. Loosen the point set attaching screw (see **Figure 29**) and insert a screwdriver in the opposing slots in the point set and and the breaker plate. Move the point set to obtain a setting of 0.019 in. (new points) or 0.016 in. (used points), measured with blade-type feeler gauge.

5. Tighten the attaching screw and remeasure the gap. Readjust if setting was moved during the tightening of the screw.

6. Reassemble the rotor and distributor cap.

7. Connect dwell meter and tachometer to engine, using manufacturer's instructions.

8. Disconnect vacuum line from distributor and plug line with sharp end of a pencil, a golf tee, or other similar pointed object.

9. Operate the engine at normal idle speed and check dwell angle. For 6-cylinder engines, angle should be between 31 and 34 degrees.

> NOTE: *If dwell angle is not within these limits, points should be reset. Check for misalignment of points or worn distributor cam lobes.*

10. Accelerate engine to 1,750 rpm. Variation in dwell angle should not exceed 3 degrees. Higher variation indicates excessive distributor wear or loose breaker plate.

11. Reconnect vacuum line to distributor.

Adjusting Dwell Angle (8-Cylinder Engines)

1. After installing point set and reassembling distributor, start engine and allow to warm up to normal operating temperature. Disconnect and plug vacuum line to distributor.

2. Attach a dwell meter and tachometer to the engine, using the manufacturer's instructions.

3. Raise window in distributor cap and insert a hex wrench in the adjusting screw head (see **Figure 31**).

4. Turn adjusting screw until a dwell angle of 29-31 degrees (30 degrees preferred) is reached.

5. If a dwell meter is not available, turn the adjusting screw clockwise until the engine begins to misfire. Then turn screw ½ turn in opposite direction. This method will give an approximate dwell angle setting, but should be used only when a dwell meter is not available.

6. Accelerate the engine to 1,750 rpm and check for variation in dwell angle. The angle should not vary more than 3 degrees. Variation in excess of this indicates excessive distributor wear or loose breaker plate.

7. Remove hex wrench and close access window in distributor cap. Unplug and replace vacuum line to distributor.

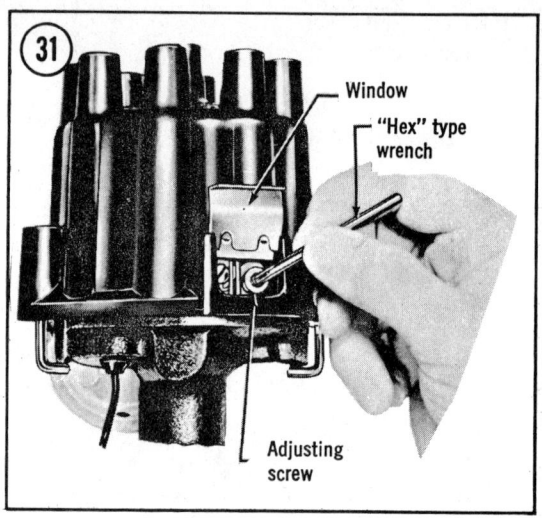

IGNITION TIMING ADJUSTMENT

Ignition timing should be checked and adjusted (if required) after point replacement and dwell angle adjustment have been completed.

1. Connect timing light in accordance with the manufacturer's instructions. **Figure 32** shows a typical hook-up.

2. Determine the timing specifications from **Table 4**.

> NOTE: Use chalk or white paint, etc., to emphasize the proper markings on the timing tab and the harmonic balancer.

3. Disconnect the vacuum line to the distributor and plug line with sharp end of a pencil or a golf tee.

4. Operate the engine at normal idle speed.

5. Aim the timing light at the timing tab (see **Figure 33**, typical).

6. Adjust timing by loosening distributor clamp bolt and rotating distributor body. Watch timing marks on timing tab and harmonic balancer. When the marks appear to stand still directly opposite each other, retighten the distributor clamp bolt. Recheck timing and readjust if required.

7. Stop engine, disconnect timing light, and reconnect vacuum hose.

HIGH ENERGY IGNITION SYSTEM

Routine maintenance is not required for the HEI system itself. If parts or components fail, they are not repairable and must be replaced. However, engine timing should be checked and the distributor components visually checked for cracks, wear, dust, moisture, burns, etc., very 18 months or 22,500 miles, whichever comes first. At the same time, the secondary wiring (spark plug wires) should be inspected, checked out with an ohmmeter, and replaced if necessary.

> NOTE: The HEI system has a larger (8mm) diameter, silicone-insulated, spark plug wires. While these gray-colored wires are more heat resistant and less vulnerable to deterioration than conventional wires, they should not be mistreated. When removing wires from spark plugs, grasp only on the boots. Twist boot ½ turn in either direction to break seal, then pull to remove.

Spark plugs also should be replaced after every 22,500 miles.

The procedure for ignition timing is identical to that given above for conventional ignition systems, except that timing light connections

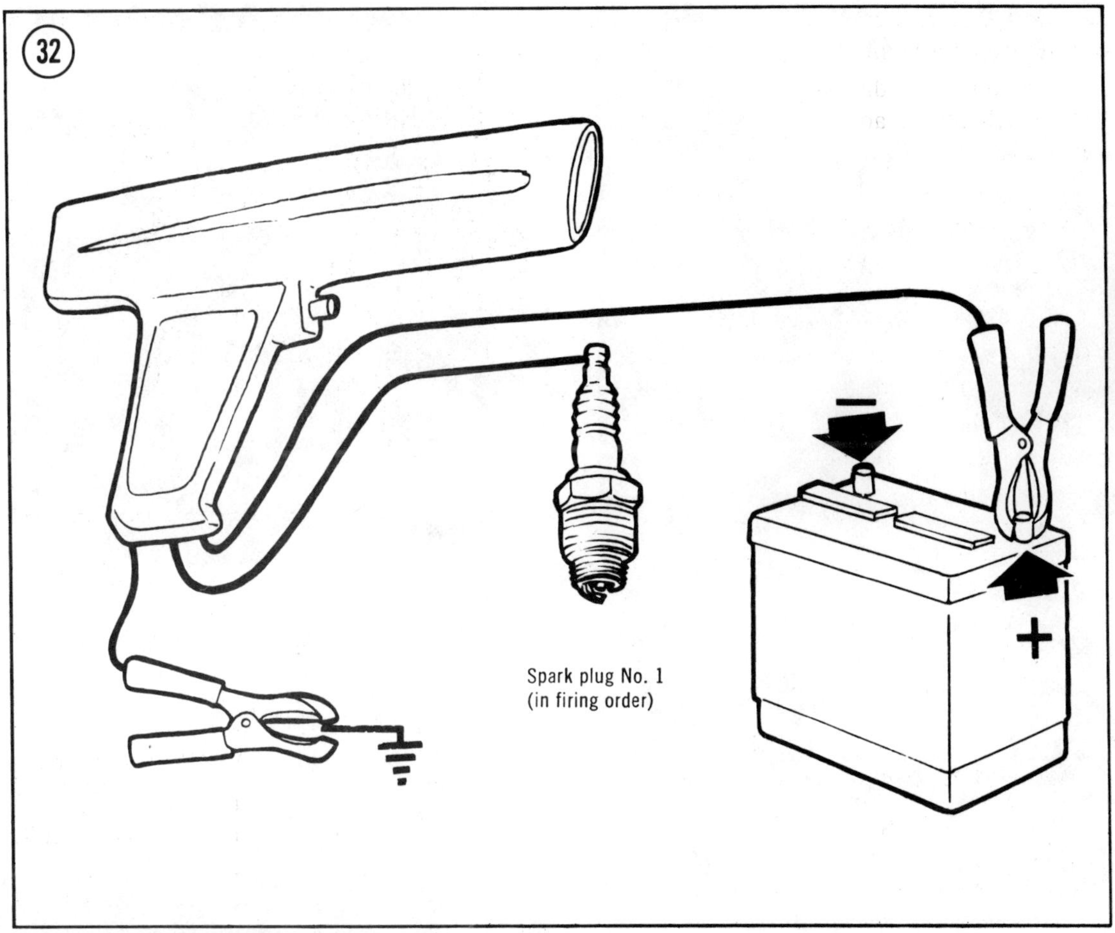

LUBRICATION, MAINTENANCE, AND TUNE-UP

32

Spark plug No. 1
(in firing order)

33

should be made in parallel using an adapter at the No. 1 spark plug wire terminal on the distributor. The distributor cap also has a special terminal marked TACH. Connect one lead of the tachometer to this terminal and the other to ground.

NOTE: *Some tachometers must connect from the* TACH *terminal to the battery positive terminal. Check the manufacturer's instructions before making connections.*

CARBURETOR ADJUSTMENT

NOTE: *Carburetors on 1971 and later models are equipped with limiter caps on idle mixture screws. These caps, which should not be altered or removed, are a part of the emission control system and restrict the richness of the fuel/air mixtures. This mixture ratio is preset at*

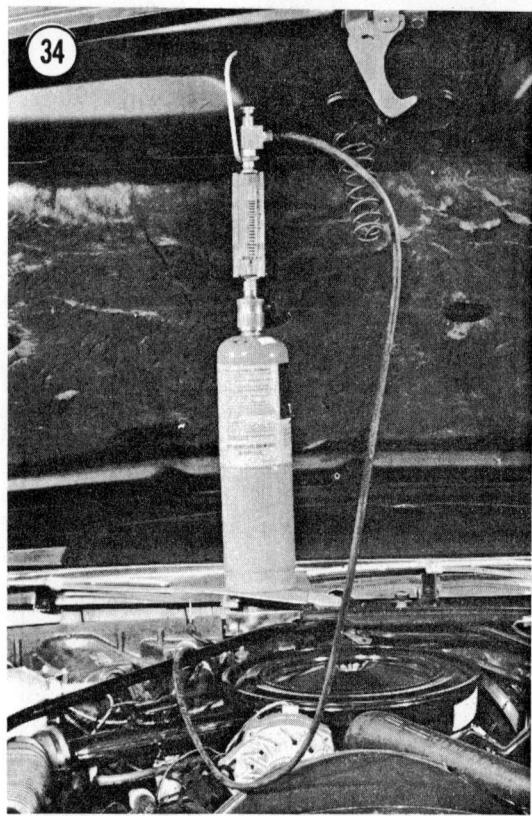

the factory and a satisfactory idle can be obtained within the range (usually ½-¾ turn) allowed by the cap. If the cap has been damaged or removed on your vehicle, a procedure is included which will permit setting the idle mixture to an acceptable ratio.

NOTE: *Before attempting to adjust idle settings on carburetor, the ignition system should be tuned and timed, using the procedures given earlier in this chapter.*

Engine idle speed is controlled by a combination of adjustments to the fuel/air mixture screw(s) and the throttle linkage. Throttle linkage adjustment is made by adjusting either an idle stop screw, an idle stop solenoid, or both.

During the period covered, Chevrolet used 1-barrel (1-bbl) carburetors on inline 6-cylinder engines, and 2- and 4-barrel carburetors on V8 engines. The V6 engines used in some late models are equipped with 2-barrel carburetors.

NOTE: *In making idle adjustments, always refer to the Vehicle Emission Control Information (VECI) decal located in the engine compartment for correct idle speeds for your vehicle. The speeds listed in Table 4 should be used only if decal is missing or cannot be read.*

ROCHESTER CARBURETORS (1975 AND LATER VEHICLES)

Propane Enrichment Idle Mixture Adjustment (All 1978-On Carburetors)

This procedure must be used on all 1978 and later cars when idle mixture adjustment is required in order to meet air pollution control requirements. Idle mixture is preset at the factory to meet these requirements, and limiter caps have been placed on the adjustment screws to prevent tampering. These caps should not be removed, and no attempt should be made to adjust the idle mixture, unless the carburetor has received a major overhaul.

In addition, most General Motors carburetors have been modified so that backing out the idle mixture adjustment screw(s) does not necessarily enrich the mixture. In view of this, the time-honored "lean drop" system (or any other system, for that matter) cannot be substituted satisfactorily for the propane enrichment method.

LUBRICATION, MAINTENANCE, AND TUNE-UP

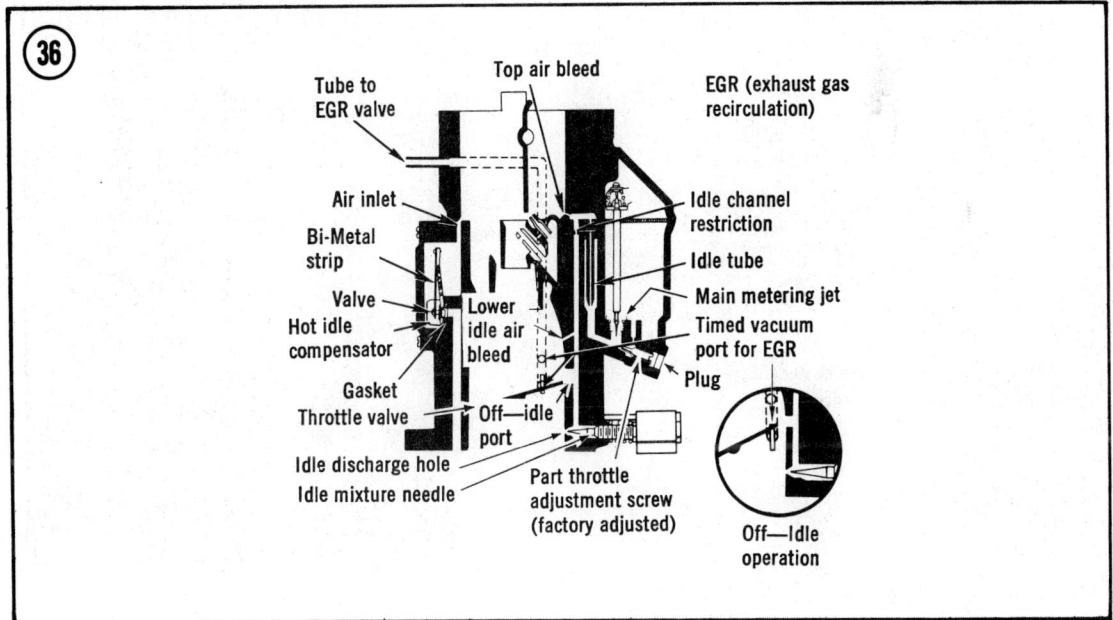

A special propane tool (General Motors J-26911) is required for this procedure. See **Figure 34**. You may be able to find a suitable substitute at an auto supply or parts store. If the tool is not available, do not try to adjust the idle mixture.

1. Set the parking brake and block the rear wheels. If the car is equipped with a vacuum brake release system, disconnect the hose at the brake and plug the hose.

2. Refer to the Vehicle Emission Control Information (VECI) decal in the engine compartment and disconnect and plug the hoses listed.

3. Connect a tachometer to the engine, following the manufacturer's instructions.

4. Disconnect and plug the vacuum advance hose at the distributor and check and adjust the timing to the specification given on the VECI decal. See *Ignition Timing Adjustment*, this chapter.

5. If the engine is cold, allow it to run until it reaches normal operating temperature. Make sure the choke is open and the air conditioner (if so equipped) is off.

6. Set the idle speed to the rpm shown on the VECI decal.

7. Remove the crankcase ventilation hose from the air cleaner and insert the propane tool hose and plug into the hose opening in the air cleaner. See **Figure 35**. The propane cartridge must be in the vertical position as shown in **Figure 34**.

8. With the engine running at idle speed, slowly open the propane control valve until maximum speed is reached with the transmission in NEUTRAL.

9. If the speed is within the enriched idle speed range given on the VECI label, no further adjustment is required. If the speed is incorrect, remove the limiter cap or plug(s) from the idle mixture screw(s). See **Figure 36** (1-bbl), **Figure 37** (2-bbl), or **Figure 38** (4-bbl).

10. Turn the mixture screw(s) in (clockwise) until lightly seated, then back out until the best idle point (maximum speed) is reached. Make sure propane is flowing during this adjustment.

11. Turn off propane and, if necessary, turn idle mixture screw(s) clockwise until the specified curb idle speed is reached.

12. Recheck the idle speed with the propane turned on. If not within the enriched idle speed specificaton, repeat Steps 10 and 11.

13. Turn the engine off and remove the propane tool. Reconnect the crankcase ventilation hose to the air cleaner.

14. Recheck the idle speed and reset to specification, if necessary, using the procedure on the VECI decal.

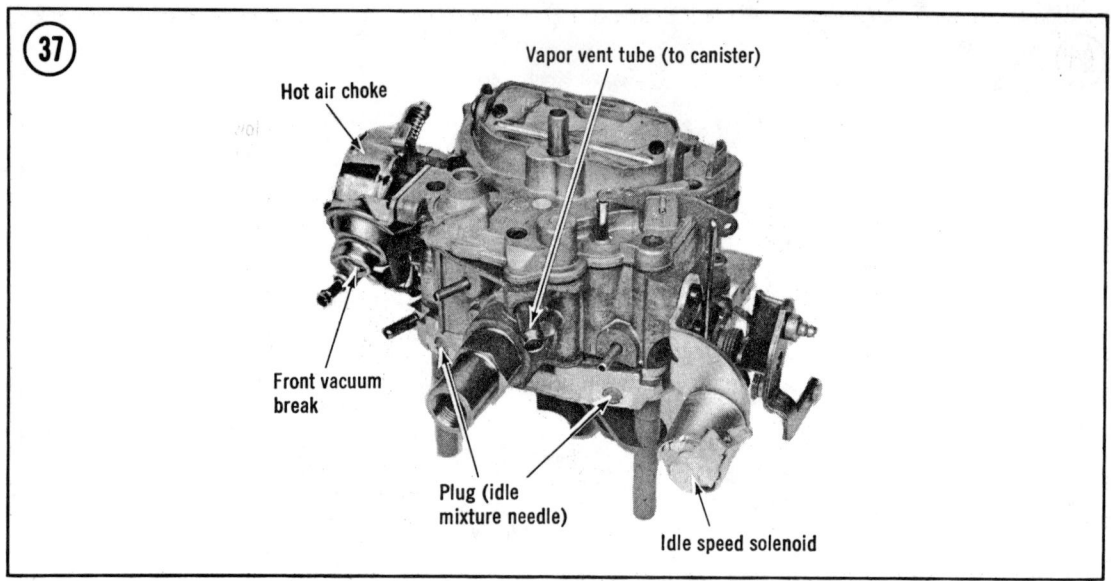

③⑦ Hot air choke / Vapor vent tube (to canister) / Front vacuum break / Plug (idle mixture needle) / Idle speed solenoid

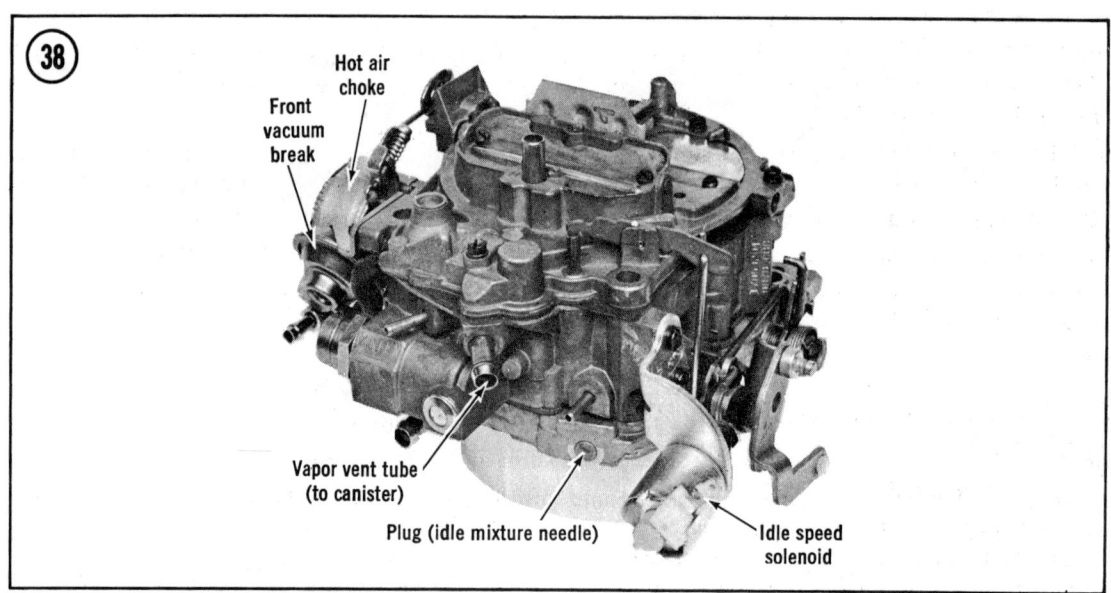

③⑧ Front vacuum break / Hot air choke / Vapor vent tube (to canister) / Plug (idle mixture needle) / Idle speed solenoid

Idle Mixture Adjustment
(1975-1977 1-BBL Carburetors)

1. Start engine and allow to reach normal operating temperature. Air cleaner should be on, choke should be fully open, and air conditioner should be off.

2. Connect a tachometer to the distributor, following manufacturer's instructions.

3. Set parking brake and block rear wheels.

4. Disconnect the fuel tank hose from vapor canister and disconnect vacuum hose from distributor and plug hose. Check VECI decal and disconnect and plug other hoses as directed.

5. Check engine timing and correct if required. Reconnect vacuum hose to distributor.

6. If the vehicle is equipped with automatic transmission, place selector in DRIVE. If equipped with manual transmission, place selector in NEUTRAL and disconnect electrical lead wire to idle stop solenoid.

7. Adjust idle stop solenoid to set idle speed to higher specified rpm.

LUBRICATION, MAINTENANCE, AND TUNE-UP

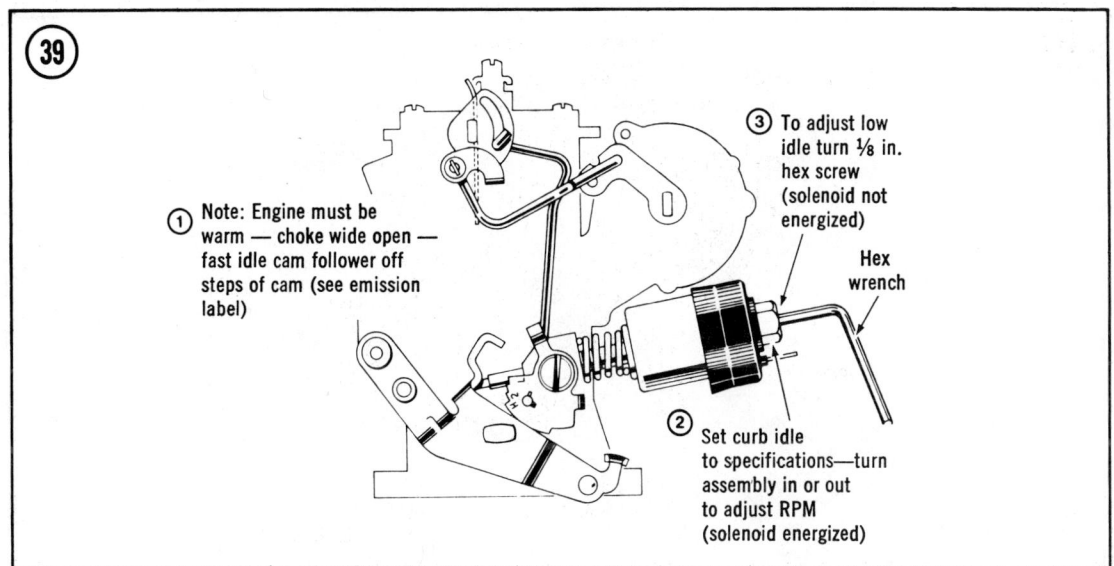

8. Turn idle mixture screw counterclockwise until maximum idle speed is obtained. Repeat Step 7, if required, to reset idle speed to higher specified rpm on VECI decal.

9. Observe tachometer and adjust idle speed to lower specified rpm on VECI decal.

10. Stop engine and remove tachometer.

11. Connect fuel tank hose to vapor canister. Reconnect electrical lead wire to idle stop solenoid on vehicles equipped with manual transmission.

Low Idle and Curb Idle Adjustment (1-BBL Carburetor)

Refer to **Figure 39**.

1. Start engine and allow it to reach normal operating temperature. Air cleaner should be installed, choke fully open, and air conditioner off. Connect a tachometer to engine, using manufacturer's instructions.

2. Set parking brake and block rear wheels.

3. Disconnect the fuel tank line from the vapor canister and disconnect the vacuum hose from the distributor.

4. Verify that timing is correctly set, then reconnect vacuum hose to distributor.

5. Adjust idle stop solenoid to obtain specified curb idle speed.

6. Disconnect electrical lead wire from idle stop solenoid.

7. Place the transmission selector in DRIVE (automatic) or NEUTRAL (manual). Turn 1/8 inch hex screw in end of solenoid to obtain the specified low idle speed.

8. Reconnect solenoid lead and crank throttle slightly. Verify that curb idle specified rpm is obtained when throttle is released.

9. Stop engine and remove tachometer.

10. Reconnect fuel tank hose to vapor canister.

Fast Idle Adjustment (1-BBL Carburetor)

Refer to **Figure 40**.

1. Verify that low and curb idle speeds are correctly set.

2. Start engine and allow it to reach normal operating speed. The air cleaner should be installed, choke should be fully open, EGR valve signal line should be removed and plugged, and air conditioner should be off. Connect tachometer to engine, using manufacturer's instructions.

3. Disconnect vacuum hose from distributor and plug hose.

4. With transmission in NEUTRAL, set fast idle cam follower so that tang is on high step of cam.

5. Bend cam follower tang to arrive at fast idle speed listed on emissions control decal.

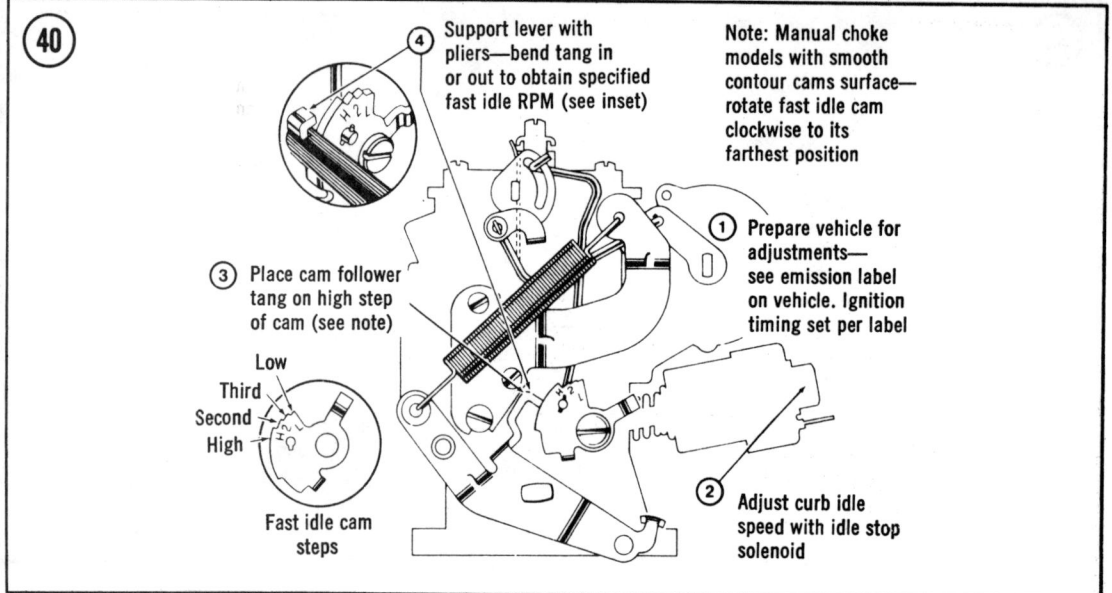

Idle Mixture Adjustment
(1975-1977 2-BBL Carburetor)

1. Start engine and allow to reach normal operating temperature. Air cleaner should be on, air conditioner should be off, and choke should be fully open. Connect tachometer to engine, using manufacturer's instructions.

2. Set parking brake and block rear wheels.

3. Disconnect the fuel tank hose from vapor canister. Disconnect the vacuum hose from distributor and plug hose. Disconnect and plug other hoses as directed by VECI decal.

4. Check engine timing and adjust if required, using the procedures given above. Reconnect vacuum hose to distributor.

5. Position the transmission selector in DRIVE (automatic transmission) or NEUTRAL (manual transmission).

> NOTE: *Obtain idle speeds from Vehicle Emission Control Information decal located in engine compartment. If decal is missing or cannot be read, specifications given in* **Table 4** *can be used.*

> NOTE: *The idle mixture screws are equipped with factory-installed caps. These caps allow acceptable adjustment and must not be removed or altered.*

6. Adjust idle speed screw to obtain the higher of 2 idle speeds given.

7. Turn idle mixture screws out equally to obtain highest possible idle speed. Reset idle speed screw to higher specified rpm if required.

8. Turn idle mixture screws in equally until the lower specified idle speed is achieved.

9. Reconnect fuel tank hose to vapor canister.

Idle Speed Adjustment
(2-BBL Carburetor)

1. Start engine and allow to warm up to normal operating temperature. Air cleaner should be installed, choke should be fully open, and air conditioner should be off. Connect tachometer to engine, using manufacturer's instructions.

2. Set parking brake and block rear wheels.

3. Disconnect fuel tank hose at vapor canister.

4. Disconnect vacuum hose from distributor and plug hose.

5. Verify that timing is properly set or adjust if required, using the procedure given above. Reconnect vacuum hose.

6. Place the transmission selector in DRIVE (automatic transmssion) or NEUTRAL (manual transmission).

7. Adjust idle speed screw to specified rpm.

8. Stop engine and remove tachometer. Connect fuel tank hose to vapor canister and remove blocks from wheels.

LUBRICATION, MAINTENANCE, AND TUNE-UP

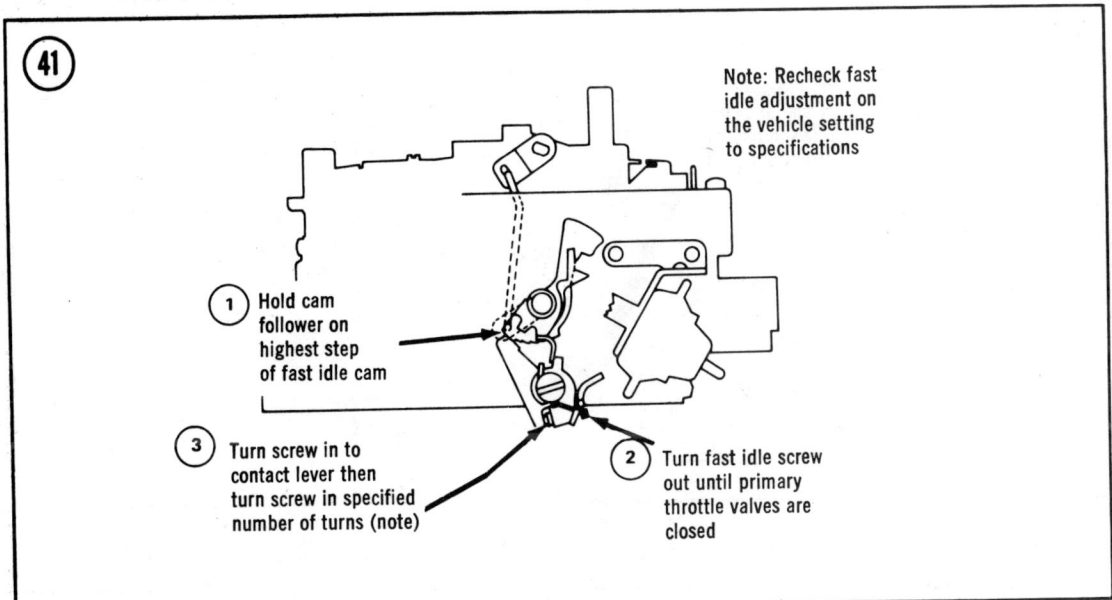

Low and Curb Idle Adjustment
(4-BBL Carburetor)

NOTE: *Obtain low and curb idle speeds from Vehicle Emission Control Information decal in engine compartment. If decal is missing or cannot be read, speeds given in Table 4 can be used.*

1. Start the engine and warm it up to normal operating temperature. The air cleaner should be installed, the choke should be opne, and the air conditioner should be turned off. Connect the tachometer to the engine, using the manufacturer's instructions.
2. Set the parking brake and block rear wheels.
3. Disconnect fuel tank hose from vapor canister. Disconnect distributor vacuum hose and plug hose.
4. Check timing, using procedure given above, and adjust as required. Reconnect distributor vacuum hose.
5. Disconnect the electrical lead at idle stop solenoid.
6. Place the transmission selector in DRIVE (automatic transmission) or NEUTRAL (manual transmission). Allow low idle screw to obtain specified low idle speed.
7. Reconnect idle stop solenoid electrical lead and open throttle slightly to extend solenoid plunger.
8. Adjust solenoid plunger screw to set specified curb idle speed.
9. Stop engine and remove tachometer. Reconnect hose to vapor canister.

Fast Idle Adjustment
(4-BBL Carburetor)

Refer to **Figure 41**.

1. Place cam follower on highest step of fast idle cam.
2. Adjust fast idle screw out until primary throttle valves are closed.
3. Turn fast idle screw into contact lever, then give screw 3 additional turns.
4. Connect tachometer to engine, start engine and check fast idle against specifications listed on Vehicle Emission Control Information decal. Readjust as required.

ROCHESTER CARBURETORS
(1972-1974 VEHICLES)

Idle Mixture Adjustment
(All Carburetors)

1. Start engine and allow to reach normal operating temperature. Air conditioner should be off, choke should be fully open, and air cleaner should be installed on engine. Parking brake should be on and rear wheels should be

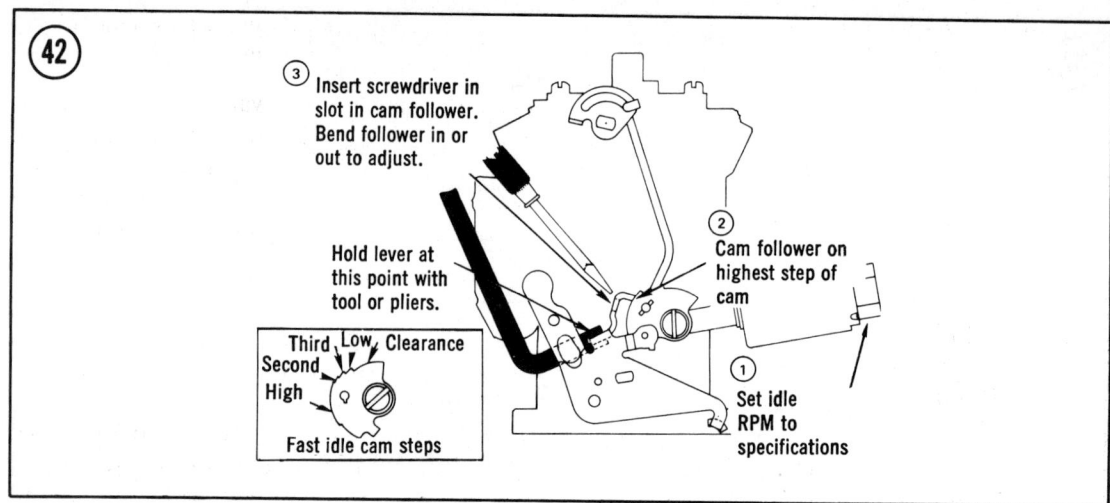

blocked. Connect a tachometer to engine using manufacturer's instructions.

2. Disconnect vacuum hose from distributor and plug hose. Verify that timing has been set, using the procedure given above.

3. Disconnect the fuel tank hose from vapor canister.

4. Adjust idle stop solenoid (turn solenoid body, using hex nut) to obtain 850 rpm (manual transmission in NEUTRAL) or 600 rpm (automatic transmission in DRIVE).

5. De-energize idle stop solenoid and use Allen wrench to adjust solenoid for 450 rpm. Energize solenoid.

6. Position fast idle lever on high step of fast idle cam. Place transmission in NEUTRAL or PARK. Set fast idle to 1,800 rpm by bending fast idle tang on throttle lever if required (**Figure 42**).

7. Reconnect vacuum hose to distributor and fuel tank line to vapor canister. Remove the tachometer.

Idle Speed Adjustment (2-BBL Carburetor)

NOTE: *Fast idle is preset in factory at approximately 1,600 rpm.*

1. Start engine and allow to reach normal operating temperature. Air conditioner should be off, choke should be fully open, and air cleaner should be installed. The parking brake should be on and rear wheels should be

blocked. Connect a tachometer to engine, using manufacturer's instructions.

2. Disconnect vacuum hose from distributor and plug hose. Verify that timing has been properly set, using the procedure given above.

3. Disconnect the fuel tank hose from vapor canister.

4. Adjust idle stop solenoid screw to obtain 900 rpm (manual transmission in NEUTRAL) or 600 rpm (automatic transmission in DRIVE).

5. De-energize the idle stop solenoid by disconnecting electrical lead. Place automatic transmission in DRIVE or manual transmission in NEUTRAL. Adjust idle cam screw (with screw on low step of cam) to obtain 400 rpm on engines with automatic transmissions and 500 rpm on 350 cu. in. engines (manual transmissions).

6. Reconnect idle stop solenoid electrical lead, distributor vacuum line, and fuel tank hose to the vapor canister. Disconnect and remove the tachometer.

Idle Speed Adjustment (4-BBL Carburetors)

1. Start engine and allow to warm up to normal operating temperature. Air conditioner should be off, choke should be wide open, and air cleaner should be installed. Set parking brake and block rear wheels. Connect tachometer to engine, using manufacturer's instructions.

2. Disconnect distributor vacuum hose and plug hose. Verify that timing has been properly set, using procedure given above.

LUBRICATION, MAINTENANCE, AND TUNE-UP

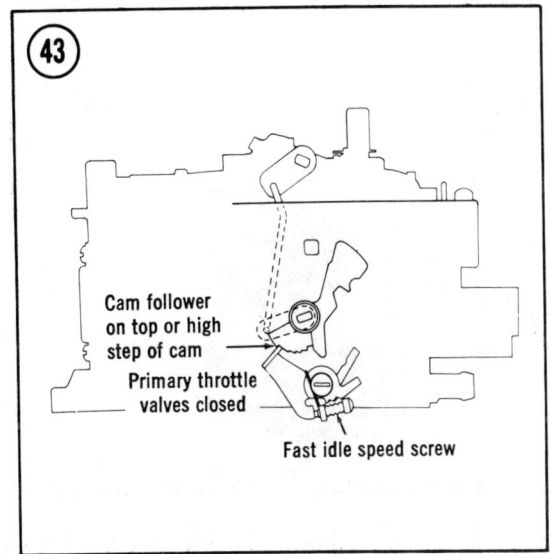

3. Disconnect the fuel tank line from vapor.
4. Adjust idle stop solenoid screws as follows:

   a. Automatic transmissions (in DRIVE) — 600 rpm

   b. Manual transmissions (in NEUTRAL) — 900 rpm

5. De-energize idle stop solenoid and adjust idle speed to low idle rpm given in **Table 4** for your engine, using low idle screw on low step of cam. Reconnect solenoid and check curb idle, readjusting if required.

6. Connect vacuum hose to distributor and place fast idle cam follower on top step of fast idle cam. Adjust fast idle speed as follows (see **Figure 43**):

   a. Automatic transmissions (in PARK) — 1,600 rpm

   b. Manual transmissions (in NEUTRAL) — 1,300 rpm

7. Reconnect fuel tank line to vapor canister. Disconnect and remove tachometer.

ROCHESTER CARBURETORS (1970-1971 VEHICLES)

Idle Mixture Adjustment (All Models)

NOTE: *The 1971 model carburetors are equipped with limiter caps on idle mixture screws. These caps, which should not be removed, do not permit adjustment of the idle mixture.*

1. Start engine and warm up to normal operating temperature (air cleaner installed, preheater valve, if so equipped, and choke valve wide open). Connect tachometer to engine, using manufacturer's instructions.

   NOTE: *Air conditioner, if so equipped, should be turned on or off, per instructions on the Tune-up decal located in engine compartment.*

2. Turn engine off and turn idle mixture screw in until it is lightly seated. Back idle mixture screw out approximately 3 turns and restart engine. Place transmission in DRIVE (automatic transmission) or NEUTRAL (manual transmission).

   NOTE: *Obtain all idle speeds from Tune-up decal in engine compartment. If decal is missing or illegible, see* **Table 4** *for specifications.*

3. Adjust idle mixture screw to obtain highest steady idle sped.

4. Adjust the idle speed screw to rpm recommended on Tune-up decal (adjust idle stop solenoid, if so equipped).

   CAUTION
   On 1971 vehicles only, do not adjust combination emissions control solenoid to set idle speed. Instead, use carburetor idle speed screw to make adjustments. See **Figure 44**.

5. If carburetor is equipped with idle stop solenoid, disconnect solenoid electrical lead and set idle speed to low rpm recommended on Tune-up decal. Reconnect solenoid electrical lead, crack throttle slightly, and recheck idle speed. Reset to recommended curb idle speed per Step 4, if required.

6. Adjust mixture screw in (lean) to obtain a 20 rpm drop (lead roll). Then turn mixture screw out ¼ turn.

7. Repeat Steps 2 through 6 for second mixture screw, if so equipped.

8. Readjust idle speed screw (or solenoid) as required to achieve specified rpm.

CHAPTER THREE

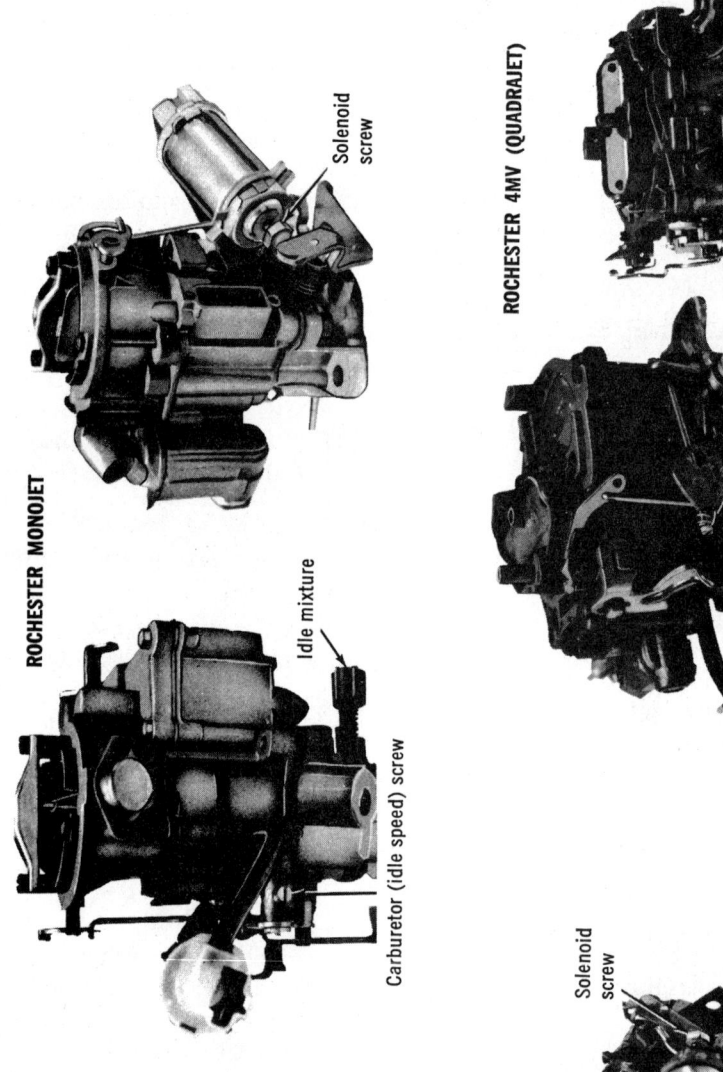

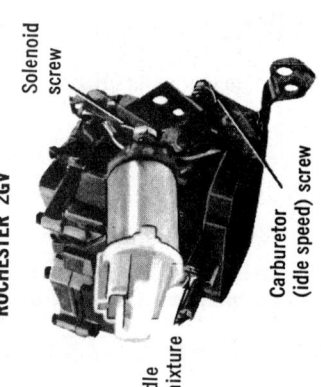

ROCHESTER CARBURETORS — 1971

LUBRICATION, MAINTENANCE, AND TUNE-UP

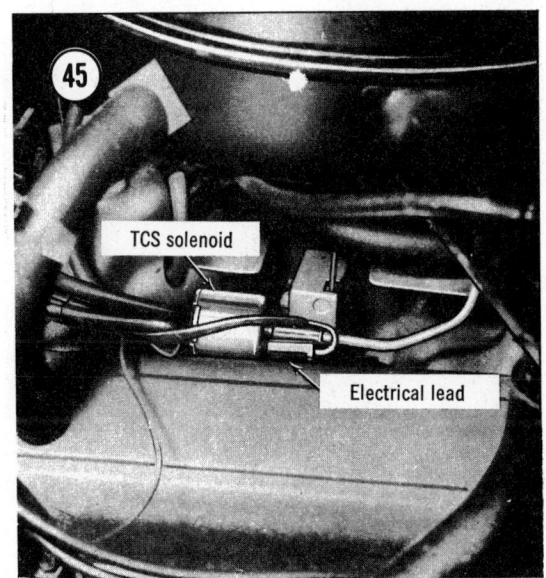

NOTE: *Disconnect electrical lead to Transmission Controlled Spark (TCS) solenoid on 1970 models so equipped. See Figure 45.*

3. Place transmission in NEUTRAL and position fast idle lever on high step (second step for 1970-1974 4-bbl carburetors) of fast idle cam.

4. Adjust fast idle to specifications (see Tune-up decal or **Table 4**) with fast idle screw (4-barrel) or by bending fast idle lever (1-barrel).

5. Reconnect electrical leads to TCS solenoid (1970 models).

6. Disconnect and remove tachometer.

CEC Solenoid Adjustment (All 1971 Models)

Refer to **Figure 46**.

CAUTION
If the CEC solenoid is used to set idle speed or is adjusted out of limits (see Table 4), decrease in engine braking may result.

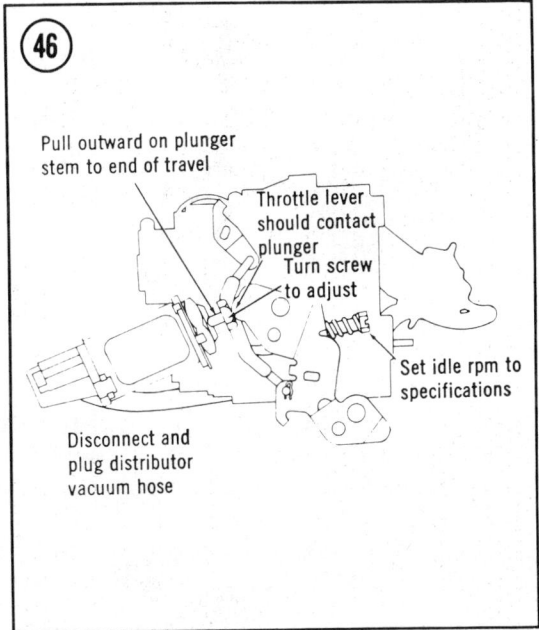

NOTE: *The CEC solenoid is adjusted only after replacement of solenoid, major overhaul of carburetor, or after throttle body is removed and replaced. Also, instructions on Tune-up sticker in engine compartment must be completed before making adjustment.*

1. Start engine and warm up to normal operating temperature. Transmission should be in NEUTRAL (manual) or DRIVE (automatic) with parking brake set and rear wheels blocked. Air conditioner should be off, distributor vacuum hose removed and plugged, and fuel tank hose removed from vapor canister. Tachometer should be connected to engine, using the manufacturer's instructions.

2. Manually extend CEC solenoid plunger to control throttle lever.

3. Adjust plunger length to obtain idle speed as specified in *CEC Solenoid* column of **Table 4**.

4. Disconnect the tachometer and reconnect distributor vacuum hose and fuel tank hose to vapor canister.

Fast Idle Adjustment (All Models)

NOTE: *No fast idle adjustment is required for 2-barrel carburetors.*

1. Start and warm up engine to normal operating temperature. Choke should be in wide open position.

2. Verify that the ignition has been tuned and timed and that curb idle has been set, using the procedure above.

CHAPTER FOUR

ENGINE

A wide range of power options has been offered. Fortunately all of these engines fall into 2 general classes — the inline 6-cylinder (L6) engine and the V6 and V8 engines. All inline sixes are remarkably close in appearance and principle, as are all V6's and V8's; and there is little difference in service procedures for each class. **Figure 1** shows a cutaway view of a typical 6-cylinder engine, while **Figure 2** shows a typical V8. The V6 is similar, except for the number of cylinders.

All engines operate on the 4 cycle principle. The crankshaft turns 2 complete rotations during one cycle. A single camshaft, operating at ½ the crankshaft speed, operates the valves through solid or hydraulic lifters, pushrods, and rocker arms. The camshaft also operates the distributor, fuel pump, and oil pump. The L6 engine has 7 main bearings and 4 camshaft bearings. The V8 has 5 main and 5 camshaft bearings. The V6 has 4 main and 5 camshaft bearing (4 camshaft bearings on the 200 and 229 CID engine).

This chapter covers removal and installation of the engine, removal and replacement of subassemblies, and inspection, adjustments and repairs of some subassemblies and components. As previously stated, the procedures for all engines are remarkably similar, and those given here are typical and general, except where specific instructions are required. Even though the illustrations usually show workbench operations, many single procedures, when not a part of a general overhaul, can be performed successfully with the engine in the car.

ENGINE REMOVAL

NOTE: *The engine and transmission should be removed as a unit.*

1. Remove hood from car.
2. Disconnect cables from battery and remove air cleaner.
3. Drain cooling system and remove radiator and shroud.
4. Remove fan blade and pulley.
5. Disconnect the following wires:
   a. Vacuum advance solenoid (if so equipped)
   b. Alternator
   c. Starter solenoid
   d. Oil pressure switch
   e. Temperature sender
   f. Ignition coil
6. Disconnect the following components:
   a. Accelerator linkage (at manifold bell crank)

6-CYLINDER ENGINE

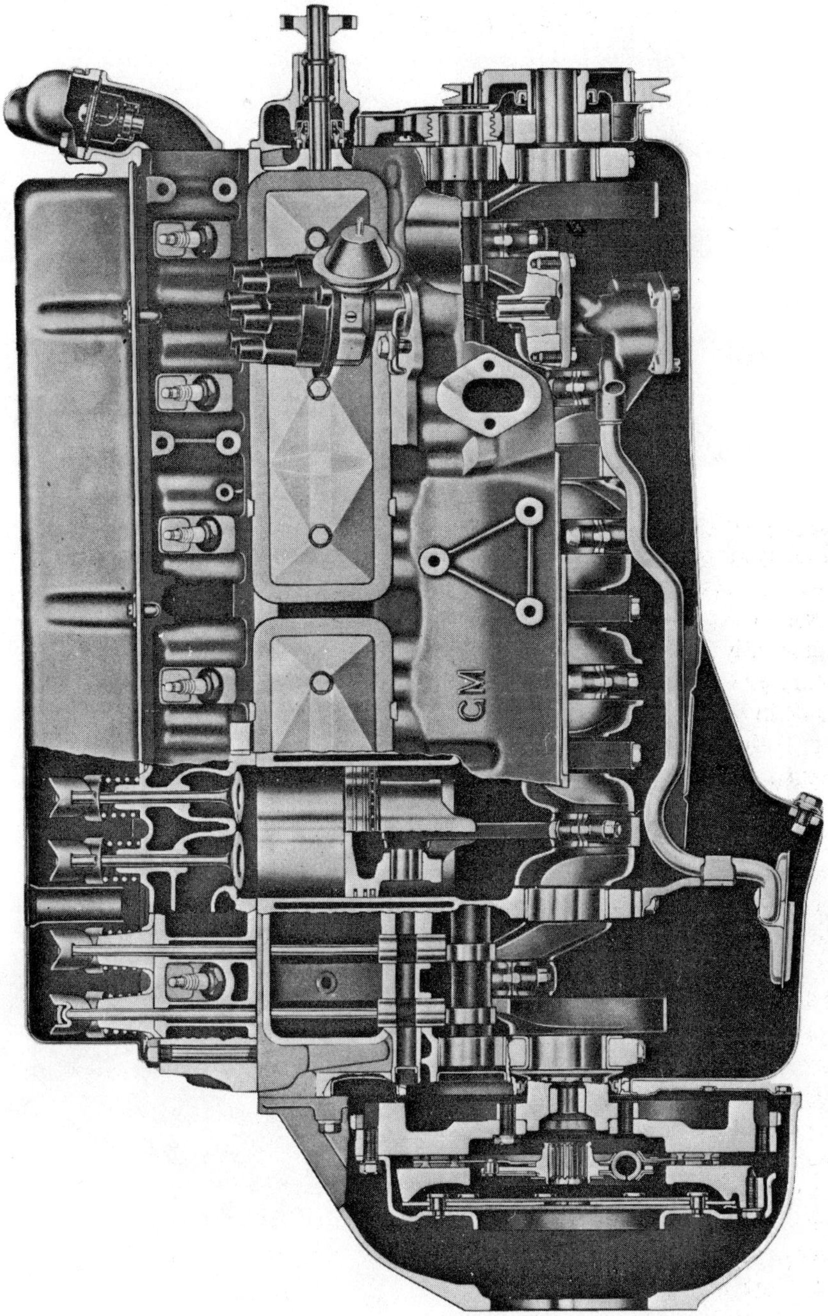

CHAPTER FOUR

② V-8 ENGINE

ENGINE

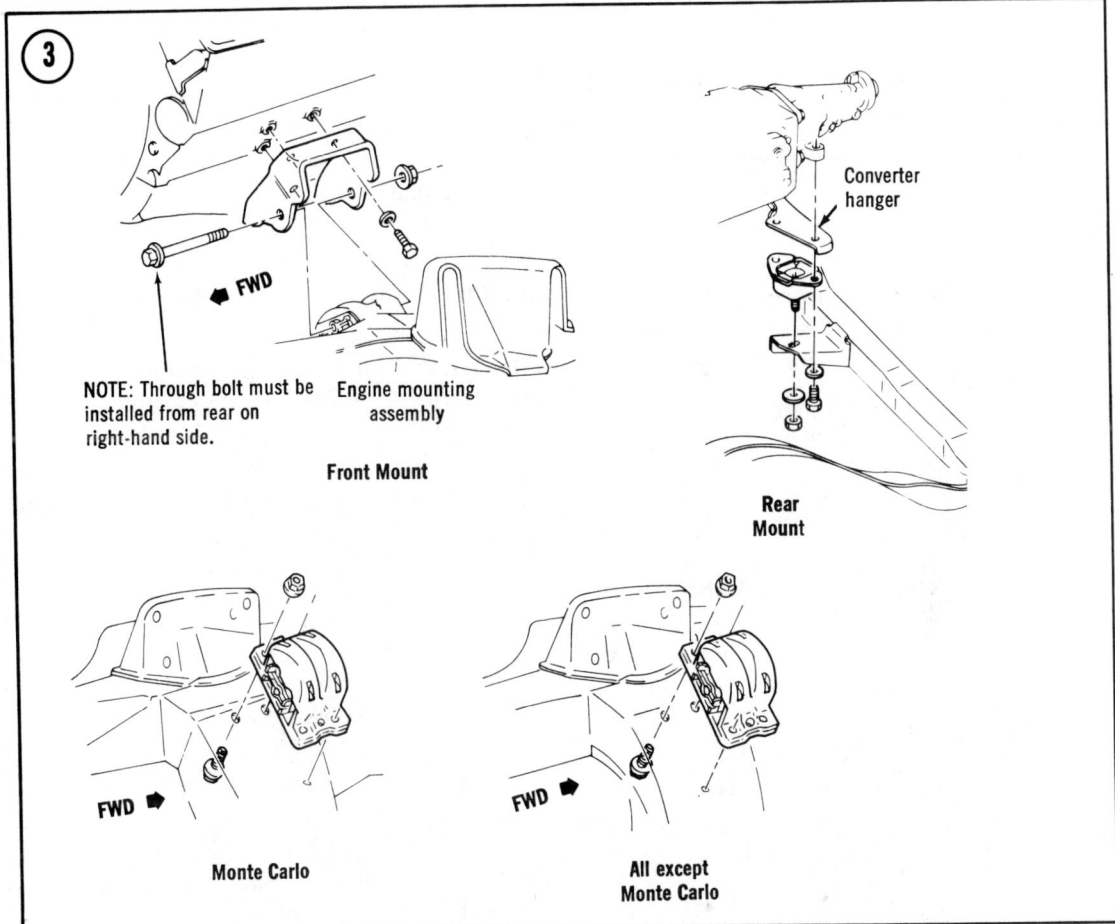

Front Mount

Rear Mount

Monte Carlo

All except Monte Carlo

    b. Fuel line (from tank at fuel pump)

    c. Exhaust pipe (at exhaust manifold flange)

    d. Vacuum line to power brake unit (at intake manifold, if equipped with power brakes)

    e. Hoses to fuel vapor canister

7. If equipped with power steering, remove pump from engine and lay to one side, taking care to avoid fluid spillage.

    NOTE: *On cars equipped with air conditioning, disconnect the compressor ground wire from mounting bracket and the electrical connector from the compressor clutch. Remove the bolts attaching the compressor to the bracket and position the compressor out of the way.*

8. Raise vehicle and drain crankcase of engine oil.

9. Remove drive (propeller) shaft.

    NOTE: *Drain transmission if a plug for propeller shaft opening in transmission is not available.*

10. Disconnect the following components:

    a. Transmission cooler lines (if so equipped)

    b. Transmission control switch (if so equipped)

    c. Shift linkage (at transmission)

    d. Speedometer cable (at transmission)

11. On vehicles with manual transmissions, disconnect clutch linkage at cross shaft, then remove cross shaft engine mount.

12. Lower vehicle and connect lifting device to engine lifting bracket.

13. Raise engine until weight is removed from front engine mounts and then remove through bolts from mounts. See **Figure 3** (typical).

14. Remove bolts holding rear mount to crossmember.

15. Using lifting device, remove engine and transmission from the car.

16A. Remove manual transmission (and clutch) from engine as follows:

    a. Remove screws from clutch housing cover plate.

    b. Remove bolts holding clutch housing to engine, and then remove transmission and clutch housing as a unit.

    NOTE: *Transmission should be supported during removal to avoid damage to clutch disc.*

    c. Remove starter and then remove clutch housing rear cover.

    d. If clutch is to be replaced, loosen clutch mounting bolts a turn at a time until all spring tension is relieved. This will prevent distortion of the clutch cover. Remove all bolts, clutch disc, and pressure plate assembly.

16B. Remove automatic transmission from engine as follows:

    a. Support engine on blocks.

    b. Remove the starter and then remove converter housing underpan.

    c. Remove bolts attaching flywheel to the converter.

    d. Support transmission on blocks.

    e. Remove bolts attaching transmission to engine.

    f. Using lifting device on engine, remove engine blocks and carefully guide engine away from transmission.

17. If engine stand is available, mount engine on stand. Otherwise, support engine on sturdy workbench.

ENGINE INSTALLATION

1. Attach lifting device to engine and remove from stand or work area.

2A. Install manual transmission (and clutch) on engine as follows:

    a. Install clutch, using procedure given in Chapter Nine.

    b. Install starter and clutch housing front cover.

    c. Install transmission and clutch housing, using the procedure given in Chapter Nine.

    d. Install and securely tighten clutch housing cover screws.

2B. Install automatic transmission on engine as follows:

    a. Using lifting device, position engine in front of transmission and align converter with flywheel.

    b. Bolt transmission to engine and then raise engine/transmission assembly and install flywheel-to-converter attaching bolts.

    c. Install starter and converter housing underpan.

3. Tilt engine and transmission and lower into vehicle. Align engine front mounts with frame supports. Install through bolts.

4. Raise rear of engine enough to install rear mount. Lower engine.

5. Remove lifting device and securely tighten through bolts (55 ft.-lb.) and rear mount bolts (30 ft.-lb.).

6. On manual transmission, install clutch cross shaft engine bracket. Then adjust and connect clutch, using the procedure given in Chapter Nine.

7. Raise vehicle and connect transmission control switch (if so equipped), speedometer cable and shift linkage.

8. Install drive shaft. Lower vehicle.

9. Connect the following:

    a. Lower steering pump (if so equipped)

    b. Power brake vacuum line

    c. Line from fuel tank to fuel pump

    d. Vapor canister hoses

    e. Exhaust pipe

    f. Accelerator linkage

10. Connect the following wires:

    a. Vacuum advance solenoid

    b. Coil

    c. Oil pressure switch

ENGINE

    d. Temperature sender
    e. Alternator
    f. Starter solenoid

11. Install pulley, fan blade and fan belt.
12. Install radiator and shroud.

> NOTE: *If car is equipped with air conditioning, reinstall the compressor on the bracket and reconnect the ground wire and electrical connector.*

13. Install and adjust hood.
14. Connect battery cables.
15. Connect and tighten all cooling system hoses and fill with coolant. Fill with engine oil and transmission fluid. Start engine and check for leaks.
16. Install air cleaner and perform any necessary adjustments. See Chapter Three for tune-up procedure.

GENERAL OVERHAUL SEQUENCE

> NOTE: *For partial disassembly of the engine, see the applicable procedure later in this chapter. For complete disassembly, follow the sequence outlined below. This sequence presumes the engine has been removed from the vehicle and mounted on an engine stand or suitable workbench.*

1. Remove the following accessories or components from the engine (if present):

    a. Air injection reactor system (and brackets)
    b. Alternator (and brackets)
    c. Accessory drive pulleys
    d. Fuel pump (and pushrod, V6 and V8 engines)
    e. Water pump and hoses
    f. Spark plug wires and distributor cap
    g. Carburetor and fuel lines
    h. Oil filter
    i. Starter
    j. Clutch pressure plate and disc
    k. Ground strap
    l. Oil dipstick and tube

2. Remove intake and exhaust manifolds.
3. Remove pushrod covers (inline engines).
4. Remove the rocker arm and shaft assemblies and remove pushrods and valve lifters. Store the removed items in racks so they can be reinstalled in the same locations from which they were removed.
5. Remove cylinder head(s).
6. Remove tortional damper.
7. Remove oil pan.
8. Remove front cover from crankcase.
9. Remove oil baffle, if so equipped.
10. Remove oil pump and screen. Remove extension shaft (V8 and 200-229 V6 engines).
11. If connecting rods and caps are not marked with cylinder number identification, mark them. Also check cylinder bores for ridges. If necessary, remove ridges.
12. Remove connecting rod caps, then remove connecting rod/piston assemblies from engine.

> NOTE: *Turn crankshaft as required to remove rods and pistons.*

> *CAUTION*
> *Use care when removing camshaft to avoid damaging bearings.*

13A. Remove camshaft from V6 and V8 as follows:

    a. Remove attaching bolts from camshaft sprocket, tap lower edge of sprocket lightly with plastic hammer to dislodge, and remove sprocket and timing chain. On 231 V6, remove camshaft and crankshaft sprockets and timing chain after aligning timing marks on sprockets.
    b. Install two $5/16$-18 bolts in camshaft sprocket bolt holes (except on 231 V6) and carefully pull out camshaft.

13B. Remove camshaft from inline engines as follows:

    a. Remove camshaft thrust plate screws, using holes in camshaft gear for access.
    b. Remove the camshaft and gear as an assembly.

14. Remove flywheel.

15. Remove caps from main bearings and remove crankshaft from cylinder block.

16. Remove rear main bearing oil seal from cylinder block and bearing cap.

17. Discard all gaskets and seals removed during the above steps.

> NOTE: *See procedures later in this chapter for further disassembly, cleaning, and inspection of the parts or subassemblies removed during this procedure.*

GENERAL REASSEMBLY SEQUENCE

> NOTE: *Use only new gaskets and seals on engine during assembly.*

1. Install crankshaft as follows:

   a. Place rear main bearing oil seal in grooves in cylinder block and bearing cap. Make certain seal lip faces toward front of engine. If seal has 2 lips, one with helix should face front of engine.

   b. Lubricate seal lips with engine oil, keeping oil off parting line surface.

   c. Install main bearings in block and bearing caps and lubricate bearing surface with engine oil.

   d. Carefully install crankshaft in cylinder block.

   e. Brush a thin coat of oil sealing compound on block mating surface (see **Figure 4**) and on corresponding surface on bearing cap.

   > CAUTION
   > *Do not allow sealer on crankshaft or seal.*

   f. Install main bearing caps, with arrows pointing to front of engine; and torque all bolts (except rear cap) to specifications (see **Tables 1 and 2** at end of chapter). Torque rear bearing cap bolts to 10-12 ft.-lb., then tap end of crankshaft, first to the rear and then to the front, with a lead hammer to line up rear main bearing and crankshaft thrust surfaces. Retorque all bolts to specifications.

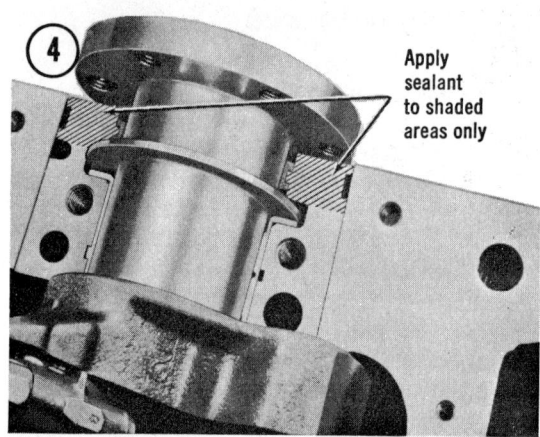

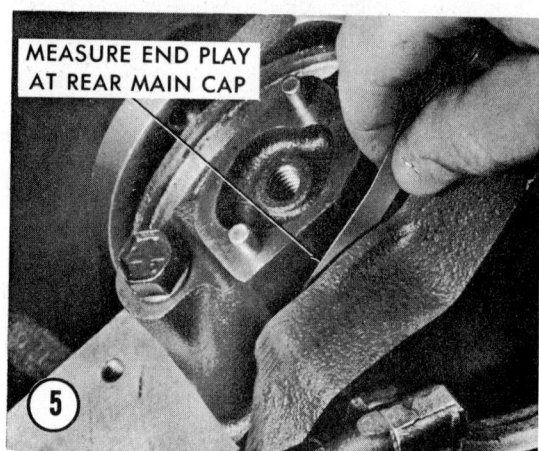

   g. Measure crankshaft end play with feeler gauge (see **Figure 5**). Force crankshaft forward as far as it will go and measure between front of rear main bearing and crankshaft thrust surface. See **Table 3**, Engine Specifications (at end of chapter), for allowable end play.

2. Install a wood block between crankshaft and cylinder block to prevent rotation and install flywheel. Torque to specifications (see **Tables 1 and 2** at end of chapter).

   > NOTE: *Align dowel holes in flywheel and crankshaft. If vehicle has automatic transmission, make certain converter attaching pads on flywheel are facing toward transmission.*

   > NOTE: *If camshaft or lifters are new, lubricate cam lobes and lifter feet with Molykote or equivalent.*

ENGINE

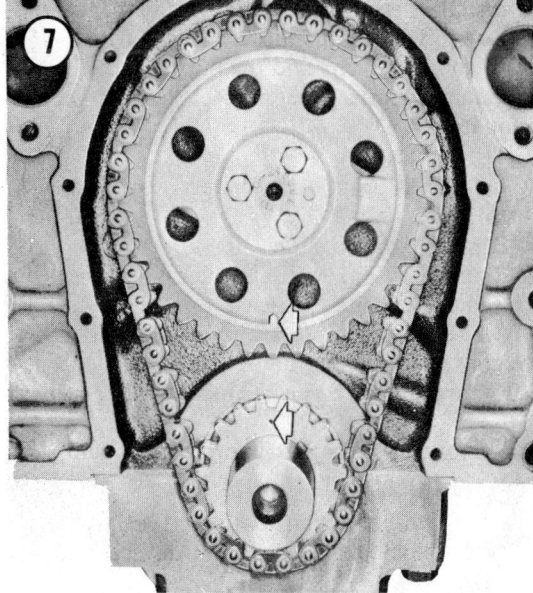

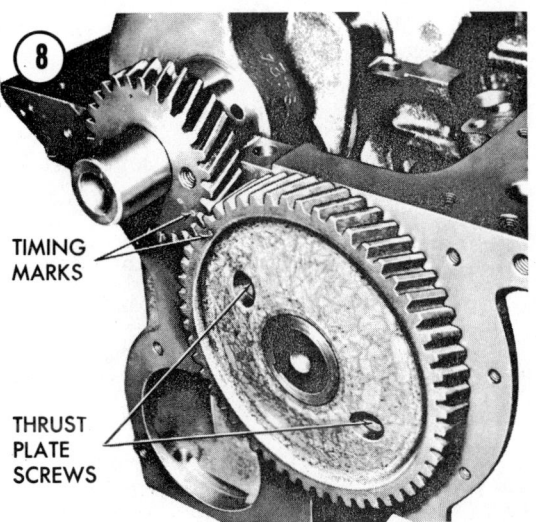

3A. Install camshaft on V6 and V8 as follows:

a. Thread 5/16-18 bolts into crankshaft bolt holes (except 231 V6), lubricate camshaft journals with engine oil, and install camshaft. See **Figure 6**, typical. Take care not to damage bearings. Remove bolts.

b. Install timing chain on camshaft sprocket. Align marks on camshaft and crankshaft sprockets (see **Figure 7**) and install chain on crankshaft sprocket. Align dowel on camshaft with hole in camshaft sprocket and install sprocket on camshaft.

CAUTION
Do not hammer on camshaft sprocket during installation, as camshaft rear expansion plug could be loosened.

c. Use mounting bolts to draw sprocket onto camshaft. Torque to specifications (see **Tables 1 and 2**, at end of chapter), and lubricate the chain with engine oil.

3B. Install camshaft on inline engines as follows:

a. Lubricate journals on camshaft with engine oil.

b. Position crankshaft and camshaft so that timing marks are aligned (see **Figure 8**) and install camshaft and gear assembly. Take care not to damage the camshaft bearings.

c. Using a dial indicator, check gear runout (see **Figure 9**). Camshaft gear runout should not exceed 0.004 in. and crankshaft gear should not exceed 0.003 in.

d. Use dial indicator to check backlash between timing gear teeth (see **Figure 10**). Backlash should be not less than 0.004 in. and not more than 0.006 in.

e. Lubricate timing gears with engine oil.

4. Install connecting rod and piston assemblies as follows:

a. Install bearings in connecting rods and caps. Lubricate bearings, pistons, piston rings, bolts and cylinder walls with engine oil.

b. Using a ring compressor to compress the piston rings, insert No. 1 piston assembly in No. 1 cylinder, using light taps with a wooden hammer handle. Hold ring compressor firmly against cylinder block until all rings have entered the bore.

CAUTION
Make certain ring gaps are properly positioned on piston and piston is properly positioned in cylinder. See Figure 11.

NOTE: *A connecting rod guide tool set will greatly help in installing piston and rod assemblies. If not available, use care to avoid damage to bearing surfaces, threads, and crankshaft journals. Masking tape over threads and sharp projections will help. However, make certain all adhesive from the tape is removed before assembling rods to the crankshaft.*

c. If installed, remove rod guide set. Install connecting rod cap and torque nuts to specifications (see **Tables 1 and 2** at end of chapter).

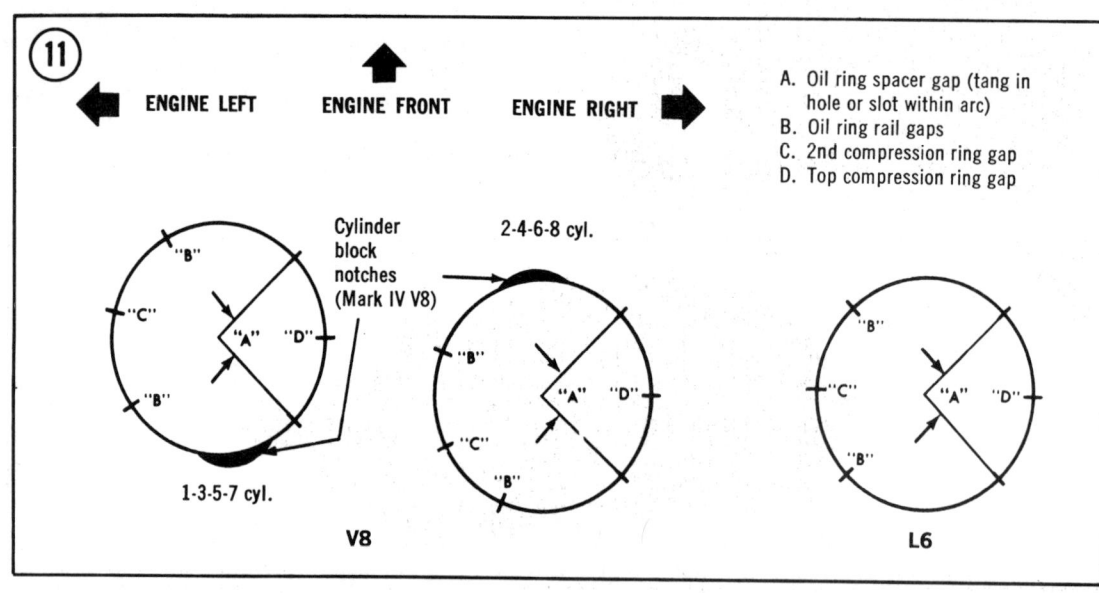

ENGINE

d. Repeat Steps b and c for remaining rod and piston assemblies.

e. Measure side clearance of each rod (see **Table 3** at the end of the chapter). Make measurement between connecting rod cap and side of crankpin (inline engines) or between connecting rod caps (V6 and V8 engines). Refer to **Figure 12 or 13**.

5. Install oil pump and screen assembly, extension shaft (V6 and V8 engines), and oil baffle, if so equipped.

6A. Install crankcase front cover on inline engines as follows:

a. Lubricate lip seal with engine oil and install crankcase front cover aligning tool (GM tool No. J-23042 or equivalent) in seal (see **Figure 14**).

b. Position gasket on cover and install cover on cylinder block. Torque to specifications (see **Tables 1 and 2** at end of chapter) and remove tool.

6B. Install crankcase front cover on V6 and V8 engines as follows:

a. Place gasket in position on dowels on cylinder block.

b. Lubricate lip seal with engine oil. Place cover in position on dowel pins. Install bolts and torque to specifications (see **Tables 1 and 2** at the end of the chapter).

7. Install oil pan as follows:

a. Place side gaskets on cylinder block sealing surfaces and apply sealer at the intersection of the end seals and side gaskets.

b. Place oil pan rear seal in groove in rear main bearing cap. Ends of seal should butt against side gaskets.

c. Place oil pan front seal in crankcase front cover. Ends of seal should butt against side gaskets.

d. Install oil pan and torque to specifications (see **Tables 1 and 2** at end of chapter).

CAUTION
The installation procedures in Step 8 must be followed, using the proper tool, or movement of the inertial weight section (assembled to hub with rubber-like material) on the hub may destroy the tuning of the torsional damper.

8A. Install drive-on torsional damper as follows:

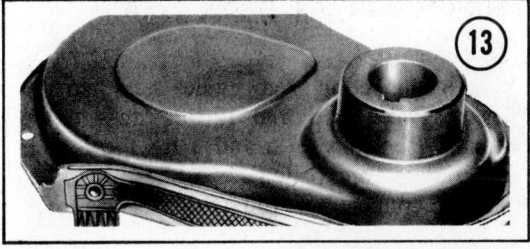

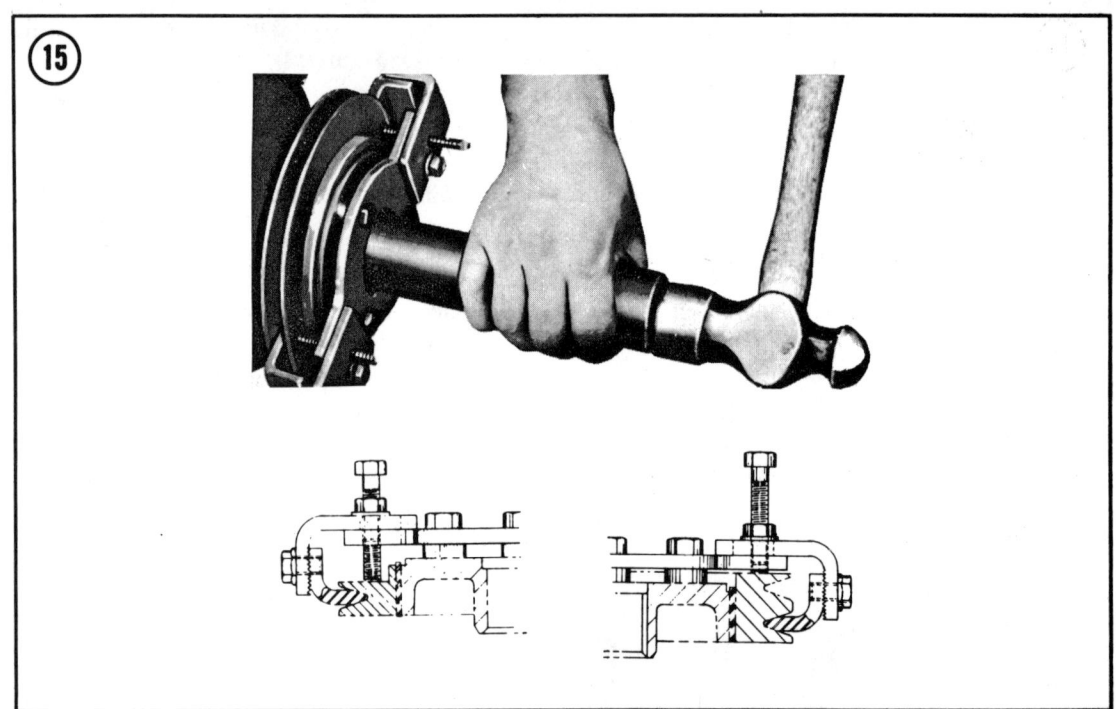

a. Coat seal contact area on damper with engine oil.

b. Attach installer tool (GM Part J-22197) to damper (see **Figure 15**). Tighten fingers of tool to keep weight from moving.

c. Position damper on crankshaft and drive into position, using hollow drift bar (GM Part J-5590), until the damper bottoms against crankshaft gear or sprocket. Remove tool.

8B. Install pull-on type torsional damper as follows:

a. Coat seal contact area on damper with engine oil.

b. Position damper over key on camshaft.

c. Using tool (GM Part J-23523), pull damper onto crankshaft (see **Figure 16**).

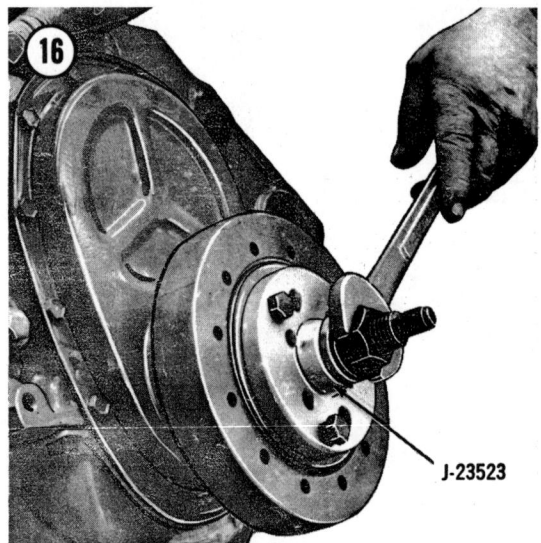

CAUTION
Bolt should be installed in crankshaft with a minimum of ½ in. thread engagement.

d. Remove tool and install retaining bolts. Torque to specifications (see **Tables 1 and 2** at end of chapter).

9. Install cylinder head(s) as follows:

CAUTION
Make certain gasket surfaces on head(s) and cylinder block are clean and free of nicks and deep scratches. Threads in the block and on cylinder head bolts must be clean, as dirty threads will affect bolt torques.

a. Apply a thin coat of sealer on both sides of steel gaskets.

ENGINE

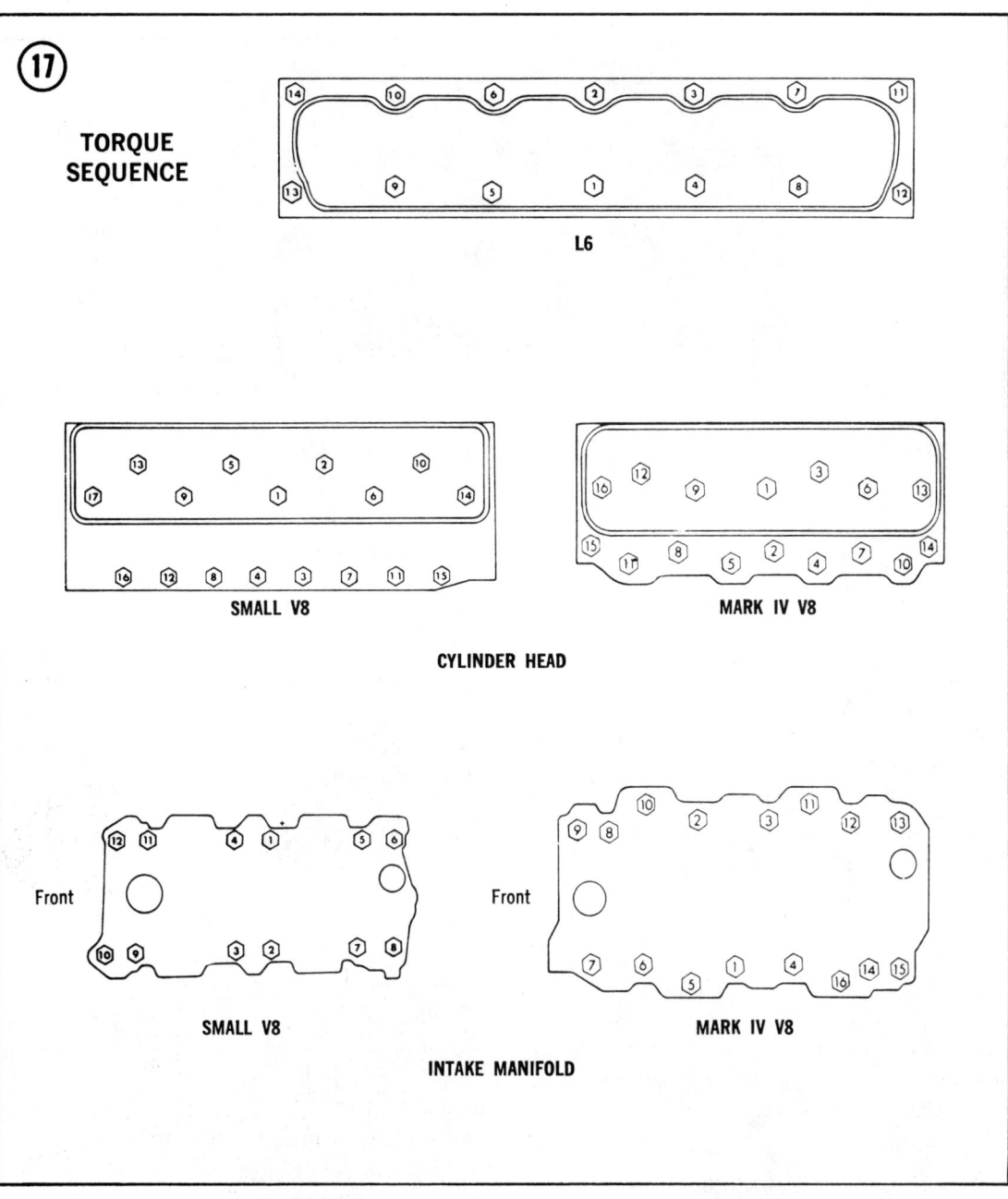

CAUTION
Too much sealer will affect the sealing quality of the gasket.

CAUTION
Do not use sealer on either side of a composite steel-abestos gasket.

b. Place gasket over dowel pins on cylinder block with bead side up.

c. Carefully lower cylinder head into place.

d. Apply sealing compound to bolt threads and install them finger-tight.

e. Following the torque sequence shown in **Figure 17**, tighten each cylinder head bolt a small amount at a time until all bolts have been tightened to specification (see **Tables 1 and 2**, end of chapter).

NOTE: *For 200 and 229 V6 engines, use a pattern similar to that shown for small V8s. Start in the center and work toward both ends, using a crisscross pattern. See* **Figure 18** *for 231 V6.*

10. Install valve lifters and pushrods, making certain they are replaced in the same locations from which they were removed.

NOTE: *If new lifters and/or rocker arms and balls are installed, coat feet of valve lifters and surfaces of rocker arm and balls with Molykote or equivalent.*

11. Install rocker arms, balls and nuts. Tighten nuts to take up all pushrod end play.

12. Install pushrod covers (inline engines).

13. Install exhaust and intake manifolds, and torque bolts in sequence shown in **Figure 17** to specifications (see **Tables 1 and 2** at end of chapter). See note following Step 9e above. See **Figure 19** for 231 V6.

14. Install the following applicable sub-assemblies or components, using the appropriate procedures given later in this chapter.

    a. Engine oil dipstick tube and dipstick
    b. Clutch pressure plate and disc
    c. Ground strap
    d. Starter
    e. New oil filter
    f. Fuel lines and carburetor
    g. Distributor
    h. Distributor cap and spark plug wires
    i. Fuel pump (and pushrod on V6 and V8 engines)
    j. Water pump and bypass hose
    k. Accessory drive pulleys and belts
    l. Alternator and bracket
    m. Air injector reactor system with brackets

15. Adjust all belts to proper tension

16. Adjust valves, using the procedure given elsewhere in this chapter.

17. Connect lifting device to engine lifting brackets and install in car, using the procedure given elsewhere in this chapter.

MANIFOLD ASSEMBLY (INLINE ENGINES)

Removal

1. Remove air cleaner and disconnect both throttle rods at bellcrank and remove throttle spring.

2. Disconnect vacuum and fuel lines from carburetor.

3. Disconnect ventilation hose from rocker arm cover and hose at vapor canister.

4. Disconnect exhaust pipe from manifold flange and discard packing.

5. Remove attaching bolts and clamps and then remove manifold assembly. Discard all gaskets.

6. Check for cracks in manifold assembly.

7. Separate manifolds by removing bolt and nuts at center.

Installation

1. Clean all gasket surfaces on manifold and cylinder head.

ENGINE

2. Check manifold exhaust port for gaps by placing a straightedge along the full length of the port faces. If at any point a gap of 0.030 in. exists, the manifold will not, in all probability, seal and the manifold should be replaced.

3. Rejoin the exhaust and intake manifolds by installing the center bolt and nuts (finger-tight).

4. Place new gasket on cylinder head manifold end studs. Place manifold assembly on cylinder head and install bolts and clamps. Torque to 25-30 ft.-lb. (inner bolts) and 20 ft.-lb. (outer bolts).

5. Torque exhaust-to-intake manifold bolt and nuts to 30 ft.-lb.

6. Using new packing, connect exhaust pipe to manifold flange.

7. Connect ventilation hose to rocker arm cover and connect hoses to vapor canister.

8. Connect vacuum and fuel lines to carburetor.

9. Reconnect throttle linkage and install throttle spring.

10. Install air cleaner, start engine, and check for manifold leaks and adjust carburetor idle speed.

INTAKE MANIFOLD (V6 AND V8 ENGINES)

Removal

NOTE: *If engine has been removed from vehicle, skip to Step 5.*

1. Remove air cleaner and drain radiator.
2. Disconnect the following:
   a. Battery cables (at battery)
   b. Radiator hoses (at manifold)
   c. Accelerator linkage (at pedal lever)
   d. Fuel line (at carburetor)
   e. All coil and temperature sender wires
   f. Power brake vacuum hose
   g. Crankcase ventilation hoses (if so equipped)
   h. Vacuum hose (at distributor and carburetor).
   i. Idle stop solenoid wire (if so equipped).

3. Remove cap from distributor and mark position of rotor. First remove distributor clamp, then remove distributor from vehicle. Position distributor cap rearward so it is clear of manifold.

4. Remove alternator upper bracket.

5. Remove coil bracket and coil.

6. Remove attaching bolts and then remove intake manifold and carburetor from vehicle. Discard all gaskets and seals.

7. If the manifold is to be replaced, transfer the following to the new manifold, using new gaskets:
   a. Carburetor (and attaching bolts)
   b. Water outlet and thermostat
   c. Heater hose adapter

d. Choke coil

e. EGR valve (if so equipped)

Installation

1. Clean all sealing and gasket surfaces on cylinder block, cylinder heads, and intake manifold.

2. Install new end seals, and side gaskets as shown in **Figures 20 and 21**. Use sealing compound around water passages.

3. Install attaching bolts and torque to specifications (see **Tables 1 and 2** at end of chapter), using the torquing pattern shown in **Figure 17** (V8) or **Figure 19** (231 V6).

4. Install coil bracket and coil.

5. Install distributor and distributor clamp. Make certain rotor points toward mark made during removal. Install distributor cap.

> NOTE: *If crankshaft was rotated while distributor was removed, see procedure later in this chapter for installation instructions.*

6. Install alternator upper bracket and adjust belt tension and tighten bolts.

7. Connect the following components:

   a. Battery cables
   b. Radiator hoses
   c. Accelerator linkage
   d. Fuel line
   e. Wires to coil
   f. Power brake vacuum hose
   g. Distributor vacuum hose
   h. Crankcase ventilation hoses

8. Fill cooling system with coolant, start engine, and check manifold and cooling system for leaks. Adjust timing and idle speed, using the procedures given in Chapter Three.

EXHAUST MANIFOLD (V6 AND V8 ENGINES)

Removal

> NOTE: *If engine is equipped with Air Injection Reactor System, remove AIR manifold and tubes, using procedure given in Chapter Five.*

1. Disconnect battery ground cable and remove pre-heater air stove from air cleaner. Remove alternator, if required.

2. Remove flange nuts and lower and support crossover (if so equipped) and exhaust pipe assembly. Disconnect EFE pipe (if so equipped).

3. Remove bolts (end bolts first) and remove exhaust manifold from engine. Discard gaskets.

Installation

1. Clean mating surfaces on manifold and cylinder heads. Install manifold and secure with center bolts.

2. Install end bolts and snug up all bolts. Torque center bolts to specifications (see **Tables 1 and 2** at end of chapter), then torque end bolts to specification.

3. Using a new gasket, install crossover and exhaust pipes to manifold flange. Connect EFE pipe, if so equipped.

ENGINE

4. Install alternator (if removed), and connect battery ground cable.

5. Install air cleaner pre-heater, and AIR manifold and tubes, if so equipped.

6. Start engine and check for leaks.

ROCKER ARM COVER (ALL MODELS)

Removal/Installation

1. Remove the air cleaner and disconnect crankcase ventilation hose(s) at rocker arm cover(s).

2. Remove all wires from clips on the rocker arm cover. Remove air injection hose from check valve of AIR pipe, if so equipped.

3. Remove attaching bolts and remove rocker arm cover. If so equipped, rotate cover out from under AIR pipe.

4. To install, clean all gasket surfaces on cylinder head and rocker cover, install new gasket, and reverse Steps 1 through 3.

VALVE MECHANISM

Removal

1. Remove rocker arm cover(s), using procedure above.

2. On inline, 200-229 V6, and V8 engines, remove rocker arm nuts, balls, arms, and pushrods. On 231 V6 engines, remove shaft and arm assemblies and pushrods. Remove Nylon rocker arm retainers with a pair of channel lock pliers to separate rocker arms from shaft.

NOTE: *Store nuts, balls, arms, and pushrods in a rack, properly identified, so they can be reinstalled in same locations from which they were removed. A rack can be built by drilling a series of 12 or 16 holes in a piece of wood and marking them with the cylinder numbers.*

Installation and Adjustment (Inline Engines)

CAUTION
Whenever new rocker arms and/or rocker arm balls are installed, coat the bearing surfaces with Molykote or equivalent.

1. Install pushrods, making certain rods seat in lifter sockets.

2. Install rocker arms, balls, and nuts. Tighten nuts until all valve lash is eliminated.

3. Adjust valves when lifter is on base circle of camshaft lobe as follows:

   a. Mark distributor housing with chalk at No. 1 and No. 6 cylinder positions, then disconnect wires from spark plugs and coil. Remove distributor cap and wires.

   b. Crank engine until distributor rotor points to No. 1 cylinder position and breaker points are open, and adjust the following valves:

   No. 1 cylinder exhaust and intake
   No. 2 cylinder intake
   No. 3 cylinder exhaust
   No. 4 cylinder intake
   No. 5 cylinder exhaust

   c. Back out adjusting nut until lash is felt at pushrod, then turn nut to eliminate all lash. Verify by checking pushrod end play while turning nut (see **Figure 22**). When all play has been removed, tighten nut one additional full turn (to center lifter plunger).

d. Crank engine until rotor points to No. 6 cylinder position and breaker points are open. Adjust the following valves:

   No. 2 cylinder exhaust
   No. 3 cylinder intake
   No. 4 cylinder exhaust
   No. 5 cylinder intake
   No. 6 cylinder intake and exhaust

4. Reinstall distributor cap and spark plug wires.

5. Install rocker arm cover, using the procedure given above.

6. Adjust idle speed.

Installation and Adjustment (V6 and V8 Engines)

> NOTE: *To adjust valves on engines equipped with mechanical valve lifters, see procedure given in Chapter Three.*

> NOTE: *If new rocker arms and/or rocker arm balls are being installed, coat bearing surfaces with Molykote or equivalent.*

1. Install pushrods in same locations from which they were removed. Verify that rods are seated in lifter sockets.

2. On 200-229 V6 and V8 engines, install rocker arms, balls, and nuts. Tighten nuts until all valve lash is removed. On 231 V6 engines, install rocker arm and shaft assemblies and tighten nuts to 30 ft.-lb.

3. Adjust valves (hydraulic lifters only) as follows:

> NOTE: *Valve lash adjustment is neither possible nor required on the 231 V6 engines.*

   a. Crank engine until mark on torsional damper is aligned with zero mark on timing tab and engine is in No. 1 firing position.

   > NOTE: *When marks are aligned as described above, engine may be in either No. 1 or No. 6 firing position (V8) or No. 4 (V6). To determine No. 1 firing*

   position, hold fingers on No. 1 cylinder valves while cranking engine. If valves are not moving as mark on damper comes near the zero mark, engine is in No. 1 firing position. If valve move, engine is in No. 6 (No. 4 on V6) firing position.

   b. With engine in No. 1 position, adjust following valves (see Step d):
      Exhaust: 1, 3, 5, 8 (V8) or 1, 5, 6 (V6)
      Intake: 1, 2, 5, 7 (V8) or 1, 2, 3, (V6).

   c. Crank engine one revolution to No. 6 firing position (No. 4 on V6) and adjust the following valves (see Step d):
      Exhaust: 2, 5, 6, 7 (V8) or 2, 3, 4 (V6)
      Intake: 3, 4, 6, 8 (V8) or 4, 5, 6 (V6)

   d. Adjust valves by backing off adjusting nut (rocker arm stud nut) until play can be felt in pushrod. Tighten nut to the point that all pushrod-to-rocker arm clearance is removed. This can be determined by rotating pushrod while tightening nut (see **Figure 23**). When pushrod does not turn easily, clearance has been eliminated. Now tighten nut one additional turn to center hydraulic lifter plunger. No other adjustment is required.

4. Install rocker arm cover(s) using new gaskets and procedure given elsewhere in this chapter.

ENGINE

HYDRAULIC VALVE LIFTERS

All inline V6 and most V8 engines are equipped with hydraulic valve lifters, which seldom require attention and do not require readjustments. Exercise care and cleanliness when handling parts, however.

Valve Lifter Removal (Inline Engine)

1. Remove valve mechanism, using procedure given above.
2. Mark No. 1 and No. 6 cylinder positions on distributor housing, then disconnect coil and spark plug wire and remove wires and distributor cap.
3. Turn engine over until distributor rotor points to No. 1 cylinder position. Disconnect primary lead at coil and remove distributor from vehicle.
4. Remove pushrod covers and discard gaskets.
5. Remove valve lifters.

NOTE: *Store valve lifters in order on a rack so they may be reinstalled in the same locations from which they were removed.*

Valve Lifter Installation (Inline Engines)

1. Install valve lifters in the engine.

NOTE: *If new valve lifters are being installed, coat the foot of each lifter with Molykote or equivalent.*

2. Install pushrod covers, using new gaskets. Torque to 80 in.-lb. (1970 models) or 50 in.-lb. (1971 and later models).
3. Install distributor and secure with clamp and bolt. Make certain rotor is pointing to No. 1 cylinder position. Connect primary lead to coil.
4. Reinstall and adjust valve mechanism and distributor cap and wires, using the procedure given above.
5. Set ignition timing and idle speed, using the procedure given in Chapter Three.

Valve Lifter Removal (V6 and V8 Engines)

1. Remove intake manifold, using procedure given above.
2. Remove valve mechanism, using procedure given above.
3. Remove valve lifters. Store lifters in a rack so they can be reinstalled in the locations from which they were removed.

Valve Lifter Installation (V6 and V8 Engines)

1. Install valve lifters.

NOTE: *If new lifters are being installed, coat feet of lifters with Molykote or equivalent.*

2. Install intake manifold, using procedure given above.
3. Install and adjust valve mechanism, using the procedure given above.

VALVE STEM OIL SEAL AND/OR VALVE SPRING (ALL ENGINES)

Removal/Installation

1. Remove rocker arm cover(s), using procedure given above.

2. Remove spark plug, rocker arm and pushrod on valve to be serviced.

3. Install air line adapter tool in spark plug hole and apply compressed air to hold valves.

> NOTE: *Tool is GM Part No. J-23590. A suitable substitute can be made by removing the porcelain and side electrode from an old spark plug (see **Figure 24**). Drive out porcelain, after cutting as shown, by tapping on center electrode. Using a $\frac{3}{8}$ in. pipe, tap, cut thread in remaining body of plug and assemble air connection as shown. A suitable tool also may be available at an auto parts dealer.*

4. Using a valve spring compressor tool (GM Part No. J-5892 or J-806.2 on 231 V6 engines) or equivalent, applied as shown in **Figure 25** or **Figure 26** (231 V6), remove valve locks, valve cap, and valve spring and damper.

5. Remove valve stem oil seal (if replacement is required).

6. On inline V6, and small V8 engines, replace as follows:

   a. Set valve spring and damper, valve shield and valve cup in place.

   b. Compress the spring and install oil seal in lower groove of stem, taking care that seal is flat and not twisted.

   > NOTE: *A light coat of oil on seal will help avoid twisting.*

   c. Install valve locks and release spring compression. Make sure locks seat properly in upper groove of valve stem.

   > NOTE: *A small amount of grease can be used to hold locks in place.*

7. On large (Mark IV) V8 engines, replace as follows:

   a. Coat new valve stem oil seal with oil and position over valve guide.

   > NOTE: *Follow seal installation procedures supplied with service kit.*

   b. Set valve spring and damper and cap in place.

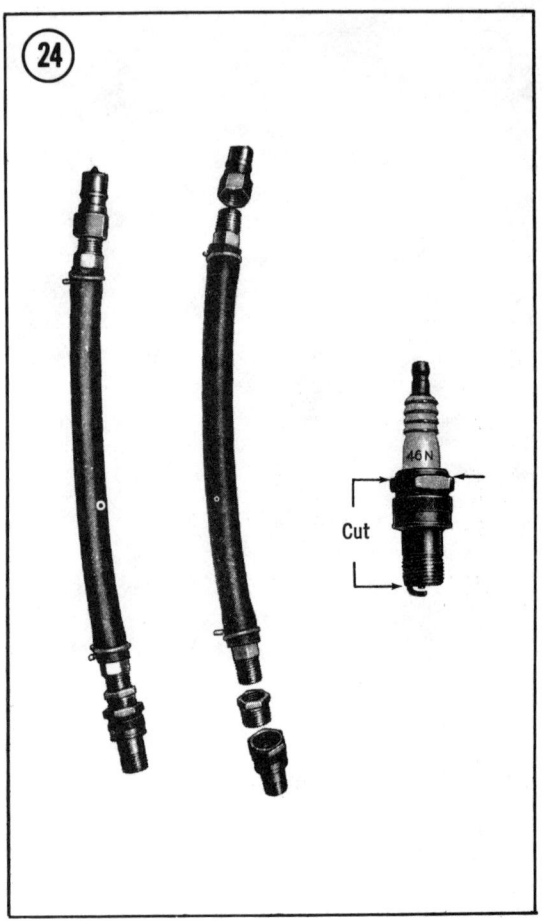

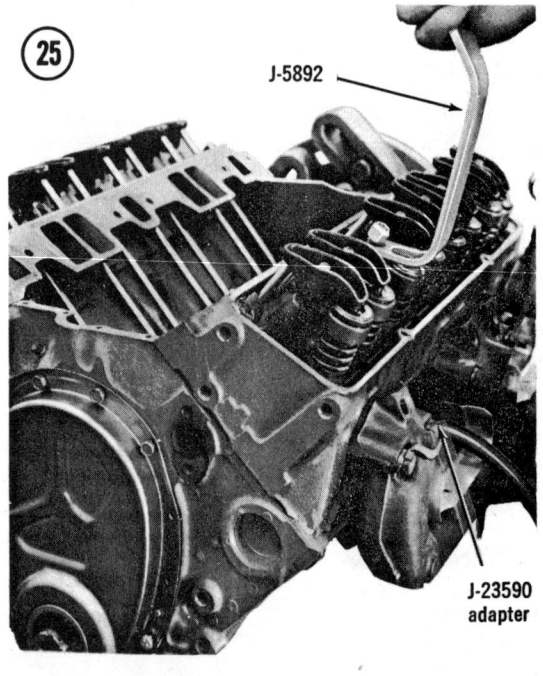

ENGINE

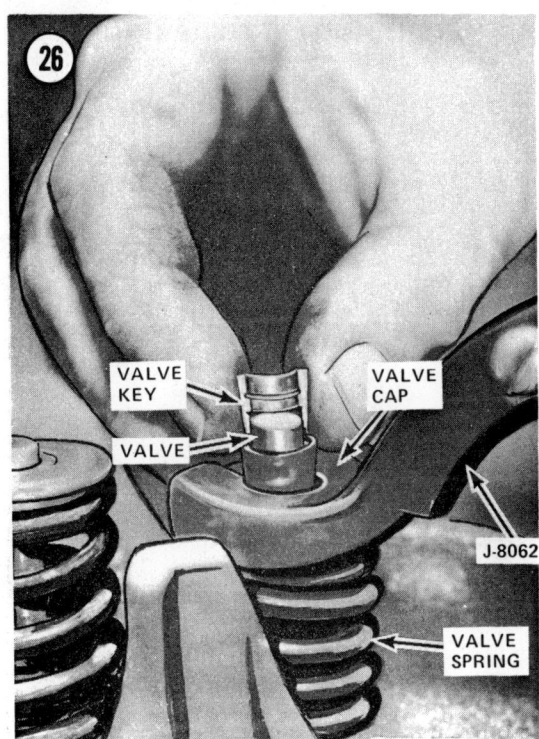

c. Compress spring and install valve locks, then release compression. Make sure lock seats properly in valve stem groove.

NOTE: *A small amount of grease can be used to hold locks in place.*

8. Remove air line adapter tool and replace spark plug.
9. Install and adjust valve mechanism, using the procedures given above.

CYLINDER HEAD

Removal

1. Drain cooling system, including block.
2. Remove intake and exhaust manifolds, using the procedures given above.
3. Remove rocker arm cover(s) and valve mechanisms, using procedure given above.
4. On inline engines, remove fuel and vacuum lines from retaining clip at water outlet and disconnect wires from temperature sending units.
5. On vehicles so equipped, disconnect Air Injection Reactor System hoses at check valve, vapor canister hose, and EGR valve.
6. On inline engines, disconnect radiator upper hose at water outlet and battery ground strap at cylinder head.
7. On inline engines, remove coil.
8. Remove the cylinder head bolts, cylinder head(s) and gasket(s). Support heads on wood blocks to prevent damage.

Installation

CAUTION
Clean gasket sealing surfaces on both head and block. These surfaces must be free of all foreign matter and nicks or heavy scratches. All threads on bolts and in cylinder blocks must be clean, as dirt will affect bolt torque. If an all-steel head gasket is used, give both sides a thin coat of sealing compound. **Do not** *use sealer on combination metal and asbestos gaskets.*

1. Place gasket in position over dowel pins with bead side up.
2. Using care, lower and guide cylinder head into position.
3. Lightly coat cylinder head bolt threads with sealing compound and install finger-tight.
4. Tighten bolts, applying a small amount of torque at a time and following the sequence shown in **Figure 17 or 18**, until the specified torque is reached (see **Tables 1 and 2** at end of chapter).
5. On inline engines, perform the following:
   a. Install coil.
   b. Connect engine ground strap and upper radiator hose.
   c. Connect fuel and vacuum lines in clip at water outlet.
   d. Connect temperature sender wires.
6. Connect vapor canister hose, if so equipped.
7. Fill cooling system and check for leaks.
8. Install intake and exhaust manifolds, using the procedures given above.
9. Install and adjust valve mechanisms, using the procedures given elsewhere in this chapter.

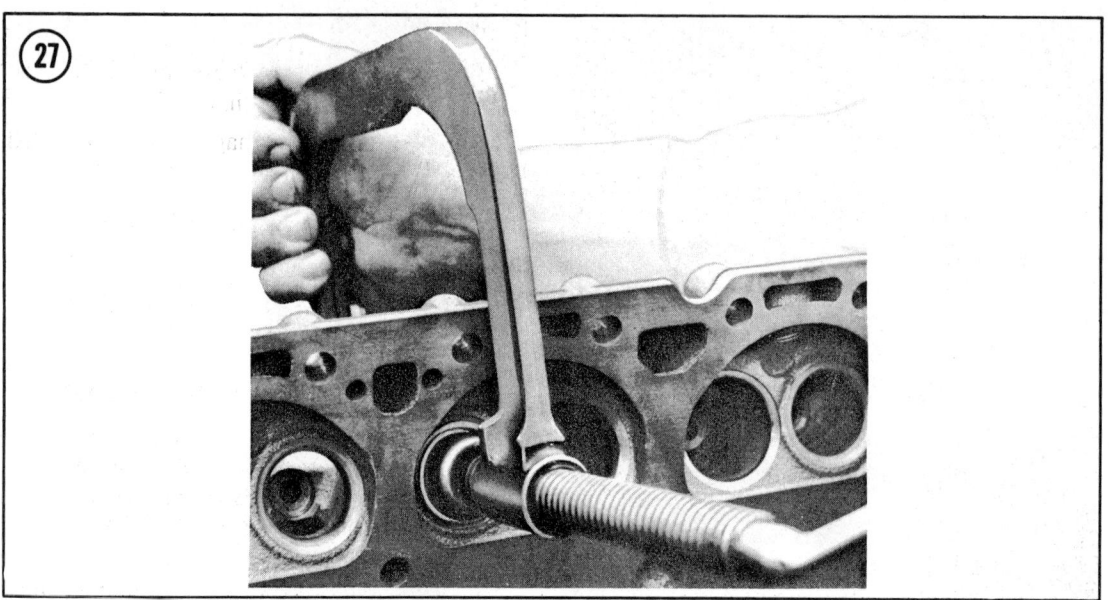

10. Install and torque rocker arm cover.
11. Connect AIR pipe, if so equipped.
12. Connect EGR valve, if so equipped.
13. Start the engine and allow it to warm up to operating temperature. Stop the engine and retorque cylinder head bolts to specification.
14. Adjust timing and idle speed, using the procedures given in Chapter Three.

Disassembly

> NOTE: *This procedure is performed after rocker arm nuts, balls, and rocker arms have been removed and the cylinder head has been removed from the engine.*

1. Using a spring compressor (GM Part No. J-8062 or equivalent), compress valve springs and remove valve keys (see **Figure 27**). Release compressor and remove rotators or spring caps, shields (if so equipped), and spring damper. Then remove oil seals and valve spring shims.
2. Remove valves from cylinder head and store in sequence in a rack so they can be replaced in their original positions.

Cleaning

1. Clean all carbon from combustion chambers and valve ports, using a wire brush (see **Figure 28**).

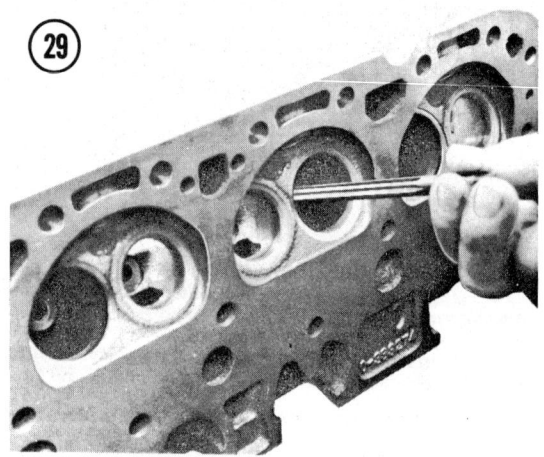

ENGINE

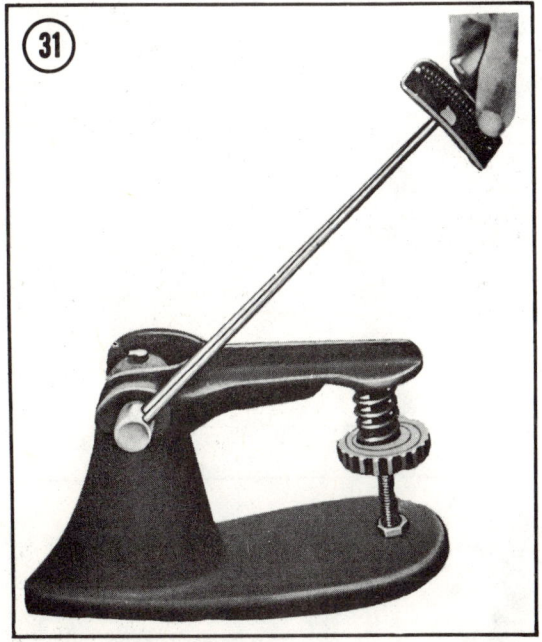

2. Clean valve guides, using a suitable tool (GM Part No. J-8101 or equivalent). Refer to **Figure 29**).

3. Clean all carbon and sludge from pushrods, rocker arms, and rocker arm guides.

4. Clean valve stems and heads, using a buffing wheel.

5. Clean all foreign material from head gasket mating surface.

Inspection

1. Inspect for cracks in exhaust ports, combustion chambers and water chambers.

2. Check valves for damaged stems, cracked faces or burned heads.

> NOTE: *Excessive valve stem clearance will result in excessive oil usage and could cause stems to break. Insufficient clearance can result in noisy operation and sticking of valves.*

3. Measure valve stem clearance as follows:

> NOTE: *Valve stem clearance measurement and valve guide reaming, if required, are tasks which should be performed by a competent automotive machinist.*

   a. Install the dial indicator as shown in **Figure 30**, so that side-to-side (crosswise to head) movement of valve stem will cause direct movement of the indicator stem. Indicator stem must contact valve stem just above valve guide.

   b. Drop valve head about $\frac{1}{16}$ inch off valve seat. Move valve stem from side to side, using light pressure, and take indicator reading. If reading exceeds specified tolerance (see **Table 3** at end of chapter), valve guide must be reamed and oversized valves installed.

4. Check valve spring tension, using a tension tester (GM Part No. J-8056 or equivalent). Refer to **Figure 31**.

> NOTE: *If a tool is not available, have tension checked by an automotive machine shop. If tool is available, compress spring to specified height (see **Table 3** at end of chapter) and measure tension. Spring should be replaced if not within 10 lb. of specified load (without dampers).*

5. Inspect rocker arm studs for wear or damage. On Mark IV engines, inspect pushrod guides for wear or damage.

Repairs

Rocker arm studs and pushrod guides:

1. On Mark IV and 350 cu. in. high performance engines, pushrod guides are related to

the cylinder head by rocker arm studs (see **Figure 32**). If necessary, replace and torque to specifications (see **Tables 1 and 2** at the end of the chapter).

> NOTE: *Before assembling rocker arm studs to cylinder head, coat threads with sealer.*

2. On inline, 220-229 V6, and small V8 engines, damaged or loose rocker arm studs should be replaced with oversized studs (available in 0.003 in. and 0.03 in. oversize). This task should be performed by an experienced automotive machinist, as holes must be reamed before the oversized studs can be installed and special tools and skills are required.

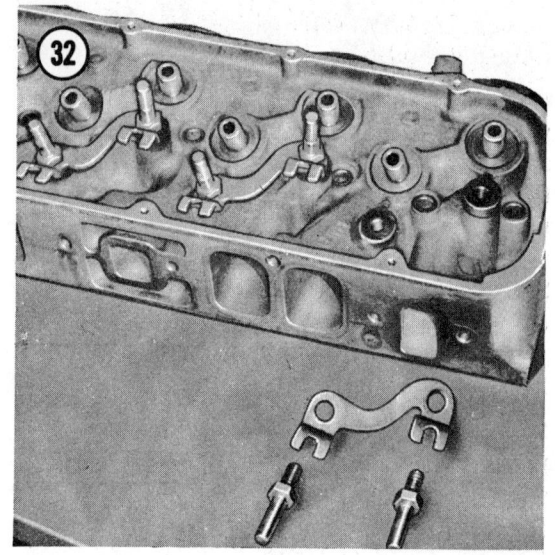

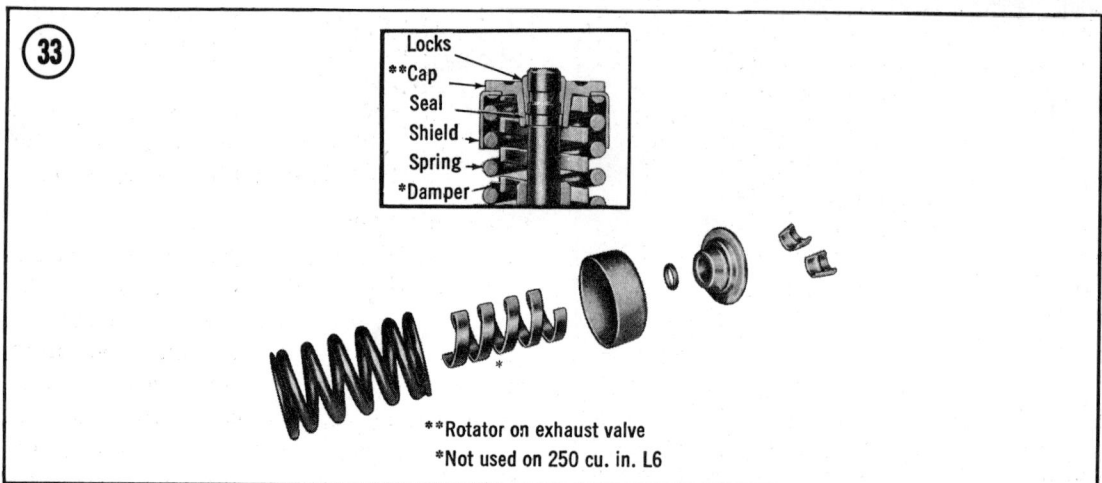

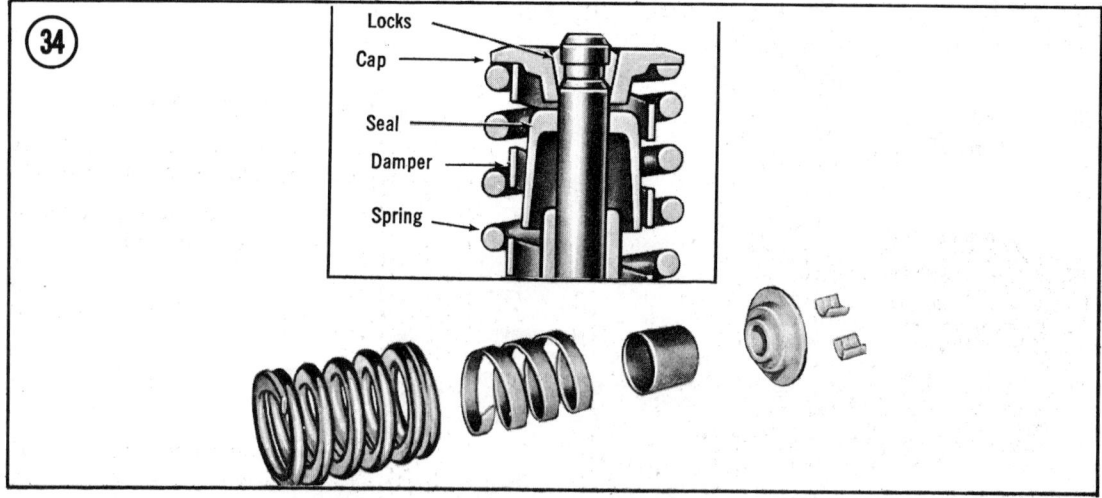

ENGINE

Valve guide bores, valve seats and valves (all models):

The reconditioning of valve guide bores, valve seats and valves requires the use of precision special equipment and should not be attempted by the home mechanic. These tasks can be performed competently and economically by a well-equipped automotive machine shop. Specifications are given in **Table 3**, end of chapter.

Assembly

1. Insert a valve in the port from which it was removed.
2. Assemble related parts, including spring as follows:

   a. For inline, 220-229 V6, and small V8 engines:

   NOTE: *If engine uses exhaust valve rotators, make certain that the shorter springs are used on exhaust valves.*

   (1) Position valve spring shim, spring, damper (if used), shield and cap or rotator on valve stem (see **Figure 33**).

   (2) Using a spring compressor tool (GM Part No. J-8062 or equivalent), compress the spring.

   (3) Install oil seal in stem lower groove. Make sure seal is not twisted.

   (4) Install valve locks and remove compressor. Make sure that lock seats in upper valve stem groove.

   b. On Mark IV V8 engines:

   (1) Install shim on valve spring seat and install new oil seal over valve and valve guide.

   (2) Install valve spring (with damper) and valve cap (see **Figure 34**).

   (3) Using a spring compressor tool (GM Part No. J-8062 or equivalent), compress the spring.

   (4) Install valve locks and release compressor. Make sure locks seat in valve stem groove.

   c. On 231 V6, install valve spring, cap, and cap retainer, using the same equipment used during disassembly. See **Figure 26**.

3. Install remaining valves.
4. Check height of installed springs against specification (see **Table 3** at the end of the chapter). A cutaway scale (see **Figure 35**) will help. Measure from shim or spring seat top to valve spring or valve spring shield top (see **Figure 36**). If height exceeds specification, install a valve seat shim (approximately $\frac{1}{16}$ in. thick). Do not shim spring to give an installed height under the specified minimum.

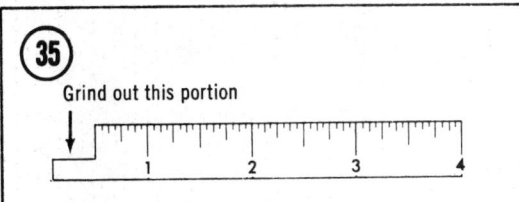

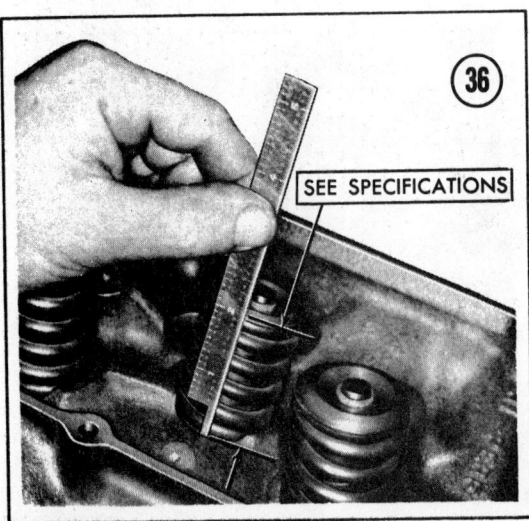

OIL PAN REPLACEMENT

1. Disconnect positive battery cable. Remove oil dipstick and tube on 200 and 229 V6 and V8.
2. Remove radiator upper mounting panel or side mounting bolts. Remove upper half of fan shroud on V8 engines.
3. Place a heavy sheet of cardboard between fan and radiator.
4. Disconnect fuel suction line at fuel pump, if so equipped.
5. Raise vehicle and drain engine oil. On 231 V6, remove exhaust crossover pipe.

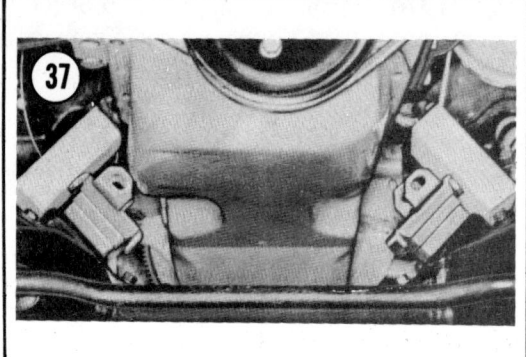

6. Disconnect and remove starter on manual transmission models or where required for flywheel or converter housing underpan.

7. Remove either flywheel underpan or converter housing underpan and splash shield. If equipped with automatic transmission cooler, disconnect cooler lines at transmission before removing converter housing underpan (if necessary).

8. If necessary (on older models), disconnect steering rod at idler lever and position steering linkage to one side for oil pan clearance.

9. Rotate crankshaft until timing mark on torsional damper points to 6 o'clock position.

10. If required for clearance, remove bolts attaching brake line to front crossmember. Move line away from crossmember.

11. Remove through bolts from front engine mounts.

12. On inline and V8 models, raise engine (using a suitable jack) and insert 2 x 4 in. blocks under engine mounts as shown in **Figure 37**. Lower engine onto blocks.

13. Remove oil pan bolts and lower and remove oil pan.

14. To install oil pan, clean all gasket sealing surfaces and, using all new gaskets and gasket sealer to hold gaskets in place, reverse Steps 1 through 13. See **Figure 38** (inline) **or** 39 (V8 engines). V6 is similar to V8. Replace coolant in cooling system (if removed), refill crankcase with engine oil, start engine, and check for leaks.

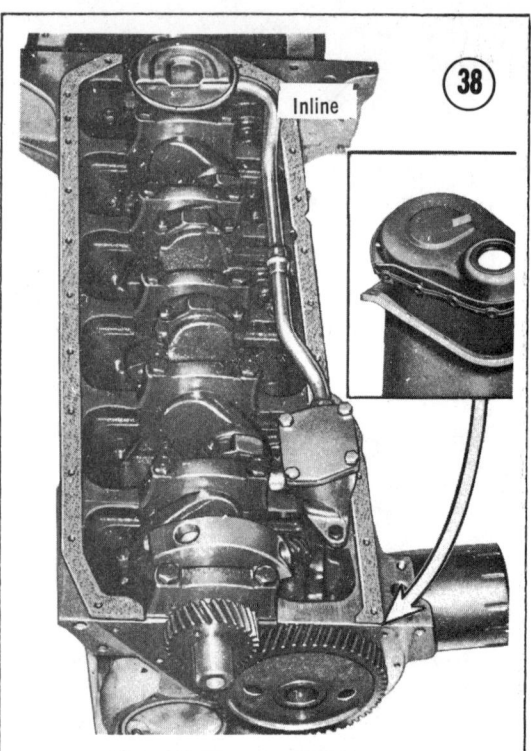

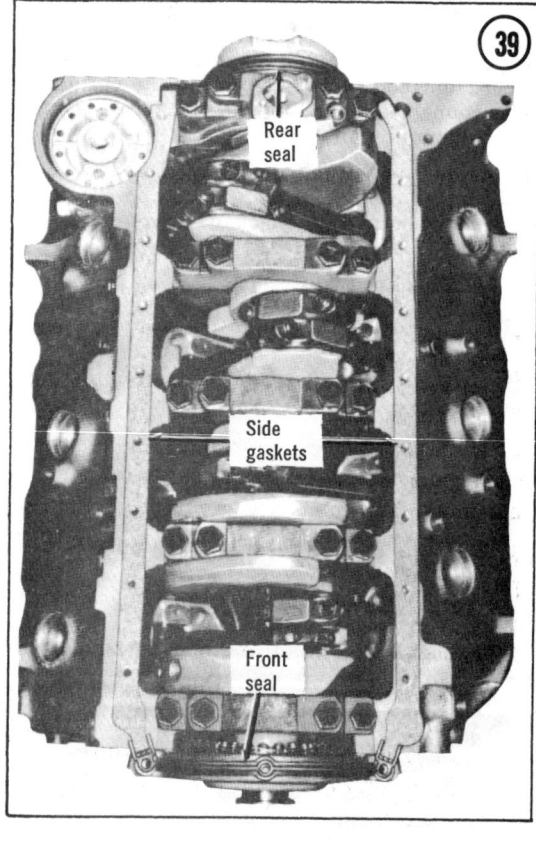

ENGINE

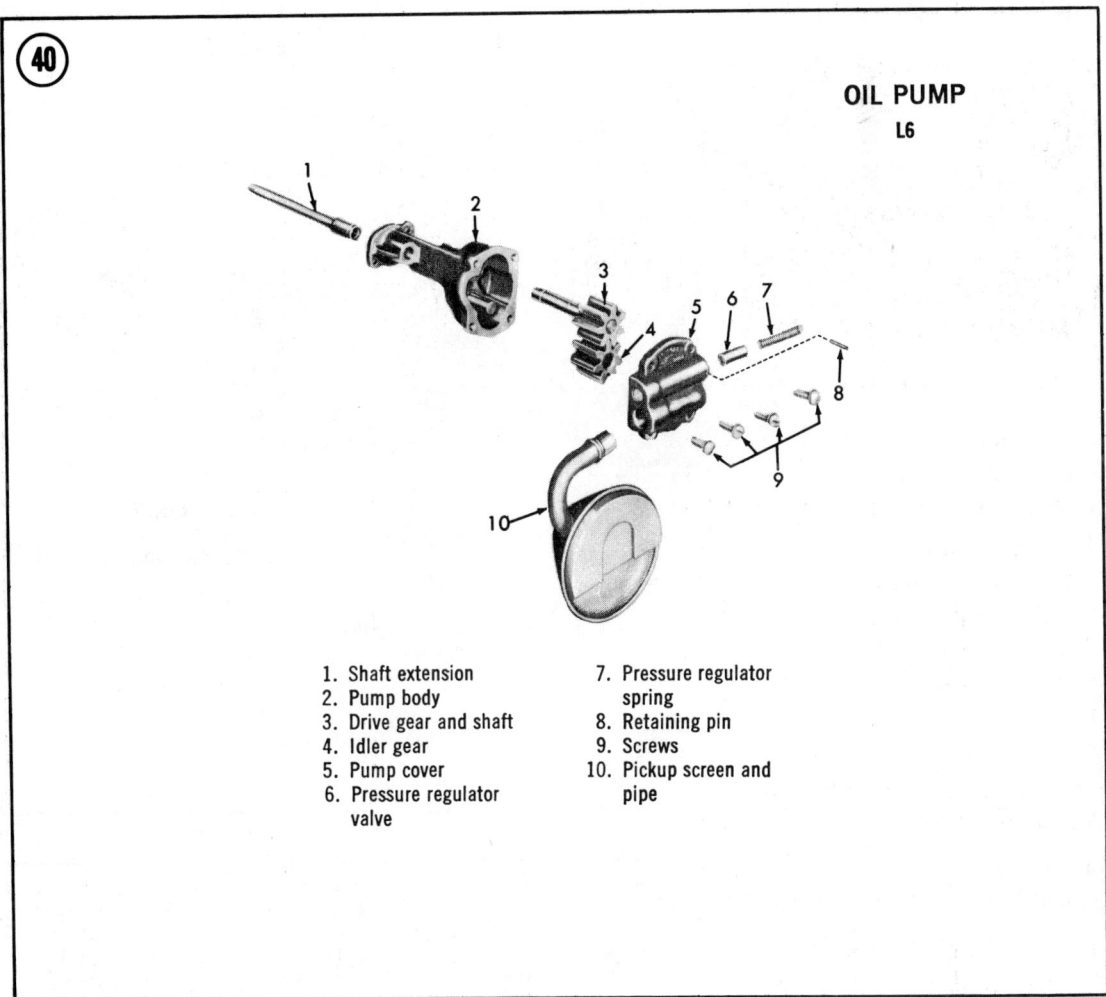

40

OIL PUMP
L6

1. Shaft extension
2. Pump body
3. Drive gear and shaft
4. Idler gear
5. Pump cover
6. Pressure regulator valve
7. Pressure regulator spring
8. Retaining pin
9. Screws
10. Pickup screen and pipe

OIL PUMP
(EXCEPT 231 V6)

Removal/Installation

1. Remove oil pan, using the procedure given above.

2. On inline engines, remove 2 flange mounting bolts, then remove pump and screen as an assembly.

3. On 200-229 V6 and V8 engines, remove pump-to-rear main bearing cap bolt and remove pump and seal.

4. To install oil pump, proceed as follows:

   a. On inline engines, align oil pump drive shaft to match with distributor tang and install oil pump to block. Position flange over lower distributor bushing. No gasket is required.

NOTE: *If oil pump does not slide easily into place, remove and realign slot with distributor tang.*

   b. On V8 engines, assemble pump and extension shaft to rear main bearing cap, aligning slot in shaft top end with tang on lower end of distributor shaft. Install bolt and torque to specification (see **Tables 1 and 2** at the end of the chapter).

NOTE: *Bottom edge of oil pump screen should be parallel to oil pan rails.*

Disassembly

Refer to **Figures 40, 41, and 42** for this procedure.

1. Remove pump cover attaching screws and pump cover. On inline engines, remove gasket.

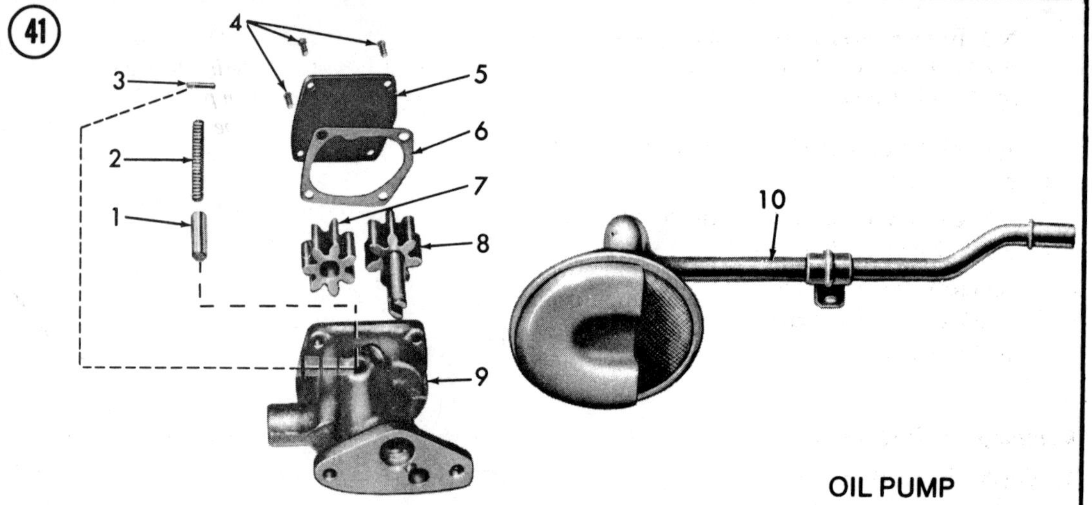

104   CHAPTER FOUR

④¹

OIL PUMP

V6 and small V8

1. Pressure regulator valve
2. Pressure regulator spring
3. Retaining pin
4. Screws
5. Pump cover
6. Cover gasket
7. Idler gear
8. Drive gear and shaft
9. Pump body
10. Pickup screen and pipe

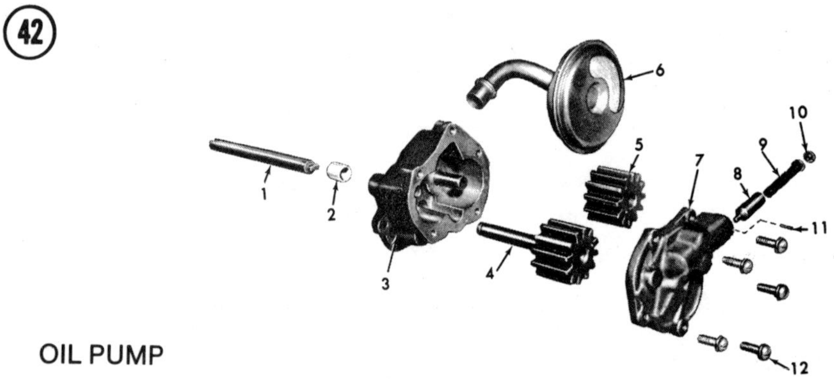

④²

OIL PUMP

MARK IV V8

1. Shaft extension
2. Shaft coupling
3. Pump body
4. Drive gear and shaft
5. Idler gear
6. Pickup-screen and pipe
7. Pump cover
8. Pressure regulator valve
9. Pressure regulator spring
10. Washer
11. Retaining pin
12. Screws

ENGINE

NOTE: *Mark gear teeth so same indexing of teeth can be duplicated when reassembling pump.*

2. Remove idler and drive gears and drive shaft from pump body.

3. Remove retaining pin, then remove pressure regulator valve and related parts.

4. If replacement is indicated, mount pump in soft-jawed vise and extract pickup screen and pipe from pump body.

Cleaning and Inspection

1. Wash all parts in solvent. Dry with compressed air (if available).

2. Inspect pump body for cracks and signs of wear.

3. Inspect gears for excessive wear or damage.

4. Check shaft of drive gear for looseness in pump body.

5. Inspect inside of cover for wear which would allow oil to leak past ends of gears.

6. Inspect pickup screen and pipe for damage.

7. Check pressure regulator valve for fit.

NOTE: *Pump body and gears cannot be replaced separately. If damage or excessive wear is present, entire pump must be replaced.*

Assembly

Refer to **Figures 40, 41, and 42** for this procedure.

CAUTION
If pickup screen and pipe assembly was removed, install new assembly. Do not attempt to install old assembly, as loss of press fit could result in loss of oil pressure.

1. To install new pickup screen and pipe assembly, mount pump in soft-jawed vise. Apply sealer to end of pipe and, using a suitable tool (GM Part No. J-8369 for inline, V6, and small V8 engines; No. J-22144 for Mark IV engines), tap pipe into place with plastic hammer.

CAUTION
Avoid twisting, shearing or collapsing pipe while installing in pump. Screen on inline engines must be parallel to bottom of oil pan when pump is installed.

2. Install pressure regulator valve and related parts with retaining pin.

3. Install drive gear and shaft in body, then install idler gear so that teeth mesh in original position and smooth side of gear is facing pump cover opening.

4. Install pump cover (using new gasket on inline engine) and torque screws to 70 in.-lb. (inline engine) or 80 in.-lb. (V6 and V8 engine).

5. Check for smooth operation by turning drive shaft by hand.

OIL PUMP (231 V6)

Pipe and Screen Assembly Removal/Installation

1. Remove the oil pan as described elsewhere in this chapter.

2. Remove the bolts attaching the pipe and screen assembly to the cylinder block and remove the assembly from the engine. See **Figure 43**.

3. Clean the assembly in solvent, paying particular attention to the screen and the pipe passage, and blow dry with compressed air.

4. Clean the gasket surface on the cylinder block flange and install the pipe and screen assembly on the flange, using a new gasket. Securely tighten the attaching bolts.

5. Install the oil pan, using a new gasket, as described elsewhere in this chapter.

Pump Removal/Installation

Figure 44 is a diagram of the lubrication system and shows the location of the oil pump and filter assembly. **Figure 45** is a cut-away view of the oil pump and filter assembly. **Figure 46** is an exploded view of the oil pump cover and relief valve assembly.

1. Remove the oil filter and then remove the oil pump cover assembly by removing the attaching screws. Slide out the oil pump gears.

43

GASKET
PIPE AND
SCREEN ASSEMBLY
CYLINDER
BLOCK

44

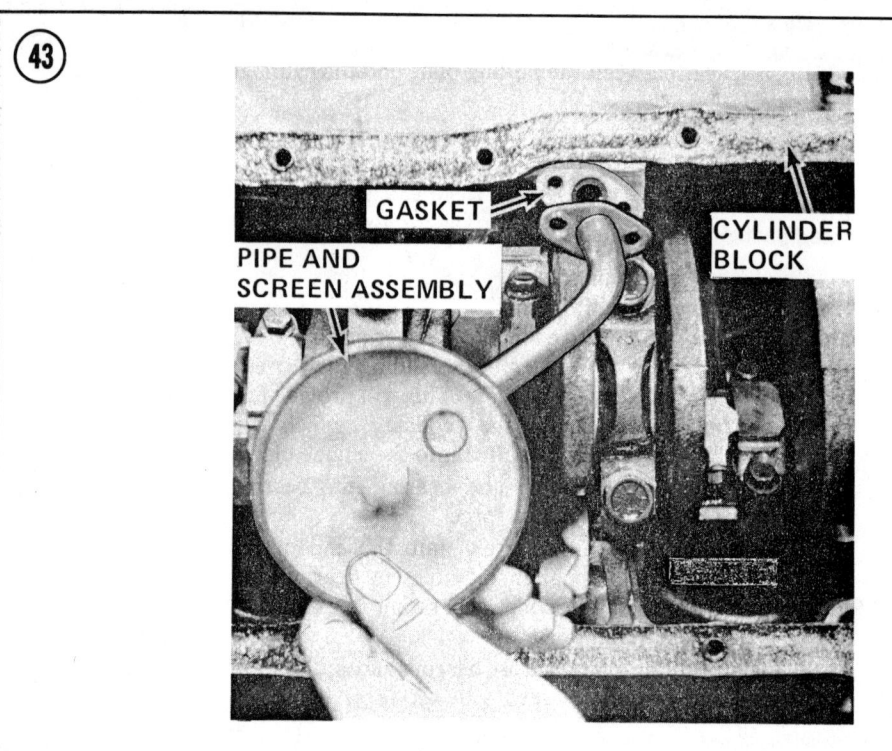

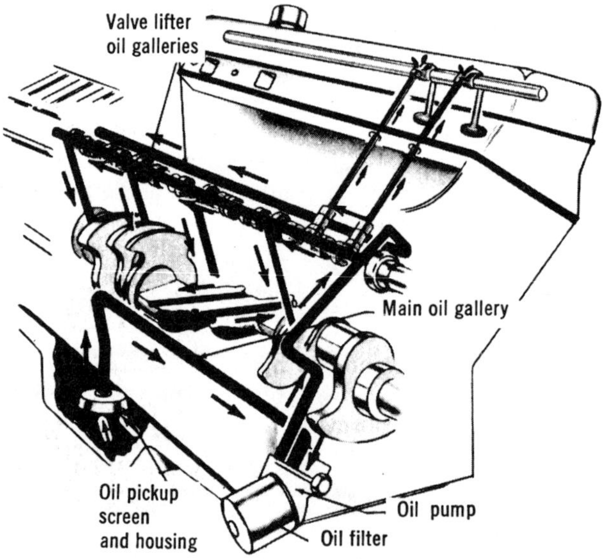
Valve lifter
oil galleries
Main oil gallery
Oil pickup
screen
and housing
Oil pump
Oil filter

ENGINE

⑤

⑥

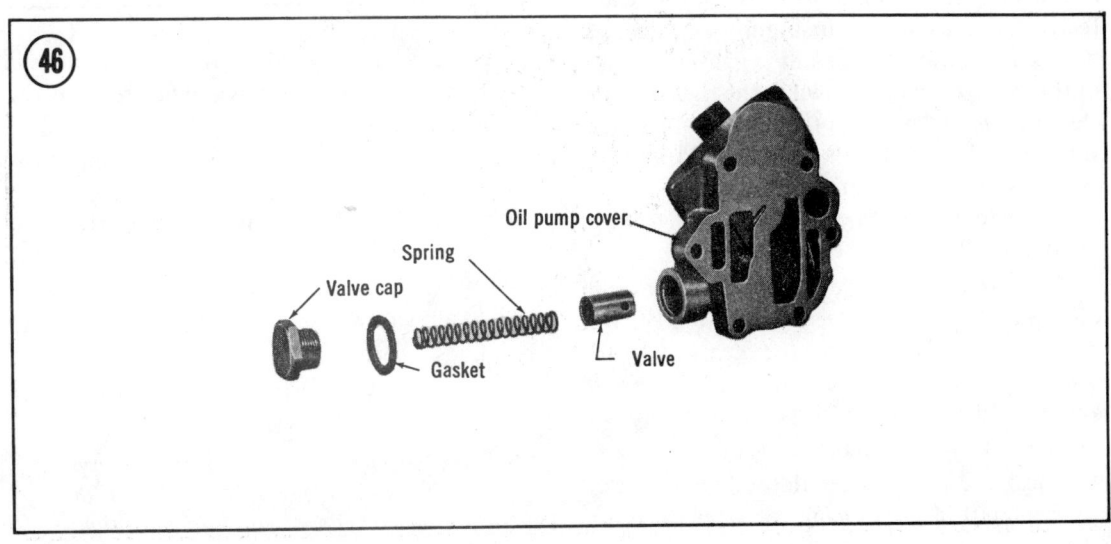
Oil pump cover · Spring · Valve cap · Gasket · Valve

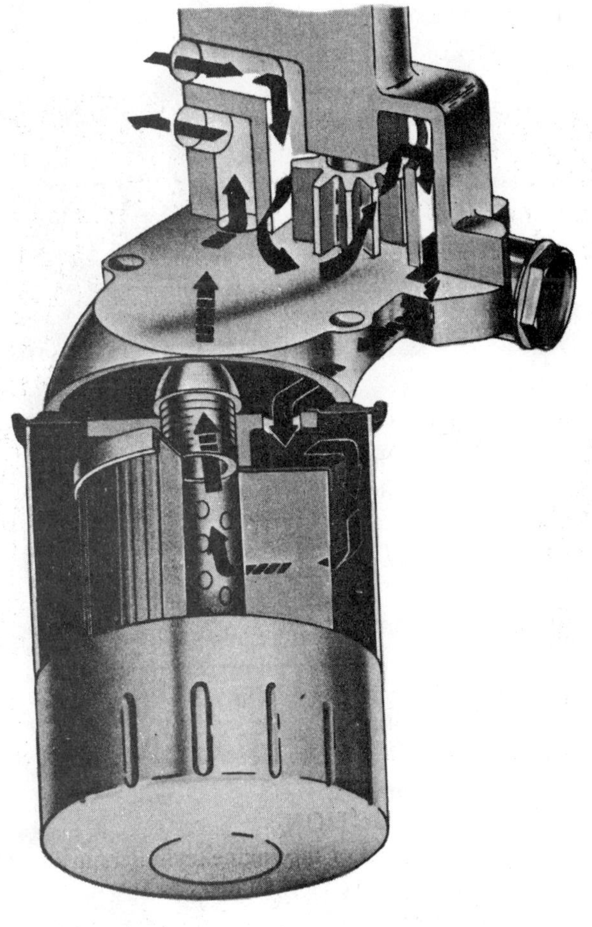

2. Clean the gears in solvent and check them for wear, scoring, or other damage. Gears in poor or doubtful condition should be replaced.

3. Remove the valve cap (**Figure 46**) and remove the spring and valve from the oil pump cover. Clean the parts in solvent and check the valve for wear, scoring, or other damage. Also check the spring for side wear or other damage. Replace parts that are questionable.

> NOTE: *The oil filter bypass valve and spring are permanently staked in place in the oil pump cover and should not be removed.*

4. Check the fit of the relief valve in its bore in the oil pump cover. It should be an easy slip fit, without side movement. If the valve fits too loosely, or does not slide easily, install a new valve and recheck the fit. If the valve still does not fit properly, replace the pump cover.

5. Inspect the bypass valve for warping or other damage. The valve should be flat and should be free of nicks and scratches. Replace the cover assembly if the bypass valve appears to be damaged.

6. Lubricate the relief valve and spring with engine oil and install them in their bore in the pump cover. Install the valve cap, using a new gasket, and tighten the cap to 35 ft.-lb., using a reliable torque wrench. Do not over-tighten.

7. Install the gears and shaft in the oil pump cavity in the timing chain cover. Check the end clearance of the gears, using a straight edge and feeler gauge as shown in **Figure 47**. Measure between the straight edge and the gasket surface of the pump cavity as shown in the figure. If the clearance is not between 0.002-0.006 in., measure the gears as shown in **Figure 48** and the cavity as shown in **Figure 49** to determine which is out of tolerance. Specifications are given in the figures. Replace the defective part(s).

8. Measure the gear side clearance as shown in **Figure 50**. If not between 0.0025-0.0050 in., measure the diameter of the gears and the diameter of the pockets as shown in **Figures 48 and 49** to determine which is out of tolerance. Replace the defective part(s).

9. Check the pump cover flatness by trying to insert a 0.001 in. feeler gauge between the cover face and a straight edge. If the clearance is 0.001 in. or larger at any point, replace the cover.

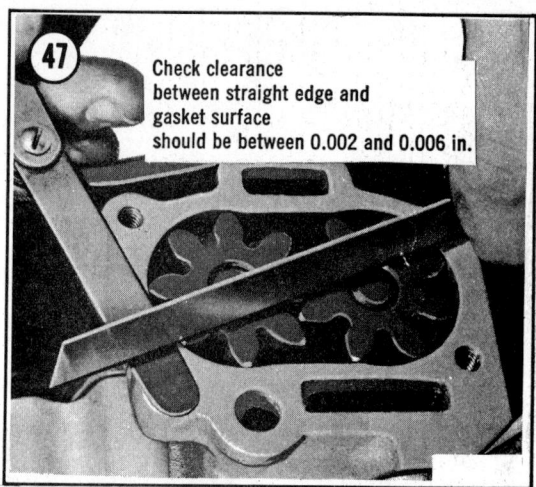

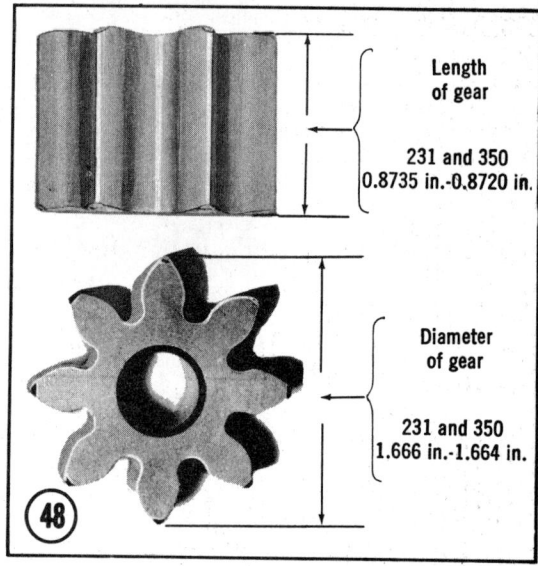

10. After all clearances are within specifications, remove the gears and pack the pump cavity with petroleum jelly (Vaseline, etc.). Do not use chassis lube or any other type of lubricant.

11. Reinstall the gears in the pump cavity, making sure that the petroleum jelly fills all spaces and that no air pockets are left.

> CAUTION
> *This step is very important. If the pump cavity is not properly packed with petroleum jelly, it could fail to prime*

ENGINE

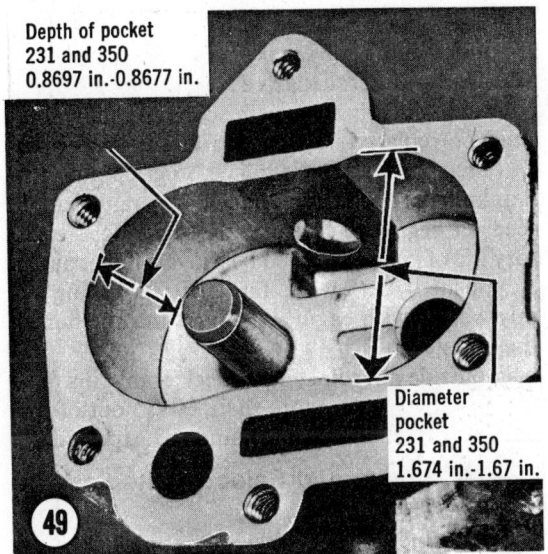

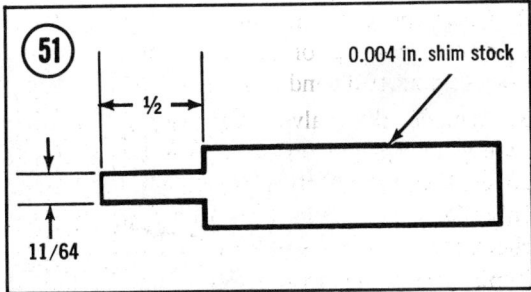

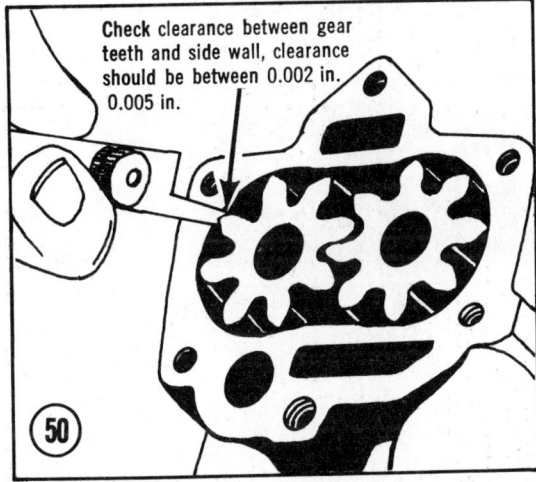

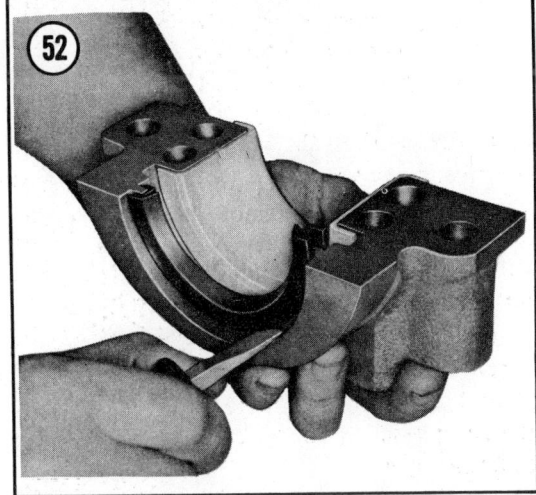

itself and serious damage to engine parts could result from the lack of lubrication.

12. Using a new gasket, install the pump cover. Tighten the attaching screws evenly and alternately to 10 ft.-lb.
13. Install a new oil filter.

REAR MAIN OIL SEAL REPLACEMENT (EXCEPT 231 V6)

NOTE: *Upper and lower seal must be replaced as a unit. Install seal with lip facing front of engine.*

The rear main bearing seal can be replaced without removing the crankshaft. However, care must be taken to protect the sealing bead on the outside diameter of the seal. An installation tool (see **Figure 51**) can be used to protect the bead.

1. Remove oil pan, using the procedure given elsewhere in this chapter.
2. Remove oil pump, using the procedure given above.
3. Remove rear main bearing cap. Use a small screwdriver to remove seal half from bearing cap (see **Figure 52**).
4. Tap one end of upper half of seal with a small hammer and brass pin punch (see **Figure 53**) until other end protrudes enough to grasp and remove with pliers.
5. Use non-abrasive cleaner to remove all traces of sealant and other foreign matter from cylinder block, crankshaft and bearing cap.
6. Inspect all components for defects on sealing surfaces, case assembly, and crankshaft.

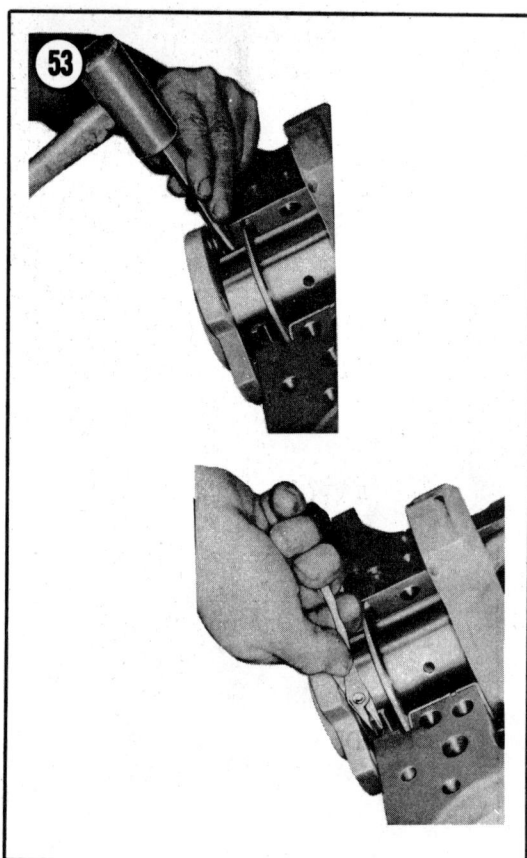

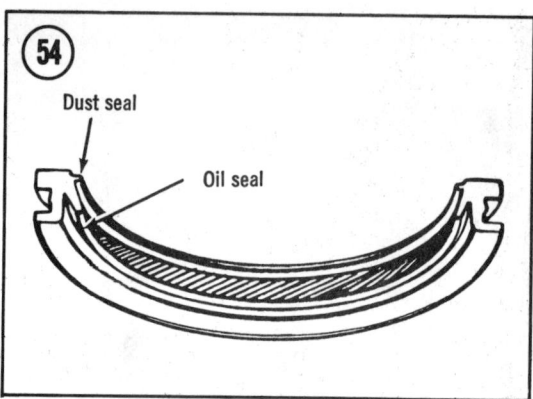

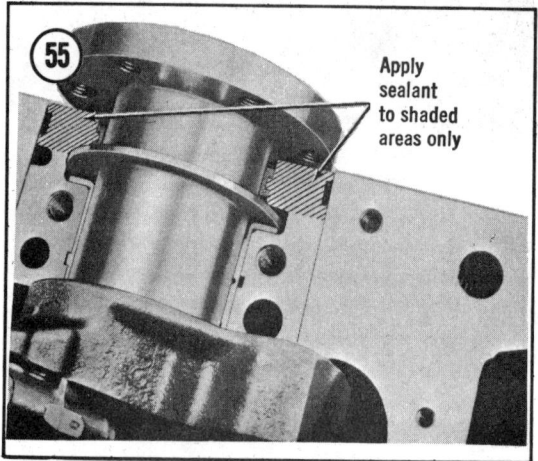

Coat seal lips and bead with engine oil, keeping oil off seal mating ends.

7. Position tool (**Figure 51**) tip between crankshaft and cylinder block seal seat.

8. Position seal between crankshaft and tip of tool so that sealing bead contacts tip of tool, making certain that seal lip points toward front of engine (see **Figure 54**).

9. Roll seal around crankshaft using tool as a "shoehorn" to protect seal bead.

CAUTION
Installation tool must remain in position until seal is positioned with both ends flush with block to protect seal bead from sharp edge of seal seat surface in cylinder block.

10. Remove tool. Use care not to remove or displace seal.

11. Again using tool as "shoehorn" install seal half in bearing cap. Feed seal into cap, using light pressure with thumb and finger.

12. Apply sealant as shown in **Figure 55** and install bearing cap. Take care to avoid getting sealant on the seal split line. Torque to specifications (see **Tables 1 and 2** at the end of the chapter).

REAR MAIN OIL SEAL REPLACEMENT (231 V6)

Replacement of this seal requires that the crankshaft be removed from the engine. However, the old seal can be repaired in some cases without crankshaft removal. Repair consists of removing the oil pan and rear main bearing cap, and packing the seal into the block with a special tool (J-21526-2 or equivalent), as shown in **Figure 56**, about $\frac{1}{4}$ to $\frac{3}{4}$ in. on both ends. Then a guide tool (J-21526-1) is installed (**Figure 57**). A piece of the old seal removed from the bearing cap is cut for each side, using a razor blade. Each piece should be about $\frac{1}{16}$

ENGINE

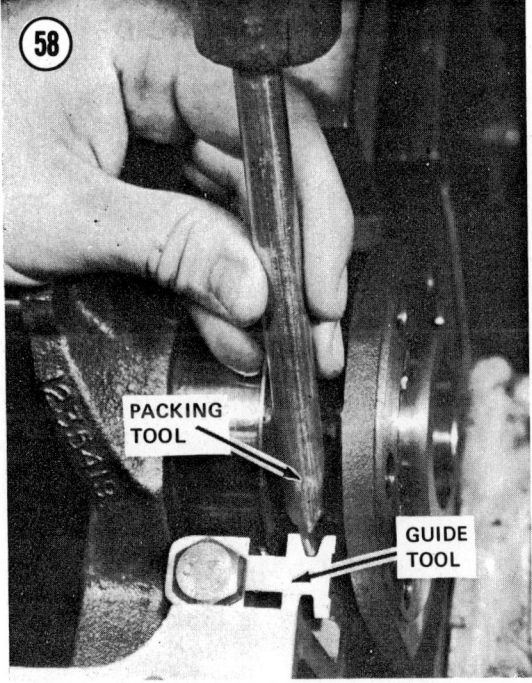

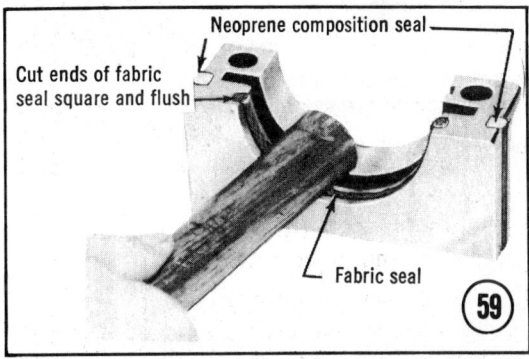

in. longer than the depth the old seal was driven into the block. Use the bearing cap as a holding fixture so the seal pieces can be cut squarely. Then use the packing tool to work the short pieces of seal into the block on both sides of the rear main bearing. See **Figure 58**. Install a new fabric seal in the bearing cap, install the bearing cap, and tighten the attaching bolts to 100 ft.-lb. Then reinstall the oil pan.

To install new seals, remove the crankshaft as described elsewhere in this chapter. Remove the old seals from the grooves in the rear main bearing cap and the cylinder block, and install new seals. Use a hammer handle or similar tool to force the fabric seal into the grooves (see **Figure 59**), and then cut the ends of the seal off squarely and flush. Install new Neoprene composition seals in the side grooves of the bearing cap (**Figure 59**). These seals, which swell when subjected to oil and heat, are a loose fit and are slightly longer than the grooves. Do not trim them or attempt to pack them. A small amount of leakage is normal when these seals are newly installed, but should disappear in a short time.

This time can be shortened by soaking the seals in oil or kerosene for a couple of minutes just prior to installation. Before installing the bearing cap, coat the surfaces shown in **Figure 60** with Silastic Sealer. Install the crankshaft and then install the bearing caps. Tighten the attaching bolts to 100 ft.-lb.

MAIN BEARINGS

Precision insert type bearings are used in all engines, and shims are not required for fitting. At any time bearing clearance is found to be excessive, a new bearing (both upper and lower halves) is required. Bearings are available in standard and 0.001, 0.002, 0.009, 0.010, 0.020, and 0.030 in. undersize. Close tolerances are obtained in production by selective fitting of rod and main bearing inserts. Thus one half of a bearing may be standard and the other half 0.001 in. undersize (this decreases clearance by 0.0005 in.). If a production crankshaft cannot be properly fitted in this manner it is ground 0.009 in. undersize on the main journals only and again is selectively fitted. If your engine has a 0.009 in. undersize crankshaft, the counterweight forward of the center main journal will be stamped with ".009," and the number 9 will be stamped on the block at the left front oil pan rail.

> NOTE: *In repair work, shimming may be required if bearing caps are replaced for any reason. Laminated shims are available for service work. Shimming is not permitted on the 231 V6 engine, however.*

Inspection

As a rule the lower bearing half shows greater wear and distress from fatigue. If inspection shows lower half is suitable for use, it can usually be assumed that the upper half also is satisfactory without removing it. If the lower half is unsatisfactory, replace both halves (never replace only one half of a main bearing).

Checking Bearing Clearance

> NOTE: *Use Plastigage or equivalent for checking bearing clearances. All bearing*

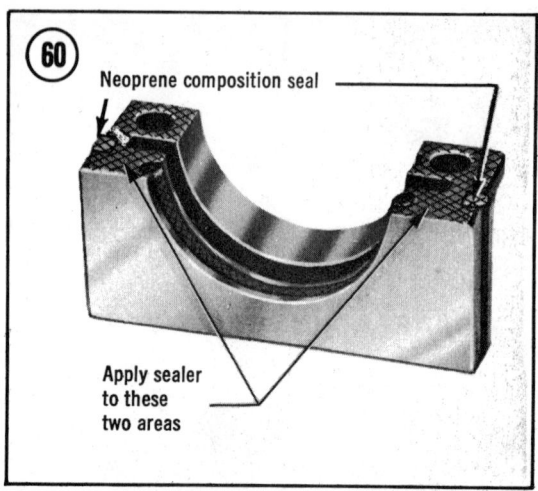

cap bolts should be torqued to specification, and the surfaces to be measured should be wiped free of oil. If engine is still installed in vehicle, support crankshaft at flywheel and torsional damper to remove clearance from upper bearing half.

1. With oil pan and pump removed (see procedures above), and starting with rear main bearing, remove cap and wipe oil from journal and bearing half in cap.

2. Place a piece of Plastigage the full width of the bearing on the journal, parallel to crankshaft (see **Figure 61**).

> CAUTION
> *Do not rotate crankshaft while Plastigage is between journal and bearing, as false reading will be obtained.*

ENGINE

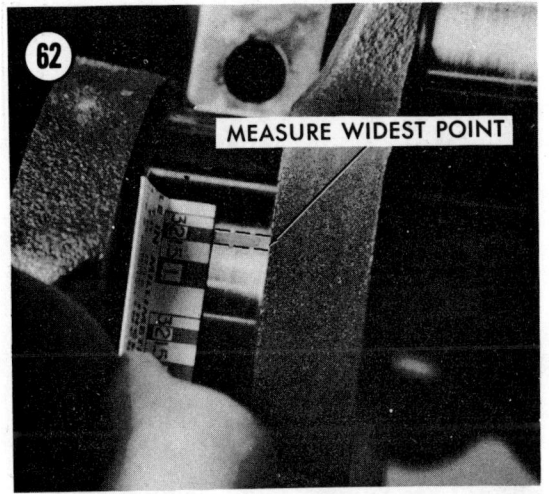

MEASURE WIDEST POINT

3. Install bearing cap and evenly torque bolts to specification (see **Tables 1 and 2** at the end of the chapter).

4. Remove bearing cap. Flattened Plastigage will be found either on journal or bearing.

5. Using the scale printed on the Plastigage package, measure width of compressed plastic at its widest point. Do not remove plastic before measurement is made (see **Figure 62**).

> NOTE: *As a rule, main bearings wear evenly and are not out-of-round. If a bearing is fitted to an out-of-round journal (0.001 in. max.), the bearing must be fitted to the maximum diameter of the journal. Otherwise rapid wear of the bearing can be expected. If the flattened plastic tapers toward the center or ends, and a difference of more than 0.001 in. is indicated, the crankshaft journal should be checked for roundness, and corrected if required, before proceeding.*

6. If clearance is within specifications, insert is satisfactory. If not, replace both halves of insert.

> NOTE: *If new bearing cap is being installed and clearance is less than 0.001 in., inspect for nicks and burrs. If none are found, use shims to obtain specified clearance.*

7. If a standard or 0.001 in. or 0.002 in. undersize bearing fails to produce proper clearance, journal must be ground for the next undersize bearing.

8. Repeat the above steps for remaining bearings, then rotate crankshaft to verify that excessive drag is not present.

9. Measure crankshaft end play (see specifications, **Table 3**, end of chapter) by forcing crankshaft to extreme front. Measure at front end of rear main bearing. See **Figure 62**.

10. Install new rear main bearing oil seal, using the procedure given above.

Main Bearing Replacement (With Crankshaft Removal)

1. Remove and inspect crankshaft.

2. Remove old bearings from cylinder block and caps.

3. Select new bearings of correct size (see *Checking Bearing Clearance* procedure above) and coat bearings with engine oil. Install in cylinder blocks and bearing caps.

4. Install crankshaft in cylinder block and install bearing caps with arrows pointing to front of engine.

5. Install bolts and torque to specification (see **Tables 1 and 2** at the end of the chapter).

Main Bearing Replacement (Without Crankshaft Removal)

1. Remove oil pan, oil pump, and spark plugs from engine (see procedures above). Remove cap from main bearing to be replaced and remove bearing from cap.

2. On inline engine crankshaft, rear main bearing journal has no oil hole. Rear main bearing upper half is removed as follows:

   a. Use hammer and small drift punch to start bearing half out of block.

   b. Grasp bearing thrust surface and oil slinger with a pair of pliers (with taped jaws) and rotate crankshaft to remove bearing half (see **Figure 63**).

   c. Select new bearing of correct size and coat with oil. Insert unnotched end between crankshaft and notched side of block.

d. Using pliers as in Step b, rotate bearing half into place. During last fourth of movement, pliers may be used on slinger only, or, if care is taken, bearing may be tapped into place with a soft drift punch.

3. All remaining main bearing journals on in-line crankshafts and all V6 and V8 crankshaft journals have oil holes. Upper bearing halves may be installed as follows:

a. Install a long cotter pin in oil hole in crankshaft journal and rotate crankshaft clockwise (viewed from front of engine) to roll out insert.

b. Select and oil proper size upper bearing (see *Checking Bearing Clearance* above) and insert unnotched end between crankshaft and notched side of block. Rotate into place and remove cotter pin from journal oil hole.

4. Oil new lower bearing insert and install in bearing cap. Install cap with arrows pointing to engine front. Torque cap bolts to specifications (see **Tables 1 and 2** at the end of the chapter).

CONNECTING ROD BEARING REPLACEMENT

CAUTION
Do not file rods or rod caps to obtain specified clearance. If clearance is excessive, new bearings are required. Bearings are available in standard and 0.001 in. and 0.002 in. undersize for use with new and used standard size crankshafts, and in 0.010 in. and 0.020 in. undersize for use with reconditioned crankshafts.

1. Remove oil pan and oil pump, using procedures given above.
2. Remove connecting rod cap and bearing, and inspect bearing for wear or damage. Do not reuse bearing if either is in evidence.
3. Clean bearings and crankpin to remove all oil.
4. Using a micrometer, measure crankpin for out-of-round or taper. If not within specifications (see **Table 3** at the end of the chapter), replace or recondition crankshaft. If within tolerances and a new bearing is to be installed, measure maximum diameter of crankpin to determine required bearing size.

CAUTION
If a new bearing is fitted to the minimum diameter of an out-of-round crankpin, rapid bearing failure will result.

5. Install new or used bearing and check clearance with Plastigage or equivalent as follows:

ENGINE

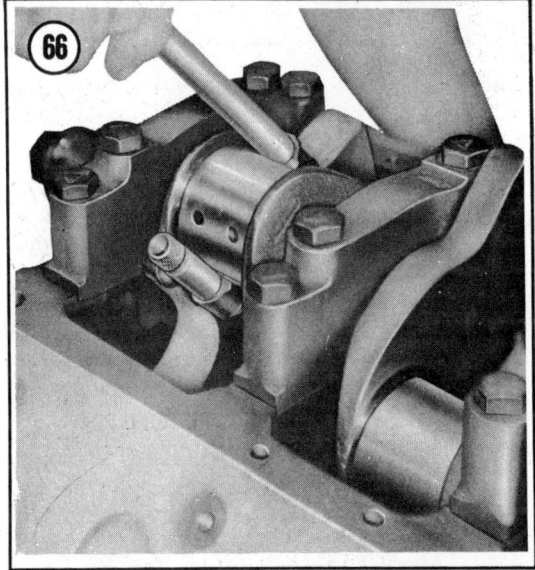

d. Remove cap and measure Plastigage at widest point, using scale on Plastigage package (see **Figure 65**).

6. If the clearance is not within specified tolerances, select and install a different, correct size bearing and remeasure clearance.

7. Coat bearing surfaces with oil, install rod and cap on crankpin, and torque nuts to specification.

8. When all bearings have been installed, tap each connecting rod lightly to assure clearance, and measure side clearances (see **Table 3**, end of chapter) as shown in **Figure 12** (inline engines) or **Figure 13** (V6 and V8 engines).

9. Replace oil pump and oil pan, using the procedures given above.

CONNECTING ROD AND PISTON ASSEMBLIES

Removal

1. Remove oil pan, oil pump, and cylinder head, using the procedures given earlier in this chapter.

2. If ridges or deposits are found in the upper end of cylinder bore, remove them as follows:

   a. Turn crankshaft until piston is at bottom of stroke and place a cloth on top of piston to catch cuttings.

   b. Use a ridge reamer to remove ridge or deposit.

   c. Turn crankshaft until piston is at top of stroke and remove cloth and all cuttings.

3. Inspect rods and bearing caps for cylinder identification markings. If none are present, mark rods and caps so they can be replaced in proper cylinders.

4. Remove bearing caps and install removal tool (GM Part No. J-5239 for $\frac{3}{8}$ in. or J-6305 for $\frac{11}{32}$ in.) and push out assembly as shown in **Figure 66**. If this tool is not available, tape ends of rod and threads of bolts to prevent cylinder wall damage and push out rod and piston assembly with a piece of hardwood or a hammer handle.

NOTE: *It will be necessary to rotate the crankshaft slightly to remove some assemblies.*

a. Place a piece of Plastigage on crankpin (parallel to crankshaft). Plastigage should cover the full width contacted by the bearing (see **Figure 64**).

b. Install bearing halves in connecting rod and bearing cap.

c. Install connecting rod and cap on crankpin and torque nuts evenly to specification (see **Tables 1 and 2** at the end of the chapter).

CAUTION
Do not turn crankshaft while Plastigage is installed. To do so will distort Plastigage which will result in incorrect measurement.

CHAPTER FOUR

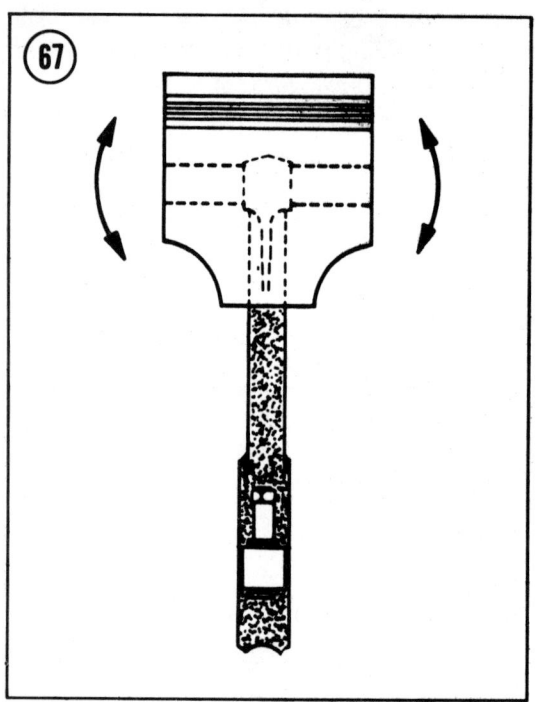

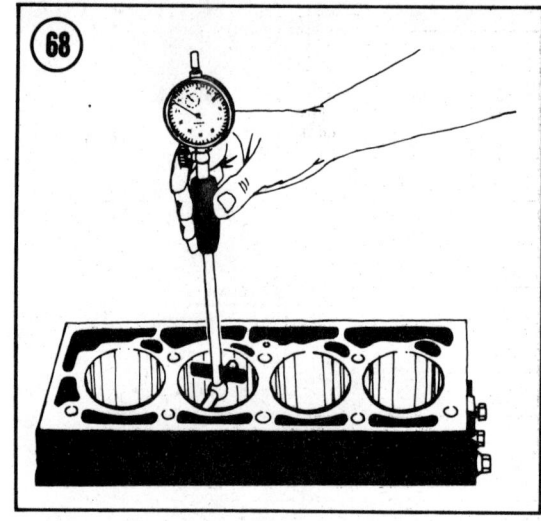

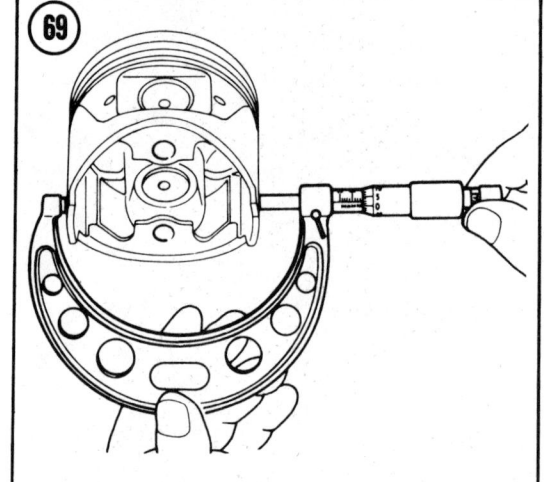

Disassembly

1. Remove piston rings with a ring expander tool.

2. Support rod tightly and rock piston as shown in **Figure 67**. Any rocking motion (do not confuse with sliding) indicates wear in the piston pin, rod bushing, pin bore, or more likely, a combination of all three. Mark the assembly for further inspection.

> NOTE: *Disassembly and assembly of the piston, piston pin, and rod requires an arbor press and special tools. These jobs are best performed by an automotive machine shop, as the use of makeshift tools could result in damage.*

Inspection

1. Clean pistons thoroughly in solvent. Scrape carbon deposits from top and grooves. Take care to avoid damage to piston.

> CAUTION
> *Do not use wire brush on piston skirts.*

2. Examine grooves for burrs, dented edges, and side wear. Pay particular attention to top compression ring groove, as it usually wears more than the others.

ENGINE

Table 4 PISTON SIZE CHART

Engine Displacement	Years	Pistons Available Standard (in.)	0.001 in.	Oversize 0.020 in.	0.030 in.
200-229-267	1978-on	3			
250-307	1970	3.8750-3.8760	3.8760-3.8770	3.8937-3.8957	3.9037-3.9057
250-307	1971	3.8750-3.8760	3.8760-3.8770	3.8935-3.8955	3.9035-3.9055
250-307	1972-on	3.8750-3.8760	3.8760-3.8770	—	3.9035-3.9055
231	1978-on	3			
305	1976-on	3.7360-3.7370	3.7370-3.7380	—	3.7390-3.7400
350	1970	4.0000-4.0010	4.0010-4.0020	4.0187-4.0207	4.0287-4.0370
350	1971 except 330 hp	3.9998-4.0008	4.0008-4.0010	4.0183-4.0203	4.0283-4.0303
350	1971 330 hp	3.9953-3.9963	3.9963-3.9973	4.0138-4.0158	4.0252-4.0272
350	1972-on	3.9998-4.0008	4.0008-4.0018	—	4.0283-4.0303
400	1973-1976	4.1241-4.1251	4.1251-4.1261	—	4.1526-4.1546
400[1]	1970 350 hp	4.1240-4.1250	4.1250-4.1260	4.1450-4.1470	4.1550-4.1570
400[1]	1970 375 hp	4.1206-4.1216	4.1216-4.1226	—	4.1525-4.1545
402	1971-72	4.1237-4.1247	4.1245-4.1257	4.1450-4.1470	4.1550-4.1570
454[2]	1970	4.2455-4.2465	4.2465-4.2475	—	4.2760-4.2780
454	1971-1975	4.2481-4.2491	4.2491-4.2501	—	4.2775-4.2795

1. Actual displacement on 1970 400s-396 cid   2. 0.060 in. oversize = 4.3060-4.3080   3. See dealer for available sizes.

3. Measure piston-to-cylinder clearance, using procedure given below.

4. If damage or wear indicate piston replacement, select a new piston as described in *Piston Clearance and Selection* procedure.

5. Measure all parts marked in Step 2 of the *Disassembly* procedure above with micrometer and dial bore gauge to determine which parts are worn (this can be done by an automotive machine shop if measuring equipment is not available). Replace piston and pin set as a unit if either or both are worn.

Piston Clearance and Selection

1. Make sure cylinder walls are clean and dry.

2. Measure cylinder bore with a telescope bore gauge. Make measurement 2½ in. from top of bore (see **Figure 68**).

3. Measure outside diameter of piston at bottom of skirt (see **Figure 69**).

4. The difference between the two readings is the piston clearance. Compare difference with the specification in **Table 4**. If greater than specification, select another piston size from **Table 3**, end of chapter.

Piston Ring Installation

> NOTE: *All compression rings have marks on their upper sides. Install these rings with marks toward top of piston. Oil control rings are made of 3 pieces, including 2 rails and a spacer.*

1. Select new rings of proper size for piston being used.

2. Insert compression ring into bore of cylinder, about ¼ in. above ring travel. Square ring with cylinder wall by tapping with a piston. Using a feeler gauge, measure the gap (see **Figure 70**) and compare to specifications (see **Table 3**, end of chapter). If gap is too large, fit

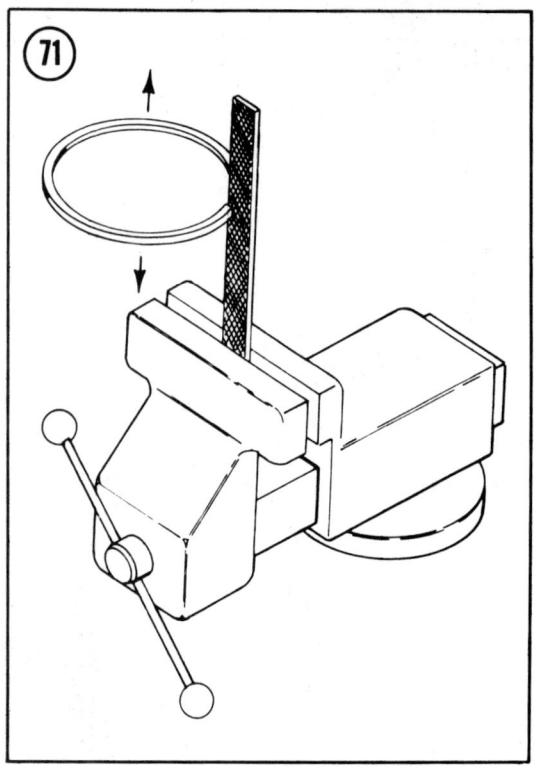
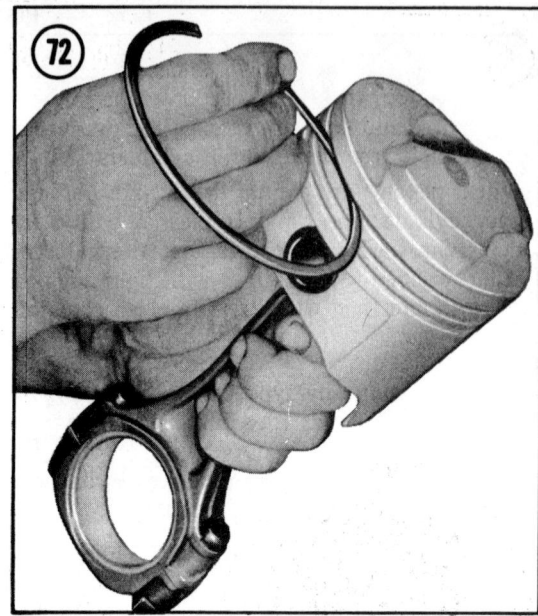
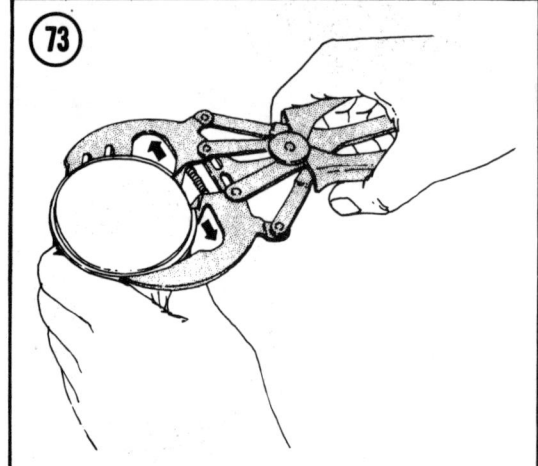
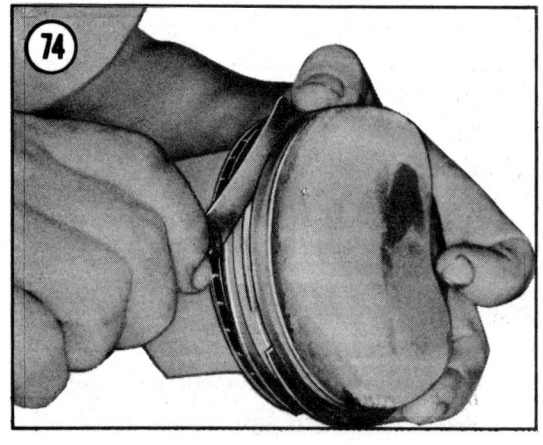

another ring. If smaller than specified, hold a small file in a vise, grip ends of ring with fingers, and carefully enlarge gap as shown in **Figure 71**.

3. Fit each cylinder with 2 compression rings, using the instructions in Step 2.

4. Roll each ring around its piston groove as shown in **Figure 72** to check for binding. Minor binding may be cleaned up with a fine cut file, if care is used.

5. Using a ring expander tool, install oil ring spacer in groove and, except on inline engines, install anti-rotational tang in oil hole. Hold spacer ends butted and install lower steel rail with gap properly located. Then install upper steel rail with gap properly located (see **Figure 11** for gap locations).

6. Flex oil ring assembly in its groove to make sure ring is free and binding does not occur at any point. Minor binding can be cleaned up with a fine cut file.

7. Install 2 compression rings, marked side up, using ring expander tool (see **Figure 73**). Make certain the rings fitted to each cylinder are installed on the proper pistons.

ENGINE

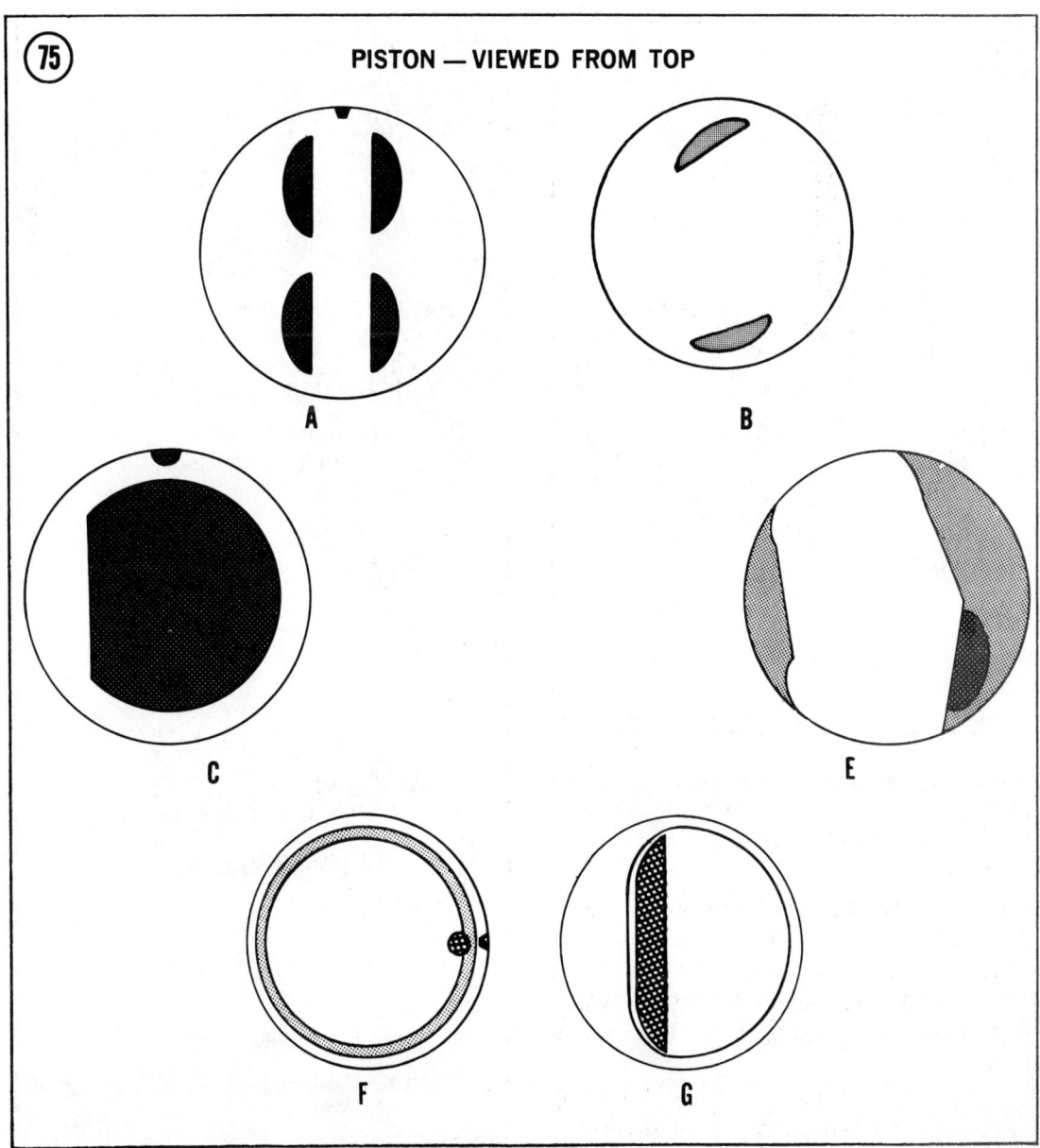

75 PISTON — VIEWED FROM TOP

8. Check side clearance of each ring as shown in **Figure 74**. Compare with specifications for your engine (see **Table 3** at the end of the chapter).

9. Verify that ring gaps are properly positioned as shown in **Figure 11**.

Piston Installation

NOTE: *Used pistons must be installed in the cylinders from which they were removed, and new pistons must be installed in the cylinders to which they were fitted.*

1. Coat cylinder walls, piston and rings with light engine oil.

2. With bearing caps removed, either install piston rod guide tool or tape bearing end of rod, including bolt threads, and have an assistant help you to slowly guide rods into place.

3. Install each piston in its correct cylinder bore. Orient rod as shown in **Figure 75** and **Table 5**, or **Figure 76** (231 V6).

Table 5   CONNECTING ROD AND PISTON RELATIONSHIP

Engine (cu. in. displacement)	Piston Type (Figure 75)	Side of Piston Aligned With Connecting Rod Bearing Tangs
250	C	Notch to front; alignment not necessary
200-229	F	Cylinders 1-3-5; left side Cylinders 2-4-6; right side
231	N.A.	Piston notches to front, rod bosses to rear on left, to front on right side
267, 305, 307, and 1970 350	A	Cylinders 1-3-5-7; left side Cylinders 2-4-6-8; right side
350 (1971-on)	A, F, or G	Cylinders 1-3-5-7; left side Cylinders 2-4-6-8; right side
396	E	All cylinders, left side
400, 402	E or F	Cylinders 1-3-5-7; left side Cylinders 2-4-6-8; right side
454	B or E	Cylinders 1-3-5-7; left side Cylinders 2-4-6-8; right side

4. Compress rings with a ring compressor tool and press pistons into bores with a wood hammer handle (**Figure 77**).

5. Guide bearing carefully onto journal. Install bearing caps and check clearances as described above. Torque nuts to specifications (**Tables 1 and 2**, end of chapter).

CAMSHAFT

Camshaft removal and installation procedures are given in Steps 13 and 3 of the *General Overhaul* and *Reassembly* sequences, respectively.

Inspection

1. Measure camshaft bearing journals for out-of-round condition, using a micrometer. Replace camshaft if out-of-round exceeds 0.001 in.

> NOTE: *If measuring equipment is not available, take the camshaft to an automotive machinist for the measurements described in this procedure.*

2. Mount camshaft in V-blocks and, using a dial indicator (**Figure 78**), check bearing journals for alignment. Replace camshaft if out of alignment more than 0.0015 in.

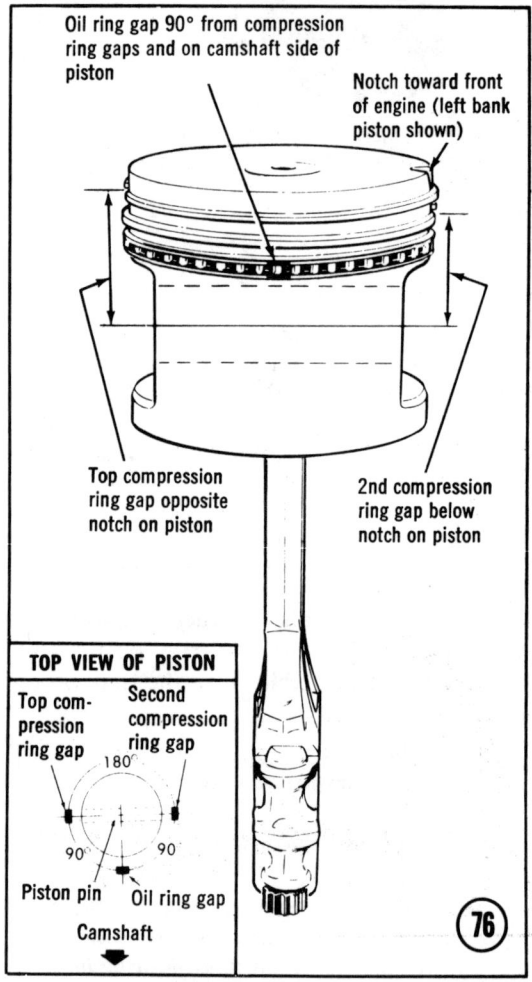

ENGINE

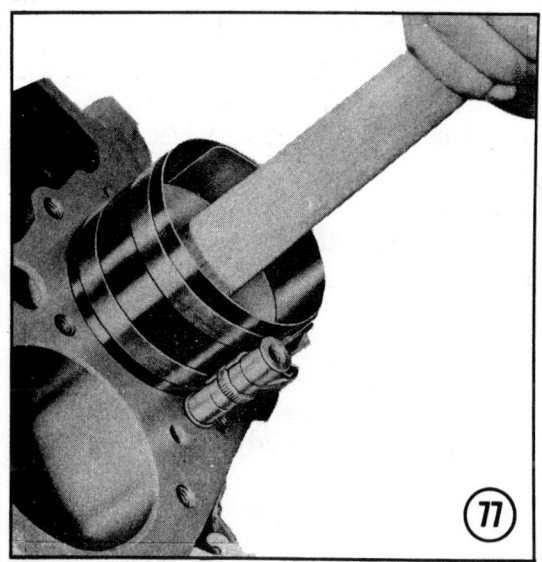

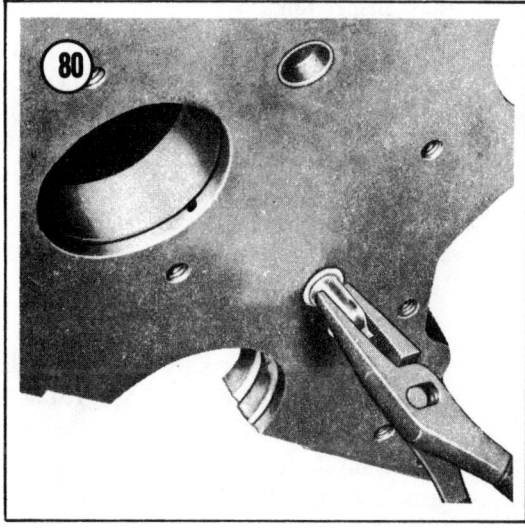

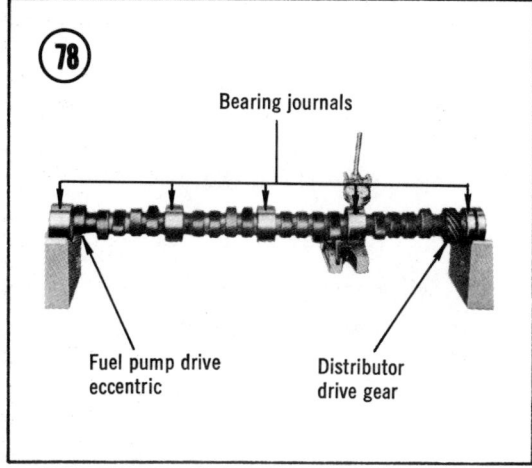

3. On inline engines, check gear and thrust plate for wear and damage. Check end play (**Figure 79**). End play should be between 0.001 and 0.005 in.

Oil Nozzle Replacement
(Inline Engines)

1. Pull out nozzle with pliers (**Figure 80**).
2. Drive new nozzle in with plastic hammer. Make sure oil hole is in vertical position.

Camshaft Gear Replacement
(Inline Engines)

Replacement of camshaft gear requires an arbor press and special tools. This job should be referred to an automotive machine shop.

Camshaft Bearing Replacement

Bearing replacement requires special tools and adequate substitutes are not easily improvised. An automotive machine shop will replace the bearings for a small fee.

CRANKSHAFT

The engine must be removed from the vehicle in order to remove the crankshaft. The procedures for crankshaft removal; bearing removal, inspection, and replacement; and installation are given in Steps 15 and 1 of the *General Overhaul* and *Reassembly* sequences, respectively.

Crankshaft Inspection

1. Using a micrometer, check crankshaft journals and crankpins for out-of-round taper, or undersize. See **Table 3** (end of chapter) for specifications.

> NOTE: *If measuring equipment is not available, take the crankshaft to an automotive machine shop to have the checks described in this procedure.*

2. Mount front and rear main journals in V-blocks and check at intermediate main journals with a dial indicator for runout. See **Table 3** (end of chapter) for specifications.

3. Replace crankshaft or have it reconditioned if out of specifications.

CYLINDER BLOCK

Cleaning and Inspection

1. Clean block with solvent and remove all foreign material from gasket surfaces.

2. Remove plugs from oil galleries and clean all oil passages.

3. Clean and inspect all water passages.

4. Check for cracks in cylinder walls, water jacket, valve lifter bores, and main bearing webs.

5. Check cylinder walls for out-of-round, taper, or excessive ridges, using a dial indicator (this job should be referred to an automotive machinist). If cylinders exceed specifications, honing or reboring will be required.

ENGINE

Table 1 ENGINE TORQUES (ALL EXCEPT 231 V6)

Bolt Size	Usage	Inline	Small V8 and 220-229 V6	Mark IV V8
1/4-20	Camshaft thrust plate	80 in. lb.	—	—
	Crankcase front cover	80 in. lb.	80 in.-lb.	80 in.-lb.
	Flywheel housing cover	80 in.-lb.	80 in.-lb.	80 in.-lb.
	Oil filter bypass valve	—	80 in.-lb.	80 in.-lb.
	Oil pan (to crankcase)	80 in.-lb.	80 in.-lb.	80 in.-lb.
	Oil pan (to front cover) 1970-1973	55 in.-lb.	—	80 in.-lb.
	Oil pan (to front cover) 1974-on	50 in.-lb.	—	55 in.-lb.
	Oil pump cover	70 in.-lb.	80 in.-lb.	80 in.-lb.
	Rocker arm cover 1970	55 in.-lb.	55 in.-lb.	50 in.-lb.
	Rocker arm cover 1971-on	45 in.-lb.	45 in.-lb.	50 in.-lb.
11/32-24	Connecting rod cap	35 ft.-lb.	—	—
5/16-18	Camshaft sprocket	20 ft.-lb.	20 ft.-lb.	20 ft.-lb.
	Clutch pressure plate	20 ft.-lb.	—	—
	Oil pan (to crankcase) 1970	125 in.-lb.	65 in.-lb.	135 in.-lb.
	Oil pan (to crankcase) 1971-1973	75 in.-lb.	65 in.-lb.	135 in.-lb.
	Oil pan (to crankcase) 1974-on	75 in.-lb.	265 in.-lb.	135 in.-lb.
	Oil pump	115 in.-lb.	—	—
	Pushrod cover 1970	80 in.-lb.	—	—
	Pushrod cover 1971-on	50 in.-lb.	—	—
	Water pump	15 ft.-lb.	—	—
3/8-16	Clutch pressure plate	—	—	35 ft.-lb.
	Distributor clamp	20 ft.-lb.	20 ft.-lb.	20 ft.-lb.
	Flywheel housing	30 ft.-lb.	30 ft.-lb.	30 ft.-lb.
	Manifold, exhaust	—	—	20 ft.-lb.[1]
	Manifold (exhaust to inlet) 1970-1972	25 ft.-lb.	—	—
	Manifold (exhaust to inlet) 1973	30 ft.-lb.	—	—
	Manifold (inlet)	—	30 ft.-lb.	30 ft.-lb.
	Manifold-to-head (outer) 1970-1971	20 ft.-lb.	—	—
	Manifold-to-head (all others) 1970-1971	30 ft.-lb.	—	—
	Manifold-to-head 1972-on	35 ft.-lb.	—	—
	Thermostat housing	30 ft.-lb.	—	—
	Water outlet 1970	20 ft.-lb.	20 ft.-lb.	20 ft.-lb.
	Water outlet 1971-on	30 ft.-lb.	30 ft.-lb.	30 ft.-lb.
	Water pump	—	30 ft.-lb.	30 ft.-lb.
3/8-24	Connecting rod cap	—	45 ft.-lb.	50 ft.-lb.
7/16-14	Cylinder head	—	65 ft.-lb.	80 ft.-lb.
	Main bearing cap 1970	65 ft.-lb.	75 ft.-lb.[2]	—
	Main bearing cap 1971-on	65 ft.-lb.	70 ft.-lb.[2]	—
	Oil pump	65 ft.-lb.	65 ft.-lb.	65 ft.-lb.
	Rocker arm stud	—	—	50 ft.-lb.

(continued)

Table 1 ENGINE TORQUES (ALL EXCEPT 231 V6) (continued)

Bolt	Usage	Inline	Small V8 and 200-229 V6	Mark IV V8
7/16-20	Flywheel	60 ft.-lb.	60 ft.-lb.	65 ft.-lb.
	Torsional damper	60 ft.-lb.	60 ft.-lb.	—
1/2-13	Cylinder head	95 ft.-lb.	—	—
	Main bearing cap (4 bolt) 1970	—	—	105 ft.-lb.
	Main bearing cap 1971-on	—	—	110 ft.-lb.
1/2-14	Temperature sending unit	20 ft.-lb.	20 ft.-lb.	20 ft.-lb.
1/2-20	Torsional damper	—	—	85 ft.-lb.
	Oil filter	Hand tight	25 ft.-lb.	25 ft.-lb.
	Oil pan drain plug	20 ft.-lb.	20 ft.-lb.	20 ft.-lb.
14mm (5/8 in.)	Spark plug 1970	25 ft.-lb.	25 ft.-lb.	25 ft.-lb.
	Spark plug 1971-on	15 ft.-lb.	15 ft.-lb.	15 ft.-lb.

1. Inside bolts on 302, 307, and 350 engines = 30 ft.-lb.
2. Outer bolts on engines with 4 bolt caps = 65 ft.-lb.

ENGINE

Table 2 ENGINE TORQUES (231 V6 ENGINE)

	Ft.-lb.
Spark plugs	20
Crankshaft bearing caps to cylinder block	100
Connecting rods	40
Cylinder head to cylinder block	80
Harmonic balancer to crankshaft	175
Fan driving pulley to harmonic balancer	20
Flywheel to crankshaft (auto. & manual)	60
Oil pan to cylinder block	14
Oil pan drain plug	30
Oil pump cover to timing chain cover	10
Oil pump pressure regulator retainer	35
Oil pressure switch to cylinder block	23
Filter assembly to pump cover	20
Timing chain cover to block	30
Water pump cover to timing chain cover	7
Thermostat housing to intake manifold	13
Intake manifold to cylinder head	45
Exhaust manifold to cylinder head	25
Fuel pump eccentric and timing chain sprocket to camshaft	22
Rocker arm cover to cylinder head	4
Rocker arm shaft to cylinder head	30
Starting motor to block	35
Distributor hold-down clamp	13
Lower flywheel housing cover (manual)	10
Lower flywheel housing cover (automatic)	4
Bellhousing	25
Automatic transmission to cylinder block	35
Timing chain dampener to cylinder block bolt	8
Bolt — special moveable timing chain dampener	12

Table 3 ENGINE SPECIFICATIONS

	250 CID L6
General	
Bore	3⅞ in.
Stroke	3.53 in.
Firing order	1-5-3-6-2-4
Cylinder bore	
Diameter	3.8775 in.
Out-of-round, new (wear limit)	(0.005 in. maximum)
Piston	
Clearance in bore	0.0015 in.
Piston rings	
Number per cylinder	3
Ring end gap	
Top	0.020 in.
2nd	0.020 in.
Oil control	0.055 in.
Ring side clearance	
Top	0.0027 in.
2nd	0.0032 in.
Oil control	0.005 in.
Piston pin	
Diameter	0.9273 in.
Clearance	
In piston	0.00025 in.
In rod	0.0016 in.
Crankshaft	
End play	0.006 in.
Main bearing journal	
Diameter (all)	2.2993 in.
Taper	0.0002 in. maximum
Out-of-round	0.0002 in. maximum
(continued)	

Table 3 ENGINE SPECIFICATIONS (continued)

	250 CID L6
Crankshaft (continued)	
Main bearing clearance (all)	0.0003-0.0029 in.
Crankpin	
Diameter	1.999-2.000 in.
Taper, new (wear limit)	0.0003 in. maximum (0.001 in.)
Out-of-round, new (wear limit)	0.002 in. maximum (0.001 in.)
Connecting rods	
Side clearance	0.007-0.016 in.
Bearing clearance	0.0007-0.0027 in.
Camshaft	
Journal diameter	1.8682-1.8692 in.
Runout	0.0015 in. maximum
Valve system	
Lifter type	Hydraulic
Rocker arm ratio	1.75:1
Valve lash (intake & exhaust)	One turn down from zero lash
Intake valve	
Face angle	45°
Seat angle	46°
Seat width	1/32-1/16 in.
Stem-to-guide clearance	0.0010-0.0027 in.
Seat runout	0.002 in. maximum
Exhaust valve	
Face angle	45°
Seat angle	46°
Seat width	1/16-3/32 in.
Stem-to-guide clearance	0.0015-0.0032 in.
Seat runout	0.002 in. maximum
Valve springs (outer)	
Free length	1.92 in.
Load @ length (lbs. @ in.)	
Closed	55-64 @ 1.66 in.
Open	180-192 @ 1.27 in.
Installed height (± 1/32 in.)	1 21/32 in.

(continued)

Table 3 ENGINE SPECIFICATIONS (continued)

231 CID V6	
General	
Bore	3.800 in.
Stroke	3.400 in.
Firing order	1-6-5-4-3-2
Cylinder bore	
Diameter	3.800 in.
Out-of-round	0.0005 in. max.
Taper	0.0005 in. max.
Piston	
Clearance in bore	0.0008-0.0020 in.
Piston rings	
Number per cylinder	3
Ring end gap	
Top	0.010-0.020 in.
2nd	0.010-0.020 in.
Oil control	0.015-0.035 in.
Ring side clearance	
Top	0.003-0.005 in.
2nd	0.003-0.005 in.
Oil control	0.0035 in. max.
Piston rings	
Diameter	0.9391-0.9394 in.
Clearance	
In piston	0.0004-0.0007 in.
In rod	0.0007-0.0017 in.
Crankshaft	
End play	0.004-0.008 in.
Main bearing journals	
Diameter (all)	2.4995 in.
Out-of-round	0.0015 in. maximum
(continued)	

ENGINE

Table 3 ENGINE SPECIFICATIONS (continued)

231 CID V6	
Crankshaft (continued)	
Main bearing clearance (all)	0.0004-0.0015 in.
Crankpin	
Diameter	2.2495-2.2487 in.
Connecting rods	
Side clearance	0.006-0.027 in.
Bearing clearance	0.0005-0.0026 in.
Camshaft	
Journal diameter	1.8682-1.8692 in.
Runout	0.0015 in. maximum
Valve system	
Lifter type	Hydraulic
Rocker arm ratio	1.55:1
Valve lash (intake & exhaust)	Non-adjustable
Intake valve	
Face angle	45°
Seat angle	45°
Seat width	1/32-1/16 in.
Stem-to-guide clearance	0.0015-0.0032 in.
Exhaust valve	
Face angle	45°
Seat angle	45°
Seat width	1/16-3/32 in.
Stem-to-guide clearance	0.0015-0.0032 in.
Valve springs (outer)	
Load @ length (lb. @ in.)	
Closed	59-69 @ 1.727
Open	162-174 @ 1.327
(continued)	

Table 3 ENGINE SPECIFICATIONS (continued)

307 CID V8		
General		
Bore	3⅞ in.	
Stroke	3.25 in.	
Firing order	1-8-4-3-6-5-7-2	
Cylinder bore		
Diameter	3.8745-3.8775 in.	
Out-of-round, new (wear limit)	0.001 in. (0.002 in.)	
Piston		
Clearance in bore	0.0005-0.0011 in.	
Piston rings		
Number per cylinder	3	
Ring end gap		
Top	0.010-0.020 in.	
2nd	0.010-0.020 in.	
Oil control	0.015-0.055 in.	
Ring side clearance		
Top	0.0012-0.0027 in.	
2nd	0.0012-0.0032 in.	
Oil control	0.055 in. maximum	
Piston pins		
Diameter	0.9270-0.9273 in.	
Clearance		
In piston (wear limit)	0.00015-0.00025 in. (0.001 in.)	
In rod	0.0008-0.0016 in.	
Crankshaft		
End play	0.002-0.006 in.	
Main bearing journal diameter	1970-71:	1972-73:
No. 1	2.4484-2.4493 in.	2.4484-2.4493 in.
Nos. 2-3-4	2.4484-2.4493 in.	2.4481-2.4490 in.
No. 5	2.4479-2.4488 in.	2.4479-2.4488 in.
Taper	0.0002 in.	0.0002 in.
Out-of-round	0.0002 in.	0.0002 in.
(continued)		

Table 3 ENGINE SPECIFICATIONS (continued)

307 CID V8		
Crankshaft (continued)		
Main bearing clearance	1970:	1971-73:
No. 1	0.0003-0.0015 in.	0.0008-0.0020 in.
Nos. 2-3-4	0.0006-0.0016 in.	0.0011-0.0023 in.
No. 5	0.0008-0.0023 in.	0.0017-0.0033 in.
Crankpin		
Diameter	2.099-2.100 in.	2.099-2.100 in.
Taper, new (wear limit)	0.0003 in. (0.001 in.)	0.0003 in. (0.001 in.)
Out-of-round, new (wear limit)	0.002 in. (0.001 in.)	0.002 in. (0.001 in.)
Connecting rods	1970:	1971-73:
Side clearance	0.008-0.014 in.	0.008-0.014 in.
Bearing clearance	0.0007-0.0028 in.	0.0013-0.0035 in.
Camshaft		
Journal diameter	1.8682-1.8692 in.	1.8682-1.8692 in.
Runout	0.015 in. maximum	0.015 in. maximum
Valve system		
Lifter type	Hydraulic	Hydraulic
Rocker arm ratio	1.50:1	1.50:1
Valve lash	One turn down from zero lash	One turn down from zero lash
Intake valve		
Face angle	45°	45°
Seat angle	46°	46°
Seat width	1/32-1/16 in.	1/32-1/16 in.
Stem-to-guide clearance	0.0010-0.0027 in.	0.0010-0.0027 in.
Seat runout	0.002 in. maximum	0.002 in. maximum
Exhaust valve		
Face angle	45°	45°
Seat angle	46°	46°
Seat width	1/16-3/32 in.	1/16-3/32 in.
Stem-to-guide clearance	0.0012-0.0029 in.	0.0012-0.0029 in.
Seat runout	0.002 in. maximum	0.002 in. maximum
Valve springs (outer)	1970-72 & 1973 intake:	1973 exhaust:
Free length	2.03 in.	1.91 in.
Load @ length (lb. @ in.)		
Closed	76-84 @ 1.70 in.	76-84 @ 1.61 in.
Open	194-206 @ 1.25 in.	183-195 @ 1.20 in.
Installed height	1 23/32 in.	1.518 in.
Damper free length	1.94 in.	1.94 in.

Table 3 ENGINE SPECIFICATIONS (continued)

	200-229 CID V6 and 267-305-350 CID V8				
	200 CID V6	229 CID V6	267 CID V8	305 CID V8	350 CID V8
General					
Bore	3.50 in.	3.74 in.	3.50 in.	3.736 in.	4.0 in.
Stroke	3.48 in.	3.48 in.	3.48 in.	3.48 in.	3.48 in.
Firing order	1-6-5-4-3-2	1-6-5-4-3-2	1-8-4-3-6-5-7-2	1-8-4-3-6-5-7-2	1-8-4-3-6-5-7-2
Cylinder bore					
Diameter	3.4995-3.5025 in.	3.7350-3.7385 in.	3.4995-3.5025 in.	3.7350-3.7385 in.	3.9995-4.0025 in.
Out-of-round, new (wear limit)	0.001 in. (0.002 in.)				
Piston					
Clearance in bore		See Table 4, Chapter 4			
Piston rings					
Number per cylinder		3			
Ring end gap					
Top		0.010-0.020 in.			
2nd		0.013-0.023 in.			
Oil control		0.015-0.055 in.			
Ring side clearance					
Top		0.0012-0.0032 in.			
2nd		0.0012-0.0027 in.			
Oil control		0.0000-0.005 in. maximum			
Piston pins					
Diameter		0.9270-0.9273 in.			
Clearance					
In piston		0.00015-0.00025 in.			
In rod		0.0008-0.0016 in.			
Crankshaft					
End play		0.002-0.006 in. (1970-75)			
Main bearing journals					
Diameter					
No. 1, 2, 3 & 4		2.4483-2.4493 in. (1970-1971)			
No. 5		2.4478-2.4488 in. (1970-1971)			
No. 1		2.4484-2.4493 in. (1972-on)			
No. 2, 3 & 4		2.4481-2.4490 in. (1972-on)			
No. 5		2.4479-2.4488 in. (1972-on)			
Taper		0.0002 in. maximum			
Out-of-round		0.0002 in. maximum			
Main bearing clearance					
No. 1		0.003-0.0015 in. (1970)			
No. 2, 3 & 4		0.0006-0.0018 in. (1970)			
No. 5		0.0008-0.0023 in. (1970)			
No. 1		0.0008-0.0020 in. (1971-on)			
No. 2, 3 & 4		0.0011-0.0023 in. (1971-on)			
No. 5		0.0017-0.0033 in. (1971-on)			
		(continued)			

Table 3 ENGINE SPECIFICATIONS (continued)

200-229 CID V6 and 267-305-350 CID V8	
Crankshaft (continued)	
Crankpin	
Diameter	2.099-2.100 in.
Taper, new (wear limit)	0.003 in. (0.001 in.)
Out-of-round, new (wear limit)	0.002 in. (0.001 in.)
Connecting rods	
Side clearance	0.005-0.014 in.
Bearing clearance (wear limit)	0.0013-0.0035 in. (0.0030 in.)
Camshaft	
Journal diameter	1.8682-1.8692 in.
Runout	0.0015 in. maximum
Valve system	
Lifter type	Hydraulic-Mechanical[1]
Rocker arm ratio	1.50:1
Valve lash	See Table 5, Chapter 3 for mechanical lifters
Intake valve	
Face angle	45°
Seat angle	46°
Seat width	1/32-1/16 in.
Stem-to-guide clearance	0.0010-0.0027 in.
Seat runout	0.002 maximum
Exhaust valve	
Face angle	45°
Seat angle	46°
Seat width	1/16-3/32 in.
Stem-to-guide clearance	0.0012-0.0029 in.
Seat runout	0.002 in. maximum
Valve springs (outer)	
Free length	2.03 (1967-72 intake/exhaust & 1973-75 intake valves)
	1.90 (1973-1975 exhaust valves)
Load @ length (lb. @ in.)	
Closed	76-84 @ 1.70 in. (1970-72 intake/exhaust & 1973-on intake valves)
Open	194-206 @ 1.25 in. (1970-72 intake/exhaust & 1973-on intake valves)
Closed	76-84 @ 1.61 in. (1973-on exhaust valves)
Open	183-195 @ 1.20 in. (1973-on exhaust valves)
Installed height	1 23/32 in.
Damper free length	1.86 in.

1. Mechanical lifters were used on 1970 350 CID 370 hp engines.

(continued)

Table 3 ENGINE SPECIFICATIONS (continued)

396 CID V8			
General			
Bore	4 3/32 in.		
Stroke	3.76 in.		
Firing order	1-8-4-3-6-5-7-2		
Cylinder bore			
Diameter	4.0925-4.0995 in.		
Out-of-round, new (wear limit)	0.001 in. (0.002 in.)		
Piston			
Clearance in bore	See Table 3, Chapter 5		
Piston rings			
Number per cylinder	3		
Ring end gap			
Top	0.010-0.020 in.		
2nd	0.010-0.020 in.		
Oil control	0.010-0.030 in.		
Ring side clearance			
Top	0.017-0.032 in.		
2nd	0.017-0.032 in.		
Oil control	0.0005-0.0065 in.		
Piston pins			
Diameter	0.9895-0.9898 in.		
Clearance			
In piston	0.00025-0.00035 in. (0.00030-0.00040, 1969 375 hp)		
In rod	0.0008-0.0016 in.		
Crankshaft			
End play	0.006-0.010 in.		
Main bearing journal			
Diameter	1967:	1968-69 exc. 375 hp:	1968-69 375 hp:
No. 1	2.7487-2.7497 in.	2.7484-2.7493 in.	2.7484-2.7493 in.
No. 2	2.7487-2.7497 in.	2.7484-2.7493 in.	2.7481-2.7490 in.
Nos. 3-4	2.7482-2.7492 in.	2.7481-2.7490 in.	2.7481-2.7490 in.
No. 5	2.7478-2.7488 in.	2.7478-2.7488 in.	2.7478-2.7488 in.
Taper	0.0002 in.	0.0002 in.	0.0002 in.
Out-of-round	0.0002 in.	0.0002 in.	0.0002 in.
(continued)			

ENGINE

Table 3 ENGINE SPECIFICATIONS (continued)

396 CID V8			
Crankshaft (continued)			
Main bearing clearance	1967:	1968-69 exc. 375 hp:	1968-69 375 hp:
No. 1	0.0004-0.0020 in.	0.0010-0.0022 in.	0.0013-0.0025 in.
Nos. 3-4	0.0009-0.0025 in.	0.0013-0.0025 in.	0.0013-0.0025 in.
No. 5	0.0013-0.0029 in.	0.0015-0.0031 in.	0.0015-0.0031 in.
Crankpin	1967:	1968-69 exc. 375 hp:	1968-69 375 hp:
Diameter	2.199-2.200 in.	2.199-2.200 in.	2.1985-2.1995 in.
Taper, new (wear limit)	0.0003 in. (0.001 in.)	0.0003 in. (0.001 in.)	0.0003 in. (0.001 in.)
Out-of-round, new (wear limit)	0.002 in. (0.001 in.)	0.002 in. (0.001 in.)	0.002 in. (0.001 in.)
Connecting rods	1967:	1968-69 exc. 375 hp:	1968-69 375 hp:
Side clearance	0.015-0.021 in.	0.015-0.021 in.	0.019-0.025 in.
Bearing clearance	0.0007-0.0028 in.	0.0009-0.0025 in.	0.0014-0.0030 in.
Camshaft		1968-69:	
Journal diameter		1.9482-1.9492 in.	
Runout		0.0015 in.	
Valve system			
Lifter type		Hydraulic (mechanical on 1968-69 375 hp)	
Rocker arm ratio		1.70:1	
Valve lash		See Table 2, Chapter 4	
Intake valve			
Face angle		45°	
Seat angle		46°	
Seat width		1/32-1/16 in.	
Stem-to-guide clearance		0.0010-0.0025 in.	
Seat runout		0.002 in. maximum	
Exhaust valve			
Face angle		45°	
Seat angle		46°	
Seat width		1/16-3/32 in.	
Stem-to-guide clearance		0.0012-0.0027 in.	
Seat runout		0.002 maximum	
Valve springs (outer)		325 hp:	Others:
Free length		2.11 in.	2.09 in.
Load @ length (lb. @ in.)			
Closed		84-96 @ 1.88 in.	94-106 @ 1.88 in.
Open		210-230 @ 1.46 in.	303-327 @ 1.38 in.
Installed height		1 7/8 in.	1 7/8 in.
Damper free length		1.94-2.00 in.	1.94-2.00 in.

Table 3   ENGINE SPECIFICATIONS (continued)

402 CID V8		
General		
Bore	4 1/8 in.	
Stroke	3.76 in.	
Firing order	1-8-4-3-6-5-7-2	
Cylinder bore		
Diameter	4.1246-4.1274 in.	
Out-of-round, new (wear limit)	0.001 in. (0.002 in. maximum)	
Piston		
Clearance in bore	See Table 3, Chapter 5	
Piston rings	1970:	1971-72:
Number per cylinder	3	3
Ring end gap		
Top	0.010-0.020 in.	0.010-0.020 in.
2nd	0.010-0.020 in.	0.010-0.020 in.
Oil control	0.010-0.030 in.	0.015-0.055 in.
Ring side clearance	1970:	1971-1972:
Top	0.0017-0.0032 in.	0.0017-0.0032 in.
2nd	0.0017-0.0032 in.	0.0017-0.0032 in.
Oil control	0.0005-0.0065 in.	0.0005-0.0065 in.
Piston pins		
Diameter	0.9895-0.9898 in.	0.9895-0.9898 in.
Clearance	All exc. 1970 375 hp:	1970 375 hp:
In piston	0.00025-0.00035 in.	0.00030-0.00040 in.
In rod	0.0008-0.0016 in.	0.0008-0.0016 in.
Crankshaft		
End play	0.006-0.010 in.	0.006-0.010 in.
Main bearing journals		
Diameter	All exc. 1970 375 hp:	1970 375 hp:
Nos. 1-2	2.7487-2.7496 in.	2.7481-2.7490 in.
Nos. 3-4	2.7481-2.7490 in.	2.7481-2.7490 in.
No. 5	2.7478-2.7488 in.	2.7473-2.7483 in.
Taper	0.0002 maximum	0.0002 maximum
Out-of-round	0.0002 maximum	0.0002 maximum

(continued)

Table 3 ENGINE SPECIFICATIONS (continued)

402 CID V8			
Crankshaft (continued)			
Main bearing clearance	1970 350 hp:	1970 375 hp:	1971-72:
No. 1	0.0007-0.0019 in.	0.0013-0.0025 in.	0.0007-0.0019 in.
Nos. 2-3-4	0.0013-0.0025 in.	0.0013-0.0025 in.	0.0013-0.0025 in.
No. 5	0.0024-0.0040 in.	0.0029-0.0045 in.	0.0019-0.0035 in.
Crankpin		All exc. 1970 375 hp:	1970 375 hp:
Diameter		2.199-2.200 in.	2.1985-2.1995 in.
Taper, new (wear limit)		0.0003 in. (0.001 in.)	0.0003 in. (0.001 in.)
Out-of-round, new (wear limit)		0.002 in. (0.001 in.)	0.002 in. (0.001 in.)
Connecting rods		All exc. 1970 375 hp:	1970 375 hp:
Side clearance		0.015-0.021 in.	0.019-0.025 in.
Bearing clearance		0.009-0.0025 in.	0.0014-0.0030 in.
Camshaft			
Journal diameter		1.9487-1.9497 in.	1.9487-1.9497 in.
Runout		0.015 in. maximum	0.015 in. maximum
Valve system		All exc. 1970 375 hp:	1970 375 hp:
Lifter type		Hydraulic	Mechanical
Rocker arm ratio		1.70:1	1.70:1
Valve lash		One turn down from zero lash	0.024-0.028 in. ex.
Intake valve			
Face angle		45°	45°
Seat angle		46°	46°
Seat width		1/32-1/16 in.	1/32-1/16 in.
Stem-to-guide clearance		0.0010-0.0027 in.	0.0010-0.0027 in.
Seat runout		0.002 in. maximum	0.002 in. maximum
Exhaust valves			
Face angle		45°	45°
Seat angle		46°	46°
Seat width		1/16-3/32 in.	1/16-3/32 in.
Stem-to-guide clearance		0.0012-0.0027 in.	0.0012-0.0027 in.
Seat runout		0.002 in. maximum	0.002 in. maximum
Valve springs (inner)			
Free length		2.12 in.	2.03 in.
Load @ length (lb. @ in.)		69-81 @ 1.88 in.	69-81 @ 1.88 in.
		228-252 @ 1.38 in.	228-252 @ 1.38 in.
Installed height		1 7/8 in.	1 7/8 in.
Valve springs (inner)			
Free length		2.06 in.	2.06 in.
Load @ length (lb. @ in.)		26-34 @ 1.78 in.	26-34 @ 1.78 in.
		81-99 @ 1.28 in.	81-99 @ 1.28 in.

Table 3 ENGINE SPECIFICATIONS (continued)

400 CID V8	
General	
Bore	4.125 in.
Stroke	3.75 in.
Firing order	1-8-4-3-6-5-7-2
Cylinder bore	
Diameter	4.1246-4.1274
Out-of-round, new (wear limit)	0.005 in. (0.001 in.)
Piston	
Clearance in bore	0.0014-0.0020 in.
Piston rings	
Number per cylinder	3
Ring end gap	
Top	0.010-0.020 in.
2nd	0.010-0.020 in.
Oil control	0.015-0.055 in.
Ring side clearance	
Top	0.0012-0.0027 in.
2nd	0.0012-0.0032 in.
Oil control	0.000-0.005 in.
Piston pins	
Diameter	0.9270-0.9273 in.
Clearance	
In piston	0.00015-0.00025 in.
In rod	0.0008-0.0016 in.
Crankshaft	
End play	0.006-0.010 in.
Main bearing journal	
Diameter (1, 2, 3 & 4)	2.6484-2.6493 in.
Diameter (5)	2.6479-2.6488 in.
Taper	0.0002 in. (0.001 in. wear limit)
Out-of-round	0.0002 in. (0.001 in. wear limit)
(continued)	

Table 3 ENGINE SPECIFICATIONS (continued)

400 CID V8	
Crankshaft (continued)	
Main bearing clearance (No. 1)	0.0008-0.0020 in.
(No. 2, 3 & 4)	0.0011-0.0023 in.
(No. 5)	0.0017-0.0033 in.
Crankpin	
Diameter	2.099-2.100 in.
Taper, new (wear limit)	0.0003 in. (0.001 in.)
Out-of-round, new (wear limit)	0.002 in. (0.001 in.)
Connecting rods	
Side clearance	0.008-0.014 in.
Bearing clearance	0.0013-0.0035 in.
Camshaft	
Journal diameter	1.9482-1.9492 in.
Runout	0.0015 in. maximum
Valve system	
Lifter type	Hydraulic
Rocker arm ratio	1.70:1
Valve lash — intake and exhaust	one turn down from zero lash
Intake valve	
Face angle	45°
Seat angle	46°
Seat width	1/32-1/16 in.
Stem-to-guide clearance	0.0010-0.0027 in.
Seat runout	0.002 in.
Exhaust valve	
Face angle	45°
Seat angle	46°
Seat width	1/16-3/32 in.
Stem-to-guide clearance	0.0012-0.0027 in.
Seat runout	0.002 in.
Valve springs (outer)	
Free length	2.03 in.
Load @ length (lbs. @ in.)	
Closed	76-84 @ 1.70 in.
Open	194-206 @ 1.25 in.
Installed height	1 23/32 in.
Damper free length	1.94 in.
(continued)	

Table 3 ENGINE SPECIFICATIONS (continued)

454 CID V8	
General	
Bore	4.25 in.
Stroke	4.0 in.
Firing order	1-8-4-3-6-5-7-2
Cylinder bore	
Diameter	4.2495-4.2525 in.
Out-of-round, new (wear limit)	0.001 in. (0.002 in.)
Piston	
Clearance in bore	0.0040-0.0050 in.
Piston rings	
Number per cylinder	3
Ring end gap	
Top	0.010-0.020 in.
2nd	0.010-0.020 in.
Oil control	0.015-0.055 in.
Ring side clearance	
Top	0.0017-0.0032 in.
2nd	0.0017-0.0032 in.
Oil control	0.0005-0.0065 in.
Piston pins	
Diameter	0.9895-0.9898 in.
Clearance	
In piston	0.00025-0.00035 in.
In rod	0.0008-0.0016 in.
Crankshaft	
End play	0.006-0.010 in.
Main bearing journal	
Diameter	
Nos. 1-2-3-4	2.7481-2.7490 in.
No. 5	2.7478-2.7488 in.
Taper, new (wear limit)	0.0002 in. (0.001 in.)
Out-of-round, new (wear limit)	0.0002 in. (0.001 in.)
(continued)	

ENGINE

Table 3 ENGINE SPECIFICATIONS (continued)

454 CID V8	
Crankshaft (continued)	
Main bearing clearance	
Nos. 1-2-3-4	0.0013-0.0025 in.
No. 5	0.0029-0.0045 in.
Crankpin	
Diameter	2.1985-2.1995 in.
Taper, new (wear limit)	0.0003 in. (0.001 in.)
Out-of-round, new (wear limit)	0.002 in. (0.001 in.)
Connecting rods	
Side clearance	0.015-0.021 in.
Bearing clearance	0.0009-0.0025 in.
Camshaft	
Journal diameter	1.9487-1.9497 in.
Runout	0.015 in. maximum
Valve system	
Lifter type	Mechanical
Rocker arm ratio	1.70:1
Valve lash	0.024 in., 0.28 in. ex.
Intake valve	
Face angle	45°
Seat angle	46°
Seat width	1/32-1/16 in.
Stem-to-guide clearance	0.0010-0.0027 in.
Seat runout	0.002 maximum
Exhaust valve	
Face angle	45°
Seat angle	46°
Seat width	1/16-3/32 in.
Stem-to-guide clearance	0.0012-0.0027 in.
Seat runout	0.002 in. maximum
Valve springs (outer)	
Free length	2.12 in.
Load @ length (lb. @ in.)	69-81 @ 1.88
	228-252 @ 1.38
Installed height	1 7/8 ± 1/32 in.
Valve springs (inner)	
Free length	2.06 in.
Load @ length (lb. @ in.)	26-34 @ 1.78 in.
	81-99 @ 1.28 in.
Installed height	1 25/32 ± 1/32 in.

CHAPTER FIVE

FUEL AND EXHAUST SYSTEMS

FUEL SYSTEM

The fuel system basically consists of a fuel tank, a mechanical fuel pump, and a carburetor, all connected by fuel lines.

A large number of carburetors are used on these cars. These are designed to meet the requirements of particular engine/transmission combinations. Therefore, carburetors that look alike are not always interchangeable. To obtain correct repair parts for your carburetor, copy the carburetor part number stamped on the carburetor body (or on a metal tag) and take it with you to your dealer.

> NOTE: *Holley carburetors made for these cars should have both a Chevrolet part number as well as their own model designation. The Chevrolet part number is the one used for ordering repair parts and it contains 7 digits.*

This section covers removal, installation, and repair and/or replacement of carburetors, fuel pumps, and fuel tanks.

ROCHESTER 1MV-1ME

The Rochester 1MV was used on 1970-1976 6-cylinder inline engines. Specifications are given in **Table 1**.

All 1977 inline 6-cylinder engines are equipped with the Rochester 1ME single-barrel carburetor. Although similar in appearance and in operation to the Rochester 1MV carburetor, this carburetor is an integral part of the emission control system and has been set to meet requirements of the U.S. Department of Health, Education, and Welfare and of certain state air pollution control agencies. Except for adjusting idle speed and idle mixture (without disturbing the limiter cap) as described in Chapter Three, no attempt should be made to adjust, repair, or overhaul this carburetor except in emergencies. In such case, the procedures in this section may be used. Except as noted, the figures apply to all carburetors, and specifications are given in **Table 1**. Externally, the main difference between the 1ME and the 1MV is that the choke coil of the former is mounted on the carburetor, rather than being separately mounted as is the case with the 1MV.

Removal

> NOTE: *Acceleration stumbling and other performance problems are sometimes caused by foreign matter in the carburetor. When removing carburetor, take care to avoid draining fuel from*

FUEL AND EXHAUST SYSTEMS

Table 1 ROCHESTER 1MV-1ME SPECIFICATIONS

Year	1970	1971	1972	1973
Engines	250	250	250[1]	250
Float level	¼ in.	¼ in.	¼ in.	¼ in.
Fast idle measurement	0.110 in.	0.100 in.	N/A	N/A
Fast idle rpm	2,400	N/A	N/A	N/A
Carb. choke rod (auto.)	0.170 in.	0.160 in.	0.125 in.	0.245 in.
Carb. choke rod (man.)	0.190 in.	0.180 in.	0.150 in.	0.275 in.
Primary vacuum break (automatic)	0.200 in.	0.200 in.	0.190 in.	0.300 in. 0.350 in.
Primary vacuum break (manual)	0.225 in.	0.230 in.	0.225 in.	0.350 in.
Choke unloader	0.350 in.	0.350 in.	0.500 in.	0.500 in.
Thermostat choke rod	1 rod diameter	N/A	N/A	N/A
Main metering jet (auto.)	0.104 in.	N/A	N/A	N/A
Main metering jet (man.)	0.104 in.	N/A	N/A	N/A
Throttle bore	1¹¹⁄₁₆ in.	N/A	N/A	N/A
Metering rod	0.070 in.	0.080 in.	0.080 in. Fed. 0.078 in. Cal.	0.080 in.
Year	1974	1975	1976	1977[3]
Engines	250[1]	250[1]	250[1]	250[1]
Float level	0.95 in.	¹¹⁄₃₂ in.	¹¹⁄₃₂ in.	⅜ in.
Carburetor choke rod (automatic)	0.230 in. Fed. 0.245 in. Cal.	0.160 in. Fed. 0.230 in. Cal.	0.090 in. Fed. 0.135 in. Cal.	0.085 in. Fed. 0.100 in. Cal.
Carb. choke rod (man.)	0.275 in.	0.275 in.	0.090 in.	0.100 in.
Primary vacuum break (automatic)	0.275 in. Fed. 0.300 in. Cal.	0.200 in. Fed. 0.275 in. Cal.	0.110 in. Fed. 0.150 in. Cal.	0.120 in. Fed. 0.110 in. Cal.
Primary vacuum break (manual)	0.350 in.	0.350 in.	0.110 in.	0.125 in.
Choke unloader	0.500 in.	0.215 in. Fed. A 0.275 in. Cal. A 0.275 in. M	0.265 in.	0.325 in. Fed. 0.225 in. Cal.
Metering rod	0.079 in. A[2] 0.072 in. M	0.080 in.	0.084 in. Fed. A 0.082 in. Fed. M 0.083 in. Cal. A	0.070 in.
Auxiliary vacuum break (automatic)	N/A	0.215 in. Fed. 0.312 in. Cal.	0.215 in.	N/A
Auxiliary vacuum break (manual)	N/A	0.312 in.	0.215 in.	N/A

1. Fed. = All vehicles except those originally sold in California.
   Cal. = Vehicles originally sold in California.
2. 0.073 in. on vehicles first sold in California.
3. IME carburetor.

A = Automatic transmission.
M = Manual transmission.

CHAPTER FIVE

①

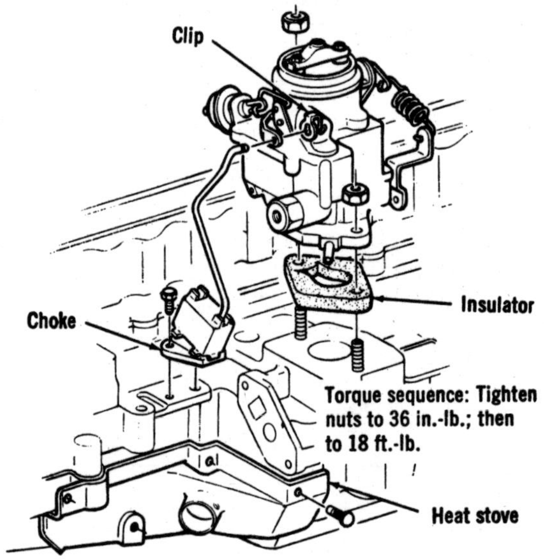

Clip

Choke

Insulator

Torque sequence: Tighten nuts to 36 in.-lb.; then to 18 ft.-lb.

Heat stove

②

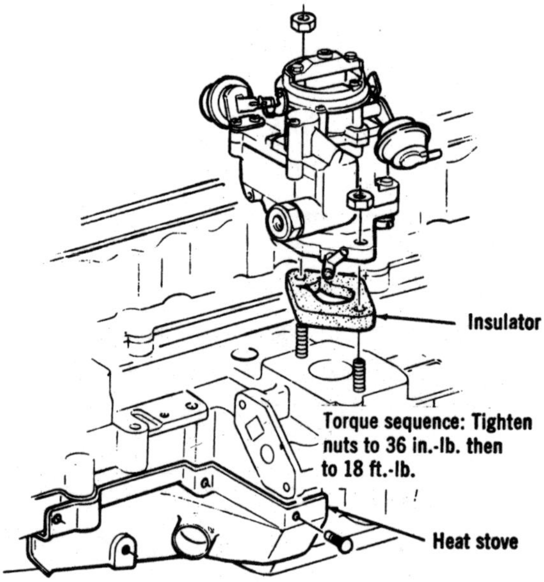

Insulator

Torque sequence: Tighten nuts to 36 in.-lb. then to 18 ft.-lb.

Heat stove

FUEL AND EXHAUST SYSTEMS

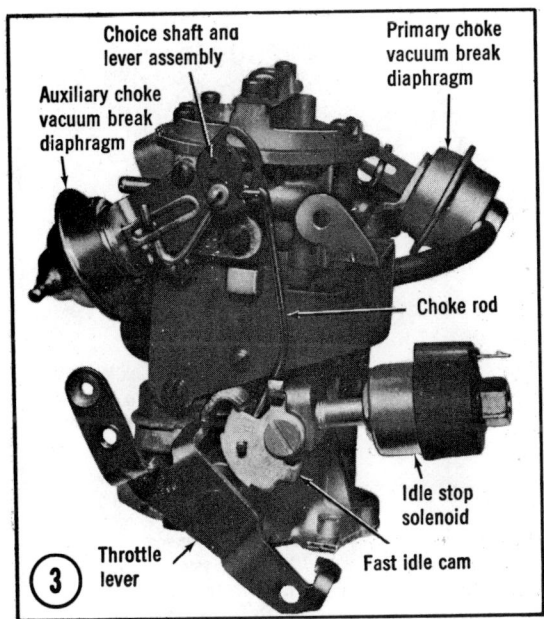

bowl so that contents of bowl can be examined for contamination when the carburetor is disassembled. Also, check fuel filter.

Refer to **Figure 1** (1MV) or **Figure 2** (1ME).

1. Remove air cleaner and gasket.
2. Disconnect fuel and vacuum lines from carburetor.
3. Disconnect choke coil rod (on 1MV only), throttle linkage, and idle stop solenoid electrical lead (at connector).
4. Remove carburetor attaching nuts and then remove carburetor and solenoid assembly.
5. Remove insulator, air cleaner bracket, and gasket from intake manifold flange.

Installation

> NOTE: *For ease of starting, fill carburetor bowl before installing carburetor. Operate throttle lever several times and verify fuel discharge from pump jets prior to installation.*

1. Verify that sealing flanges on carburetor and intake manifold are clean.
2. Install gasket, air cleaner bracket, and insulator over manifold studs.
3. Install carburetor over manifold studs and install and securely tighten attaching nuts.
4. Install and tighten vacuum and fuel lines at carburetor.
5. Connect accelerator linkage to carburetor.
6. Connect choke coil rod (1MV only) and idle stop solenoid electrical connector.

Disassembly

Air Horn (1MV)

1. Note position of choke rod carefully and remove choke rod. On early models this is done by removing retaining screw from upper choke lever and removing lever from shaft and rod from lever. On later models, remove fast idle cam and remove choke rod from cam.

2. On 1975-1976 models, remove auxiliary vacuum break diaphragm assembly from air horn by removing 2 attaching screws (**Figure 3**). Remove diaphragm plunger from link which is permanently attached to choke lever. Lever and link can be removed as a unit if desired. See *Disassembly* procedure below.

3. Remove air horn-to-bowl attaching screws (3 short and 3 long screws) and lockwashers.

4. On 1975-1976 models, remove primary vacuum break diaphragm unit from air horn. Remove vacuum break hose assembly and link from slot in diaphragm plunger stem.

Remove air horn by lifting straight up. Invert air horn and place on a clean workbench.

5. On 1971 and earlier models, remove vacuum break diaphragm cover (2 screws), hold choke wide open, and push upward on eyelet on choke valve until looped end of rod slides off wire lever attached to choke valve. Then remove diaphragm and plunger rod (**Figure 4**).

6. If desired, choke valve, vacuum break lever (if so equipped), and the choke shaft can be removed by first removing thermostatic coil lever from end of choke shaft (one screw). Then remove choke valve (2 screws) and choke shaft from air horn.

> NOTE: *No further disassembly of air horn is required.*

Air Horn (1ME)

1. Remove the hose from the vacuum break diaphragm and then remove the vacuum break

CHAPTER FIVE

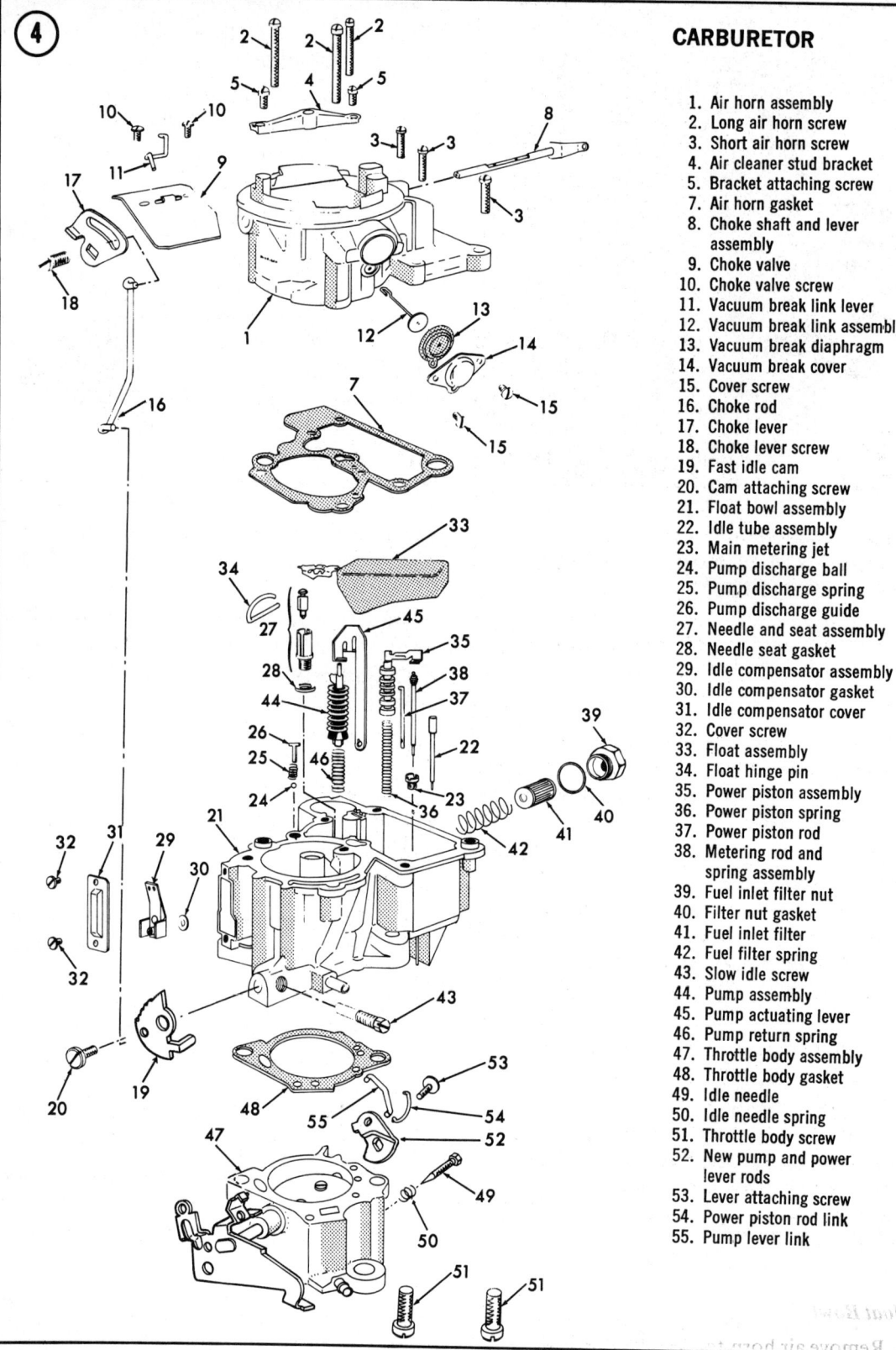

CARBURETOR

1. Air horn assembly
2. Long air horn screw
3. Short air horn screw
4. Air cleaner stud bracket
5. Bracket attaching screw
7. Air horn gasket
8. Choke shaft and lever assembly
9. Choke valve
10. Choke valve screw
11. Vacuum break link lever
12. Vacuum break link assembly
13. Vacuum break diaphragm
14. Vacuum break cover
15. Cover screw
16. Choke rod
17. Choke lever
18. Choke lever screw
19. Fast idle cam
20. Cam attaching screw
21. Float bowl assembly
22. Idle tube assembly
23. Main metering jet
24. Pump discharge ball
25. Pump discharge spring
26. Pump discharge guide
27. Needle and seat assembly
28. Needle seat gasket
29. Idle compensator assembly
30. Idle compensator gasket
31. Idle compensator cover
32. Cover screw
33. Float assembly
34. Float hinge pin
35. Power piston assembly
36. Power piston spring
37. Power piston rod
38. Metering rod and spring assembly
39. Fuel inlet filter nut
40. Filter nut gasket
41. Fuel inlet filter
42. Fuel filter spring
43. Slow idle screw
44. Pump assembly
45. Pump actuating lever
46. Pump return spring
47. Throttle body assembly
48. Throttle body gasket
49. Idle needle
50. Idle needle spring
51. Throttle body screw
52. New pump and power lever rods
53. Lever attaching screw
54. Power piston rod link
55. Pump lever link

FUEL AND EXHAUST SYSTEMS

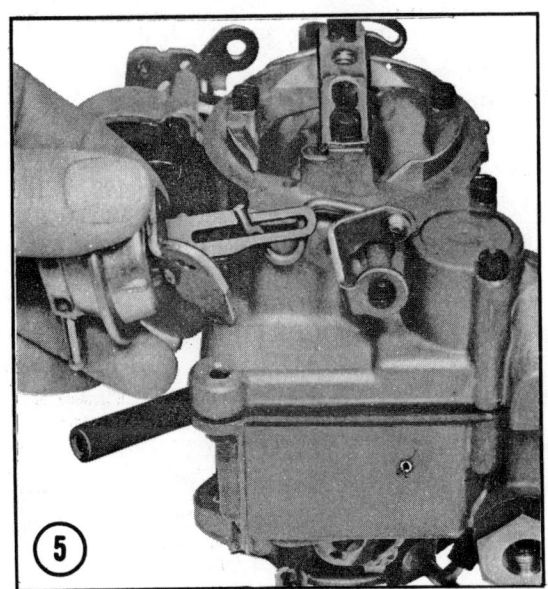

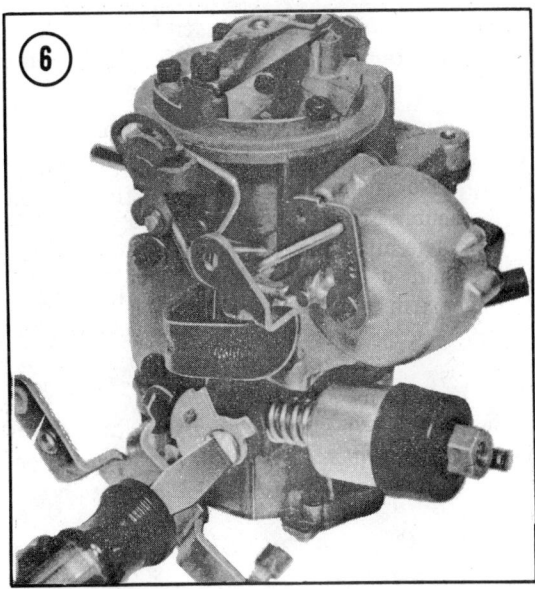

diaphragm from the air horn by removing 2 attaching screws. Remove the plunger stem from the link. See **Figure 5**.

> NOTE: *Do not remove the vacuum break lever from the choke shaft unless choke shaft replacement is required.*

2. Remove the attaching screw and then remove the fast idle cam. See **Figure 6**. Then remove the choke rod from the lever on the end of the choke shaft.

3. Remove the choke coil assembly from the choke housing (3 screws).

4. If choke coil housing replacement is required, remove the housing from the float bowl (3 screws). Otherwise, do not remove it.

5. Remove the 4 attaching screws and remove the air horn by lifting and twisting backward to disengage the choke coil lever link from the lever. Invert the air horn and place it on a clean workbench.

6. If required, remove the choke valve from the choke shaft (2 screws). Then remove the retaining screw and remove the vacuum break lever from the choke shaft. Then remove the choke shaft from the air horn.

Float Bowl

1. Remove air horn-to-float bowl gasket.

2. Lift up on float hinge pin and remove float from bowl. Remove hinge pin from float arm.

3. Remove float needle. On 1974 and earlier models, remove float needle seat and gasket, taking care to avoid damage to needle seat.

4. Remove fuel filter element, spring, and gasket by removing fuel inlet fitting.

5. Using long-nosed pliers, remove "T" guide and pump discharge spring. Invert bowl and remove spring and ball.

6. Remove accelerator pump plunger and metering rod power piston drive assemblies as follows:

   a. Remove actuating lever from throttle shaft (one screw). See 52, **Figure 4** for a typical example.

   b. Remove upper end of drive link from lower end of power piston rod.

   c. Rotate drive link from keyhole in actuating lever.

   d. Remove actuator lever from lower end of pump link in same manner.

   e. Push down on accelerator pump and remove actuator link by rotating until tang is aligned with slot in lever, then remove link.

7. Lift out power piston metering rod assembly and drive rod. Then remove spring from power piston cavity.

8. Remove pump plunger assembly from float bowl. Remove pump return spring from pump well.

9. Remove main metering jet from bottom of fuel bowl.

10. On 1975-1977 models, remove float needle seat and gasket.

11. On 1974 and earlier models, remove the screws (3) from throttle return spring bracket and remove bracket, hot idle compensator, and gasket from recess in bowl.

12. On 1975-1976 models, remove 2 screws from idle compensator cover (automatic transmission only) and remove cover, compensator, and seal.

> NOTE: *If desired, idle adjustment screw and fast idle cam (idle stop solenoid on 1975-1977 models) can be removed at this time. No further disassembly of float bowl is required.*

Throttle Body

1. With bowl inverted, remove attaching screws and remove throttle body and gasket from bowl.

2. No further disassembly is required unless idle mixture needle is damaged or idle channels need cleaning. If necessary, cut tang from plastic limiter cap to remove needle. Do not replace limiter cap when reinstalling needle.

> NOTE: *Because of the close tolerances involved, do not remove throttle valve or shaft.*

Cleaning and Inspection

1. Clean all metal parts and castings in carburetor solvent.

> **CAUTION**
> *Do not immerse any rubber or plastic parts, or electrical parts (solenoids, etc.) in carburetor cleaners.*

2. When clean, blow out all passages with compressed air. Do not clean jets or passages with drills or wires.

3. Inspect idle mixture needle for damage or wear. Replace if present.

4. Examine float needle and seat for wear. Replace with a new factory matched set if worn.

5. Inspect all sealing surfaces for damage.

6. Examine rods and holes in levers for excessive wear. Replace if worn.

7. Check fast idle cam for wear or damage.

8. Check levers and valves for binds and other damage.

9. Replace filter element.

Assembly

Float Bowl (1974 and Earlier)

1. If removed, install fast idle cam to float bowl boss (one screw) on 1974 and earlier models.

2. If removed, install seal in recess in idle compensator cavity and install idle compensator (2 screws). Securley tighten screws.

3. Install throttle return spring bracket (3 screws) and tighten securely.

4. Install pump return spring in pump well, making sure it is properly seated.

5. Install pump plunger assembly in pump well. Slide must protrude through bottom of bowl. Push down on slide and install drive link in hole in lower end of shaft. Drive link ends should point toward carburetor boss. Tang on link upper end holds link to slide.

6. Attach power piston and pump actuating lever to lower end of link. Lever projection must point down.

7. Install power piston in its cavity.

8. Install end of actuating rod into groove on side of power piston. Install metering rod assembly and actuating rod into float bowl. Metering rod end must enter jet orifice.

9. Install drive link in actuating lever keyhole.

10. While holding assembly down in bowl, install power piston drive link upper end in hole in power piston actuating rod lower end.

> NOTE: *Align "D" hole in actuating lever with flats on throttle shaft and install lever on shaft. Retain by securely tightening screw in end of shaft. Check operation of components installed for free operation before proceeding.*

FUEL AND EXHAUST SYSTEMS

11. Install idle tube in float bowl cavity.

12. Install discharge ball, spring, and retainer. Retainer must be flush with top of bowl casting.

13. Install fuel filter element and spring, using new gasket. Tighten inlet nut securely.

> NOTE: *Filter open end should face inlet nut hole.*

14. Install float needle seat and gasket. Carefully avoid damage to seat while tightening securely. Install float needle into seat.

15. Insert straight part of float hinge pin into float arm and install float into bowl.

Float Bowl (1975-1976)

1. If removed, install idle stop solenoid.

2. If so equipped, install seal into recess in idle compensator cavity and install compensator assembly. Install cover (2) screws and tighten securely.

3. Install main metering jet in fuel bowl and tighten securely.

4. Install float needle seat and gasket.

5. Install idle tube. Make sure tube is flush with bowl casting.

6. Install pump ball, spring, and "T" guide in pump discharge hole.

7. Press on "T" guide until it is flush with casting (**Figure 3**).

8. Install fuel filter element, spring, and gasket and tighten inlet nut securely.

9. Install accelerator pump return spring and install power piston return spring into bowl piston cavity.

10. Install right angle end of actuating rod into slot in power piston, and install piston, metering rod and actuating rod into float bowl. Make certain end of actuating rod enters hole in bowl and metering rod is located in jet orifice.

11. Install pump plunger in pump well so actuating lever protrudes through bottom of bowl. Press down on pump lever and install drive link in slot in lower end of shaft. Ends of drive link must point in toward carburetor bore. Tang on link upper end holds link to pump shaft (**Figure 3**).

12. Install pump link lower end in throttle shaft actuator lever.

13. Install power piston actuator link (curved end) into throttle actuator lever. End should face outward. End has tang to retain link to lever.

14. Hold power piston assembly down and slide upper end of curved power piston actuator link into lower end of power piston actuator rod.

15. Install actuating lever on end of throttle shaft (align flats on lever with flats on shaft) with retaining screw and tighten securely.

16. Install float needle valve into seat.

17. Install hinge pin into float arm and then install the assembly into float bowl.

Air Horn (1MV)

1. Install choke shaft assembly and choke valve into air horn, if removed. Align valve and tighten retaining screws. Stake screws securely or use Loctite or equivalent.

2. Carefully install air horn to float bowl with 3 long and 3 short screws and lockwashers. Tighten securely.

> NOTE: *Install primary choke vacuum break diaphragm assembly under 2 short screws next to thermostatic coil lever. Connect link to plunger stem and install lever to end of choke shaft with retaining screws. Tighten securely.*

3. Install hose to vacuum break diaphragm and tube on float bowl.

4. Install choke rod in slot in upper choke lever (end of rod points away from air horn).

5. Install lower end of choke rod into fast idle cam. Cam steps should face fast idle tang on throttle lever. Install cam to boss on float bowl and securely tighten screw.

6. On 1975-1976 models, install auxiliary vacuum break diaphragm link attached to choke lever to slot in diaphragm plunger stem. Then install auxiliary diaphragm unit to air horn and tighten screws securely.

7. If removed, install idle stop solenoid.

Air Horn (1ME)

1. Install the choke shaft in the air horn and then install the choke valve (if removed). Align the choke valve with the air horn wall and then securely tighten and stake the 2 retaining screws.

2. Apply Loctite or similar retaining compound to threads of retaining screw and install the vacuum break lever on the choke shaft.

3. Engage the choke coil lever link into the notched hole in the choke lever on the choke housing and install the air horn on the float bowl. See **Figure 7**.

> NOTE: *Carefully twist and lower the air horn onto the float bowl.*

4. Install the 3 long and one short attaching screws, using lockwashers.

5. Install the vacuum break diaphragm assembly, connecting the diaphragm assembly link to the slotted diaphragm plunger stem. Then tighten the 2 tapered-head attaching screws to properly locate the assembly. Then tighten all air horn screws in the sequence shown in **Figure 8**.

6. If removed, install the choke housing.

7. Install the fast idle cam and the fast idle cam link-to-upper choke lever assembly, making sure the numbers on the cam face outward.

8. Install the choke coil assembly in the choke housing, using 3 retainers and attaching screws.

9. Install the choke vacuum diaphragm hose to the tube on the diaphragm and connect to the vacuum tube on the bowl.

Adjustment

Float Level

Refer to **Figure 9** for this procedure.

1. Hold float retaining pin in place and push down on float arm at outer end.

2. Using adjustable "Y" scale, measure from top of float at index point on toe to float bowl gasket surface (gasket removed). See **Table 1** for specifications.

3. Bend float up or down, as required, at float arm junction to adjust.

Metering Rod

1. On 1975-1977 models, open the throttle valve, slide metering rod out of holder and remove from main metering jet.

2. Back out idle stop solenoid (if so equipped) or idle speed screw so throttle will close. Rotate fast idle cam so follower does not contact cam steps.

3. With throttle valve completely closed, apply pressure to power piston top and hold piston down against its stop.

4. While holding pressure as described in Step 3, swing metering rod holder over flat surface of bowl next to carburetor bore until metering

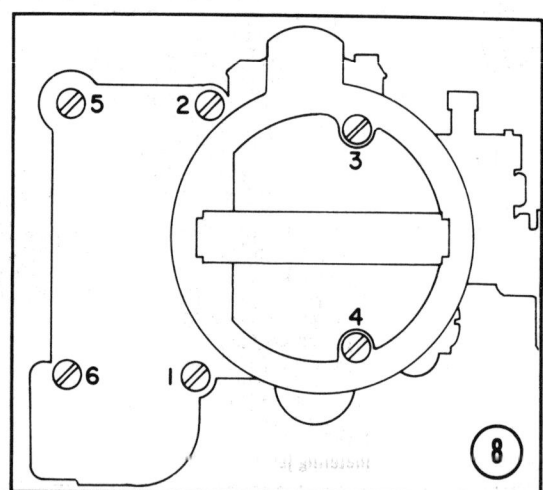

FUEL AND EXHAUST SYSTEMS

rod rests lightly against inside edge of bowl **(Figure 10)**.

5. Insert specified gauge (see **Table 1** for specifications) between bowl casting and metering rod holder lower surface. Gauge should have a slide fit.

6. Carefully bend metering rod up or down, as required, to adjust.

7. On 1975-1977 models, install metering rod and spring assembly. Install rod in jet, then in hanger.

8. Install air horn gasket on float bowl, carefully sliding slit portion over metering rod holder. Align gasket with dowels and press firmly into place.

Fast Idle Cam

Refer to **Figure 11** (1MV) or **Figure 12** (1ME).

1. Check fast idle speed, using procedure in Chapter Three, and adjust as necessary.

2. Set fast idle cam follower tang firmly on cam second stop **(Figure 11)**.

3. Rotate and hold choke valve toward closed position by using light force on choke coil rod.

4. Insert specified gauge **(Table 1)** between lower edge of choke and inside wall of air horns.

5. Bend cam-to-choke rod as required to obtain specified clearance.

6. Verify that cam-to-choke rod does not bind or meet interference.

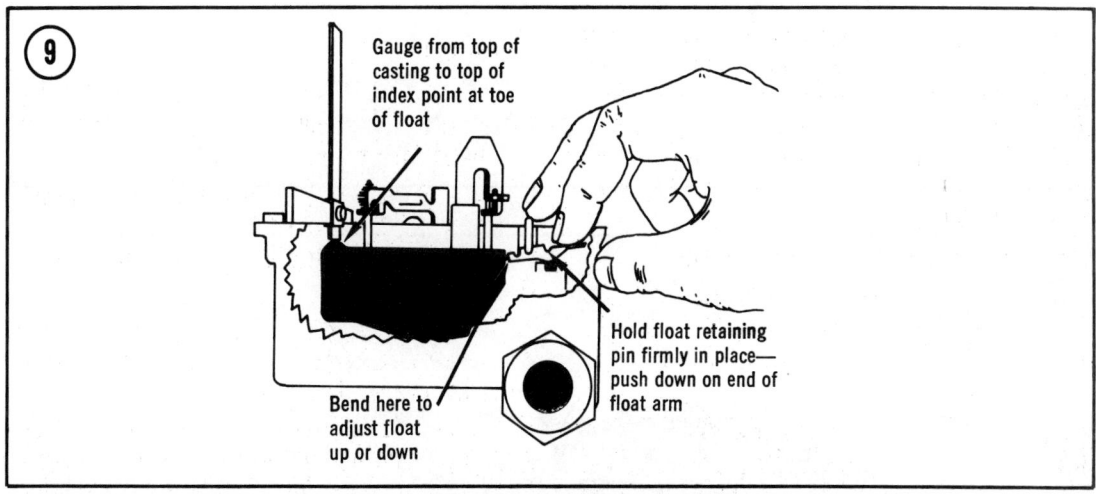

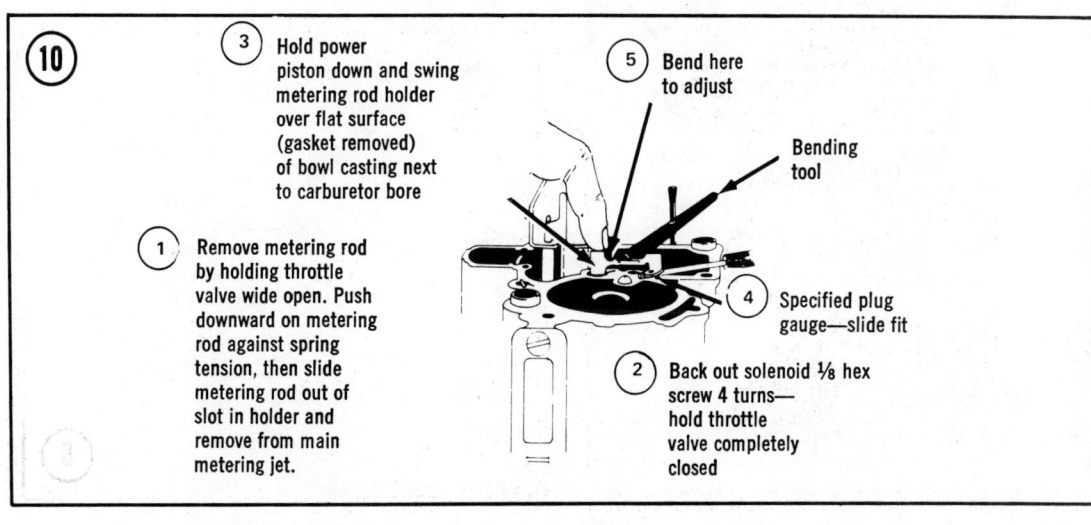

CHAPTER FIVE

11

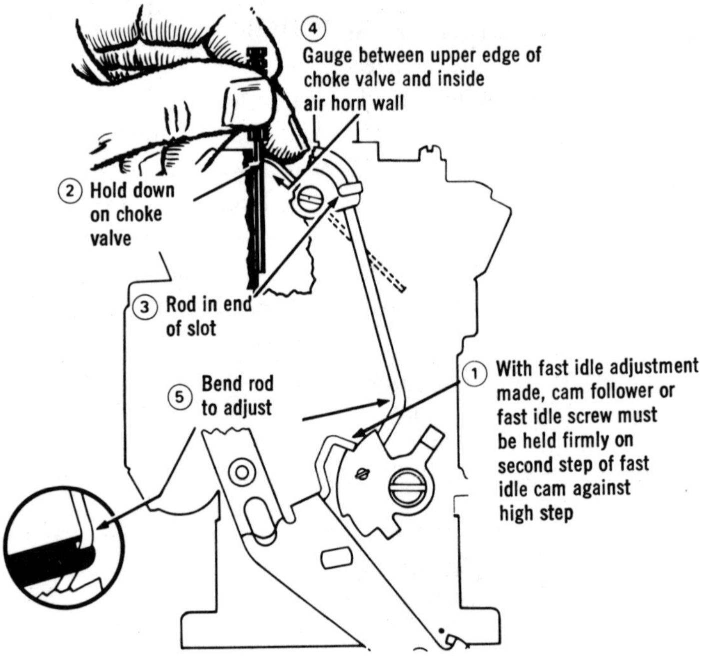

④ Gauge between upper edge of choke valve and inside air horn wall

② Hold down on choke valve

③ Rod in end of slot

⑤ Bend rod to adjust

① With fast idle adjustment made, cam follower or fast idle screw must be held firmly on second step of fast idle cam against high step

12

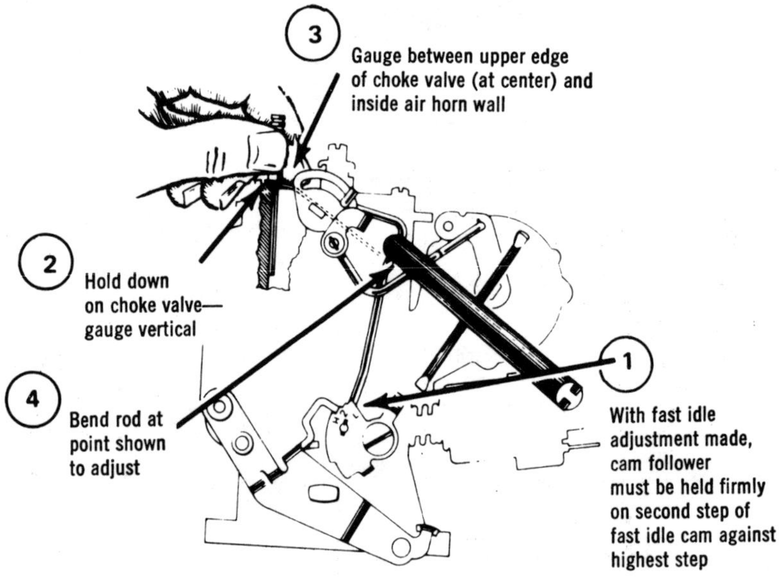

③ Gauge between upper edge of choke valve (at center) and inside air horn wall

② Hold down on choke valve—gauge vertical

④ Bend rod at point shown to adjust

① With fast idle adjustment made, cam follower must be held firmly on second step of fast idle cam against highest step

FUEL AND EXHAUST SYSTEMS

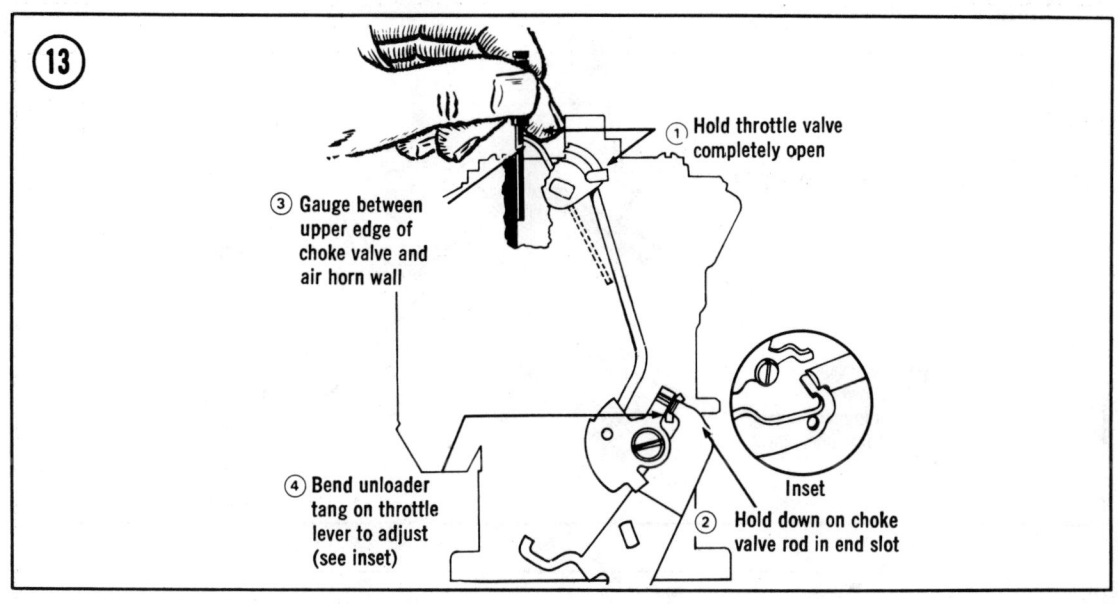

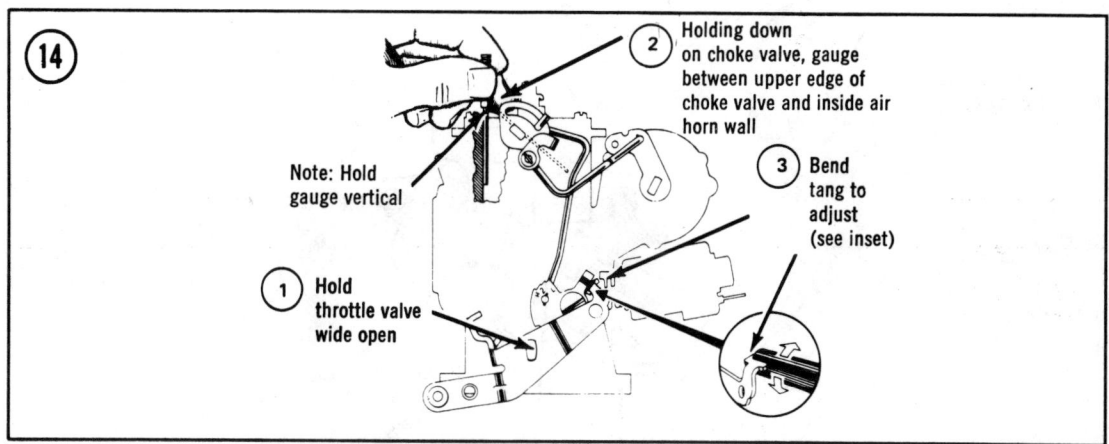

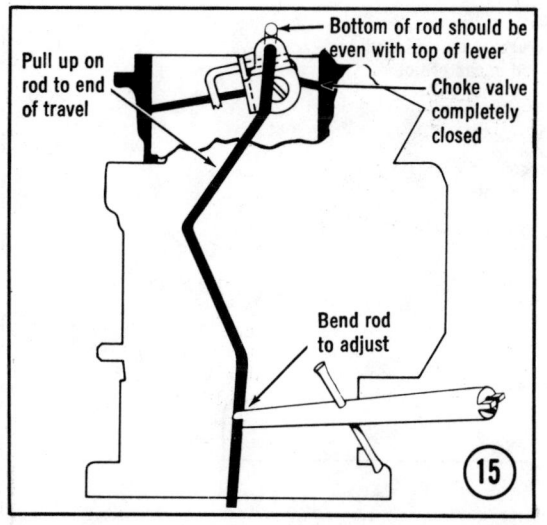

Choke Unloader

Refer to **Figure 13** (1MV) or **Figure 14** (1ME).

1. Hold down on choke valve by using light force on choke coil lever.

2. Rotate throttle valve to wide open position.

3. Insert gauge (**Table 1**) between upper edge of choke valve and inside wall of air horn.

4. If adjustment is required, bend tang on throttle lever.

Choke Coil Rod (1MV)

Refer to **Figure 15**.

1. Pull choke coil rod up to end of travel to completely close choke valve.

16

2 Hold choke valve completely closed
3 0.120 in. plug gauge must pass through hole in lever and enter hole in casting
1 Place cam follower on highest step of fast idle cam
4 Bend link to adjust

17

6 Bend rod to adjust
5 Place gauge between upper edge of choke valve and air horn wall
4 Push up on choke coil lever rod in end of slot
2 Use outside vacuum source to seat diaphragm
Note: Plug purge bleed hole with masking tape over end cover. Remove after adjustment.
3 Diaphragm plunger in until it seats
1 Place cam follower on highest step of fast idle cam

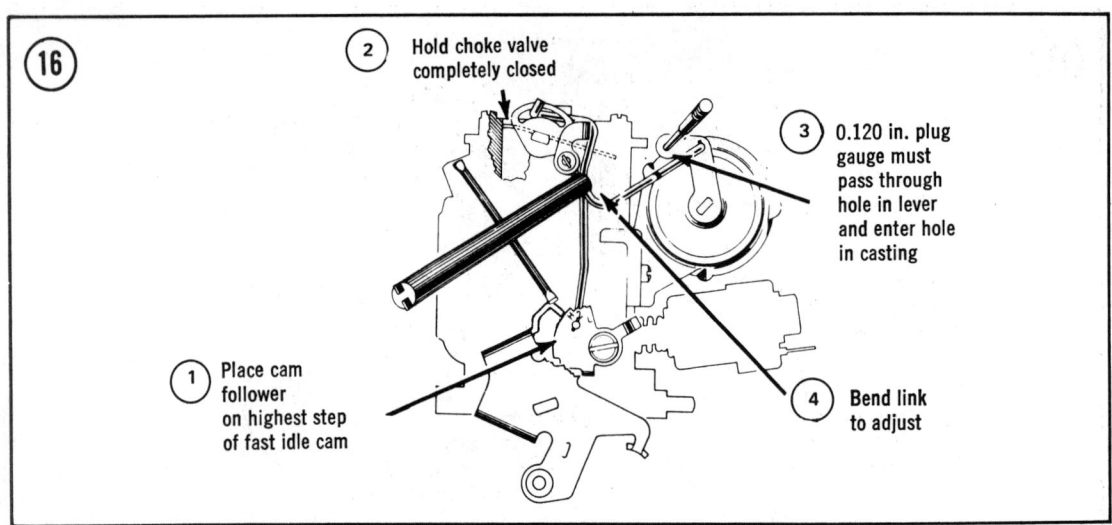

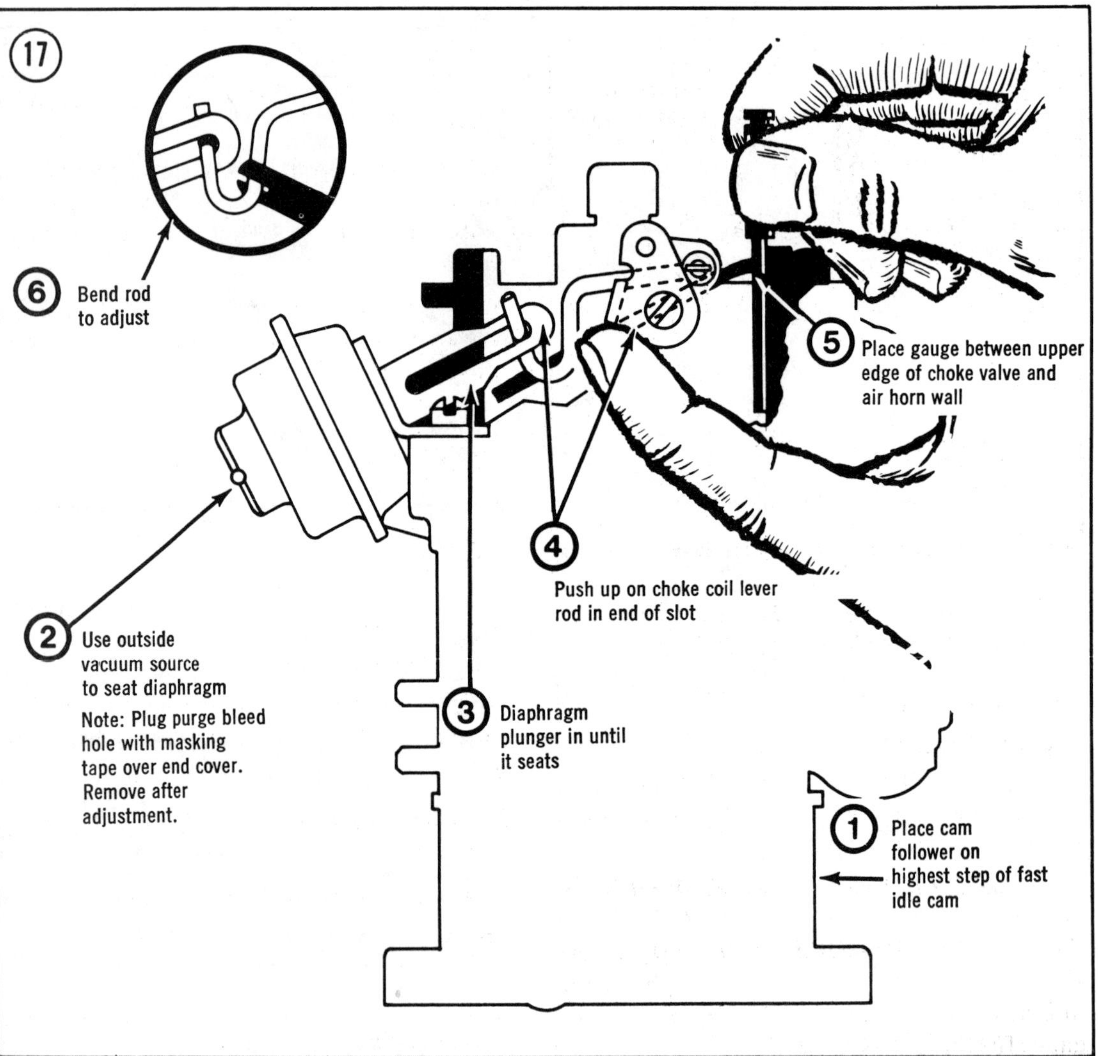

FUEL AND EXHAUST SYSTEMS

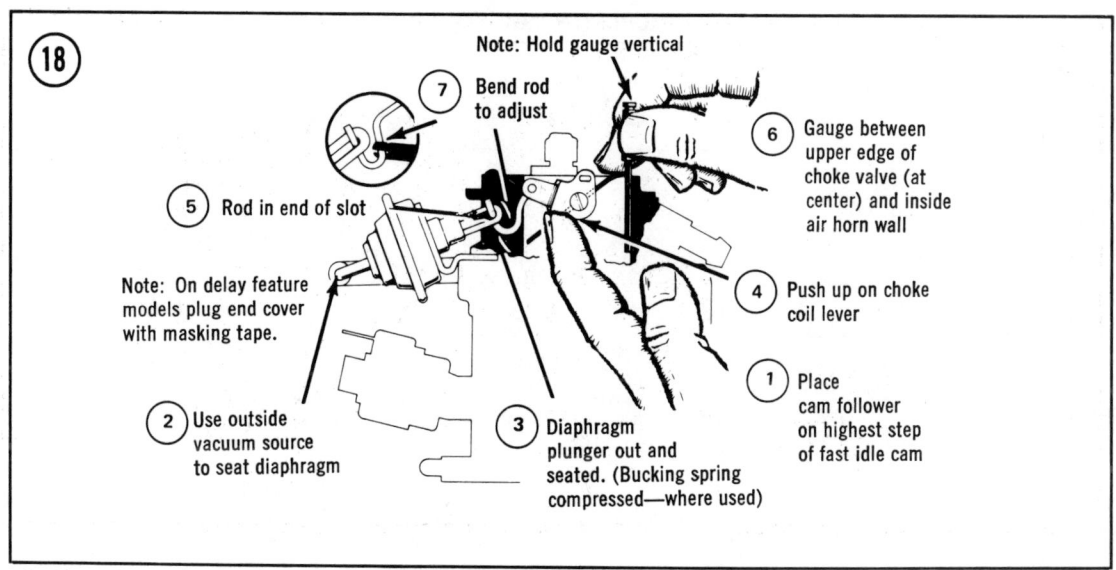

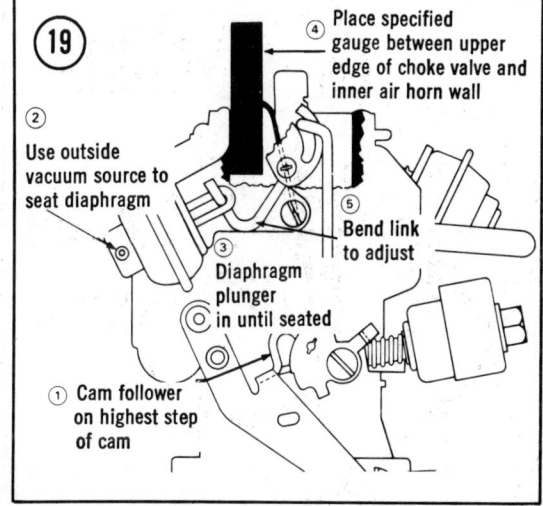

Primary Vacuum Break

Refer to **Figure 17** (1MV) or **Figure 18** (1ME).

1. Apply an outside vacuum source to primary vacuum break diaphragm until plunger is fully seated. Push up on choke coil lever rod in end of slot.

2. Insert specified gauge (**Table 1**) between lower edge (upper edge on 1ME) of choke valve and inside wall of air horn.

3. Make adjustment, if required, by bending vacuum break rod.

4. Verify that no binding or interference is present.

2. If bottom of rod is not even with top of lever, adjust as required by bending rod.

Auxiliary Vacuum Break

Refer to **Figure 19**.

1. Apply an outside vacuum source to auxiliary vacuum break diaphragm until plunger is fully seated.

Choke Coil Lever (1ME)

Refer to **Figure 16**.

1. Place the cam follower on the highest step of the fast idle cam, and hold the choke valve completely closed.

2. Insert a 0.120 in. plug gauge through the hole in the choke coil lever and attempt to insert it into the hole in the casting. If the gauge enters the hole, adjustment is OK. If not, bend the link at the point shown in **Figure 16** until the gauge fits through both holes.

2. Position cam follower on highest step of fast idle cam.

3. With vacuum break diaphragm fully seated, insert specified gauge (**Table 1**) between upper edge of choke valve and inner wall of air horn.

4. Bend link between vacuum break and choke valve to adjust, if required.

Table 2 ROCHESTER 2GV SPECIFICATIONS

Year	1970		1971	
Engine	307	350	307	350
Float level	27/32 in.	23/32 in.	13/16 in.	23/32 in. M 25/32 in. A
Float drop	1 3/4 in.	1 3/8 in.	1 3/4 in.	1 3/8 in.
Idle vent	0.020 in.	0.020 in.	N/A	N/A
Fast idle measurement	One turn	One turn	N/A	N/A
Fast idle rpm	2,200-2,400	2,200-2,400	N/A	N/A
Choke rod	0.060 in.	0.085 in.	0.075 in. M 0.040 in. A	0.100 in.
Vacuum break (man.)	0.125 in.	0.215 in.	0.110 in.	0.180 in.
Vacuum break (auto.)	0.100 in.	0.200 in.	0.080 in.	0.170 in.
Unloader	0.160 in. M 0.215 in. A	0.275 in. M 0.325 in. A	0.215 in.	0.324 in.
Main metering jet	N/A	0.059 in. M 0.060 in. A	N/A	N/A
Throttle bore	1 7/16 in.	1 11/16 in.	N/A	N/A
Accelerator pump	1 1/8 in.	1 17/32 in.	N/A	N/A
Thermostatic choke rod	1 rod dia.	1 rod dia.	N/A	N/A
Pump rod	N/A	N/A	1 3/64 in.	1 5/32 in.

Year	1972		1973	1974	1975
Engine	307	350	350	350	350
Float level	25/32 in.	23/32 in.	21/32 in.	19/32 in.	19/32 in.
Float drop	1 31/32 in.	1 9/32 in.	1 9/32 in.	1 9/32 in.	1 9/32 in.
Choke rod	0.075 in. M 0.040 in. A	0.100 in.	0.150 in.	0.200 in. M 0.245 in. A	0.200 in. M 0.245 in. A
Vacuum break (man.)	0.110 in.	0.180 in.	0.080 in.	0.130 in.	0.130 in.
Vacuum break (auto.)	0.080 in.	0.170 in.	0.080 in.	0.140 in.	0.140 in.
Unloader	0.215 in.	0.325 in.	0.215 in.	0.250 in. M 0.325 in. A	0.250 in. M 0.325 in. A
Pump rod	1 5/16 in.	1 1/2 in.	1 5/16 in.	1 7/16 in.	1 9/32 in. M 1 3/16 in. A

A = Automatic transmission.   M = Manual transmission.

FUEL AND EXHAUST SYSTEMS

Table 3 ROCHESTER 2GC SPECIFICATIONS

	1975 350 CID	1976 305 CID	1976 350 CID
Float level	21/32 in.	9/16 in.	9/16 in.
Float drop	31/32 in.	1 9/32 in.	1 9/32 in.
Pump rod	1 5/8 in.	1 21/32 in.	1 21/32 in.
Choke rod			
(fast idle cam)	0.400 in.	0.260 in.	0.260 in.
Vacuum break	0.130 in.	0.130 in.	0.140 in.
Unloader	0.350 in.	0.325 in.	0.325 in.
	1977-1978 305 CID	1977-1978 350 CID	1978 231 CID
Float level	19/32 in.	21/31 in.	7/16 in.
Float drop	1 9/32 in.	1 9/32 in.	1 5/32 in.
Pump rod	1 21/32 in.	1 21/32 in.	1 5/8 in.
Choke rod			
(fast idle cam)	0.260 in.	0.260 in.	0.080 in.
Vacuum break	0.130 in.	0.140 in.	0.150 in.
Unloader	0.325 in.	0.325 in.	0.140 in.

ROCHESTER 2GV-2GC-2GE

NOTE: *Carburetors on late models (1975 and later) are a part of the emissions control system. Service by the home mechanic should be limited to adjusting idle speed and idle mixture as described in Chapter Three. Other service should be referred to an expert.*

The 2GV, 2GC and 2GE carburetors are basically the same, except the automatic choke coil is located on the exhaust manifold in the 2GV series, and is attached to the carburetor in the 2GC and 2GE series. See **Tables 2 and 3** for specifications. The 2GE carburetor is used on some 1978 V6 engines.

Removal/Installation

1. Remove air cleaner.
2. Disconnect vacuum and fuel lines.
3. On 1974 and earlier models, remove choke coil rod. On 1975 and later models, remove fresh air and choke hoses from choke system.
4. Disconnect accelerator linkage.
5. If equipped with automatic transmission, disconnect downshift linkage.
6. Remove all vacuum and electrical connectors, including idle stop solenoid wiring (if so equipped).
7. Remove attaching nuts or bolts and remove carburetor.
8. Remove gasket or insulator.
9. To install carburetor, clean all gasket sealing surfaces and, using new gasket or insulator, reverse Steps 1 through 8.

Disassembly

Refer to **Figure 20** which shows the air horn.

1. Remove fuel filter nut, gasket, filter, and spring.
2. Remove spring clip and disconnect lower end of pump rod from throttle lever (**Figure 21**). Remove upper end of pump rod from pump lever by rotating rod out of hole in lever.
3. Remove hose connecting tube on throttle body to tube on vacuum break diaphragm unit. Remove attaching screws and remove vacuum break diaphragm assembly, detaching link assembly from choke shaft lever.
4. Remove retaining screw and then remove vacuum break lever from end of choke shaft.
5. On 2GV models, remove choke coil lever from choke vacuum diaphragm.
6. On 2GC-2GE models, remove intermediate choke rod from vacuum break lever and from lever on thermostatic coil housing.

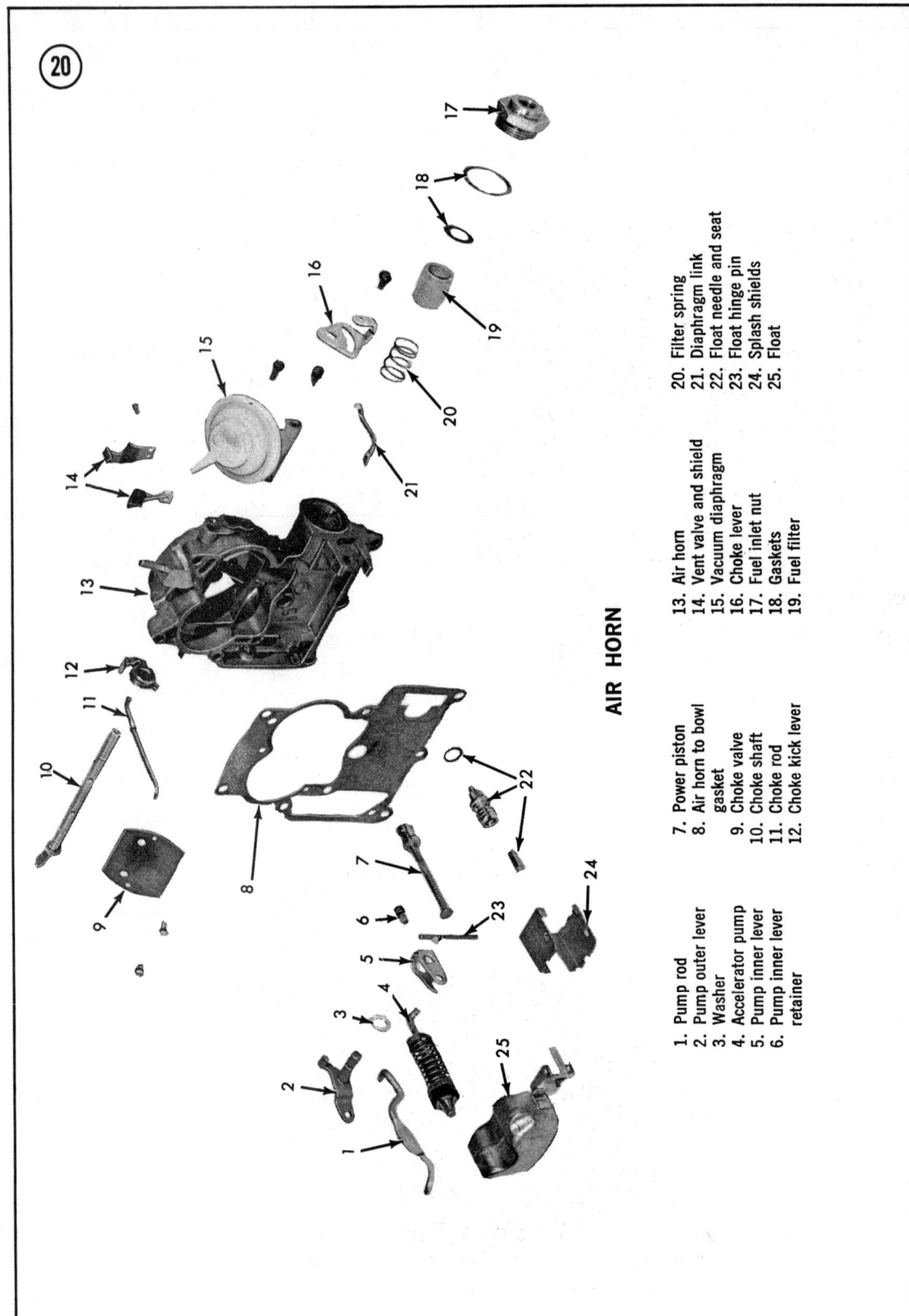

FUEL AND EXHAUST SYSTEMS

7. Remove fast idle cam. Remove cam from rod by rotating cam.

8. On 2GV models, hold choke open and rotate choke rod upward and remove from lever on choke shaft. On 2GC-2GE models, rod cannot be removed until air horn is removed from bowl.

9. Remove attaching screws and lift air horn from bowl.

10. Place air horn on flat surface and remove float hinge pin, float, gasket, splash shield (if used), and float needle.

11. Remove float needle seat and gasket (use a wide blade screwdriver).

12. Depress stem of power piston and allow piston to snap free (repeat several times, if required). Remove piston, using care to avoid bending stem.

13. Remove setscrew and remove pump plunger assembly and inner pump lever from pump shaft. See **Figure 22**.

CAUTION
Do not remove pump plunger stem from inner pump lever unless the plunger requires replacement. On some models, the stem must be broken before it can be removed and on others it is rotated out. Bending of the inner lever could cause binding which could hold the throttle open.

NOTE: *If plunger is to be reused, place it in a container of kerosene or gasoline until it is required for reassembly. Do not clean plunger or other rubber or plastic parts in carburetor solvent.*

Spring clip

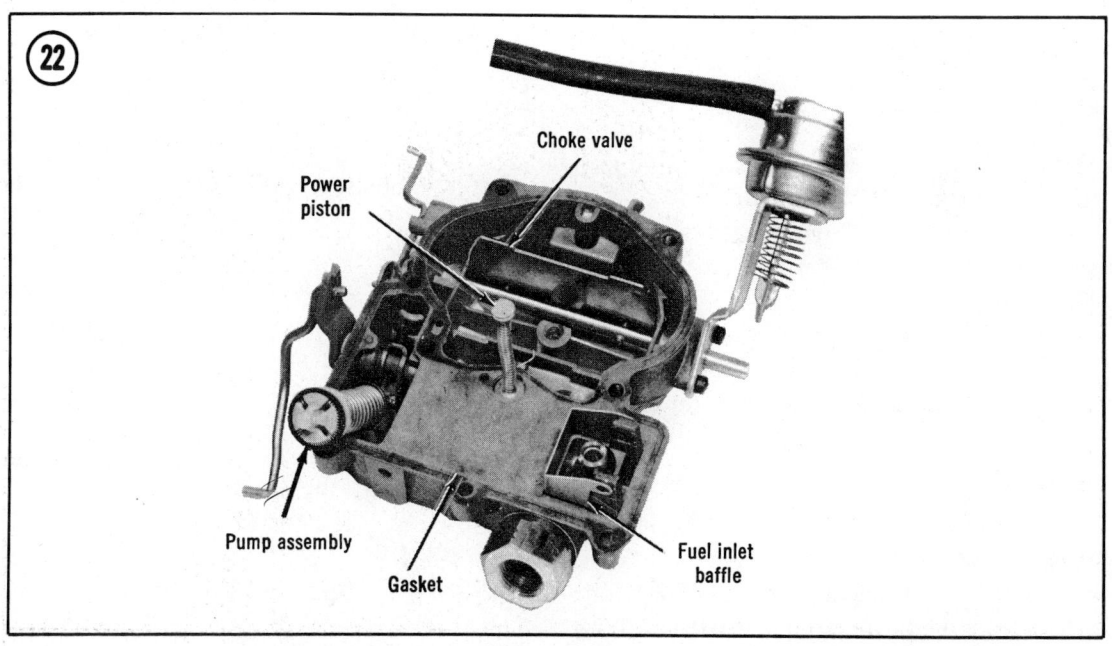

Power piston

Choke valve

Pump assembly

Gasket

Fuel inlet baffle

14. Remove gasket from air horn and remove fuel inlet baffle, if so equipped.

15. Remove choke valve from shaft (2 screws) and remove shaft from air horn.

> NOTE: *If screws are staked, file off staked ends.*

16. On 2GC-2GE models, remove fast idle cam rod and lever from choke shaft.

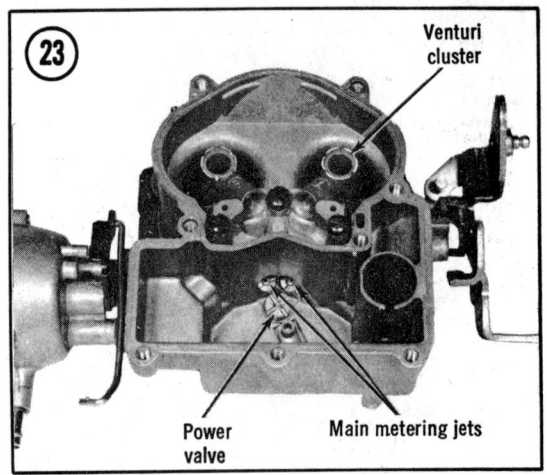

Float Bowl

1. Remove pump plunger return spring and invert bowl to remove aluminum check ball. See **Figure 23**.

2. Remove main meterinng jets, power valve, and gasket.

3. Remove venturi cluster and gasket (3 screws). Remove plastic main well inserts. See **Figure 24**.

4. Remove pump discharge spring retainer with long-nosed pliers (**Figure 25**). Remove spring and check ball from passage.

5. Invert the carburetor and remove bowl-to-throttle body retaining screws and separate the 2 units.

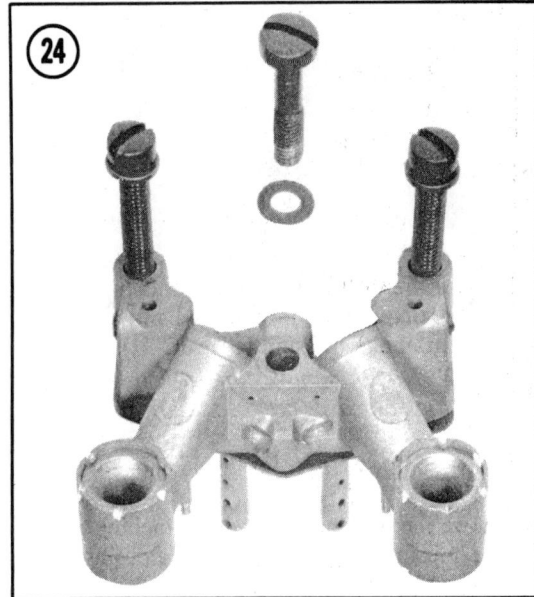

Throttle Body

> NOTE: *On GV models, further disassembly of throttle body is not recommended, unless idle mixture screws require replacement or cleaning.*

1. Remove idle mixture screws. If equipped with limiter caps, cut off cap tang.

2. On GC-GE models, remove choke cover (3 screws) and remove thermostatic coil. See **Figure 26**.

> CAUTION
> *Do not remove cup baffle beneath cover as coil distortion could result.*

3. Remove baffle plate from choke housing (**Figure 26**).

4. Remove choke housing and gasket (2 screws) from throttle body.

5. Remove intermediate choke lever (one screw) from intermediate choke shaft. Remove inner lever and shaft, and rubber dust seal from choke housing.

> CAUTION
> *Do not attempt to remove throttle valves, as relation to idle discharge orifices may be destroyed.*

Cleaning and Inspection

1. Wash all metal parts in carburetor solvent and blow dry, taking care to blow out all passages.

> CAUTION
> *Do not use drills or wire to clean jets and passages.*

FUEL AND EXHAUST SYSTEMS

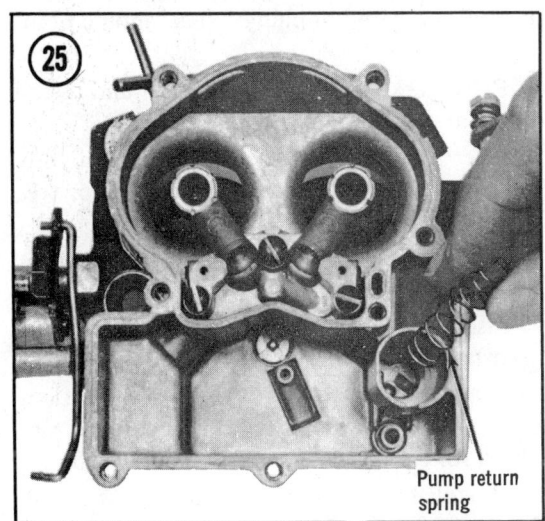

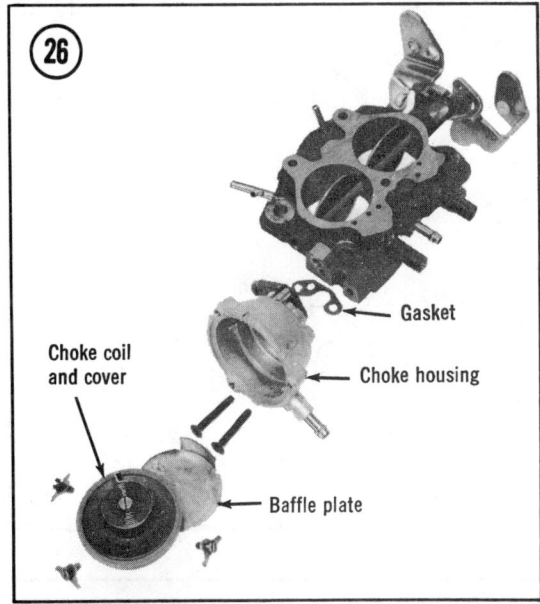

3. On 2GC-2GE models, proceed as follows:

   a. Install new rubber dust seal in choke housing cavity. Seal lips should face toward carburetor body.

   b. Install inner lever and shaft in choke housing.

   c. Install intermediate lever on shaft and securely tighten retaining screw.

   d. Install new gasket on choke housing and position housing on throttle body. Install and securely tighten retaining screws.

   e. Using the procedure given below, adjust intermediate choke rod so that with choke valve closed, lever in choke housing is aligned with gauge.

   f. Using new gaskets, install choke coil and cover with end of coil below plastic tank on inner choke housing lever. Index cover, using procedure below, and install and securely tighten cover attaching screws. See **Figure 26**.

4. Using new gasket, install float bowl to throttle body assembly (3 screws) and securely tighten. Make certain holes in gasket are aligned with holes in casting.

Float Bowl

1. Install main metering jets. See **Figure 23**.

2. Install power valve and gasket in float bowl and securely tighten with slotted screwdriver.

3. Place aluminum inlet check ball in pump well hole and install pump return spring. Press with finger to center spring.

4. Drop steel ball into pump discharge hole and install spring and retainer.

2. Inspect all parts for wear and damage and replace as necessary.

Assembly

   NOTE: *Use new gaskets in reassembly.*

Throttle Body

1. Install idle speed screw and spring, if removed.

2. Install idle mixture needles and springs, if removed. Seat needles lightly and back out 4 turns.

CAUTION
Do not interchange aluminum and steel balls during assembly. Steel ball is larger and heavier of the two.

5. Install plastic inserts in main fuel wells and make sure they are properly seated in recesses.

6. Install venturi cluster and gasket. Fit center screw with gasket to avoid pump discharge leakage, and securely tighten all screws.

Air Horn

See **Figure 22**.

1. On GC models, install upper choke rod lever and collar assembly on choke shaft. On GV models, install choke kick lever on choke shaft with tang on kick lever toward lever on shaft.

2. Install choke shaft in air horn. On GC-GE models, install shaft from throttle lever side of carburetor. On GV models, install shaft from opposite side.

3. Center choke valve on shaft with lettered or numbered side up and install with 2 screws. Securely tighten screws and lightly stake in place (or use Loctite, or equivalent, on threads). Check valve for free movement.

4. If removed, install outer pump shaft and lever assembly, making sure plastic washer is in place.

5. Install accelerator pump plunger to inner lever (if removed) and install retainer (provided in repair kit). End of plunger shaft should point inward.

6. Place gasket on float needle seat and install seat in air horn. Tighten securely.

7. Insert power piston into air horn and lightly stake retainer to casting. Check piston for free movement.

8. Install fuel inlet baffle or splash shield, as appropriate.

9. Install new air horn gasket, indexing gasket to casting dowel.

10. Install float needle, then install float assembly on air horn and insert hinge pin. Check for free movement of float and needle.

11. Install fuel filter, spring, gasket, and inlet fitting and securely tighten.

12. Install vacuum break diaphragm onto air horn. Part number on the diaphragm bracket should face outward. Position rod in diaphragm plunger and attach to choke coil lever.

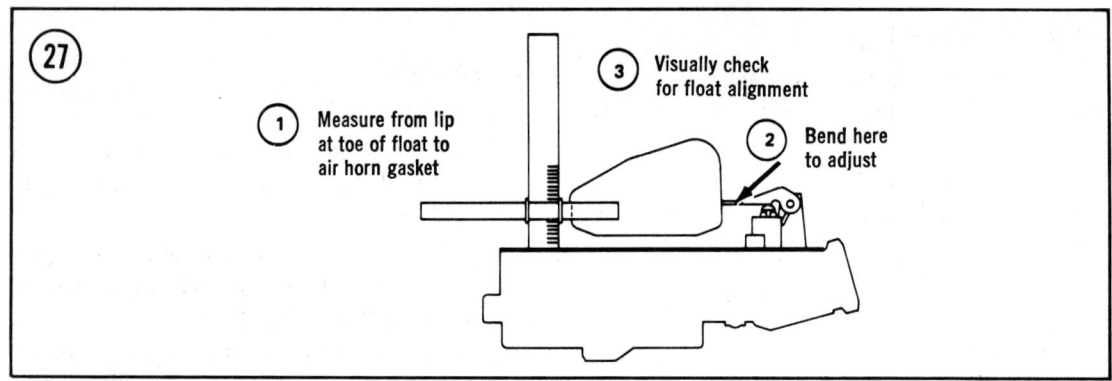

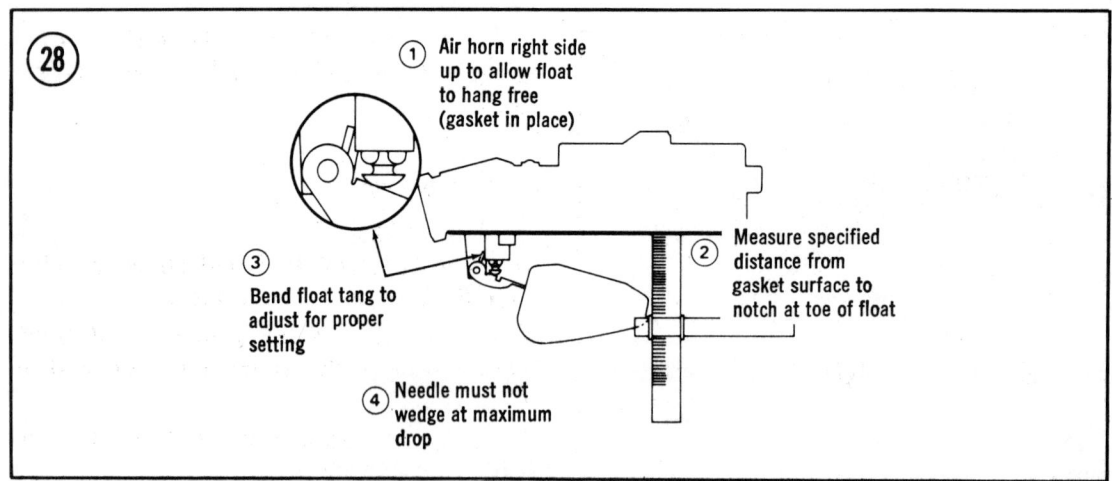

FUEL AND EXHAUST SYSTEMS

Then install choke coil lever on end of choke shaft, with tang facing outward. Install and securely tighten retaining screw.

13. On GC-GE models, install choke rod in upper choke lever, rotating rod to align square end of rod with slot in lever.

Float Level

1. Invert air horn and verify that gasket is in place and needle valve is seated. Measure distance from lip of float toe to air horn gasket. See **Figure 27**.

2. Adjust float to specifications (**Tables 2 and 3**) by bending float arm at point shown.

Float Drop

1. Hold air horn upright so float hangs free and measure distance from lip on float toe to air horn gasket. See **Figure 28**.

2. Adjust to specifications (**Tables 2 and 3**) by bending tang at rear of float hanger.

Air Horn-to-Float Bowl

1. Position air horn on bowl, making sure accelerator pump plunger is inserted in pump well and moves freely.

2. Install lockwashers and screws and securely and evenly tighten, using the pattern shown in **Figure 29**.

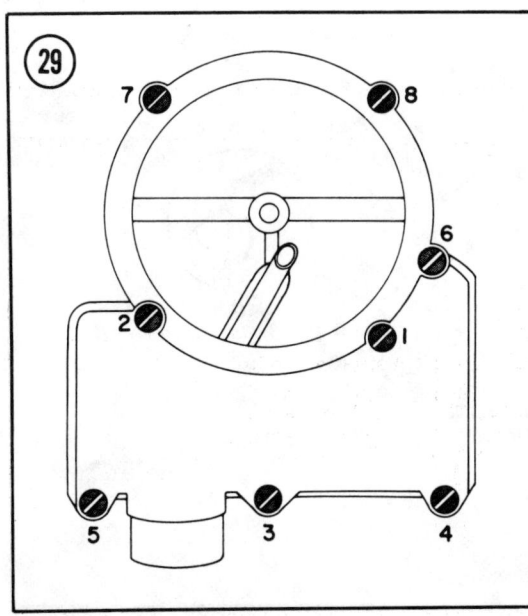

3. Install fast idle cam to lower end of choke rod (2GC-2GE); or fast idle cam rod to choke kick lever and fast idle cam (2GV). Fasten cam to bowl (part number or identification facing outward) with attaching screw and tighten securely. Check for free movement.

4. Install pump rod in upper pump lever and secure rod to throttle lever with spring clip. See **Figure 21**.

5. Install vacuum hose between vacuum break diaphragm and fitting on throttle body.

Pump Rod

Refer to **Figure 30**.

1. Back out idle speed screw and hold throttle valve completely closed.

2. Measure from top of air horn ring to top of pump rod, as shown in **Figure 30**. See **Tables 2 and 3** for specifications.

3. If adjustment is required, bend rod as shown in **Figure 30**. Reset idle using the procedure given above.

Fast Idle Cam

Refer to **Figure 31**.

1. Turn idle speed screw in until it contacts low step of fast idle cam, and then turn screw in one complete turn.

2. Place idle speed screw on step 2 of fast idle cam, with screw resting against high step as shown in inset No. 1, **Figure 31**.

3. Place gauge between upper edge of choke valve and inside air horn wall. See **Tables 2 and 3** for specifications.

4. If required, adjust by bending choke lever tang as shown in inset No. 2 or No. 3, **Figure 31**.

Choke Unloader

Refer to **Figure 32**.

1. Place choke valve toward closed position while holding throttle valve wide open.

2. Place specified gauge (**Tables 2 and 3**) between the upper edge of the choke valve and air horn wall.

3. If clearance adjustment is required, bend tang on throttle lever.

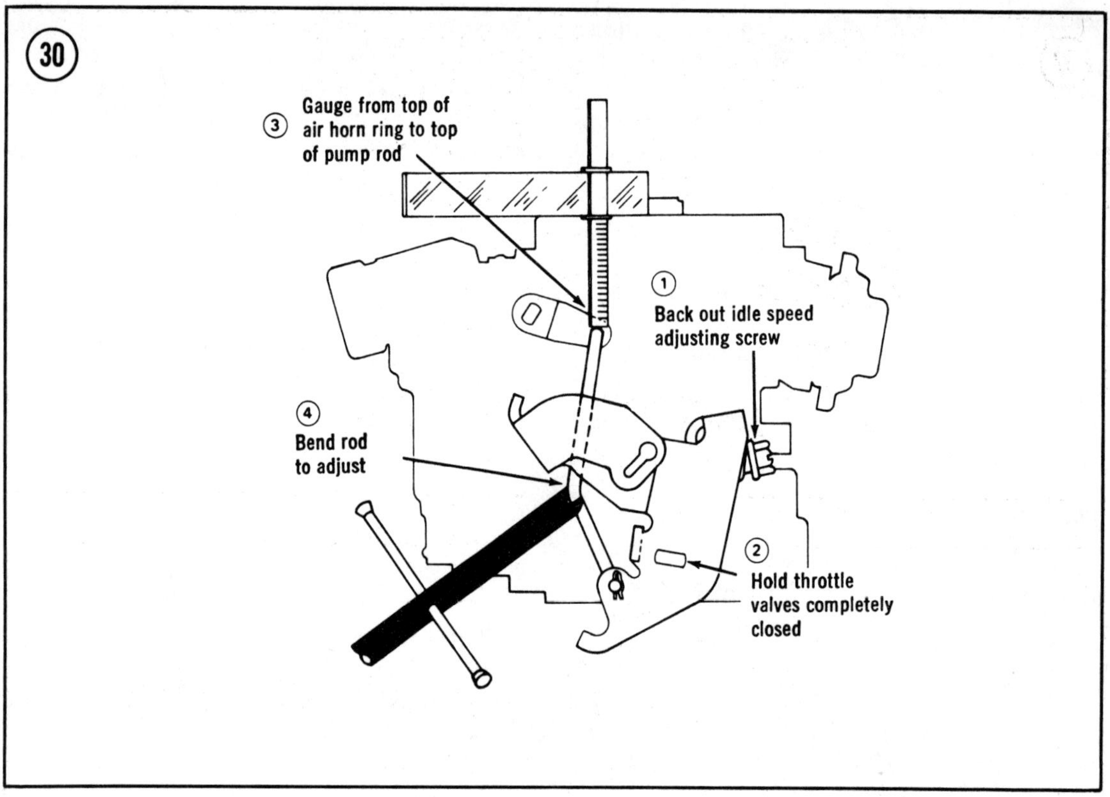

③ **30**

③ Gauge from top of air horn ring to top of pump rod

① Back out idle speed adjusting screw

④ Bend rod to adjust

② Hold throttle valves completely closed

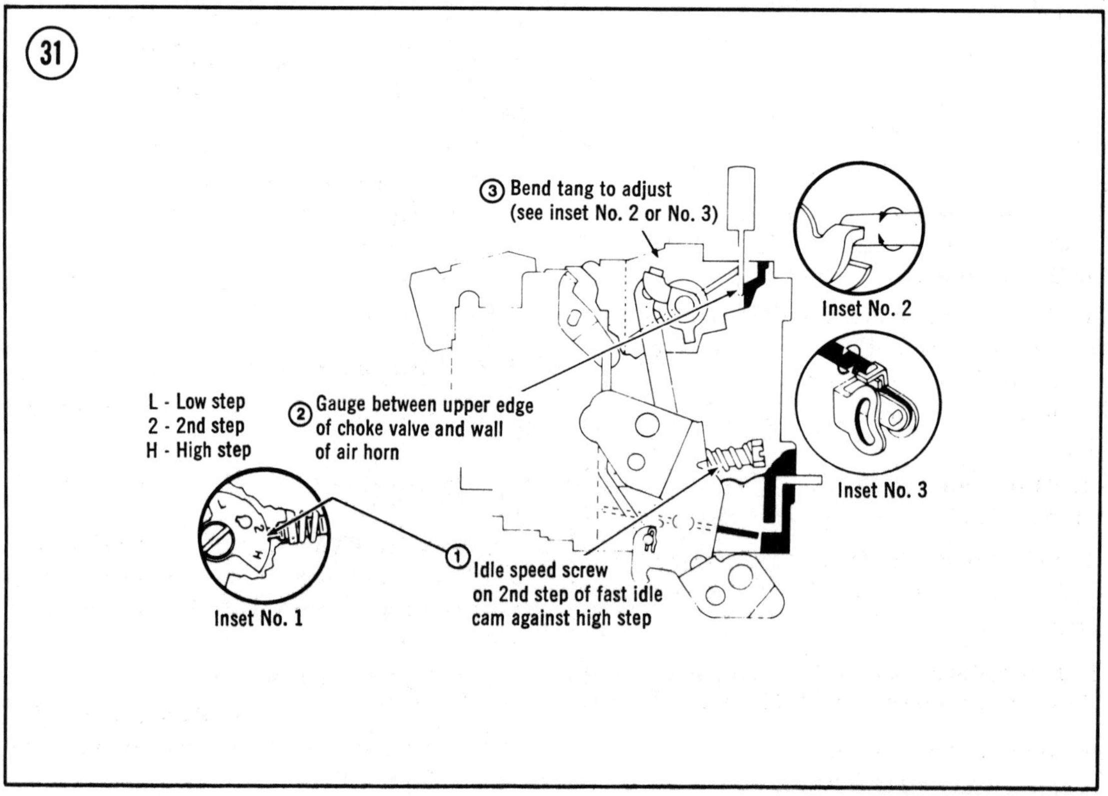

③ **31**

③ Bend tang to adjust (see inset No. 2 or No. 3)

Inset No. 2

L - Low step
2 - 2nd step
H - High step

② Gauge between upper edge of choke valve and wall of air horn

Inset No. 3

Inset No. 1

① Idle speed screw on 2nd step of fast idle cam against high step

FUEL AND EXHAUST SYSTEMS

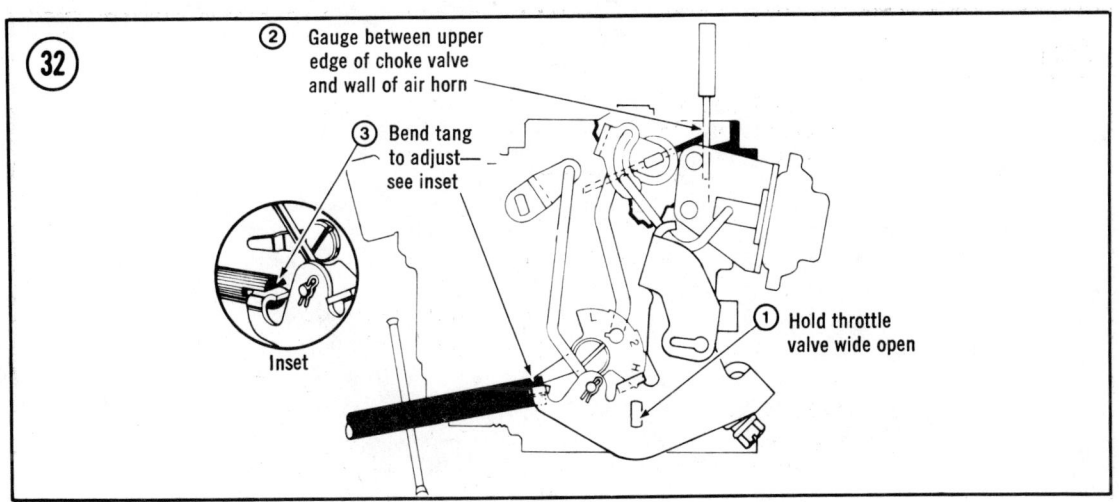

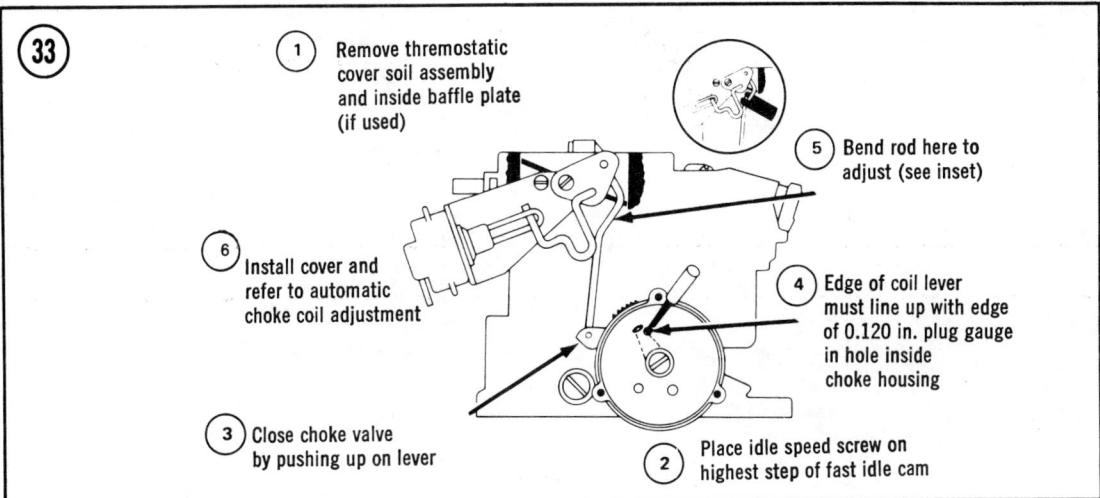

Intermediate Choke Rod

1. Remove thermostatic cover, coil assembly, gasket, and inside baffle plate (**Figure 33**) by removing 3 attaching screws.

2. Position idle speed screw on high step of fast idle cam.

3. Close choke by pushing up on intermediate choke lever.

4. Verify that edge of coil lever inside choke housing lines up with edge of 0.120 in. plug gauge.

5. If adjustment is required, bend intermediate choke rod as shown in inset, **Figure 33**.

6. Replace thermostatic cover, coil assembly, gasket, and inside baffle plate and secure with 3 attaching screws. Adjust coil using procedure given below.

Automatic Choke Coil

1. Place idle speed screw on high step of fast idle cam.

2. Loosen thermostatic choke coil cover retaining screws.

3. Rotate choke cover against coil tension until choke valve begins to close. Continue rotating until index mark lines up with index point on choke housing.

4. Tighten 3 retaining screws.

5. Place specified gauge (**Tables 2 and 3**) between upper edge of choke valve and inner air horn wall.

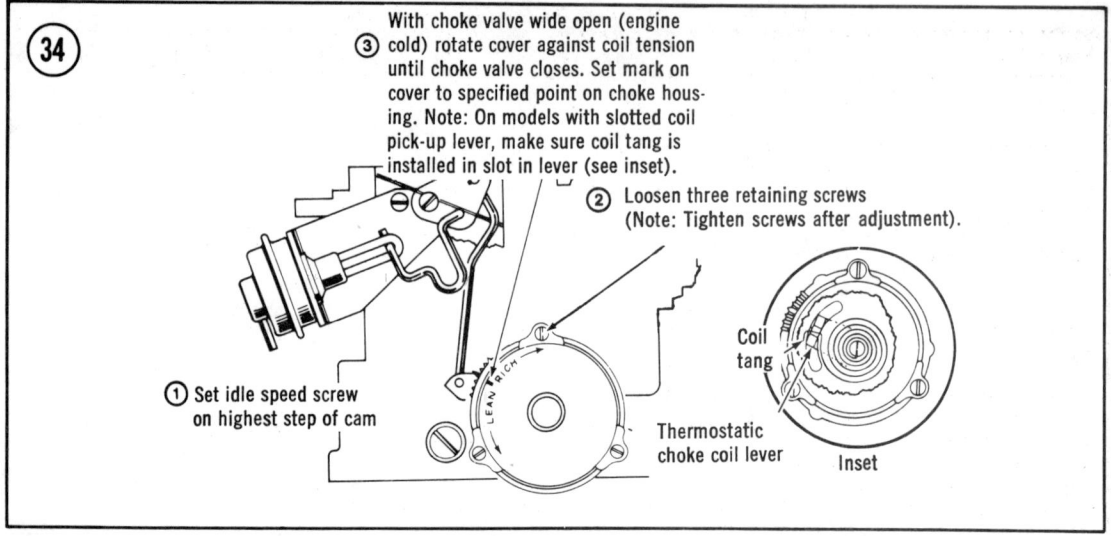

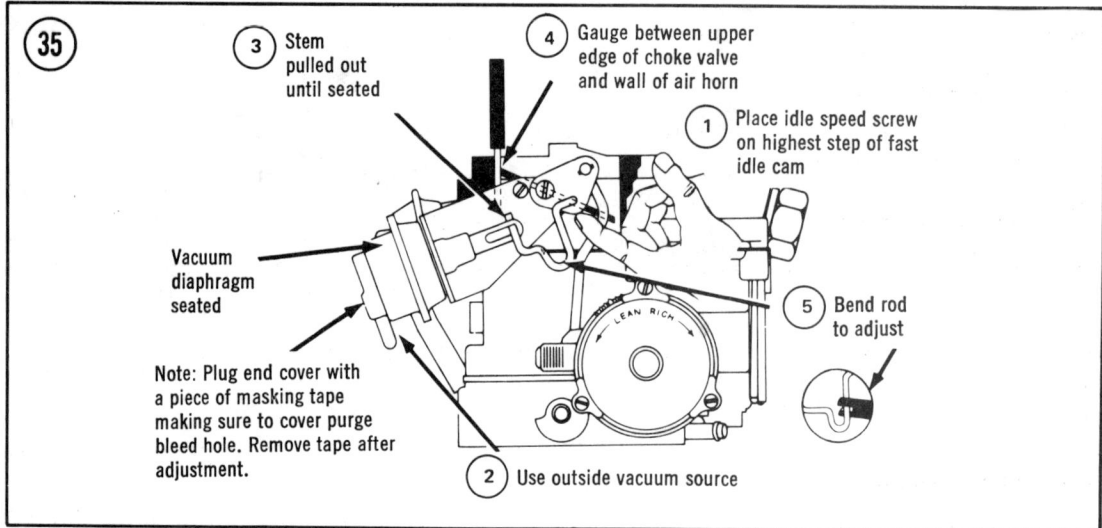

6. Bend vacuum break rod if adjustment is required. See **Figure 34**. Remove tape from bleed hole and contact vacuum hose.

Vacuum Break

Refer to **Figure 35**.

1. Use an outside vacuum source to seat vacuum break diaphragm.

2. Use a piece of tape to cover vacuum break bleed hole so diaphragm will not bleed down.

3. Locate idle speed screw on fast idle cam high step.

4. Hold choke coil lever (inside choke housing) toward closed choke position.

5. Gauge between upper edge of choke valve and air horn wall. See **Tables 2 and 3** for specifications.

6. Bend vacuum break rod to adjust. Remove tape and reconnect vacuum hose.

ROCHESTER M2M

The Rochester Model 210 (M2M) carburetor is a single-stage, two-barrel, downdraft carburetor that is essentially the primary side of the M4MC Quadrajet carburetor. This carburetor, used on some 1979-1980 305 cid engines, is a part of the emission control system and service should be limited to the procedures provid-

FUEL AND EXHAUST SYSTEMS

Table 4 ROCHESTER 4MV SPECIFICATIONS

Year	1970		1971		1972		1973	1974
Engine	350	402	350	402	350	402	350/454	350/400/454
Float level	1/4 in.	1/4 in.	1/4 in.	1/4 in.	1/4 in.	1/4 in.	7/32 in.[1]	1/4 in.[2]
Accelerator pump	5/16 in.	5/16 in.	—	—	3/8 in.	3/8 in.	13/32 in.	13/32 in.
Fast idle measurement	2 turns	2 turns	Second step of fast idle cam	Second step of fast idle cam	Second step of fast idle cam	Second step of fast idle cam	High step of fast idle cam	High step of fast idle cam
Fast idle rpm	2400 rpm	2400 rpm	2200 rpm	2200 rpm	1350M[3] 1500A[3]	1350M[3] 1500A[3]	1300M[3] 1600A[3]	1300M[3] 1600A[3]
Choke rod	0.100 in.	0.100 in.	0.100 in.	0.100 in.	0.100 in.	0.100 in.	0.430 in.	0.430 in.
Vacuum break (manual transmission)	0.275 in.	0.275 in.	0.260 in.	0.275 in.	0.215	0.250 in.	0.250 in.	0.230 in.
Vacuum break (automatic transmission)	0.245 in.	0.245 in.	0.260 in.	0.275 in.	0.215 in.	0.250 in.	0.250 in.	0.230 in.
Unloader	0.450 in.	0.450 in.	—	—	0.450 in.	0.450 in.	0.450 in.	0.450 in.
Main metering jet	0.076 in.	0.078 in.	—	—	—	—	—	—
Primary throttle bore	1 3/8 in.	1 3/8 in.	—	—	—	—	—	—
Secondary throttle bore	2 1/4 in.	2 1/4 in.	—	—	—	—	—	—
Air valve dashpot	0.020 in.	0.020 in.	0.020 in.	0.020 in.	0.020 in.	0.020 in.	—	—

1. 1/4 in. on 454   2. 3/8 in. on 454   3. Without vacuum advance

ed in Chapter Three. Other repairs and adjustments should be referred to your Chevrolet dealer or a reputable repair shop.

Removal/Installation

NOTE: *When removing the carburetor, take care not to spill the fuel in the fuel bowl. Contents of the bowl then can be examined to provide possible clues as to carburetor trouble.*

1. Remove the air cleaner and gasket and then remove the electrical wire from the solenoid, if so equipped.
2. Identify and mark all fuel and vacuum hoses so they may be returned to their original positions and then remove the hoses from the carburetor.
3. Identify and mark all electrical connections so they may be reconnected in their original positions and then disconnect the connectors.
4. Disconnect the accelerator linkage.
5. On models with automatic transmissions, disconnect the downshift cable at the carburetor.
6. Disconnect the cruise control linkage, if so equipped.
7. Remove the attaching bolts and then remove the carburetor and insulator.
8. Installation is the reverse of these steps. Tighten attaching bolts to 145 in.-lb. Check and adjust idle speed as required, using the procedures given in Chapter Three.

ROCHESTER 4MV

Removal/Installation

This is a 4-barrel, 2-stage carburetor which is easily adapted for small to large engines without design changes. See **Table 4** for specifications.

The fuel bowl is centrally located to avoid fuel slosh. The float needle valve is pressure balanced to permit use of a small single float.

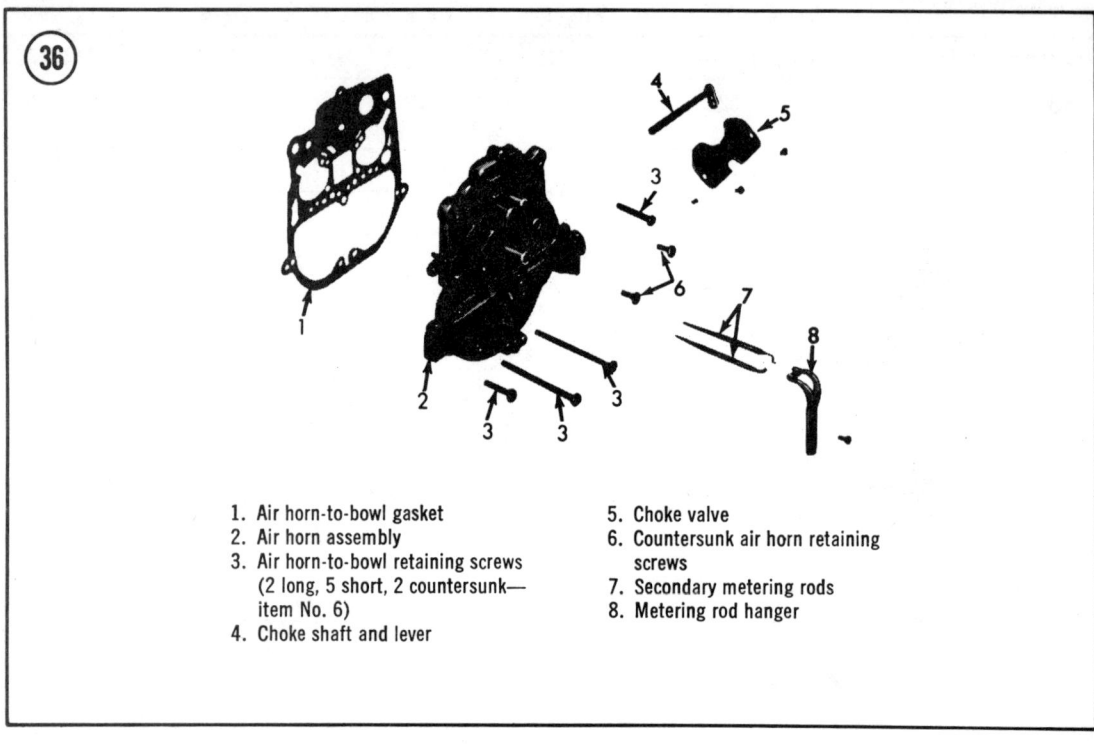

1. Air horn-to-bowl gasket
2. Air horn assembly
3. Air horn-to-bowl retaining screws (2 long, 5 short, 2 countersunk—item No. 6)
4. Choke shaft and lever
5. Choke valve
6. Countersunk air horn retaining screws
7. Secondary metering rods
8. Metering rod hanger

The primary side has small bores and a triple venturi for fine fuel control in the idle and economy ranges. The secondary side has large bores and an air valve for high air capacity.

1. Remove air cleaner, gasket, and stud.
2. Disconnect vacuum, fuel, and choke pipes at carburetor.
3. On automatic models, disconnect transmission control rod from throttle lever.
4. On models with positive crankcase ventilation (PCV), disconnect ventilation hose at valve on carburetor base.
5. Disconnect accelerator rod and throttle return spring at carburetor.
6. Remove mounting nuts and lift carburetor from manifold.
7. Reverse procedures to install. Be sure carburetor base and manifold flange are clean.

Disassembly

Air Horn

Refer to **Figure 36**.

1. Remove idle vent valve.
2. Disconnect choke rod from upper choke shaft lever and remove choke rod.
3. Disconnect and remove pump rod from the pump lever.
4. Remove 7 screws around air horn and lift straight up. Leave air horn gasket on bowl. Do not damage or remove 2 small main well air bleed tubes protruding from air horn.
5. Hold air valve wide open. Tilt secondary metering rods, then slide from holes in hanger.

Float Bowl and Throttle Body

Refer to **Figures 37 and 38**.

1. Remove pump plunger from pump well.
2. Lift air horn gasket from dowels on secondary side of bowl, then remove gasket from around power piston and primary metering rods.
3. Remove pump return spring.
4. Remove plastic filler over float valve.
5. Remove power piston and primary metering rods. See **Figure 39**.
6. Disconnect tension spring at top of each primary metering rod. Rotate each rod and remove from hanger on power piston.
7. Remove float assembly.
8. Remove pull clip and fuel inlet needle.

FUEL AND EXHAUST SYSTEMS

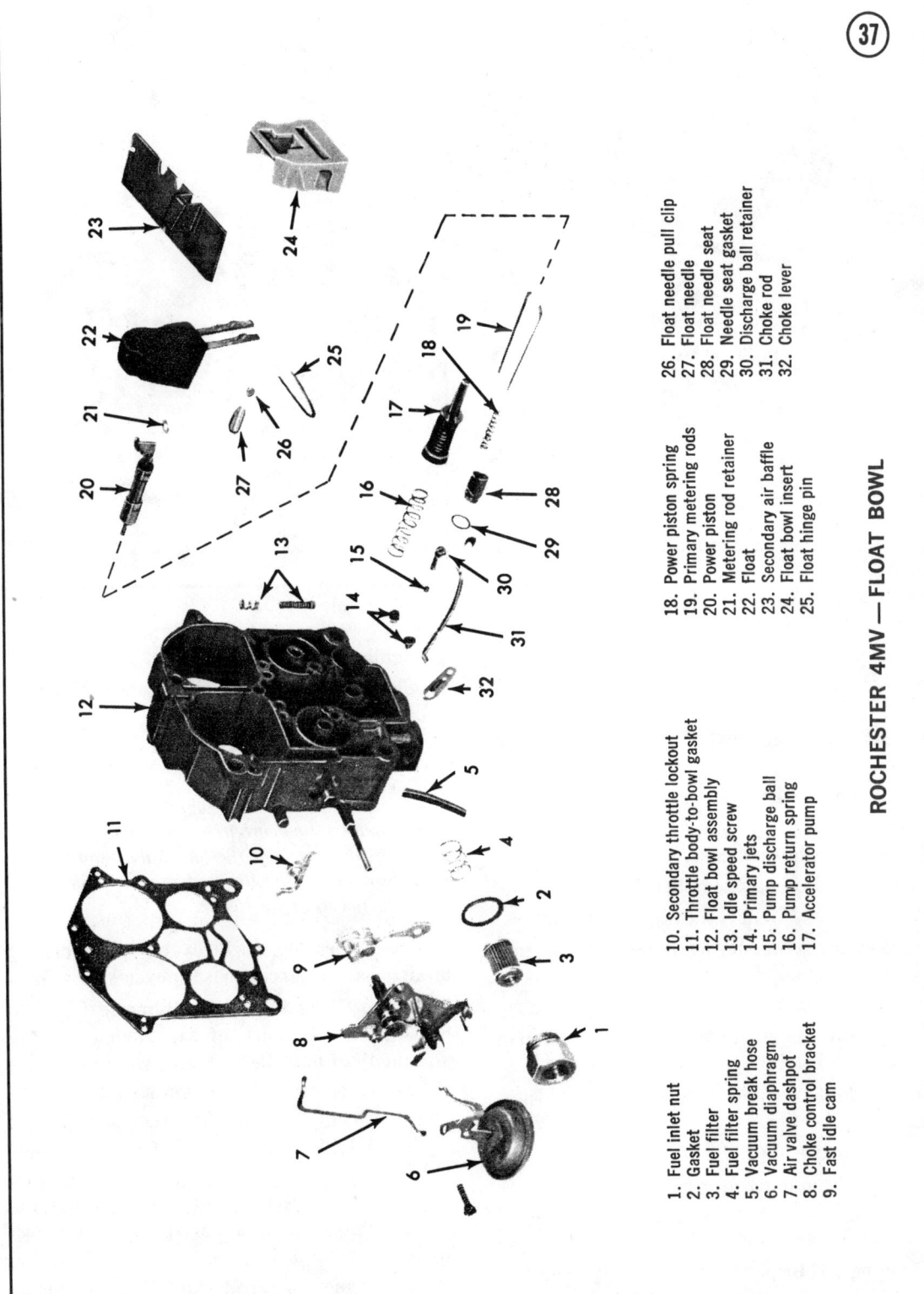

ROCHESTER 4MV — FLOAT BOWL

1. Fuel inlet nut
2. Gasket
3. Fuel filter
4. Fuel filter spring
5. Vacuum break hose
6. Vacuum diaphragm
7. Air valve dashpot
8. Choke control bracket
9. Fast idle cam
10. Secondary throttle lockout
11. Throttle body-to-bowl gasket
12. Float bowl assembly
13. Idle speed screw
14. Primary jets
15. Pump discharge ball
16. Pump return spring
17. Accelerator pump
18. Power piston spring
19. Primary metering rods
20. Power piston
21. Metering rod retainer
22. Float
23. Secondary air baffle
24. Float bowl insert
25. Float hinge pin
26. Float needle pull clip
27. Float needle
28. Float needle seat
29. Needle seat gasket
30. Discharge ball retainer
31. Choke rod
32. Choke lever

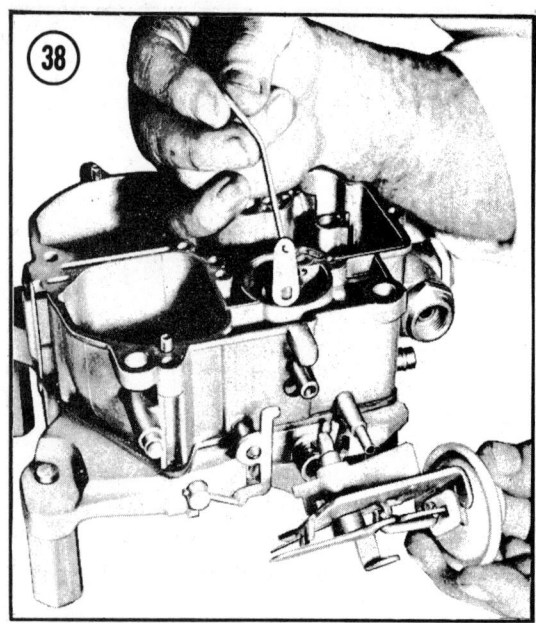

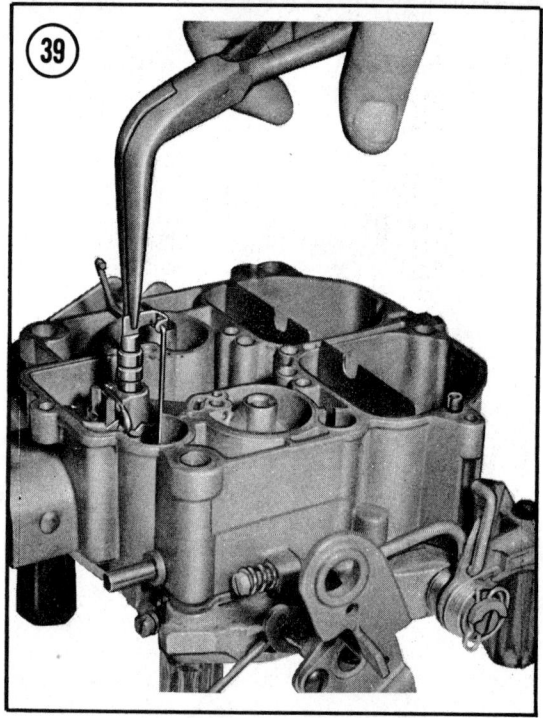

9. Remove fuel inlet needle seat with a *wide* blade screwdriver.

10. Remove primary metering jets.

CAUTION
Do not remove the secondary metering discs.

11. Remove pump discharge check ball retainer and check ball.

12. Remove baffle from secondary side of bowl.

13. Remove vacuum hose.

14. Remove choke assembly from bowl.

15. Remove secondary lockout link from bowl.

16. Remove vacuum break rod and vacuum break.

17. Remove fast idle cam.

18. Remove lower choke rod and lever from inside float bowl well.

19. Remove fuel inlet filter nut, gasket, filter, and spring.

20. Remove throttle body from bowl.

21. Remove pump rod from throttle lever.

22. Remove idle mixture screws and springs.

Cleaning and Inspection

The most frequent causes of carburetor trouble are dirt, gum, and water in the carburetor. Carefully clean and inspect all parts before assembling carburetor.

1. Clean all carburetor castings and metal parts in carburetor cleaner.

CAUTION
Do not immerse any rubber parts, plastic parts, diaphragms, or pump plungers in carburetor cleaner. The Delrin cam on the air valve shaft, however, will withstand normal cleaning in carburetor cleaner.

2. Check bearing surfaces of all operating levers, shafts, and castings for excessive wear.

3. Check floats for dents. Immerse floats in hot water. If bubbles appear, float is leaking and the float *must* be replaced. Do not attempt to solder the hole. This increases float weight and causes high fuel level.

4. Check pump plunger leather for cracks and creases.

5. Check float needles and seats for burrs and ridges. If present, replace needle and seat; never replace either alone.

6. Check metering rods and jets for bends, burrs, ridges, or other distortion. Replace rod or jet if either is damaged.

FUEL AND EXHAUST SYSTEMS

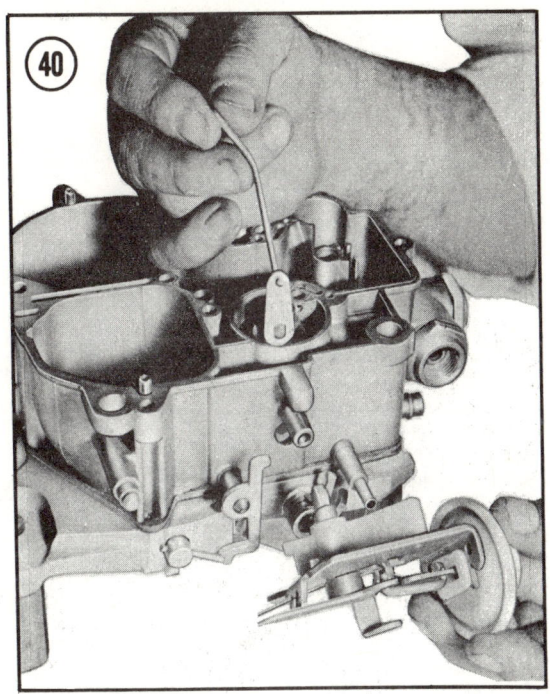

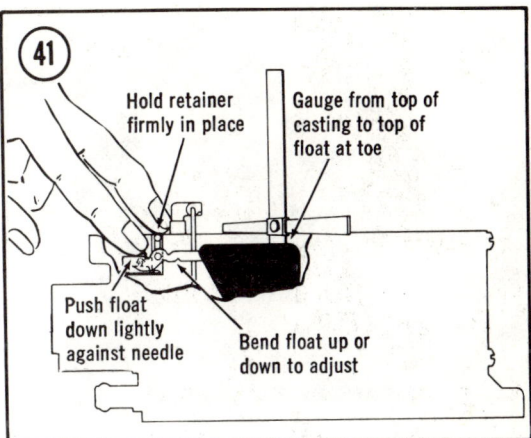

7. Inspect edges of throttle valves for nicks or gouges. Replace if necessary.

8. Check all springs for obvious distortion or bends.

9. Make certain all gasket mating surfaces of choke housing, top cover, and carburetor body are free of burrs, gouges, deep scratches, or other irregularities which could prevent a good seal.

Assembly

Float Bowl and Throttle Body

See **Figures 37 and 38**.

1. Install idle mixture needles and springs. Screw in until they seat *lightly*, then back them out 2 turns.

2. Install pump rod in throttle lever.

3. Install throttle body with new gasket.

4. Install fuel inlet filter spring, filter, new gasket, and nut.

5. Install vacuum break and rod.

6. Install fast idle cam on vacuum break assembly.

7. Connect choke rod to choke rod actuating lever. Hold choke rod with grooved end pointing inward. Position lever in well of float bowl and install choke assembly so that shaft engages with hole in actuating lever. See **Figure 40**. Install retaining screw. Remove choke rod from lever temporarily.

8. Install vacuum hose to bowl and vacuum break.

9. Install baffle in secondary side of bowl with notches toward top.

10. Install pump discharge check ball and retainer in passage next to pump well.

11. Install primary main metering jets.

12. Install fuel inlet needle seat and gasket. Use *wide* blade screwdriver.

13. Install fuel inlet needle. Secure with the pull clip.

14. Install float.

15. Measure from top of float bowl gasket surface (gasket removed) to top of float at toe. See **Figure 41**. Proper measurement is listed in **Table 4**.

16. Bend float as necessary to adjust.

17. Install power piston spring in well.

18. Install main metering rods to power piston assembly.

19. Install power piston assembly in well with main metering rods positioned in jets. Secure with retainer.

20. Install plastic filler over float needle.

21. Install pump return spring in pump well.

22. Install air horn gasket over bowl.

23. Press power piston down firmly to ensure correct alignment.

24. Install pump plunger in pump well.

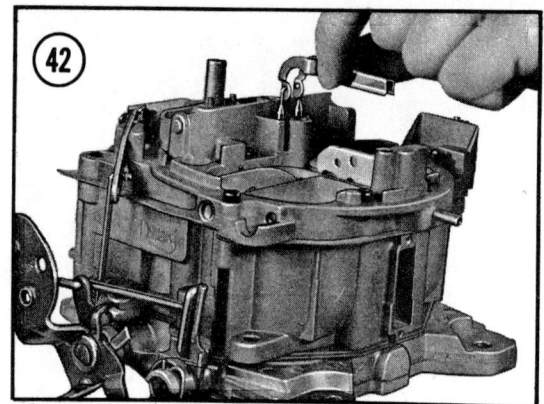

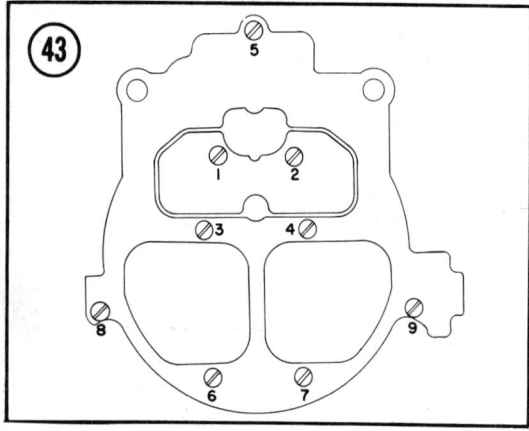

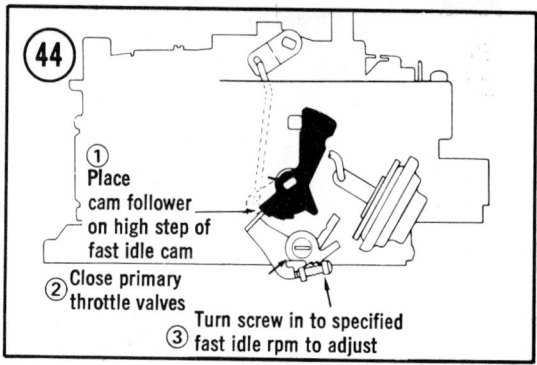

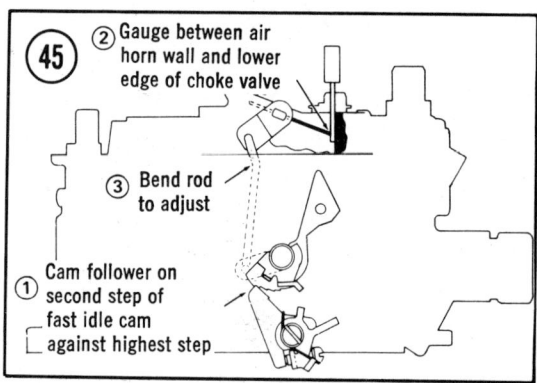

NOTE: *Steps 3 through 6 adjust choke rod. See* **Figure 45**.

Air Horn

1. Install secondary metering rods. Refer to **Figure 42**.
2. Place air horn on float bowl carefully. Secure with 7 screws. Tighten screws in sequence shown in **Figure 43**.
3. Install idle vent rod in pump lever.
4. Connect pump rod in pump lever and secure with spring clip.
5. Connect choke rod in lower choke lever. Secure upper end with spring clip.
6. Install idle vent valve.

Fast Idle

Refer to **Figure 44**.

1. Close primary throttle valves completely.
2. With cam follower over the high step of the fast idle cam, turn fast idle screw in or out until it just touches lever. Screw in 3½ turns from this point.

3. Place cam follower on second or highest step of fast idle cam (see **Table 4**).
4. Rotate choke valve toward closed position by pushing down on vacuum break tang.
5. Measure between lower edge of choke valve and wall. See **Table 4** for dimensions.
6. Bend choke rod if necessary to adjust.

Vacuum Break

Refer to **Figure 46**.

1. Hold choke valve closed with a rubber band. Cam follower must be on highest step of fast idle cam.
2. Hold vacuum break diaphragm stem against its seat so vacuum link is at end of slot.

Secondary Lockout

Refer to **Figure 47**.

1. With choke valve wide open, rotate vacuum break lever *clockwise* toward open choke position.

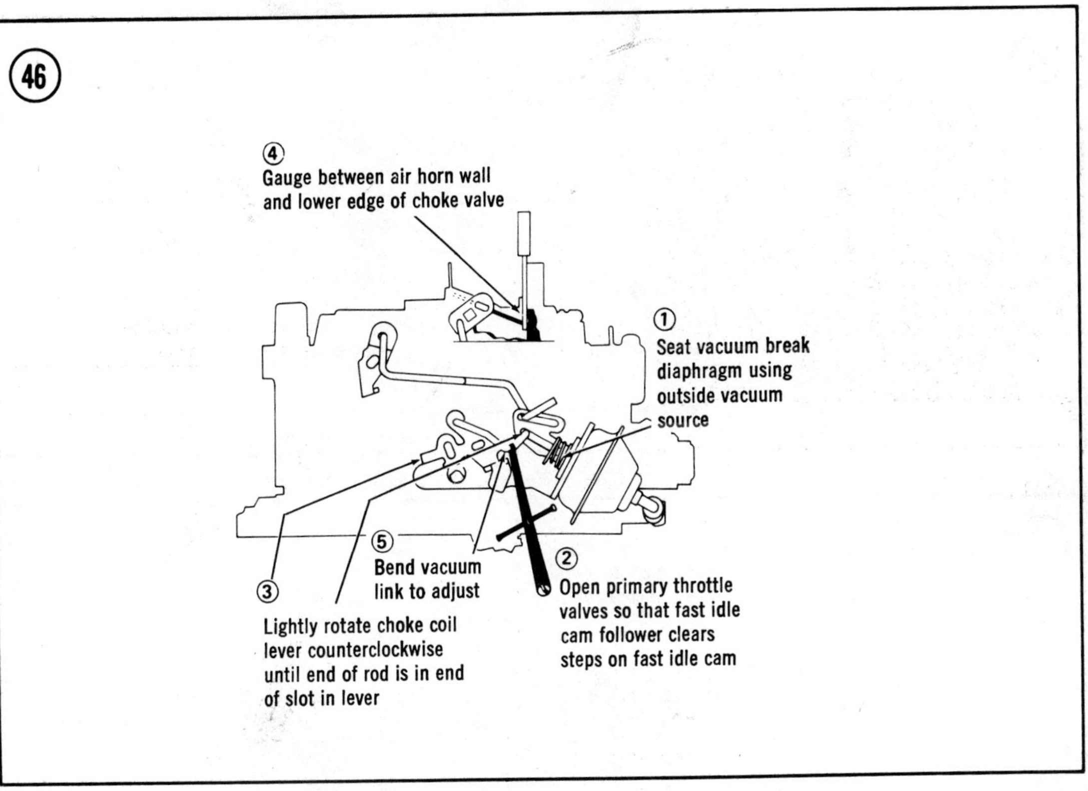

FUEL AND EXHAUST SYSTEMS

46

④ Gauge between air horn wall and lower edge of choke valve

① Seat vacuum break diaphragm using outside vacuum source

⑤ Bend vacuum link to adjust

② Open primary throttle valves so that fast idle cam follower clears steps on fast idle cam

③ Lightly rotate choke coil lever counterclockwise until end of rod is in end of slot in lever

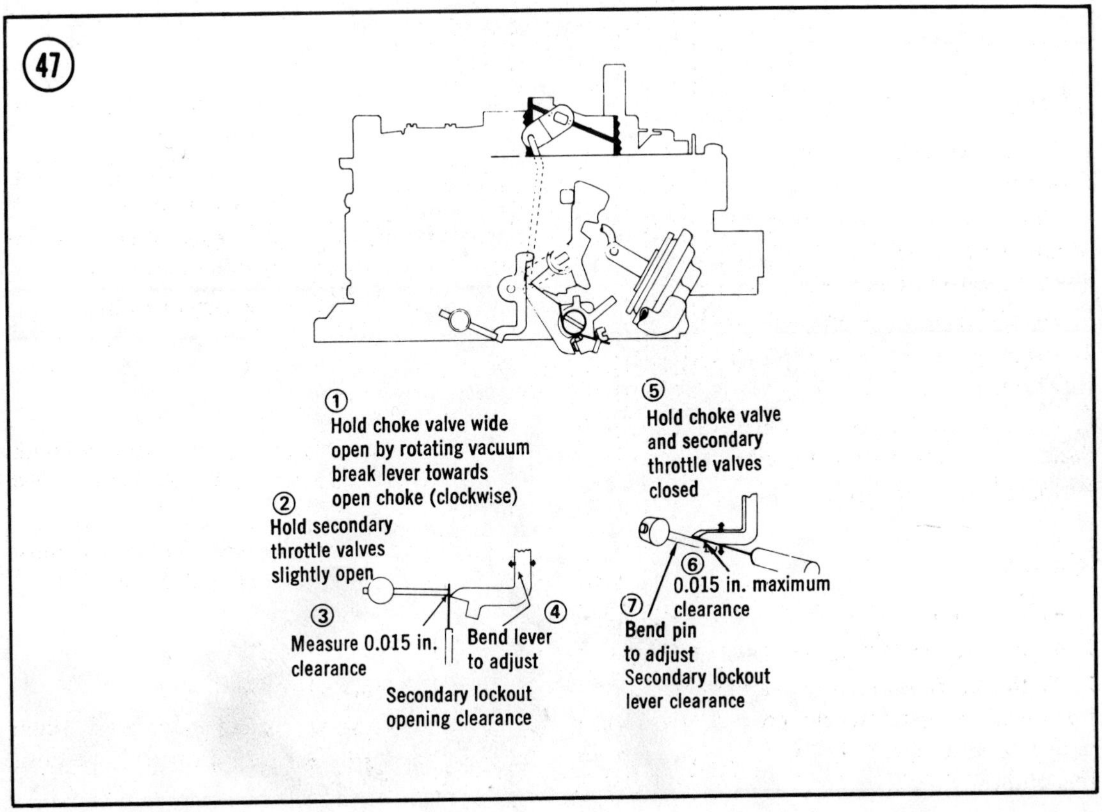

47

① Hold choke valve wide open by rotating vacuum break lever towards open choke (clockwise)

② Hold secondary throttle valves slightly open

③ Measure 0.015 in. clearance

④ Bend lever to adjust Secondary lockout opening clearance

⑤ Hold choke valve and secondary throttle valves closed

⑥ 0.015 in. maximum clearance

⑦ Bend pin to adjust Secondary lockout lever clearance

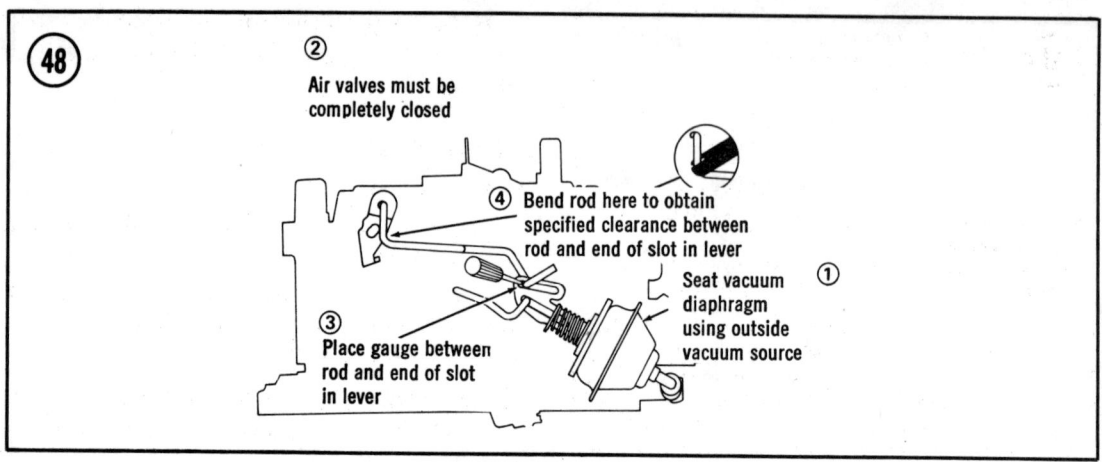

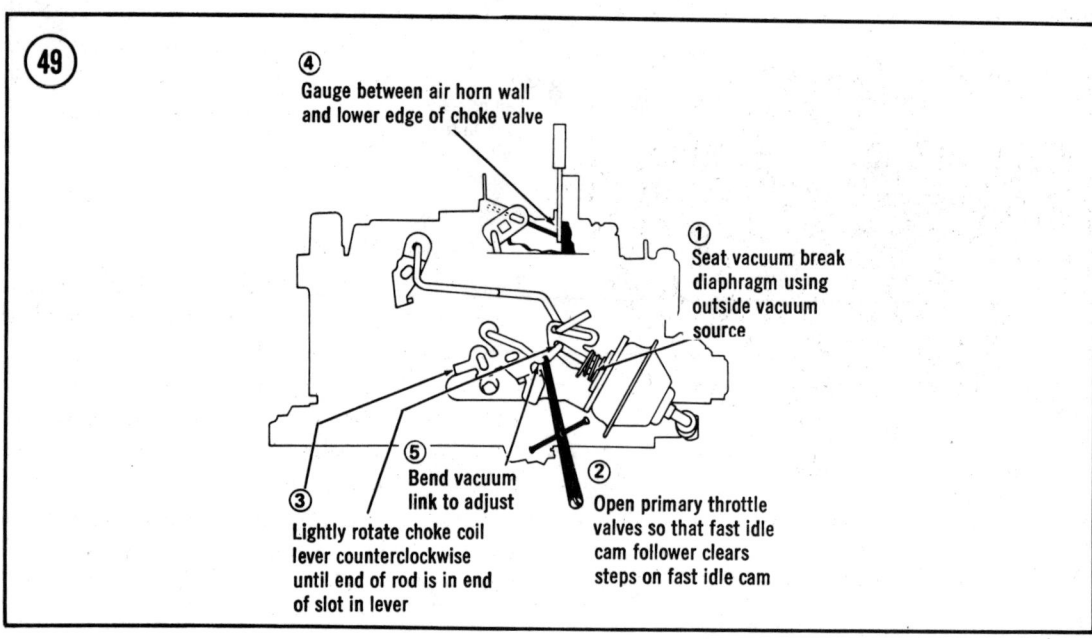

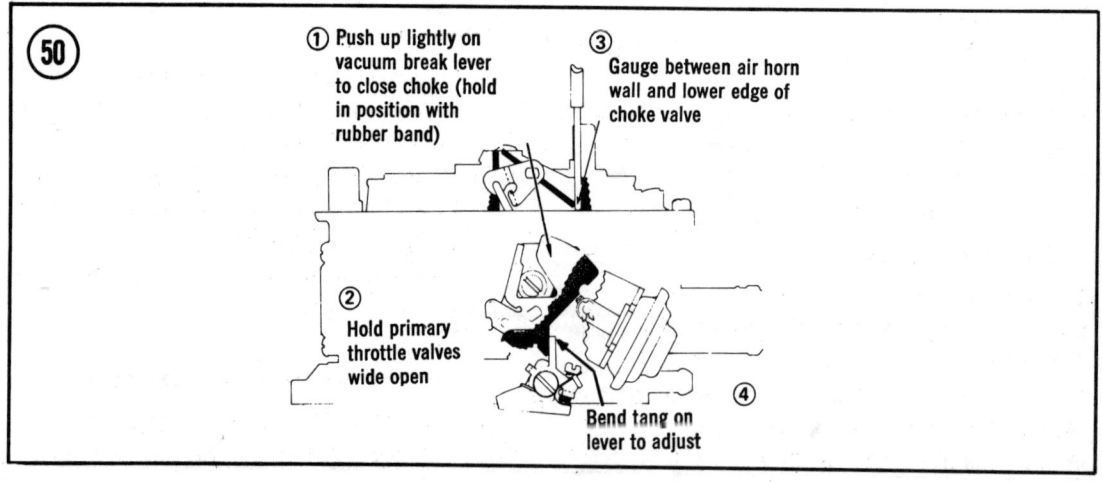

FUEL AND EXHAUST SYSTEMS

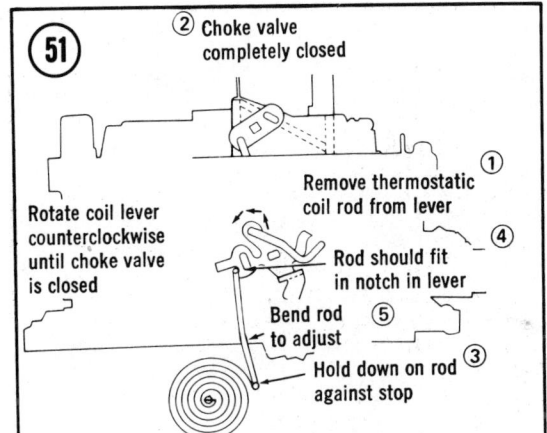

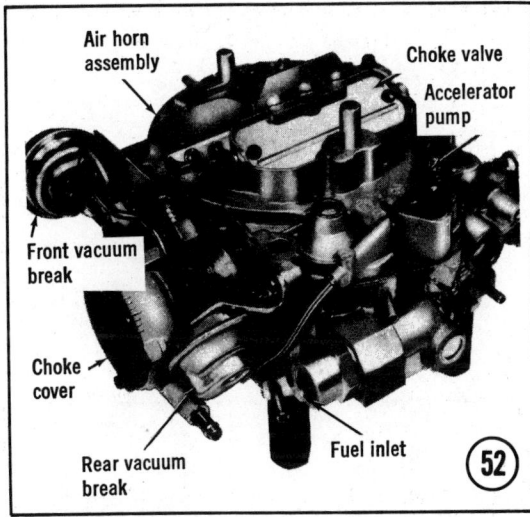

3. If the gauge is not a snug fit, bend the rod at the indicated point to obtain the specified clearance.

Vacuum Break

Refer to **Figure 49**.

1. Use a vacuum pump to seat the vacuum break diaphragm.

2. Open the primary throttle valves far enough so the fast idle cam follower is clear of all steps on the fast idle cam.

3. Rotate the choke coil lever counterclockwise until the end of the rod is in the end of the slot in the lever.

4. Insert a gauge of specified size (see **Table 4**) between the air horn wall and the lower edge of the choke valve. If the gauge is not a snug fit, bend the vacuum link at the point shown in **Figure 49** to adjust as required.

Unloader

Refer to **Figure 50**.

1. Close the choke by pushing lightly on the vacuum break lever and use a rubber band to hold the choke in this position.

2. Hold the primary throttle valve wide open and insert a gauge between the air horn wall and the lower edge of the choke valve. See **Table 4** for gauge size.

3. If the gauge does not fit snugly, bend the tang on the lever at point shown in **Figure 50** to adjust as required.

Choke Coil Rod

Refer to **Figure 51**.

1. Remove the thermostatic coil rod from the lever and turn the coil lever counterclockwise to completely close the choke valve.

2. Hold the choke coil rod against the stop. The end of the rod should fit in the notch in the lever as shown in **Figure 51**.

3. If required, bend the rod at the point shown in **Figure 51** to adjust.

ROCHESTER M4MC-M4MCA

The M4MC-M4MCA model Quadrajet carburetors (**Figure 52**) are used on some 1975 and

2. Bend lever to provide minimum 75% contact.

3. With choke valve wide open, rotate vacuum break lever clockwise toward open position.

4. Bend lever to provide clearance specified in **Figure 47**.

5. With choke valve closed, bend lever to provide clearance specified in **Figure 47**.

Air Valve Dashpot

Refer to **Figure 48**.

1. Apply an outside vacuum source (vacuum pump, etc.) to seat the vacuum diaphragm of the dashpot.

2. Make sure air valves are completely closed, then insert a 0.020 in. gauge between the rod and the end of the slot as shown in **Figure 48**.

Table 5 HOLLEY 4150 SPECIFICATIONS

Year	1970	1971
Engine Displacement	402, 454	454
Float level, primary*	0.350 in.	Float centered
Float level, secondary*	0.500 in.	Float centered
Accelerator pump	0.015 in.	0.015 in.
Idle vent	0.085 in.	—
Fast idle adjustment	0.025 in.	0.025 in.
Fast idle rpm	2,200	2,200
Choke vacuum break	0.300 in.	0.350 in.
Unloader	0.350 in.	0.350 in.
Thermostat choke rod	1 rod diameter	1.320 in. ±0.015 in.
Secondary stop	—	½ turn open
Primary main metering jet, automatic	No. 70	No. 70
Primary main metering jet, manual	No. 70	No. 70
Secondary main metering jet	No. 76	No. 76
Primary throttle bore	1 11/16 in.	1 11/16 in.
Secondary throttle bore	1 11/16 in.	1 11/16 in.

*Use sight plug for final setting

later model V8 engines. These carburetors were developed from the 4MV series.

All M4MC series carburetors are an integral part of the emission control system. Adjustment and repair by the home mechanic, other than adjustment of idle speed and idle mixture as described in Chapter Three, is not recommended. This is a job for your dealer or a carburetor shop equipped with the precision equipment and skills necessary for working with this complicated device. Removal and installation procedures are essentially the same as those for the 4MV, this chapter.

HOLLEY 4150 CARBURETOR

These 4-barrel, 2-stage carburetors consist of 7 subassemblies: throttle body, main body, primary and secondary fuel bowls, primary and secondary metering bodies, and the secondary throttle operating assembly. **Table 5** lists specifications.

Disassembly

The carburetor is disassembled into subassemblies; then each subassembly is disassembled separately. Refer to **Figures 53A and B** for the initial disassembly.

1. Loosen fuel inlet fitting, fuel bowl sight plugs, and needle/seat assembly lock screws.

2. Remove 4 primary fuel bowl screws and remove bowl, metering body, splash shield, and gaskets. See **Figure 54**.

3. Remove fuel tube if so equipped.

4. Remove 4 secondary fuel bowl screws. Remove fuel bowl.

5. Remove metering body and gasket.

6. Disconnect secondary throttle operating rod at throttle lever.

7. Remove secondary throttle operating assembly and gasket from main body.

8. Disconnect vacuum break hose at vacuum break.

FUEL AND EXHAUST SYSTEMS

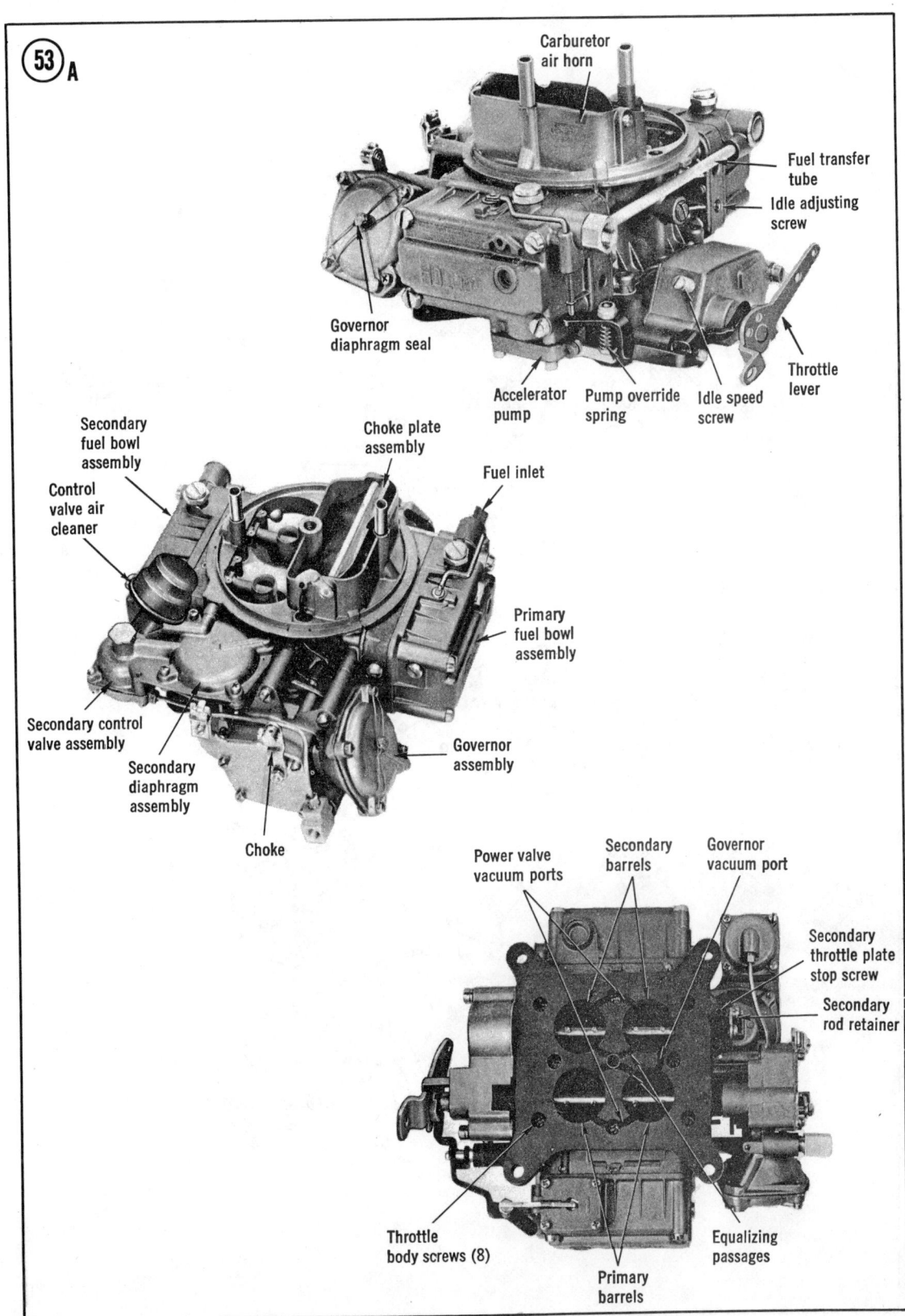

CHAPTER FIVE

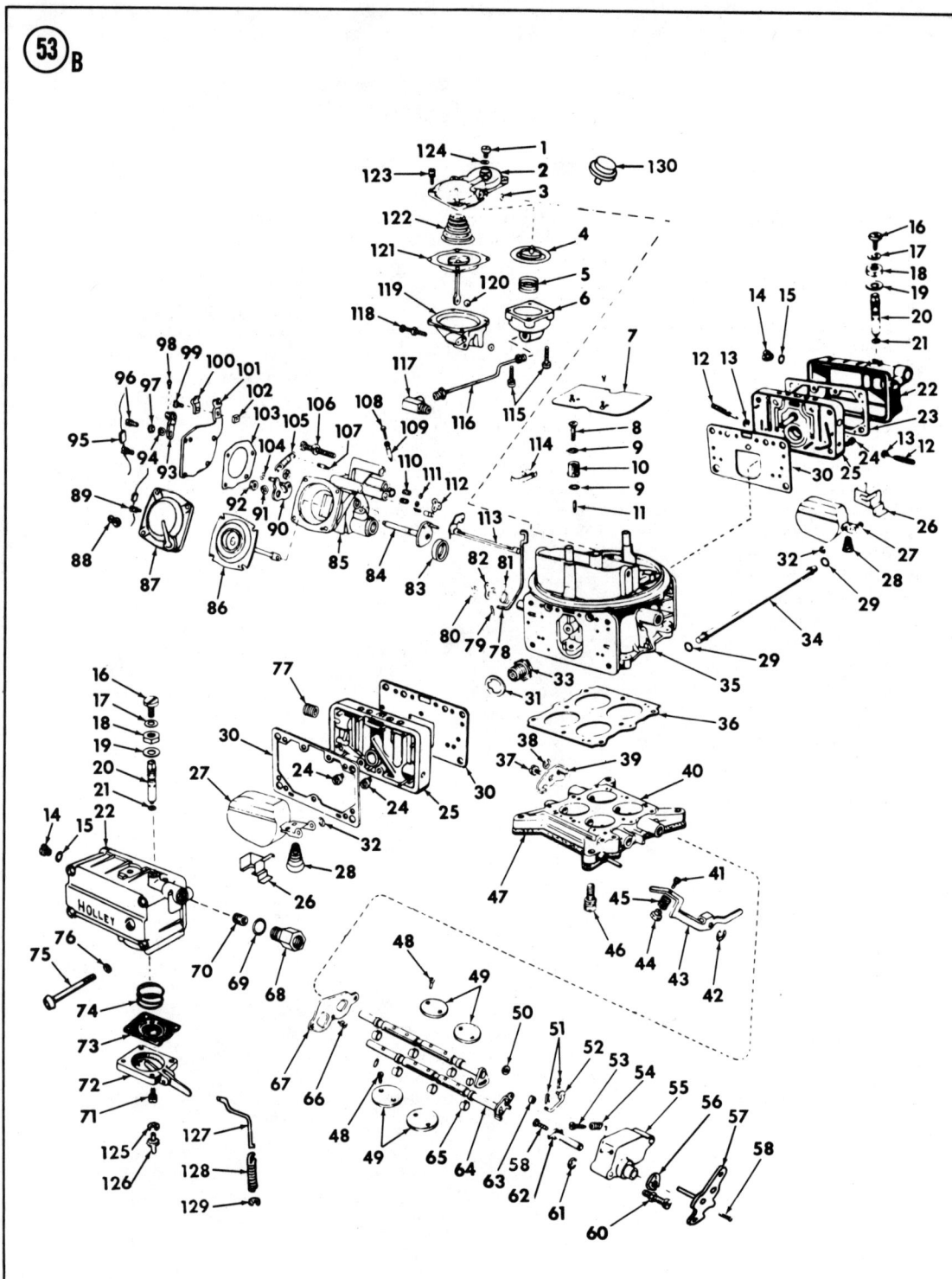

HOLLEY CARBURETOR

1. Secondary control valve plug
2. Secondary control valve cover
3. Gasket
4. Secondary control valve diaphragm
5. Secondary control valve spring
6. Secondary control valve housing
7. Choke plate
8. Nozzle screw
9. Nozzle gasket
10. Pump discharge nozzle
11. Pump nozzle valve
12. Idle adjusting needle
13. Idle needle seal
14. Sight plug
15. Sight plug gasket
16. Valve seat screw
17. Valve seat screw gasket
18. Valve seat adjusting nut
19. Valve seat gasket
20. Fuel inlet valve
21. Fule inlet valve seal
22. Fuel bowl
23. Fuel bowl gasket
24. Main jet
25. Metering body
26. Baffle plate
27. Assembly float
28. Float spring
29. Fuel tube O-ring
30. Metering body gasket
31. Power valve gasket
32. Float shaft retainer
33. Power valve
34. Fuel tube
35. Main body
36. Main body gasket
37. Screw
38. Retainer
39. Lever
40. Throttle body
41. Pump lever adjusting screw
42. Retainer
43. Pump operating lever
44. Pump lever adjusting screw nut
45. Pump lever screw spring
46. Throttle body to main body screw
47. Carburetor mounting gasket
48. Throttle plate screw
49. Throttle plate
50. Throttle rod washer
51. Throttle rod retainer
52. Throttle connect rod
53. Throttle stop screw
54. Throttle stop screw spring
55. Throttle operating housing
56. Accelerator pump operating cam
57. Throttle shaft lever
58. Throttle shaft lever screw
59. Throttle operating lever screw
60. Throttle operating housing screw
61. Retainer
62. Throttle operating lever
63. Throttle rod spacer
64. Throttle shaft
65. Throttle shaft bushing
66. Throttle shaft housing plate screw
67. Throttle shaft housing plate
68. Fuel inlet fitting
69. Fuel inlet fitting gasket
70. Filter screen
71. Accelerator pump cover screw
72. Accelerator pump cover with lever
73. Accelerator pump diaphragm
74. Accelerator pump diaphragm spring
75. Float bowl screw
76. Float bowl screw gasket
77. Plug
78. Choke rod
79. Choke rod retainer
80. Choke lever retainer
81. Choke rod spring
82. Choke rod lever
83. Throttle shaft seal
84. Fast idle cam shaft
85. Governor housing with plug
86. Governor diaphragm assembly
87. Governor diaphragm cover
88. Governor cover screw
89. Governor cover screw wire seal
90. Governor lever assembly
91. Governor lever nut washer
92. Governor lever nut
93. Choke lever
94. Choke lever washer
95. Seal wire
96. Governor housing cover screw
97. Choke lever nut
98. Choke lever swivel screw
99. Choke cable clamp screw
100. Choke cable clamp
101. Governor housing cover
102. Choke cable clamp screw nut
103. Governor housing cover gasket
104. Retainer
105. Governor spring
106. Governor housing screw
107. Governor spring post
108. Fast idle screw
109. Fast idle pin
110. Governor bypass jet
111. Governor passage seal
112. Governor housing gasket
113. Choke shaft with lever
114. Choke rod seal
115. Secondary control valve housing screw
116. Fuel control valve pipe
117. Tee
118. Control valve housing screw
119. Control valve housing
120. Check ball
121. Control valve diaphragm
122. Control valve diaphragm spring
123. Control valve cover screw
124. Secondary control valve plug gasket
125. Vent valve retainer
126. Vent valve
127. Vent valve actuating rod
128. Vent valve rod spring
129. Vent valve rod spring retainer
130. Control valve air cleaner

CHAPTER FIVE

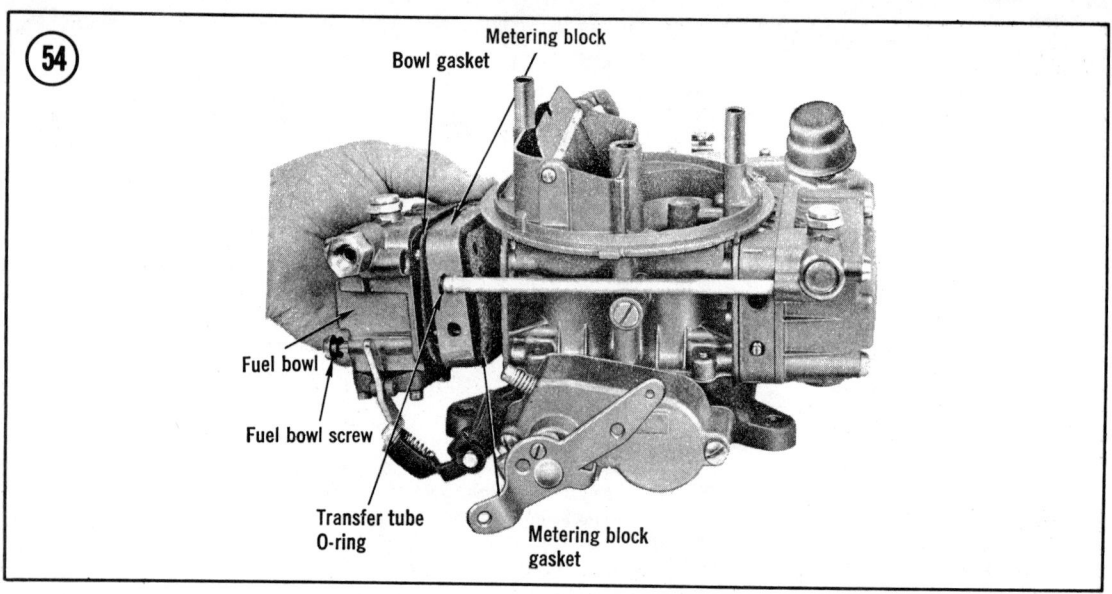

54. Metering block, Bowl gasket, Fuel bowl, Fuel bowl screw, Transfer tube O-ring, Metering block gasket

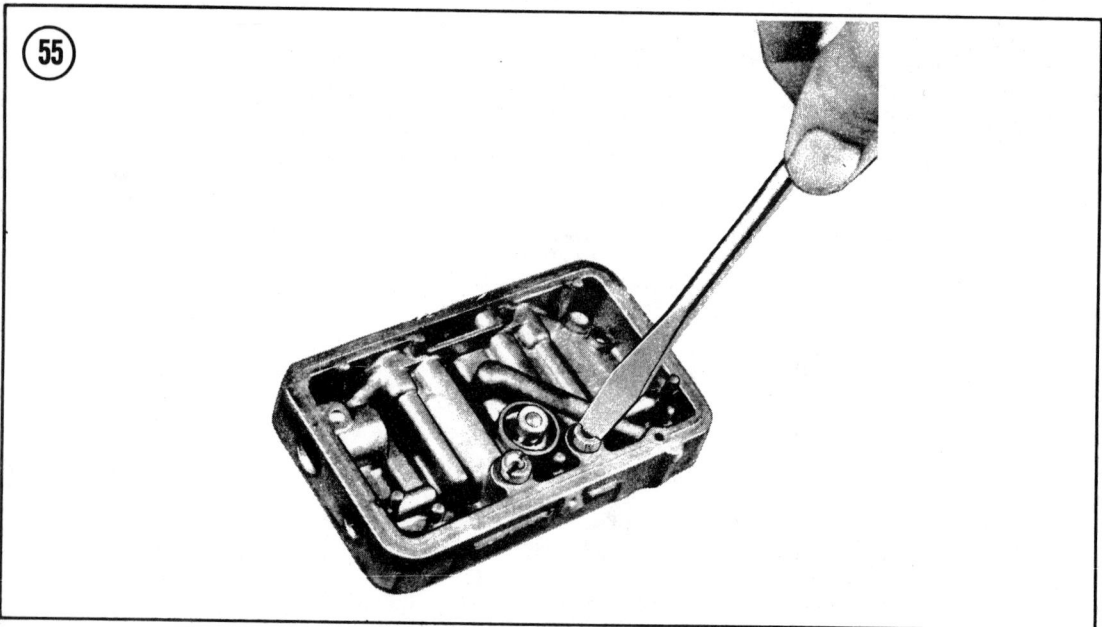

55.

9. Remove throttle body screws and separate throttle body from main body.

Fuel Bowl

Refer to **Figure 55**.

1. Remove float hinge pin retainer. Slide float from bowl.
2. Turn adjusting nut on needle and seat assembly counterclockwise. Remove assembly.
3. Remove fuel level check plug and gasket.
4. Remove inlet fitting, fuel filter, spring, and gaskets.
5. On primary bowl only, remove air vent assembly (if so equipped). Remove pump diaphragm screws, cover assembly, diaphragm, and spring.

Metering Body

1. Remove main metering jets with a *wide* blade screwdriver. See **Figure 56**.

FUEL AND EXHAUST SYSTEMS

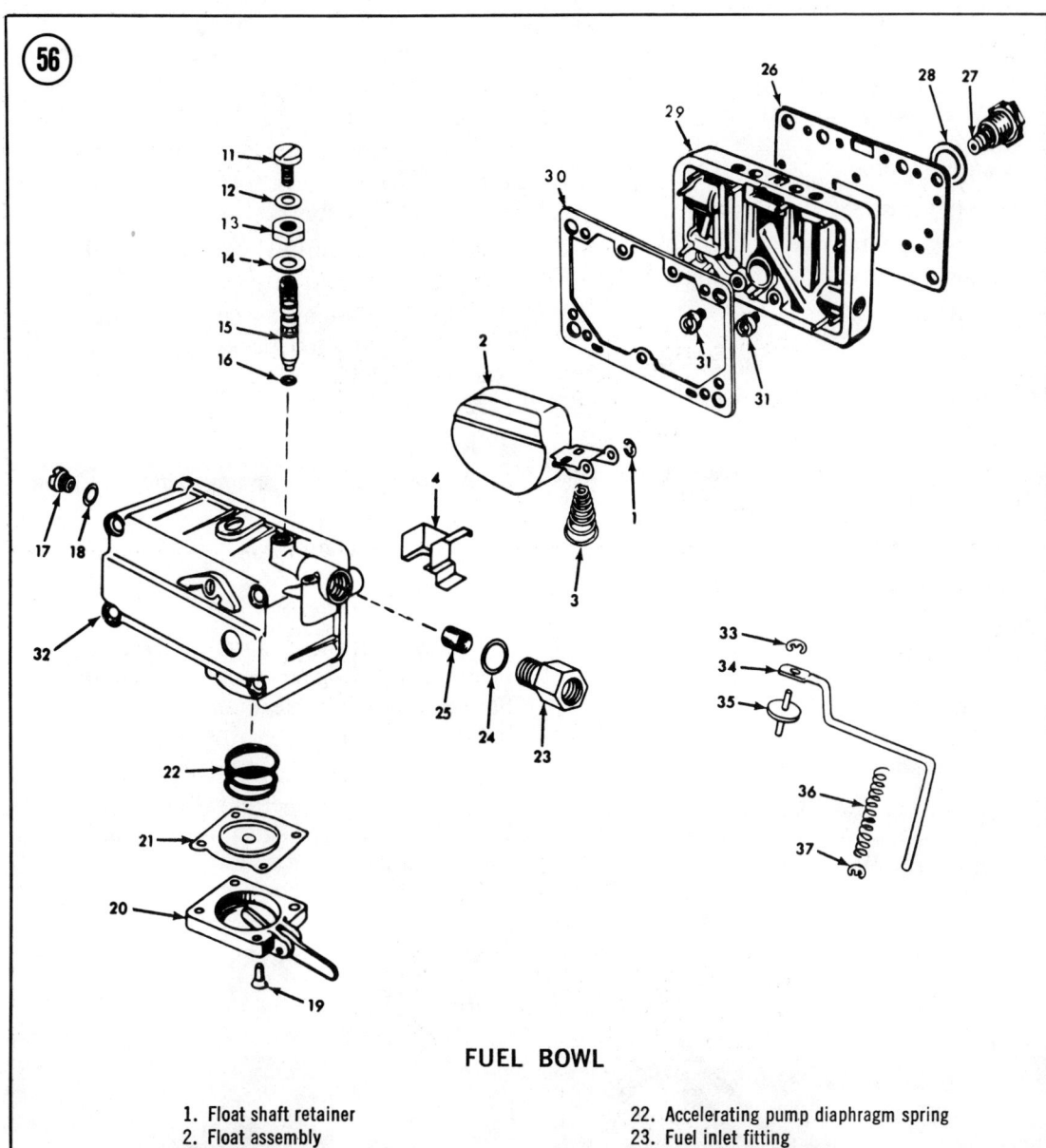

FUEL BOWL

1. Float shaft retainer
2. Float assembly
3. Float spring
4. Baffle plate
11. Lock screw, fuel valve seat
12. Fuel valve seat screw gasket
13. Fuel valve seat adjusting nut
14. Fuel valve seat adjusting nut gasket
15. Fuel valve seat assembly
16. Fuel valve seat O-ring
17. Fuel level check plug
18. Fuel level check plug gasket
19. Fuel pump cover screw
20. Accelerating pump cover assembly
21. Accelerating pump diaphragm
22. Accelerating pump diaphragm spring
23. Fuel inlet fitting
24. Fuel inlet fitting gasket
25. Filter fuel inlet screen
26. Metering body gasket
27. Power valve assembly
28. Power valve gasket
29. Metering body
30. Fuel bowl gasket
31. Main jets
32. Fuel bowl
33. Vent valve retainer
34. Vent valve actuating rod
35. Vent valve
36. Actuating rod spring
37. Actuating rod spring retainer

2. Remove vacuum fitting, idle mixture screws, and seals.

3. Remove the power valves with a 1 in., 12 point socket. See **Figure 57**.

Secondary Throttle

Refer to **Figure 58**.

1. Remove diaphragm cover.
2. Remove spring and diaphragm.
3. Remove the air filter.

Main Body

Refer to **Figure 59**.

1. Remove choke vacuum break. Disconnect link at choke lever.
2. Remove choke lever and fast idle cam.
3. Remove pump discharge nozzle and gasket. Invert body to remove pump discharge needle valve.

Throttle Body

Refer to **Figure 60**.

NOTE: *Normally, the throttle body should not be disassembled unless parts are worn or damaged.*

1. Remove pump operating lever assembly. Disassemble the spring, bolt, and nut only if necessary.
2. Remove idle speed screw and spring.
3. Remove diaphragm lever from secondary throttle shaft and fast idle cam lever from primary throttle shaft.
4. Disconnect throttle connecting link from shaft levers.
5. File off staked ends of throttle plate screws. Scribe lines on the plates along the shafts. Remove screws and plates.
6. Slide shafts out.
7. Remove secondary throttle stop screw.
8. Remove accelerator pump cam from throttle lever.
9. Remove vacuum break hose from body.

Cleaning and Inspection

The most frequent causes of carburetor problems are dirt, gum, and water in the carburetor. Carefully clean and inspect all parts before assembling carburetor.

1. Clean all parts except choke housing and accelerator pump plunger in solvent. Special carburetor cleaners, available from any auto parts supplier, work best.

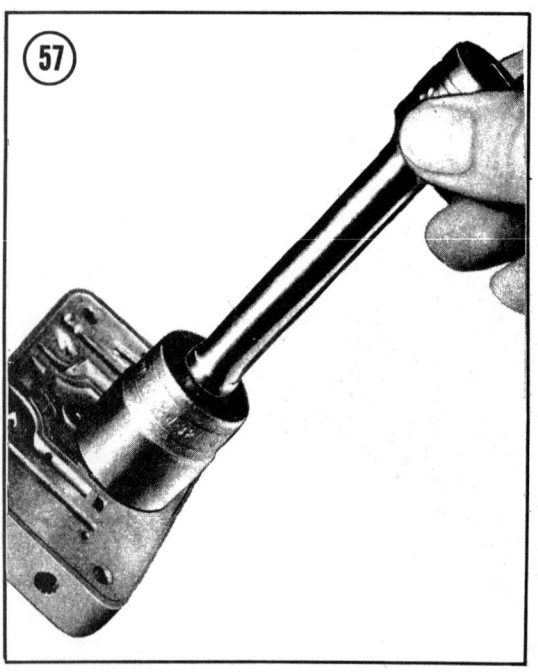

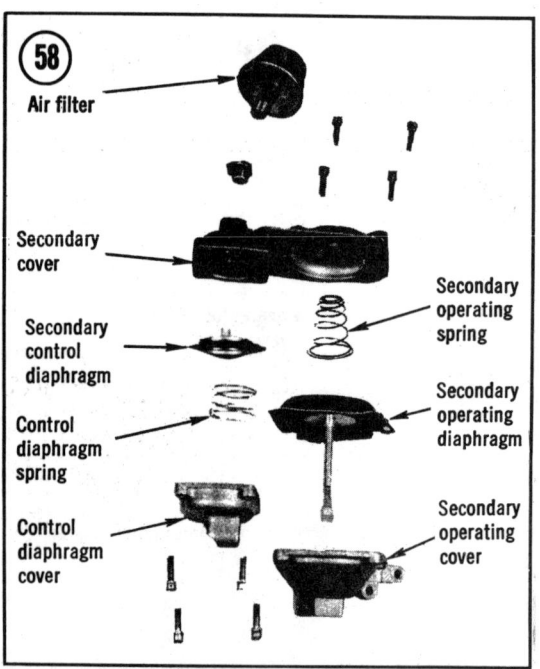

FUEL AND EXHAUST SYSTEMS

2. Check bearing surfaces of all operating levers, shafts, and castings for excessive wear.

3. Check floats for dents. Immerse floats in hot water. If bubbles appear, float is leaking and the float must be *replaced*. Do not attempt to solder the hole. This increases float weight and causes high fuel level.

4. Check pump diaphragm for damage.

5. Check float needles and seats for burrs and ridges. If present, replace needle and seat; never replace either alone.

6. Inspect edges of throttle valves for nicks or gouges. Replace if necessary.

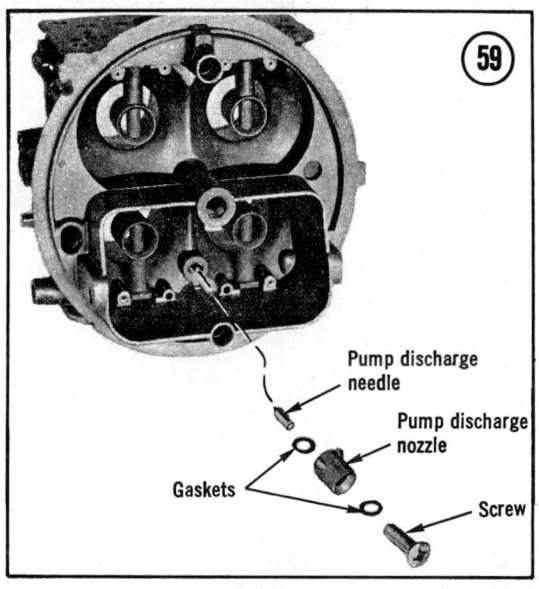

7. Check the secondary throttle operating diaphragm for free operation and leakage. To do this, push diaphragm arm up and cover vacuum passage with thumb. Diaphragm should stay up. Remove thumb and diaphragm rod should move down steadily.

8. After washing in solvent, clear all passages in metering bodies and main body with compressed air.

Assembly

The assembly procedure consists of first assembling each subassembly, making adjustments, then assembling subassemblies together.

Throttle Body

Refer to **Figure 60**.

1. Install secondary throttle stop screw.

2. Install throttle shafts in body. Roll new plastic bushings on shaft to shape them for easier installation.

3. Install throttle valve on shaft with identification numbers down. Do not tighten screws.

4. Center throttle valves on shaft by holding valves closed. Tighten screws; use Loctite to prevent loosening.

5. Install throttle connecting link to throttle shaft levers.

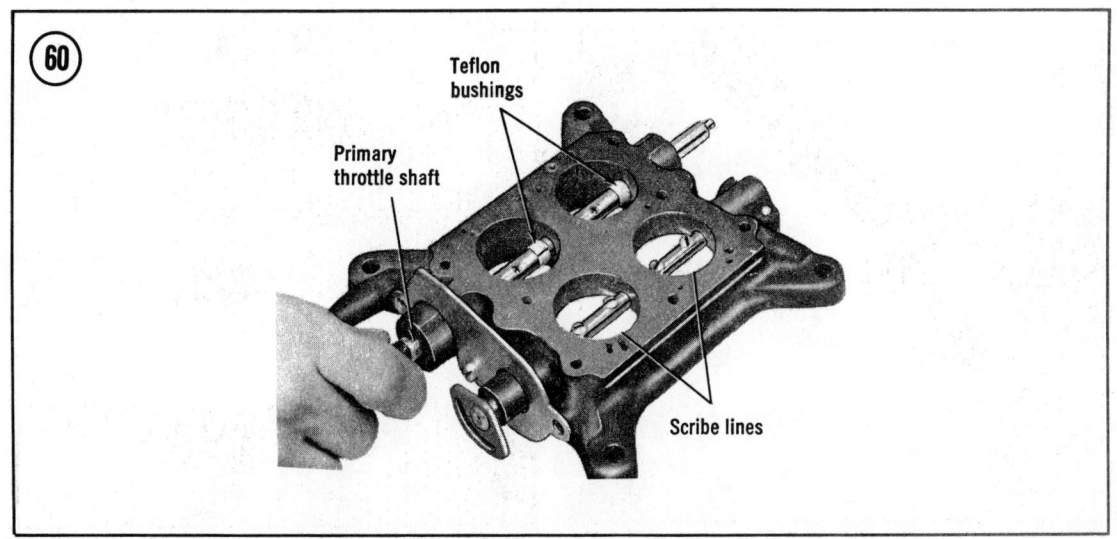

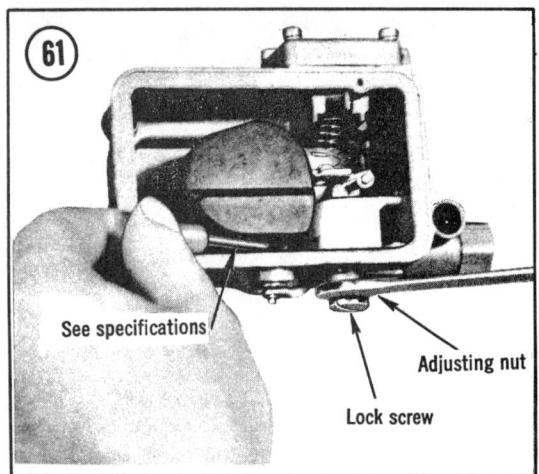

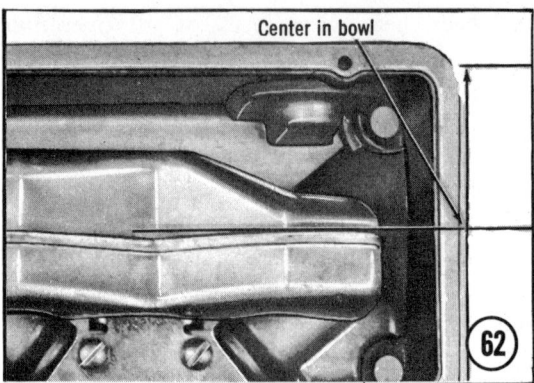

6. Install fast idle cam lever on primary throttle shaft. Install diaphragm lever on secondary throttle shaft.

7. Install idle speed screw and spring. Turn idle speed screw clockwise until it contacts throttle lever. Turn additional 1 ½ turns.

8. Install accelerator pump cam on throttle lever.

9. Assemble and install pump operating lever assembly.

10. Install vacuum break hose on fitting.

Main Body

Refer to **Figure 59**.

1. Install pump discharge valve.

2. Install pump discharge nozzle. Use new gaskets.

3. Install choke rod and seal.

4. Install choke shaft and connect upper end of choke rod.

5. Install choke valve on shaft. Hold valve closed while tightening screws to keep valve centered. Use Loctite on screws to prevent loosening.

6. Install choke lever and fast idle cam.

7. Connect vacuum break link to choke lever, then install vacuum break.

Secondary Throttle

Refer to **Figure 58**.

1. Install diaphragm assembly and spring in housing.

2. Install the diaphragm cover and tighten securely.

Metering Body

Refer to **Figures 56 and 57**.

1. Install power valve with a new gasket. Tighten securely with a one inch 12-point socket. Install main metering jets with a wide blade screwdriver.

2. On primaries only, install idle mixture screws. Turn all the way in *lightly* to seat them, then back off one turn.

Fuel Bowl

Refer to **Figure 55**.

> NOTE: *Steps apply to primary and secondary bowls unless otherwise specified.*

1. Install sight plugs with new gaskets.

FUEL AND EXHAUST SYSTEMS

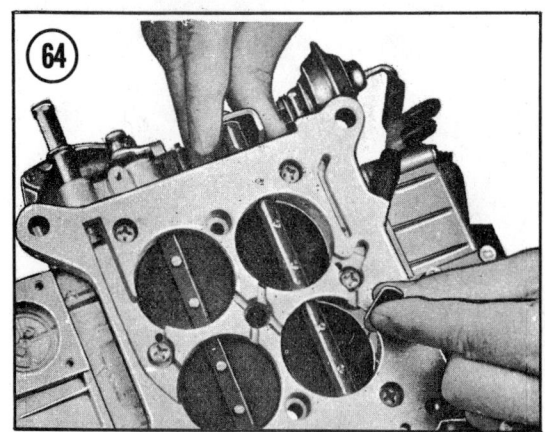

2. Install inlet needle and seats with new gaskets. Leave locknuts loose.

3. Install inlet fitting, fuel filter spring, and gaskets.

4. Assemble spring to float, slide float into bowl and install float hinge pin retainer.

5. Install inlet baffle.

6. On primary bowl only, install pump spring and diaphragm. Install air vent valve assembly.

7. Invert each fuel bowl. On 1970 models, turn adjustable needle seat until top of float is 0.350 in. (primary float) or 0.500 in. (secondary float) from top of fuel bowl. See **Figure 61**. On 1971 models, center both floats. See **Figure 62**.

Final Assembly

1. Invert main body. Install throttle body with new gasket.

2. Install secondary throttle operating assembly on main body with new gasket.

3. Install secondary metering body onto main body with new gasket. Secure 4160 metering body with 6 screws.

4. Install secondary fuel bowl with gaskets under screw heads.

5. Install primary metering body onto main body with new gasket.

6. Lubricate O-rings and install on very ends of fuel tube (if so equipped). O-rings will roll into position when installed. Install fuel tube end into secondary bowl inlet.

7. Install splash shield, new gasket, and primary fuel bowl. Align fuel tube into inlet (if so equipped).

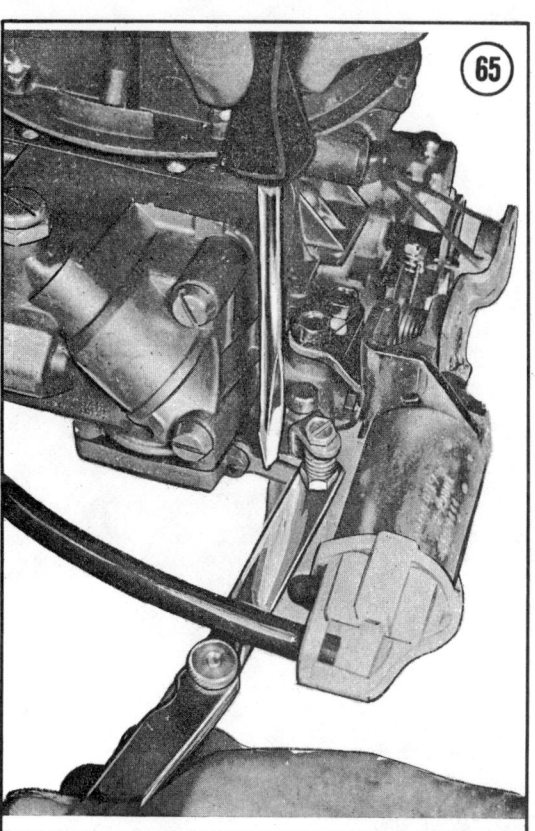

8. Align pump lever under operating lever duration spring. Install fuel bowl retaining screws with new gaskets under heads.

Adjustment

Secondary Throttle Valve Stop

1. Back off stop screw (**Figure 63**) until throttle plates are fully closed.

2. Turn stop screw in until it just touches throttle lever. Turn in additional ½ turn.

Fast Idle Cam

1. Open throttle slightly. Close fast idle lever against top step of cam.

2. Check clearance between throttle plate and bore as shown in **Figure 64**. The clearance should be 0.025 in.

3. Bend fast idle lever to adjust.

Accelerator Pump

1. Hold throttle lever wide open with rubber band. See **Figure 65**.

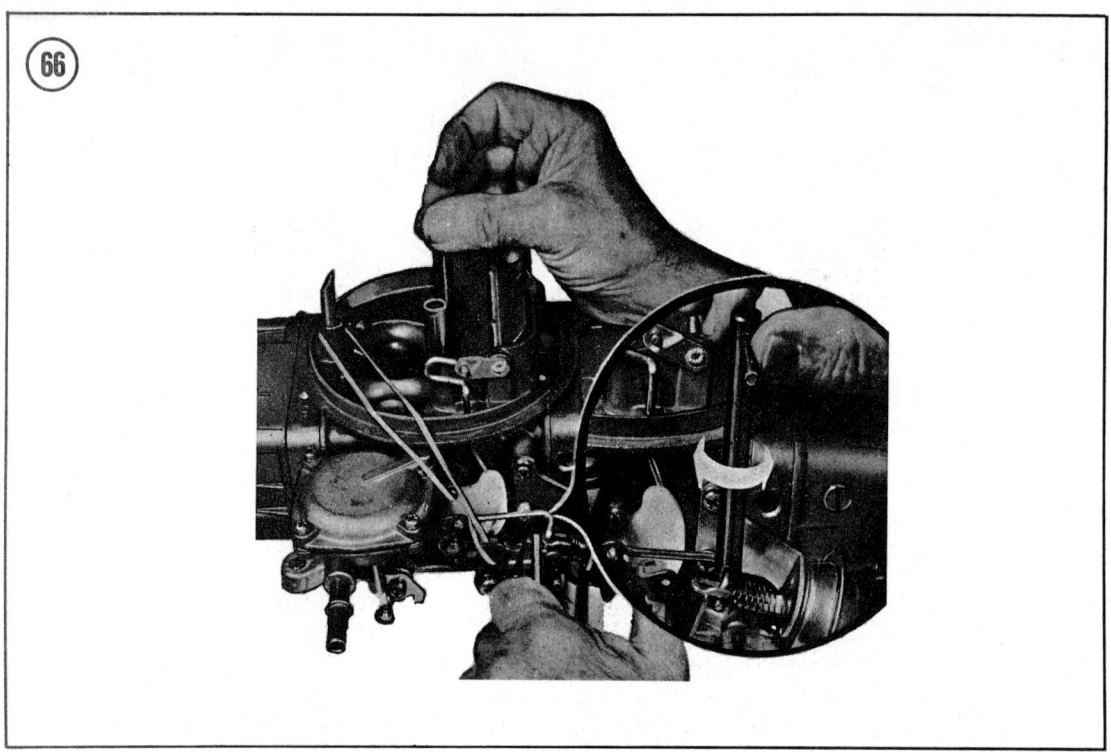

2. Hold pump lever fully down.

3. Measure clearance between spring adjusting nut and arm of pump lever. Clearance should be 0.015 in. See **Figure 65**.

4. Turn nut or screw as required to adjust.

Vacuum Break

1. Hold choke valve closed with a rubber band.
2. Hold vacuum break in against stop.
3. Measure distance between lower edge of choke valve and main body. See **Figure 66**. Proper distance is shown in specifications.
4. Bend vacuum break link to adjust.

Choke Unloader

Refer to **Figure 67**.

1. Hold throttle lever wide open with rubber band.
2. Hold choke valve toward closed position against unloader tang of throttle shaft.
3. Measure opening between lower edge of choke valve and main body. See specifications for proper measurement.
4. Bend choke rod (**Figure 67**) to adjust.

FUEL PUMP

All engines use a non-serviceable fuel pump which cannot be repaired or rebuilt; the pump is replaced as a unit.

Removal/Installation

1. Disconnect fuel inlet and outlet pipes at fuel pump and fuel cover.
2. Remove mounting bolts. Remove pump and gasket.

NOTE: *The following 2 steps are not necessary unless operating pushrod requires service (V8 engines).*

3. Remove adapter mounting bolts. Remove adapter and gasket from block.
4. Remove pushrod from block.
5. Installation is the reverse of these steps.

FUEL TANK

Removal/Installation

1. Remove ground cable from battery.

FUEL AND EXHAUST SYSTEMS

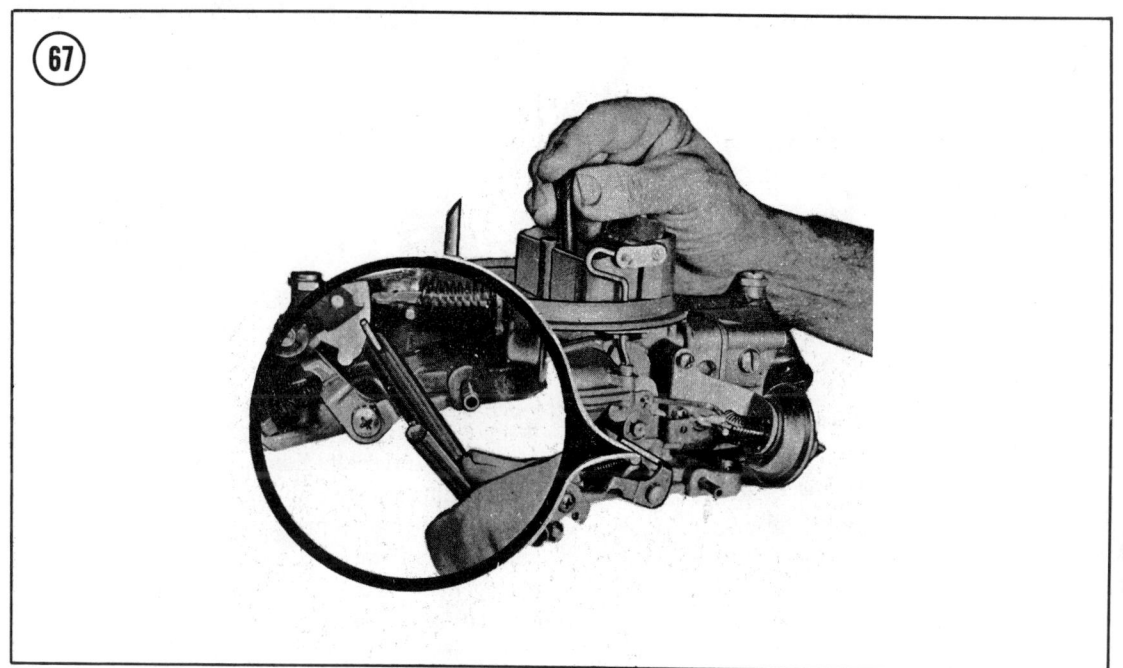

2. Disconnect meter wire at rear harness connector. Push out grommet and thread meter wire through trunk floor.
3. Raise vehicle and drain fuel tank.
4. Remove gauge ground wire screw from floor pan.
5. Disconnect fuel line at gauge pickup line.
6. Remove vent hoses.
7. Remove tank strap bolts and lower tank.
8. Reverse procedure to install tank.
9. Lower vehicle.

EXHAUST SYSTEMS

The exhaust system used in practically all vehicles consists of an exhaust pipe (or pipes in dual systems), mufflers, and tailpipes. Resonators, located between the muffler and the rear of the vehicle, are used on some models. In a typical installation, the exhaust pipes are constructed in one piece and are clamped to the muffler inlet. Tailpipes, except in installations having resonators, are also one piece and are, as a rule, welded to the muffler outlets. The following procedures are for typical installations and should be modified as required if the installation does not agree with the description given above.

MUFFLER AND TAILPIPE

Removal
1. Raise vehicle.
2. Remove clamp(s) at muffler inlet(s).
3. Remove tailpipe-to-hanger screws.
4. Disengage muffler(s) from exhaust pipe(s).
5. On single exhaust system, remove muffler and tailpipe. On dual exhaust system, cut tailpipe(s) from muffler(s), then remove muffler(s).

Installation
1. Connect new muffler(s) to exhaust pipe(s).
2. Install tailpipe(s) to new muffler and hanger.

> NOTE: *Do not tighten or fasten any screws or clamps at this time.*

3. Check complete length of exhaust system for clearance and adjust as required.
4. Tighten all clamps, screws, and attachments, and weld tailpipe to muffler, if required.
5. Lower vehicle.

EXHAUST PIPE

Removal

1. Raise vehicle.
2. Remove nuts attaching exhaust pipe to exhaust manifold.
3. Remove clamp attaching exhaust pipe to muffler.
4. Remove exhaust pipe from muffler and then from vehicle.

Installation

1. Insert exhaust pipe into muffler and loosely install clamp.
2. Using new gaskets, loosely attach exhaust pipe to exhaust manifold with nuts.
3. Check all clearances between exhaust manifold and muffler.
4. Tighten clamp and nuts.
5. Lower vehicle.

CATALYTIC CONVERTER

Exhaust systems on 1975 and later models have a catalytic converter between the front exhaust pipe and the tailpipe. This is an emission control device which reduces hydrocarbon and carbon monoxide pollutants in the exhaust gases. The converter contains beads which are coated with a catalytic material containing

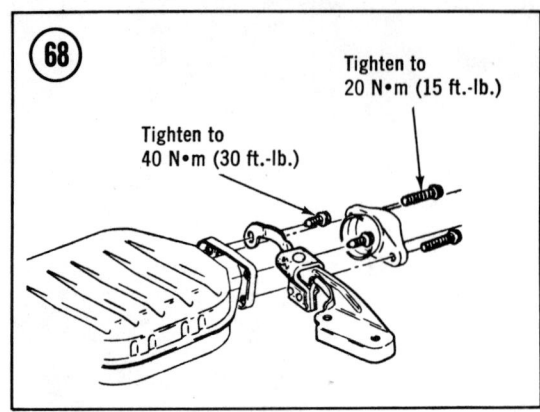

platinum and palladium. The converter does not require periodic maintenance. However, it may become necessary to replace either the converter or catalytic beads. Procedures for replacement are given below. Replacement of the catalytic beads requires expensive special tools, making it more economical to have the job done by a dealer or qualified garage.

Removal/Installation

Refer to **Figure 68** for this procedure.

1. Raise vehicle.
2. Disconnect converter at front and rear.
3. Remove converter.
4. Reverse procedure for installation.

CHAPTER SIX

EMISSION CONTROL SYSTEMS

Harmful emissions are minimized by a number of systems, depending on year:

a. Positive crankcase ventilation (PCV)
b. Controlled combustion system (CCS)
c. Air injection reaction (AIR)
d. Combination emission control system (CEC)
e. Fuel evaporation control system (ECS)
f. Exhaust gas recirculation system (EGR)
g. Carburetor calibration
h. Distributor calibration
i. Catalytic converter
j. Early full evaporation system (EFE)

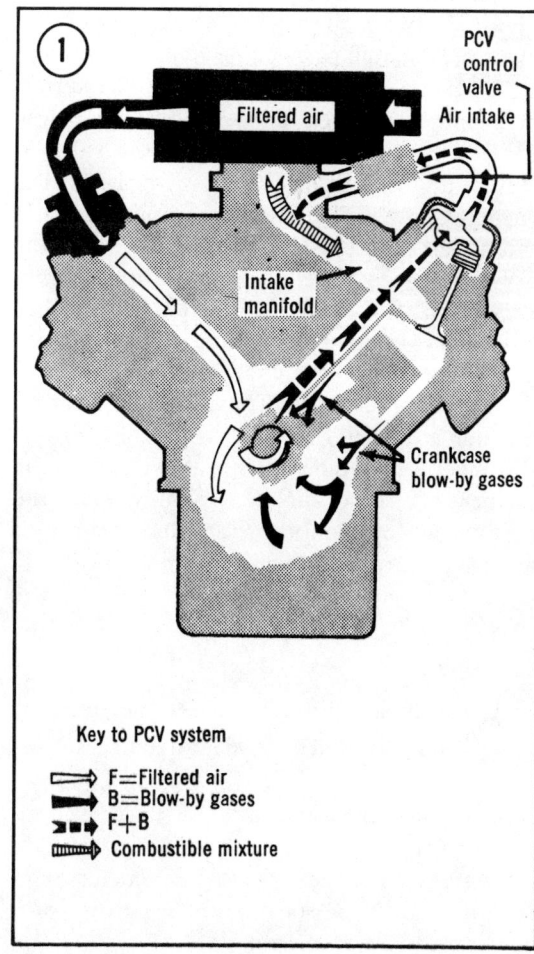

POSITIVE CRANKCASE VENTILATION

All vehicles have positive crankcase ventilation. Clean air is drawn from the air cleaner; the oil filler cap is not vented. The fresh air scavenges emissions (e.g., piston blow-by) from the crankcase and manifold vacuum draws the emissions into the carburetor. Eventually they can be reburned in the normal combustion process. **Figure 1** shows the closed system.

Either a PCV valve or fixed metered orifice mounted on the carburetor controls the volume of flow from crankcase to manifold. The valve or orifice tends to clog and must be removed and cleaned or replaced periodically. See Chapter Three. **Figure 2** shows a cross section of a PCV valve.

CONTROLLED COMBUSTION SYSTEM

The controlled combustion system on 1970 and later models includes a special thermostatically controlled air cleaner, special calibrated carburetor and distributor, and a higher temperature thermostat. See **Figure 3**.

The thermostatically controlled air cleaner maintains air to the carburetor at 100°F or more. The air cleaner includes a temperature sensor, vacuum motor, and control damper assembly. See **Figure 4**.

Operation of the air cleaner is shown in **Figure 5**. When the engine is off, absence of manifold vacuum permits the vacuum motor to close off the hot air pipe. When engine is running and underhood temperatures are below 85°F, the temperature sensor bleed valve is closed and manifold vacuum operates the vacuum motor. The vacuum motor closes off the underhood air supply and the carburetor draws the much hotter air from the hot air pipe. Between 85°F and 128°F, the temperature sensor air bleed is partially open. The vacuum motor opens both the underhood air inlet and the hot air tube inlet. The resulting blend is maintained around 100°F or more. Finally, if the underhood temperature is above 128°F, the temperature sensor air bleed is fully open, the vacuum motor cannot operate and carburetor air is drawn from underhood air inlet only.

Inspection

1. Check all heat pipe and hose connections.
2. Check for kinked or deteriorated hoses.
3. Remove air cleaner cover and install thermometer as close as possible to sensor. Install cover without wing nut. Lift cover after temperature stabilizes and read thermometer; temperature must be below 85°F before proceeding. Put cover back in place.

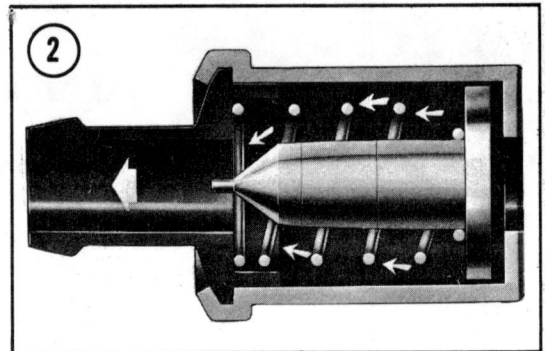

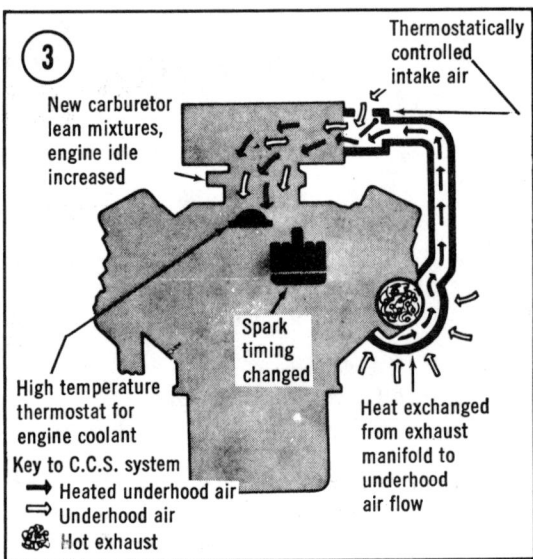

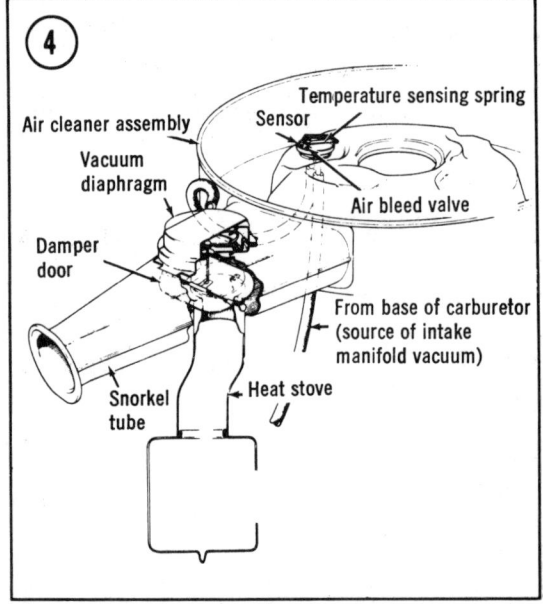

EMISSION CONTROL SYSTEMS

⑤

View A—Engine off

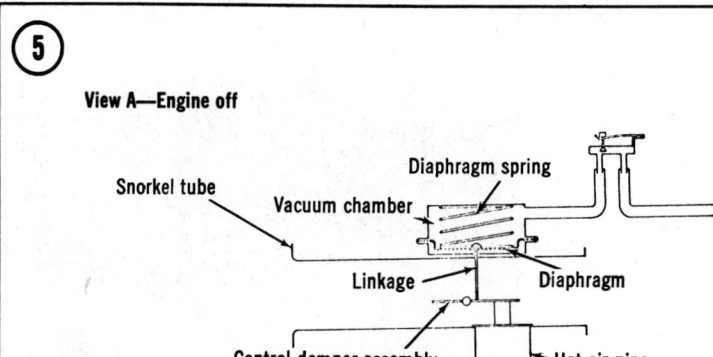

View B—Underhood temperature below 85°F

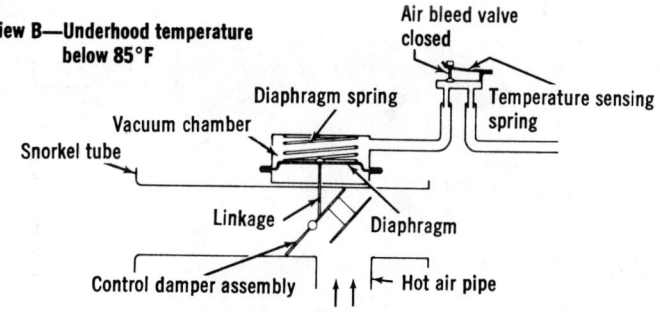

View C—Underhood temperature above 128°F

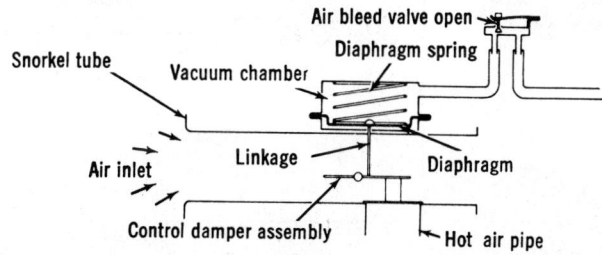

View D—Underhood temperature between 85°F and 128°F

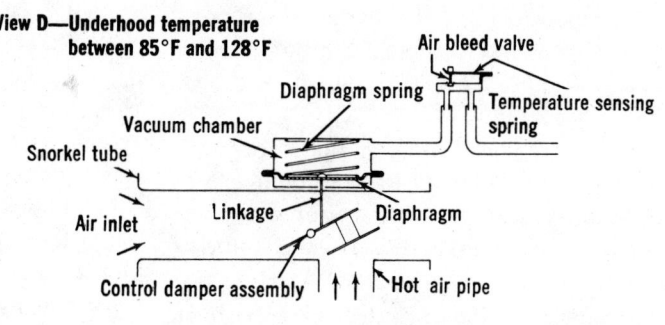

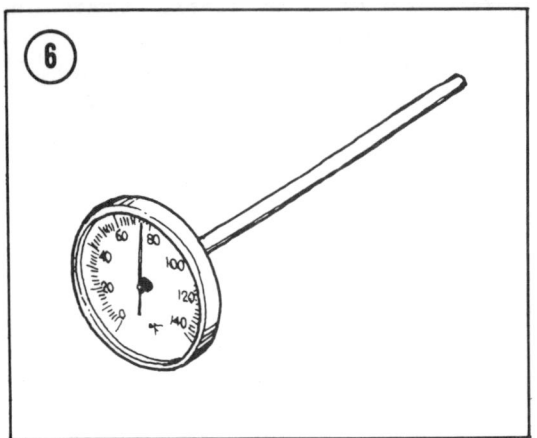

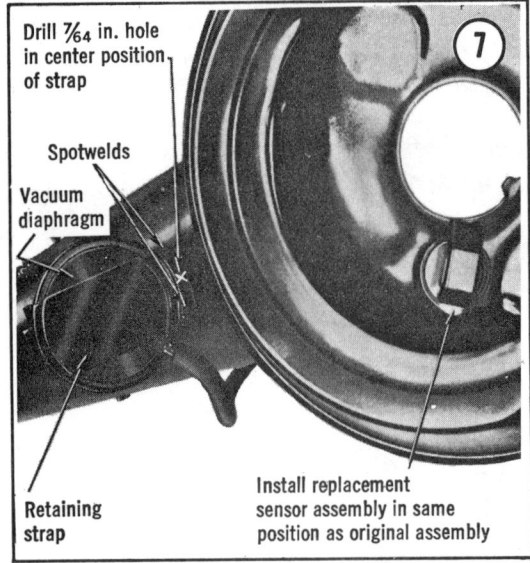

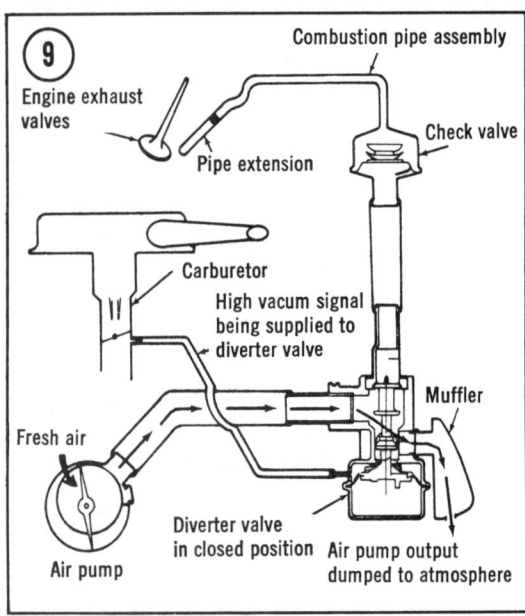

NOTE: *Use a relatively fast acting thermometer such as a photographic darkroom thermometer. See* **Figure 6**. *These are available for about $8.*

4. With engine off, observe damper door position through snorkel opening. Snorkel passage should be completely open (**Figure 5**, View A). If not, check for binds in linkage.

5. Start the engine. With the air temperature below 85°F, snorkel passage should close. When damper door begins to open passage, remove air cleaner cover and observe temperature, which should be between 85-115°F.

6. If damper door does not close completely or open at the right temperature, check the vacuum motor as described below.

Vacuum Motor Checking

1. Turn engine off. Disconnect vacuum hose between sensor and vacuum motor at sensor.

2. Suck on vacuum hose (or otherwise apply at least 9 in. Hg of vacuum). Damper door should completely close snorkel passage. If not, check for vacuum leak at other end of hose.

3. With vacuum applied, bend or tightly clamp hose to trap vacuum in motor. Damper door must remain in position. If not, the vacuum motor leaks and must be replaced.

4. If vacuum motor is good, yet system does not work properly, replace temperature sensor.

EMISSION CONTROL SYSTEMS

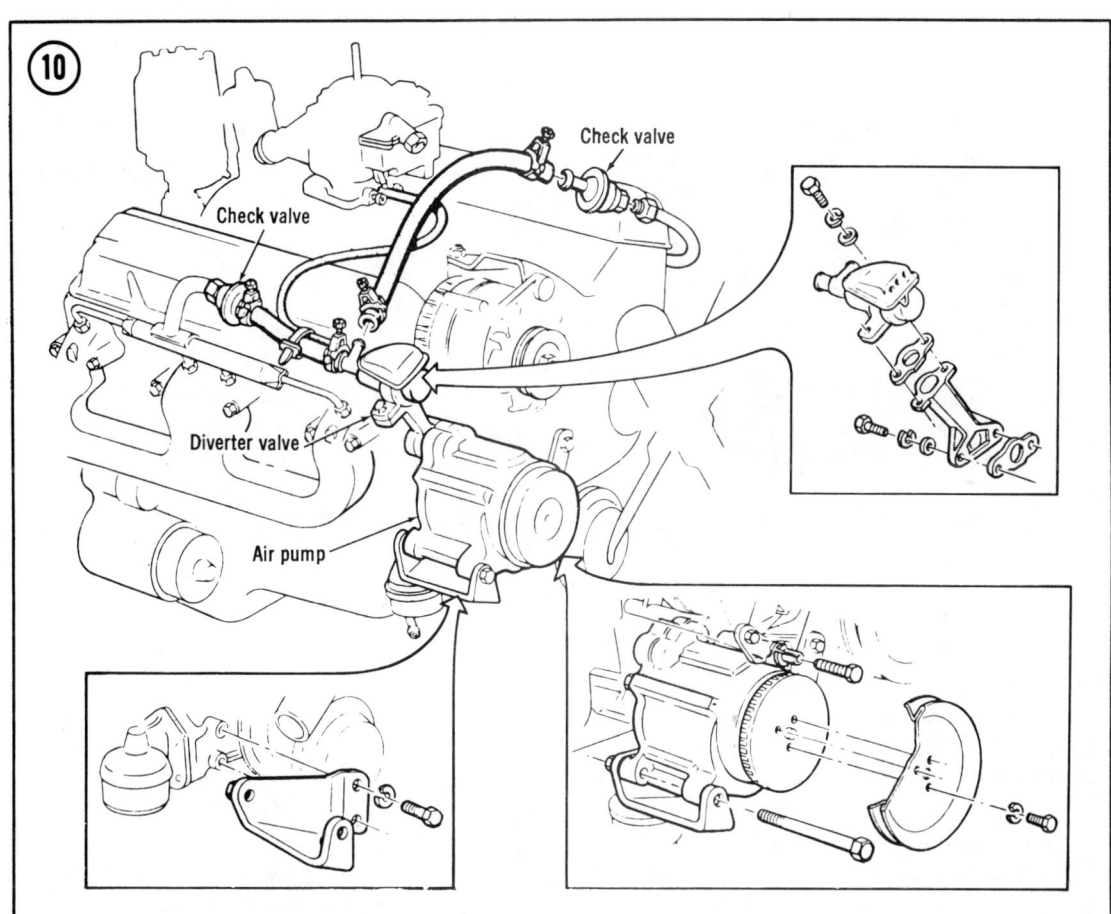

Vacuum Motor Replacement

Refer to **Figure 7** for the following procedure.

1. Remove air cleaner from engine.
2. Drill out spot welds fastening motor retaining strap to snorkel tube.
3. Unhook vacuum motor from damper door.
4. Drill $7/64$ in. hole in snorkel tube at center of vacuum motor retaining strap.
5. Connect vacuum motor to damper door.
6. Fasten retaining strap to air cleaner with sheet metal screw.
7. Install air cleaner and check operation of vacuum motor.

Temperature Sensor Replacement

1. Remove air cleaner from engine and disconnect vacuum hose at sensor.
2. Pry up sensor clip tubes. See **Figure 8**.

NOTE: *Observe position of sensor. New sensor must be installed in same position.*

3. Remove clip and sensor from air cleaner.
4. Install sensor and gasket in air cleaner in exactly the same position as the old one.
5. Press clip on sensor. Hold sensor by its sides only; do not touch control mechanism.
6. Install air cleaner and connect vacuum hoses.

AIR INJECTION REACTOR SYSTEM

The air injection reactor system reduces air pollution by oxidizing hydrocarbons and carbon monoxide as they leave the combustion chamber. See **Figures 9 and 10**.

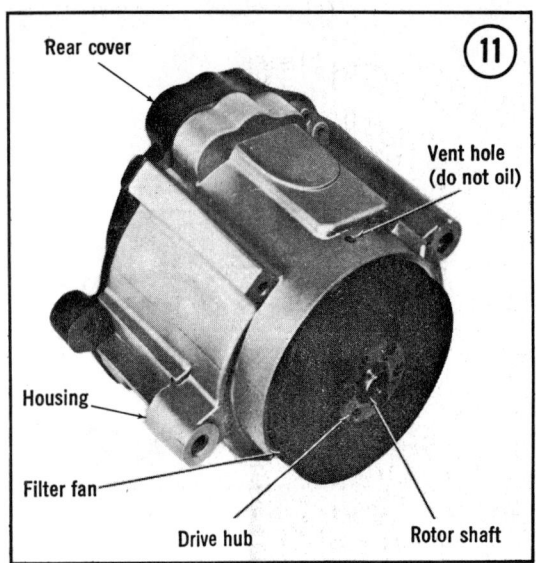

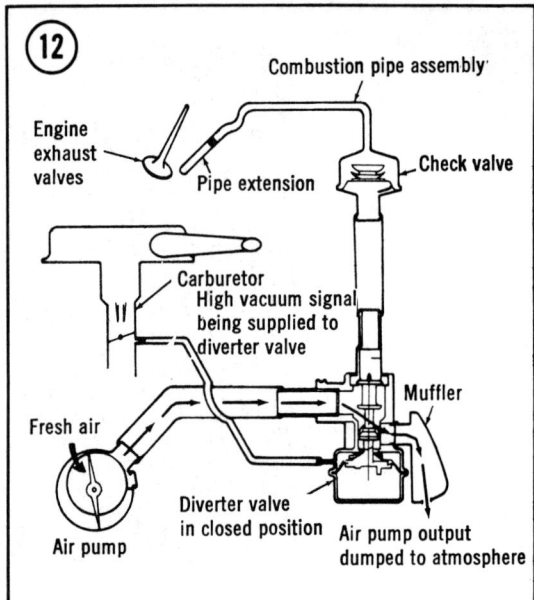

The air injection pump (**Figure 11**), driven by the engine, compresses filtered air and injects it at the exhaust port of each cylinder. The fresh air mixes with the unburned gases in the exhaust and promotes further burning.

Backfire is prevented by cutting the fresh air to the exhaust system and diverting the pump output to the atmosphere. See **Figure 12**.

The check valves prevent exhaust gases from entering and damaging the air pump if the pump becomes inoperative, e.g., from a drive belt failure. Under normal conditions, the pump delivers sufficient air pressure to prevent exhaust gases from entering the pump.

The air injection reactor system also depends on a special calibrated carburetor, distributor, and other related components.

Air Injection Pump Removal/Installation

1. Disconnect hoses at pump.

2. Compress sides of pump belt tightly together to hold pump pulley bolts.

3. Loosen pump mounting bolt and pump adjustment bracket bolt. Swing pump until drive belt can be removed.

4. Remove pump pulley.

5. Remove pump mounting bolts and remove the pump.

6. Installation is the reverse of these steps.

7. Adjust belt tension (Chapter Three).

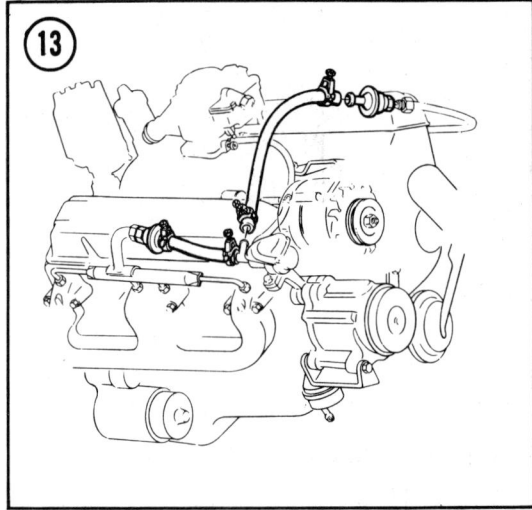

Air Injection Tube Removal/Installation

Refer to **Figure 13**.

1. Remove carbon from tubes.

2. Work tubes out of exhaust manifold. Use a penetrating oil, such as Liquid Wrench.

3. Install tubes in manifold. Use anti-seize compound on threads.

CAUTION
Straight pipe threads are used at the exhaust manifold. Do not use ¼ in. tapered pipe threads.

EMISSION CONTROL SYSTEMS

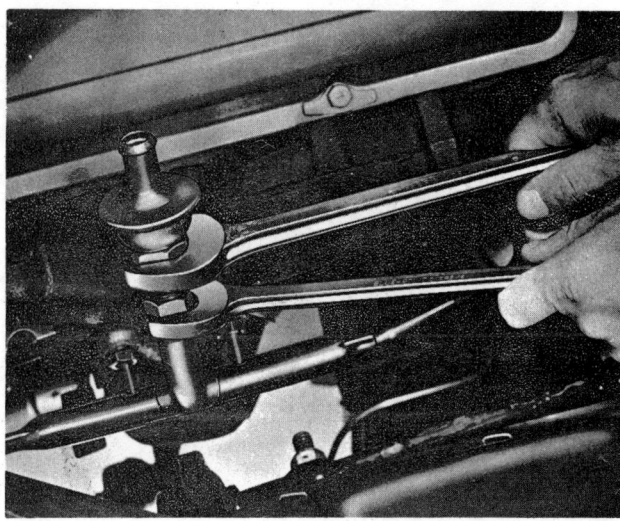

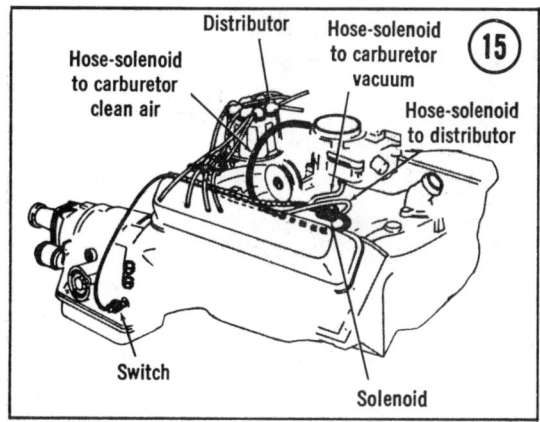

Check Valve Replacement

Refer to **Figure 14** for following procedure.

1. Disconnect pump outlet hoses from valve.
2. Unscrew check valve from air manifold. Be careful not to bend the manifold.
3. Installation is the reverse of these steps.

Air Hose Replacement

To remove any hose or tube, note routing, then remove hose or tube. Install new hose or tube and tighten connection.

CAUTION
Air reactor system hoses are made from special materials to withstand high temperatures. Do not use substitutes.

COMBINED EMISSION CONTROL SYSTEM

This system reduces exhaust emissions by permitting vacuum spark advance while in high gear only. It also prevents dieseling common to emission controlled engines.

The system consists of an electrically operated solenoid which shuts off the vacuum line between the carburetor and the distributor. See **Figure 15**. A switch on the transmission detects when the transmission is in a high gear. **Table 1** explains when vacuum is present and when it is shut off.

Figure 16 is a more detailed diagram of the 1970-1971 system. When the solenoid is not energized, vacuum to the distributor vacuum advance unit is shut off. The distributor is vented to atmosphere through a filter at the opposite end of the solenoid. When the solenoid is energized, the vacuum ports uncover and the plunger shuts off the clean air vent.

The solenoid performs another function besides that of a vacuum switch. When idling in a low gear, e.g., during high gear deceleration with throttle closed, the solenoid is energized and the plunger is extended. This provides a higher idle rpm for reduced hydrocarbon emissions during high gear deceleration.

Two switches and 2 relays control the solenoid. When the transmission is in low gear,

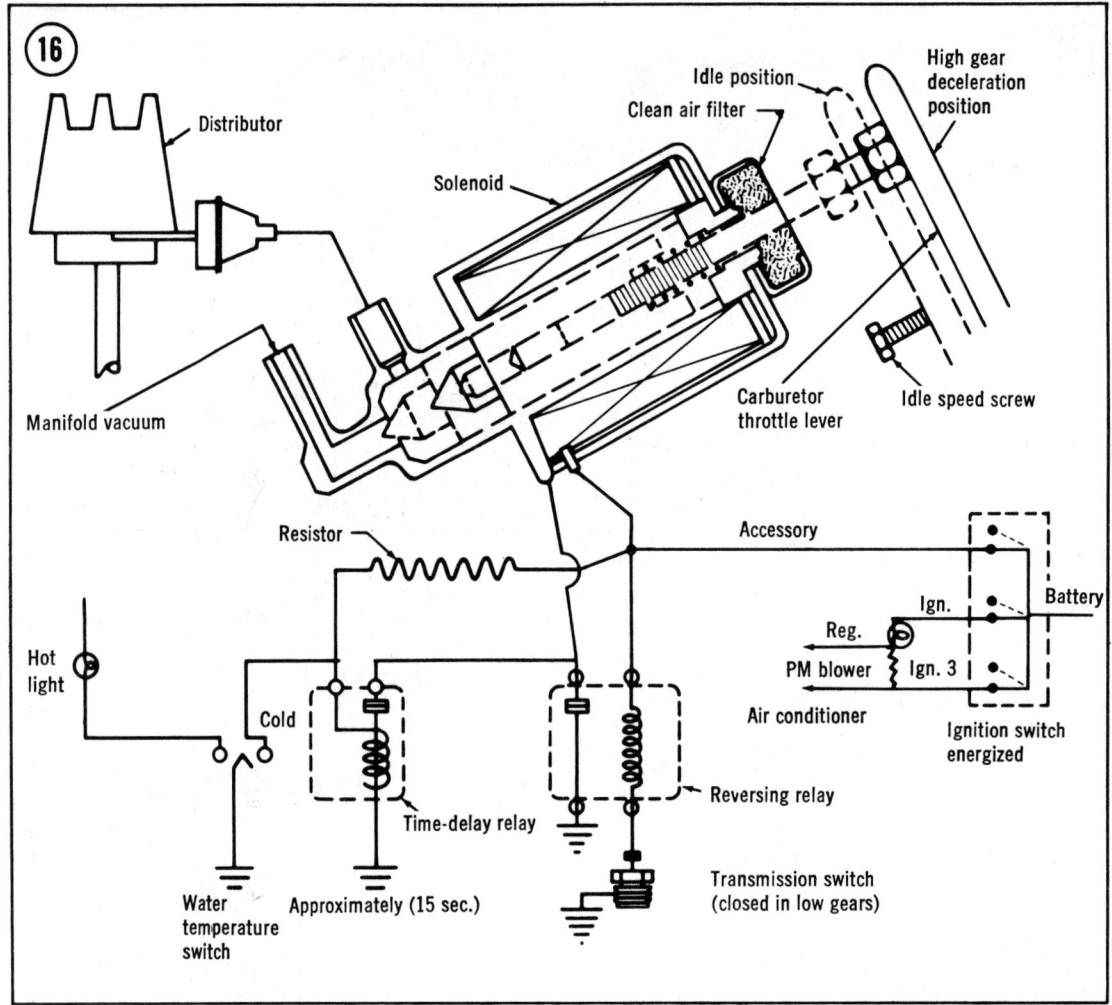

the transmission switch contacts are closed and the reversing relay contacts are open and the solenoid is de-energized. When the transmission shifts to high gear, the transmission switch contacts open, the reversing relay de-energizes, and the reversing relay contacts close. The solenoid energizes.

Two other circuits can energize the solenoid. The time delay relay holds its contacts closed for about 15 seconds after the ignition is turned on. The voltage developed across the resistor energizes the solenoid. Full vacuum during this time improves acceleration and eliminates stalling after a start. Finally, the water temperature switch provides an override when the temperature is below 82°F. The switch closes to ground below this temperature and energizes the solenoid.

The combined emission control system also provides methods to prevent dieseling which is a problem with emission controlled engines. One method is a by-product of the much lower "curb" idle rpm which occurs when the CEC solenoid is de-energized. The engine runs at such a low rpm it cannot diesel.

On air-conditioned vehicles with automatic transmission, the throttle is open more with the engine idling, and the engine tends to diesel if the air conditioner compressor happens to be off. To prevent this, a solid state timer engages the air conditioner clutch for 3 seconds after the ignition is turned off. The additional compressor load stops the engine quicker, reducing its tendency to diesel. **Figure 17** shows the location of components in the 1970-1971 system.

EMISSION CONTROL SYSTEMS

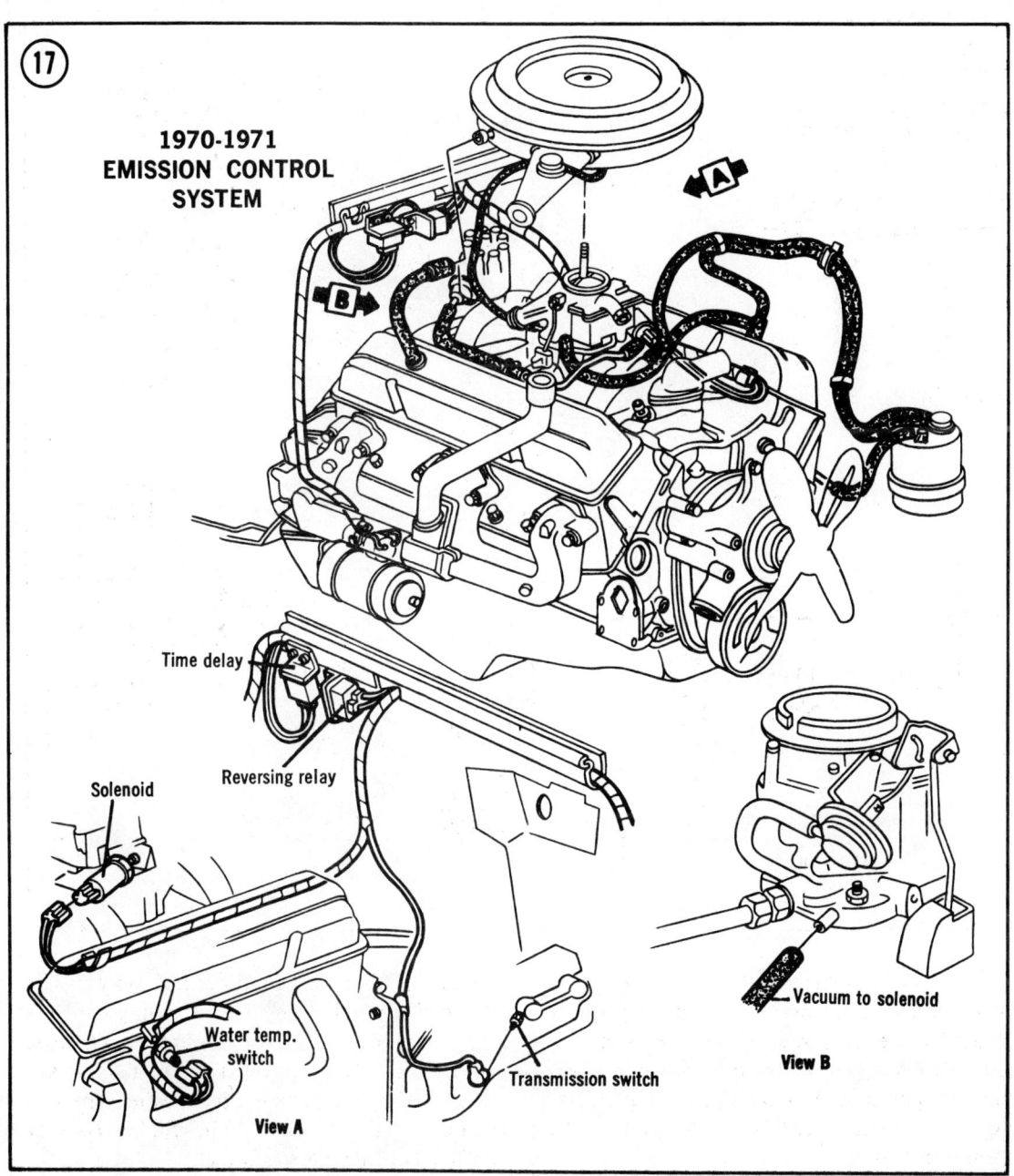

⑰ 1970-1971 EMISSION CONTROL SYSTEM

Table 1 CEC SYSTEM OPERATION

Transmission	Gear						
	Park	Neutral	Reverse	1st	2nd	3rd	4th
3-speed	—	O	O	O	O	X	—
4-speed	—	O	O	O	O	X	X
Turbo-Hydramatic	O	O	X	O	O	X	—

X—vacuum   O—no vacuum

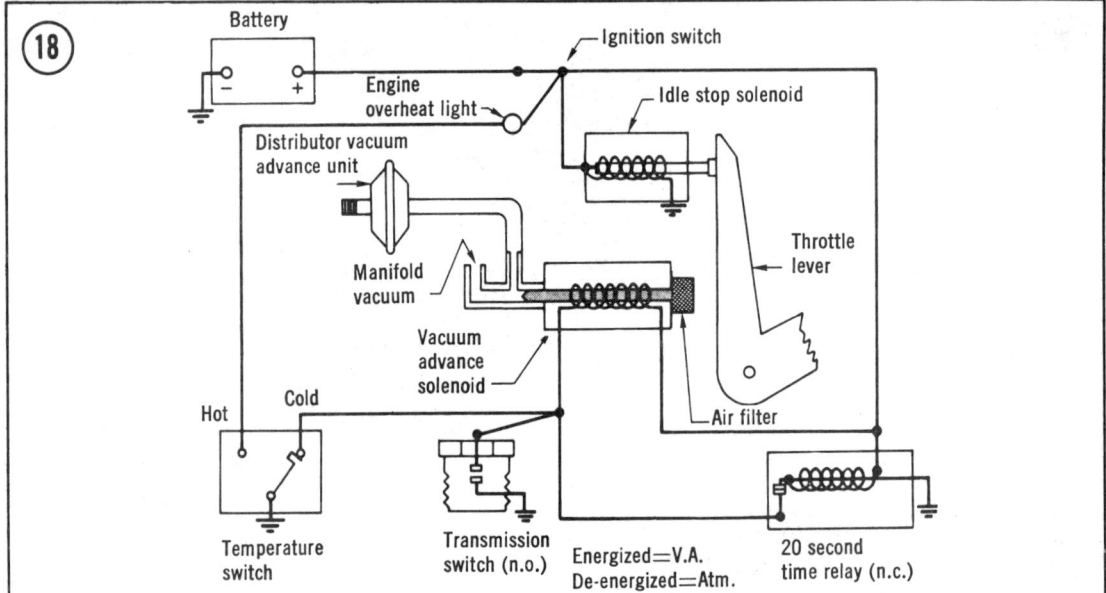

The system on 1972-1974 L6 250 and V8 350 cu. in. engines is similar to the earlier system with the following exceptions:

a. The transmission switch is open in low gears, eliminating the need for reversing relay.
b. A separate idle stop solenoid closes throttle completely when ignition is turned off to prevent dieseling.
c. The water temperature override operates below 82°F as before, but also operates above 232°F.
d. No time delay relay operates during startings.
e. A time delay prevents energizing vacuum advance solenoid for 23 seconds after shifting to high gear.

Figure 18 is a simplified schematic of the L6 250 and V8 350 cu. in. systems. When the ignition is turned on, the idle stop solenoid cracks the throttle open to idle. If the engine cools and temperature is below 82°F, the solenoid operates. Vacuum advance when the engine is cold improves acceleration and helps minimize stalling. If the coolant is above 82°F, the solenoid does not energize.

When engine coolant rises above 82°F, the solenoid can energize only when transmission is in high gear and the 20-second time delay

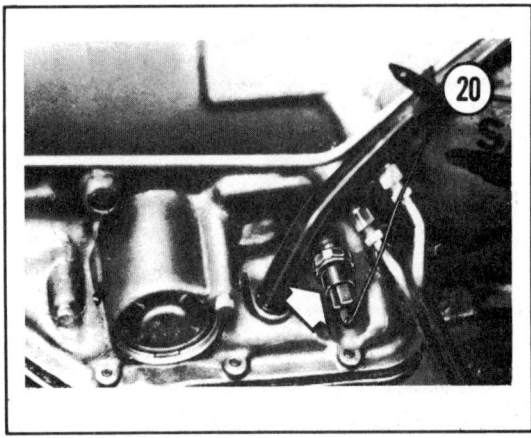

EMISSION CONTROL SYSTEMS

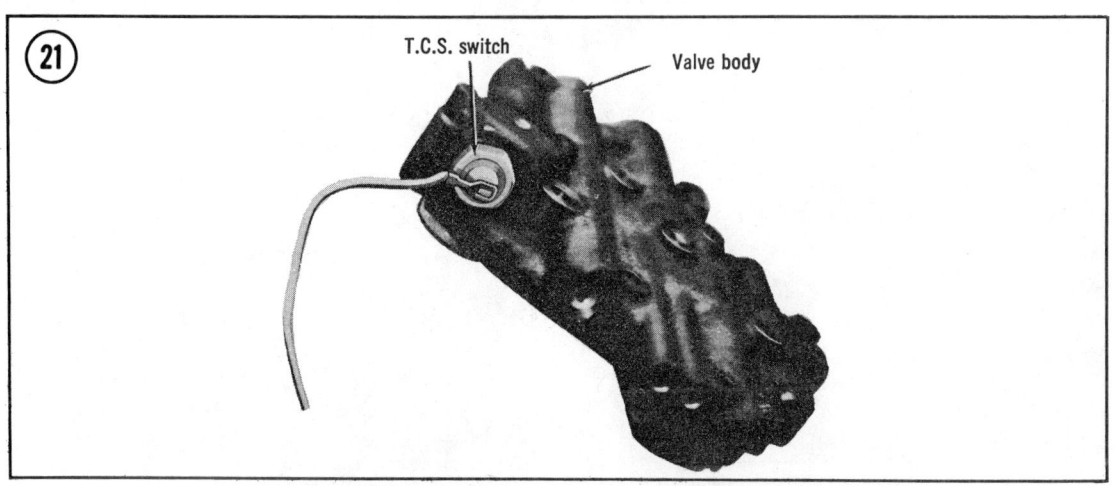

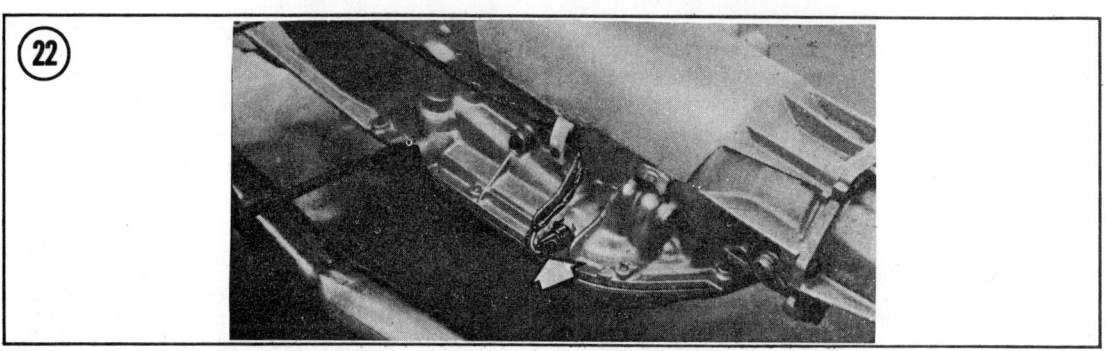

period has passed. In low gear, the transmission switch opens, de-energizing the solenoid. When the transmission is shifted to high gear, the transmission switch closes, but the time delay relay holds its contacts open for 20 seconds, preventing the solenoid from energizing. After 20 seconds, the solenoid energizes. If the transmission downshifts, even momentarily, from high gear, the time delay relay will still prevent energizing the solenoid for 20 seconds.

If coolant temperature exceeds 232°F, regardless of gear, the temperature switch overrides the system and operates the solenoid to supply vacuum advance.

Transmission Switch Replacement (1970-1971)

Refer to **Figure 17**.
1. Disconnect electrical wire.
2. Unscrew switch.
3. Screw in new switch and connect wire.
4. Test switch as described above.

Transmission Switch Replacement (1972-1974 Manual and Turbo Hydra-matic 350)

Figures 19 and 20 show location of the switch on the manual and Turbo Hydra-matic 350 transmission, respectively.
1. Disconnect electrical wire.
2. Unscrew switch.
3. Screw in new switch and connect wire.
4. Test switch as described above.

Transmission Switch Replacement (1972-1974 Turbo Hydra-matic 400)

The switch is located internally (**Figure 21**). The transmission oil pan must be removed to reach the switch. Chapter Nine explains pan removal.

A wire connects the switch to the externally mounted detent solenoid TCS connector. See **Figure 22**. Fortunately, the switch can be tested

CHAPTER SIX

externally. Start engine and place transmission in reverse. If CEC solenoid does not energize, remove the wire to the switch from the terminal on the transmission and ground it. If solenoid energizes, switch is faulty and must be replaced.

CEC Solenoid Replacement (1970-1971)

See *Disassembly* procedure for your carburetor in Chapter Five.

Vacuum Advance Solenoid Replacement (1972-1974)

On 350 engines, the solenoid is located on the right rear portion of the intake manifold. See **Figure 23**. On the 1972 and 1973 L6 engines, the solenoid is located on the carburetor. See **Figure 24**. On 1974 L6 engines, the solenoid is bracket mounted to ignition coil (**Figure 25**).

1. Disconnect vacuum hoses and electrical wires from solenoid.
2. Remove solenoid from bracket.
3. Install new solenoid. Connect vacuum hoses and wires.

Idle Stop Solenoid Replacement (1972-1974)

The idle stop solenoid is mounted on the carburetor. See **Figure 26**. Refer to *Disassembly* procedure in Chapter Five, for your carburetor.

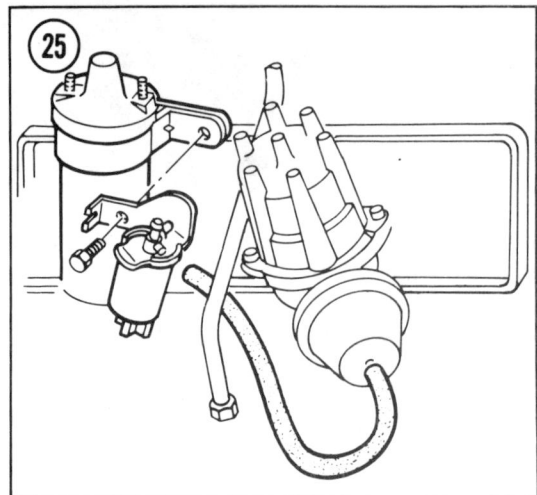

Relay Replacement (1970-1971)

Refer to **Figures 17 and 18**.

1. Disconnect cable from relay.
2. Unscrew relay bracket.
3. Install new relay and reconnect cable.

Time Delay Relay (1972-1974)

The time delay relay is located on the instrument panel reinforcement immediately behind the console instrument cluster assembly (or on the cowl vertical wall on L6 engines).

1. Disconnect cable from relay.

EMISSION CONTROL SYSTEMS

FUEL EVAPORATION CONTROL SYSTEM

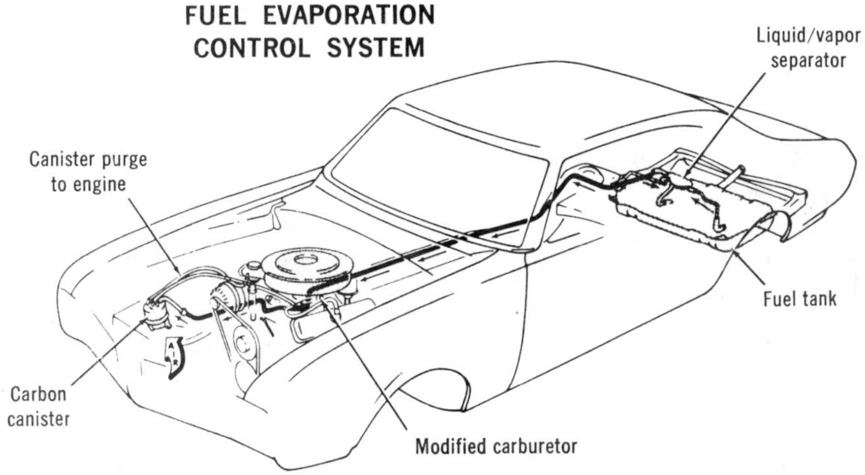

2. Unscrew relay bracket.
3. Install new relay and reconnect cable.

Water Temperature Switch Replacement (1970-1974)

The water temperature switch is located on the left cylinder head on 1970-1971 V8 engines (**Figure 18**) and on the right cylinder head on 1972-1974 V8 engines (**Figure 27**). On L6 engines, the switch is on the left side of the cylinder head.

FUEL EVAPORATION CONTROL SYSTEM

All 1972 and later vehicles (1970 and later in California) are equipped with a fuel evaporation control system which prevents release of fuel vapor into the atmosphere.

Refer to **Figure 28**. Fuel vapor from the fuel tank passes through the liquid/vapor separator to the carbon canister. The carbon absorbs and stores the vapor when the engine is stopped. When the engine runs, manifold vacuum draws

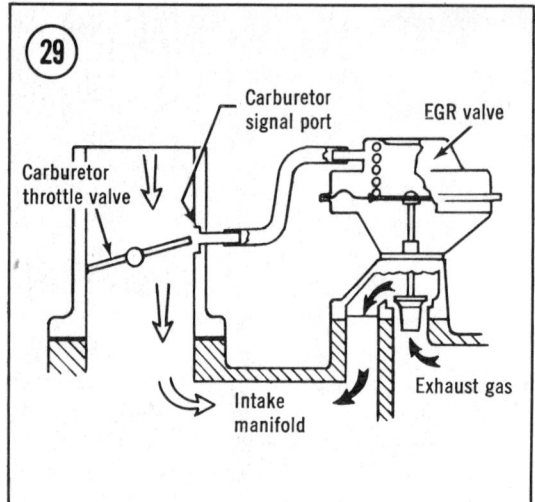
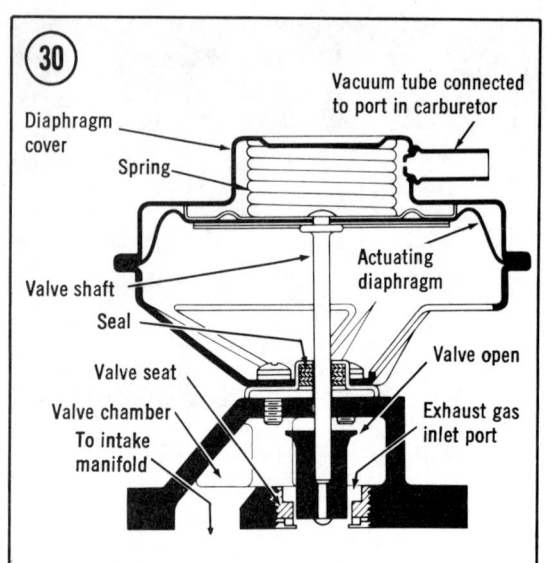

the vapor from the canister. Instead of being released into the atmosphere, the fuel vapor takes part in the normal combustion process.

There is no preventive maintenance other than replacing the filter on the bottom of the carbon canister every 12,000 miles and checking tightness and condition of all lines connecting the parts of the system.

EXHAUST GAS RECIRCULATION (1973 AND LATER)

The Exhaust Gas Recirculation (EGR) system is used to reduce emission of nitrogen oxides (NOX). Relatively inert exhaust gases are introduced into the combustion process to slightly reduce peak temperatures. This reduction in temperature reduces the formation of NOX.

The exhaust gases are introduced into the intake manifold by way of an EGR valve. See **Figure 29**. This shut off and metering valve operates on vacuum from the intake manifold via a signal port in the carburetor. On 1974 and later models, a thermal vacuum switch cuts off vacuum to the EGR valve until water temperature reaches 100-130°F. At idle speed, recirculation is not required. Thus, the carburetor signal port is located above the throttle valve and vacuum to the EGR valve diaphragm is cut off at idle speeds. This causes the EGR to close, halting the introduction of exhaust gas to the intake manifold. As the throttle valve is opened, the signal port is again exposed to manifold vacuum. This actuates the EGR valve diaphragm, which opens the valve and allows exhaust gas to be metered (through an orifice) into the intake manifold. See **Figure 30**.

On L6 engines, the EGR valve is located on the intake manifold next to the carburetor. See **Figure 31**. On V8 engines, the valve is located externally on the right side of the intake manifold next to the rocker arm cover. See **Figure 32**.

EGR Valve Replacement

1. Disconnect vacuum line at top of valve.
2. Remove bolt and clamp securing valve to manifold, then remove valve from manifold.
3. Reassemble valve to manifold, using new gasket, and torque bolt to 25 ft.-lb. Bend lock tab up over bolt head. Replace vacuum line.

EGR Valve Cleaning

CAUTION
Do not wash valve assembly in solvents or degreaser. Permanent damage could result.

1. Use wire brush or wire wheel to clean valve base and remove exhaust deposits from mounting surface.
2. A spark plug cleaner (sandblaster) can be used to clean valve seat and pintle. Insert valve and pintle into machine and blast for 30 seconds.

EMISSION CONTROL SYSTEMS

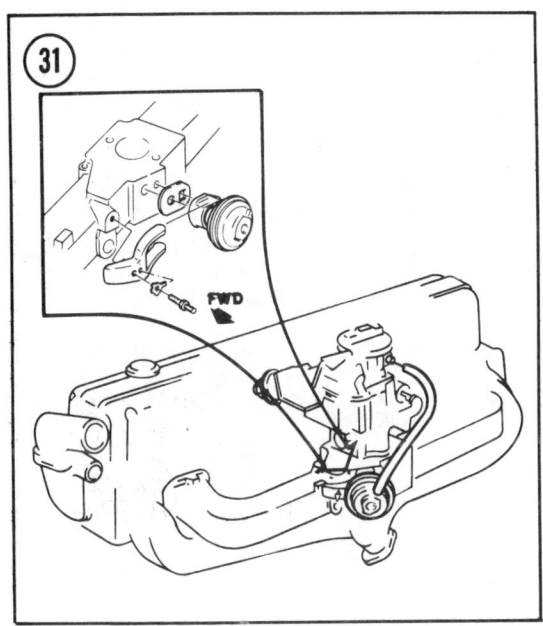

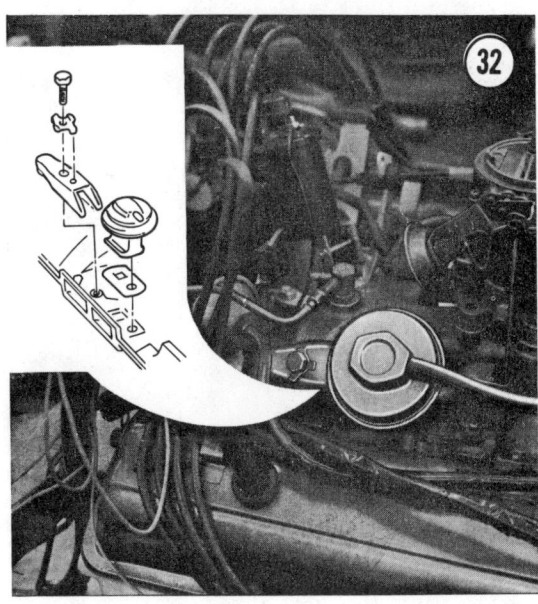

NOTE: *Many service stations have this type of spark plug cleaner and will clean the valve for a small fee.*

3. Compress the diaphragm spring so the valve is fully unseated and repeat sandblasting for 30 seconds.
4. Make sure all exhaust deposits have been removed. Repeat cleaning if required.
5. Use compressed air to remove all abrasive material from the valve.

Thermal Vacuum Switch Replacement (1974)

1. Disconnect vacuum lines and remove switch from thermostat housing.
2. Apply a sealer to threads and install switch in thermostat housing. Torque to 15 ft.-lb.
3. Rotate head of switch as required for proper hose routing and install vacuum hoses.

NOTE: *Thermal vacuum switch is non-repairable. If defective, replace.*

CARBURETOR CALIBRATION

In addition to providing the engine with a combustible mixture of air and fuel, the carburetor is also calibrated to maintain proper emission levels. The idle, off-idle, power enrichments, main metering, and accelerating pump systems all are calibrated to provide the best possible combination of performance, economy, and exhaust emission control.

Calibration is especially critical on 1975 and later models, and the tasks involved require special skills and test equipment the home mechanic is not likely to have. Except for the adjustments of idle speed and idle mixture described in Chapter Three, carburetor work on late models should be entrusted to an expert.

DISTRIBUTOR CALIBRATION

Distributor calibration consists of adjusting initial timing, centrifugal advance, and vacuum advance to obtain the best combination of engine performance, fuel economy, and exhaust emission level control. Timing specifications for each engine are given on the Vehicle Emission Control Information (VECI) decal located in the engine compartment. See *Ignition Timing Adjustment*, Chapter Three, for instructions on how to set the initial timing. Adjustment or repair of the centrifugal and vacuum spark advance systems on 1975 and later cars should be left to an expert.

CATALYTIC CONVERTER

Catalytic converters, used since 1975, reduce air pollutants by promoting further burning of

CHAPTER SIX

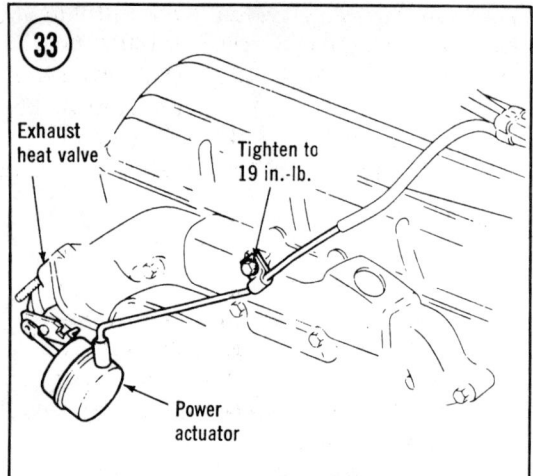

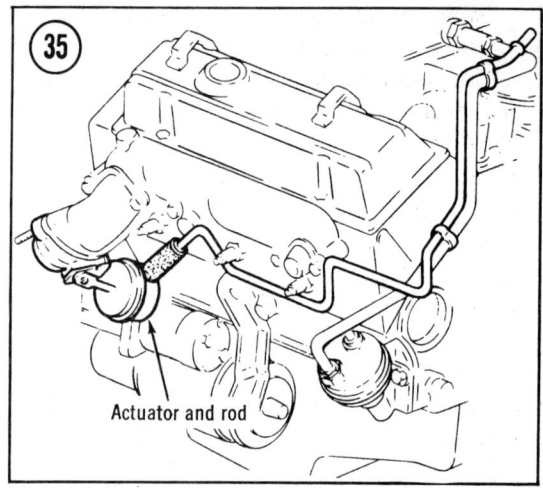

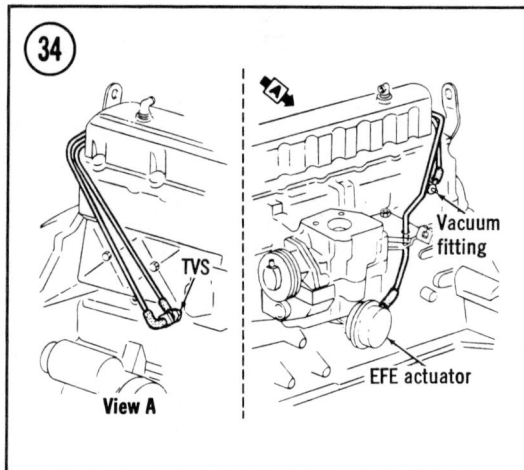

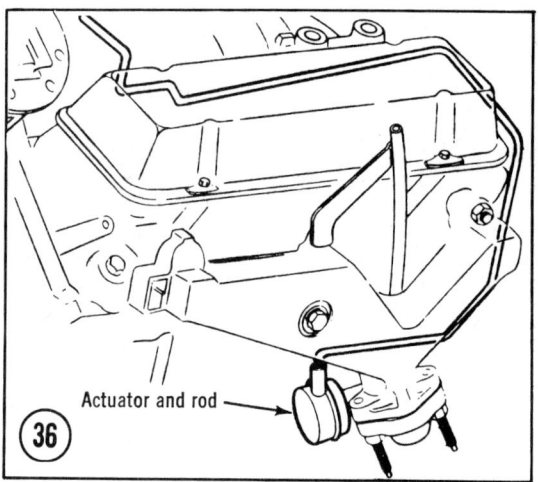

the exhaust gases. The converter is located in the exhaust line ahead of the muffler and contains a material coated with platinum and palladium. Both of these materials are catalysts, and cause the reduction of hydrocarbons and carbon monoxide by burning as the exhaust gases pass over them. The catalytic converter should be checked for general condition at the same time the remainder of the exhaust system is checked (see *Periodic Maintenance*, Chapter Three). See *Catalytic Converter*, Chapter Five for replacement procedure.

EARLY FUEL EVAPORATION SYSTEM

The early fuel evaporation (EFE) system, used on 1975 and later models, provides heating to the intake manifold while the engine is cold, to promote vaporization of the fuel. This helps cut down on choke time and also promotes more thorough burning of the fuel. This, in turn, reduces the amount of pollutants released into the air. The heart of the system is the EFE valve, which controls the amount of heat (exhaust gas) directed under the intake manifold. The valve is controlled by a thermostatic switch in the cooling system that applies or removes vacuum to the EFE valve according to the temperature of the engine coolant. Use the following procedure to check operation of the EFE system.

EFE System Check

Refer to **Figure 33** (V8 engines), **Figure 34** (L6 engines), **Figure 35** (200-229 V6 engines), or **Figure 36** (231 V6 engines).

EMISSION CONTROL SYSTEMS

1. With engine cold, place transmission in NEUTRAL or PARK, apply handbrake, and start engine. Observe movement of EFE actuator rod and heat valve. Valve should move to closed position.

2. If valve does not close, remove vacuum hose from actuator and check for presence of vacuum. If vacuum is present, replace the actuator. If no vacuum is present, remove vacuum input hose at thermovacuum switch (TVS) and check for vacuum. If vacuum is present, replace TVS. If no vacuum is present, check for damaged vacuum hose and replace as required.

3. If valve closes, allow engine to warm up until coolant temperature reaches 180°F (V6 and V8) or 150°F (L6). Exhaust heat valve should move to open position. If not, remove hose from actuator and check for vacuum. If no vacuum is present, replace actuator. If vacuum is present, replace TVS.

NOTE: *When replacing* TVS *on V8 engines, drain coolant below level of coolant outlet housing. Apply a soft setting sealant to replacement switch threads before installation, and torque to 120 in.-lb.*

CHAPTER SEVEN

COOLING, HEATING, AND AIR CONDITIONING SYSTEMS

All vehicles use pressurized cooling systems, sealed with a pressure type radiator cap. The higher operating pressure of the system raises the boiling point of the coolant. This increases the efficiency of the radiator.

COMPONENTS

Cooling system components are the radiator, pressure cap, water pump, thermostat, fan, thermostatic fan clutch (on 1972 and later models so equipped), and associated hoses and water passages.

Radiator

The cross-flow type radiator, used on all models, cools the coolant fluid by allowing it to flow from side to side through the radiator and transferring heat to the air passing through the radiator. See **Figure 1**.

Radiator Cap

The pressure-type radiator cap (**Figure 2**) allows the cooling system to operate at higher than atmospheric pressure, thus raising the boiling point of the coolant. This permits operation of the vehicle with coolant temperatures between 245°F and 260°F without vaporization of the coolant through boiling. The pressure cap has a pressure relief valve which allows excessive pressure to be vented. It also has a vacuum valve which opens to relieve the vacuum created when the system cools.

Fan/Fan Clutch

The cooling fan increases cooling system efficiency by drawing air through the radiator. Two types of fans are used on these cars. One is a fixed drive fan which rotates at water pump speed. The other type has a thermostatically-controlled clutch which ensures adequate cooling at reduced engine speeds and eliminates overcooling, excessive noise, and power loss at high speeds. See **Figure 3**.

Thermostat

The thermostat (**Figure 4**) is basically a heat-controlled valve in the cooling system, located in the engine water passage outlet to the radiator. When the engine and coolant are cold, the valve remains closed, preventing the circulation of coolant through the radiator. When the engine and coolant approach normal operating temperature, the thermostat opens to allow coolant to flow to the radiator for cooling.

COOLING, HEATING, AND AIR CONDITIONING SYSTEMS

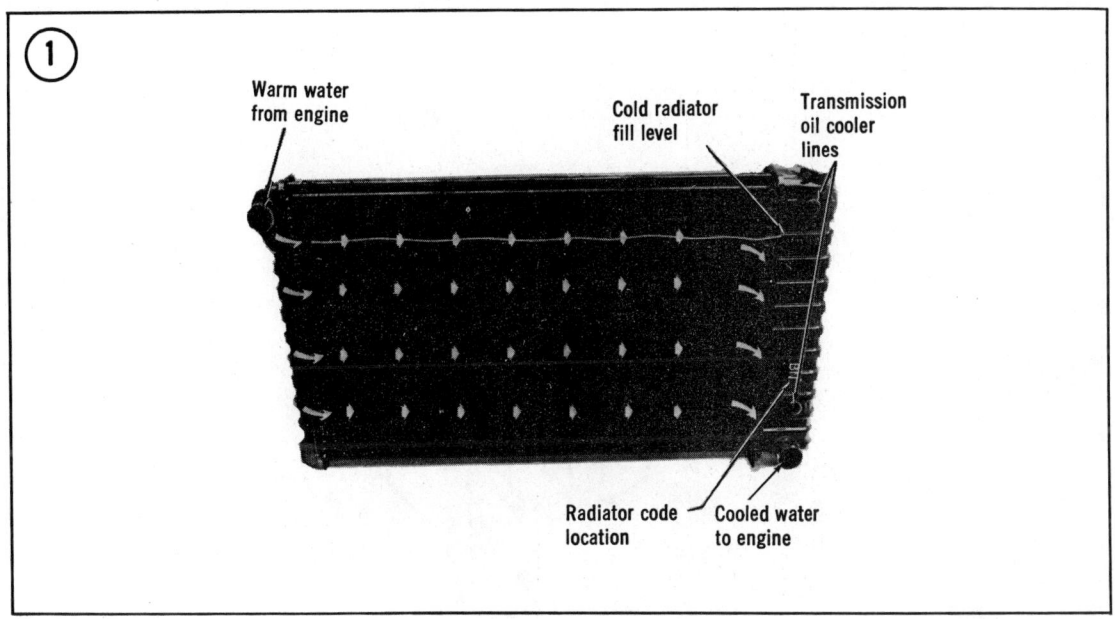

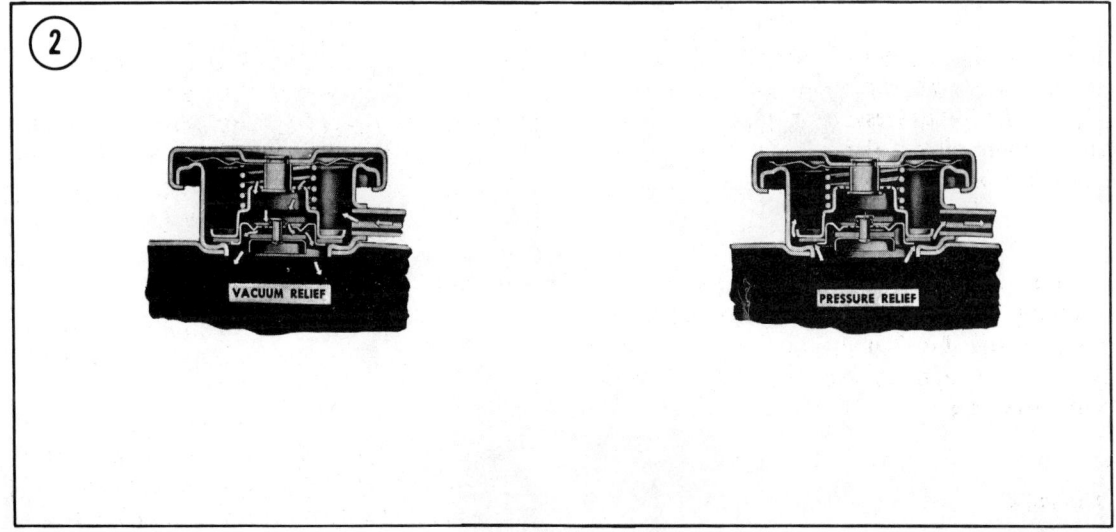

Coolant

Only ethylene glycol-based coolant meeting the requirements of GM Specification 1899-M should be used.

Coolant Recovery System

This system, which has a translucent reservoir containing a supplemental supply of coolant, catches the coolant that escapes through the radiator cap pressure relief valve. When the engine cools, the fluid is drawn back into the radiator by vacuum. See **Figure 5**.

Water Pump

The centrifugal vane impeller water pump in the cooling system has sealed bearings and requires no periodic maintenance. The pump inlet is connected to the bottom of the radiator by a hose. The pump causes water to circulate through the water passages in the engine block and cylinder heads, where engine heat is transferred to the water. The engine and head passages are connected and terminate at the water outlet of the thermostat housing, which is connected to the top of the radiator by a hose.

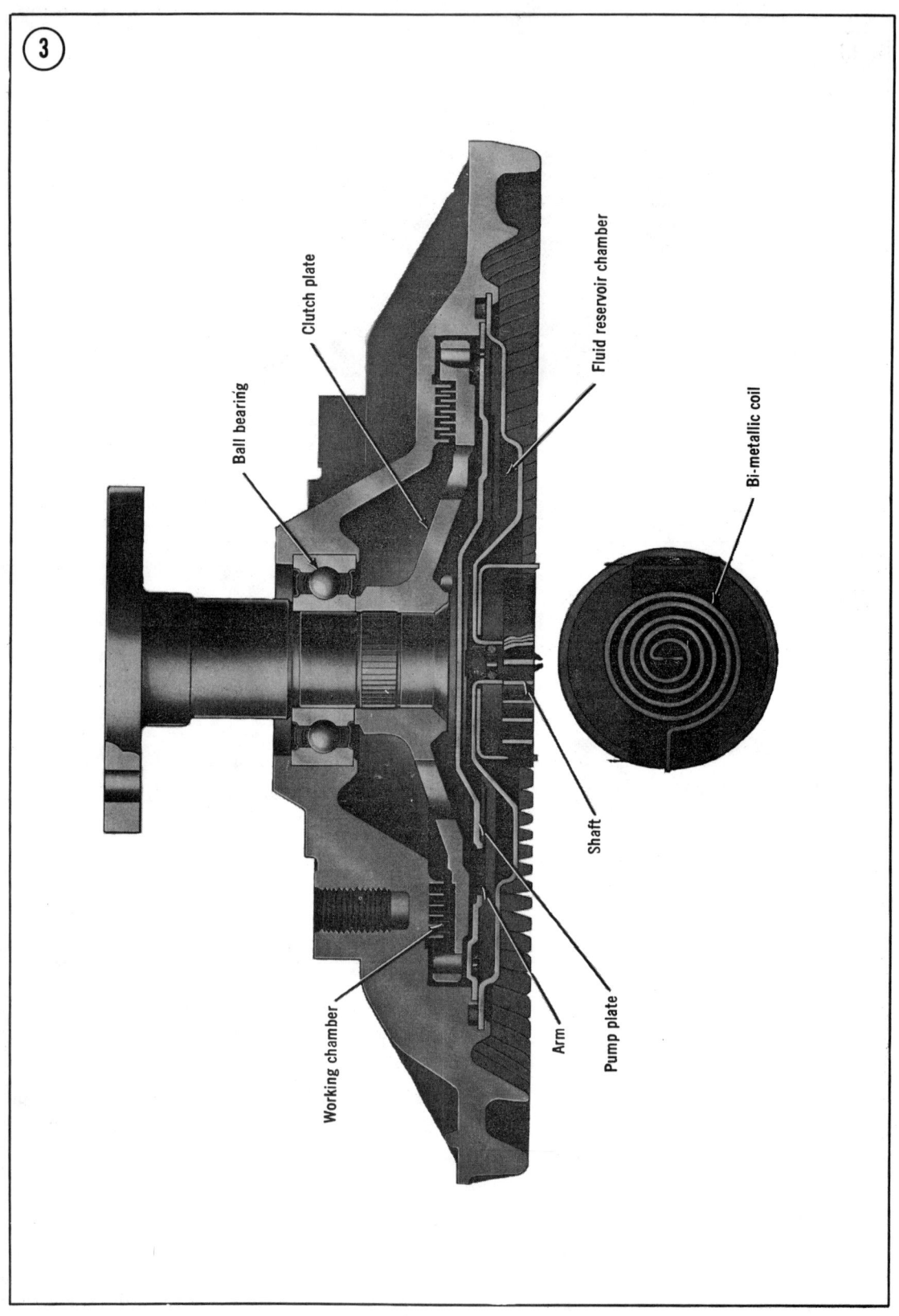

COOLING, HEATING, AND AIR CONDITIONING SYSTEMS

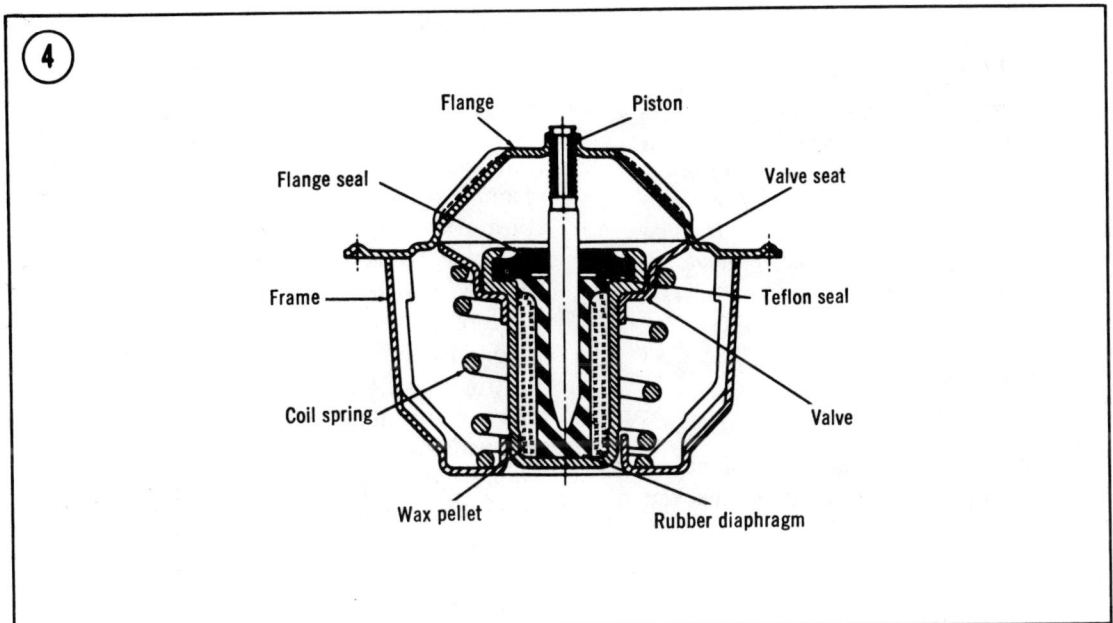

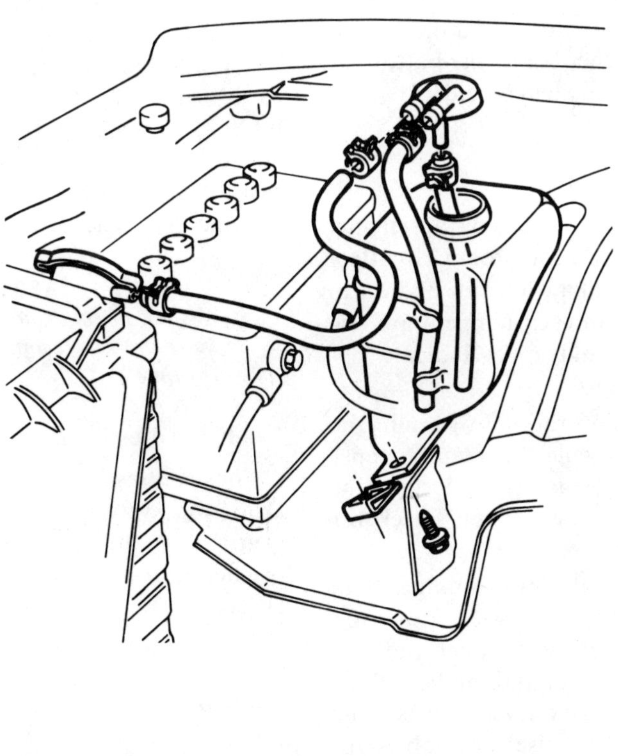

CHAPTER SEVEN

MAINTENANCE

Coolant Level

> NOTE: *On vehicles equipped with coolant recovery systems the coolant is checked by observing the liquid in the recovery system reservoir. The radiator cap should not be removed. If additional coolant is required it should be added to the recovery reservoir. Level should be at* COLD FULL *mark when engine is cool or at* HOT FULL *mark when engine is hot.*

If vehicle does not have a coolant recovery system, coolant level should be checked only when engine is cool.

> CAUTION
> *Removal of radiator cap from a hot engine, especially if an air conditioner has been in use, can result in coolant "blow out." This results in coolant loss and could cause injuries to bystanders.*

Coolant level in cross-flow radiators should be maintained at 3 inches below the bottom of the filler neck when the system is cold.

Cooling System Checks

1. Check the radiator by warming up engine. Turn engine off and feel radiator. Cross-flow radiator should be hot along left side and warm along right side with an even temperature rise from right to left. Cold spots indicate obstructed radiator sections.
2. Check water pump operation by running the warmed up engine while squeezing upper radiator hose. If a pressure surge is felt, the water pump is functioning. If not, check for plugged vent hole in pump.
3. Check for exhaust leaks into cooling system by draining coolant until level is just above top of cylinder head. Disconnect upper radiator hose and remove thermostat and fan belt. Start engine and accelerate engine several times while observing coolant. If level rises or bubbles appear, chances are that exhaust gases are leaking into the cooling system.

Periodic Maintenance

The concentration of antifreeze in the coolant should be maintained to provide protection to $-20\,°F$, regardless of expected temperatures. This is required for protection from corrosion and for proper temperature indicator light operation.

Every 2 years the cooling system should be drained and back flushed with clear water. If required, the system should be cleaned with a good cleaning solution (follow manufacturer's instructions). Back flush the system by removing radiator upper and lower hoses and replacing radiator cap. Attach a lead-away hose to upper radiator opening and a length of garden hose to lower opening. If an air/water flushing gun is available, connect it to garden hose and connect gun water hose to water supply and air hose to compressed air supply. Allow water to fill radiator, then apply air in short bursts. Repeat process until water coming from lead-away hose runs clear. If gun is not available, connect garden hose to pressurized water supply and back flush until water from lead-away hose runs clear. Then the system should be refilled with ethylene glycol-based coolant/water solution sufficient to provide protection to at least $-20\,°F$. GM Cooling System Inhibitor and Sealer, or equivalent, should also be added at this time to retard the formation of rust or scale. This inhibitor should also be added every fall thereafter.

> CAUTION
> *Alcohol or methanol coolants or plain water are not recommended for use at any time.*

Fan Belt Adjustment

Loosen bolts at Delcotron mounting and pull Delcotron away from engine until desired tension (75 ±5 lb. used, or 125 ±5 lb. new) is reached (use strand tension gauge). Tighten all Delcotron bolts securely and repeat tension check. Readjust if required.

Thermostat Check and Replacement

1. Drain coolant until level is slightly below thermostat housing base.
2. Remove upper radiator hose.

COOLING, HEATING, AND AIR CONDITIONING SYSTEMS

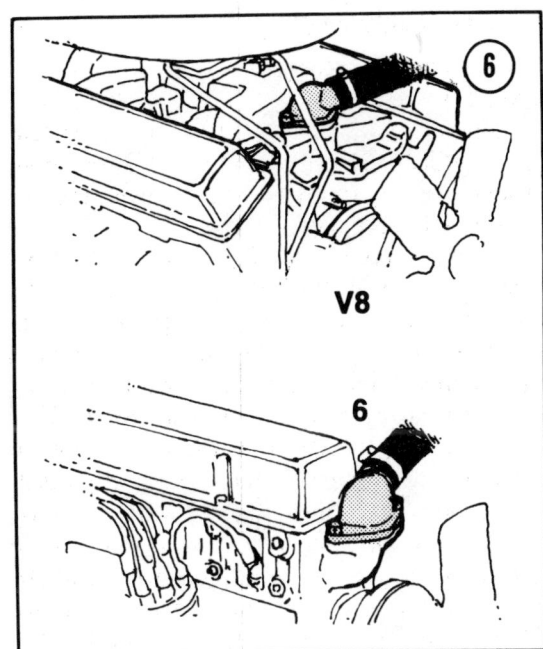

3. Remove thermostat housing bolts and then remove water outlet and gasket from thermostat housing. See **Figure 6**.

4. Remove and inspect thermostat valve for condition.

5. Test thermostat valve as follows:

   a. Place thermostat in a 33% solution of glycol heated to 25°F above the temperature stamped on the thermostat.

   b. Submerge thermostat and agitate liquid. Valve should open fully.

   c. Remove valve and place in another 33% glycol solution heated to 10°F under the temperature stamped on thermostat.

   d. Thermostat should close completely when completely submerged and liquid is agitated.

6. If thermostat fails the above test it should be replaced. If OK, reinstall in housing, using a new gasket. Tighten bolts securely.

7. Replace radiator hose and refill with coolant.

Belts and Hoses

All engine belts and hoses should be inspected for wear and damage at each oil change period. Belts which show signs of wear, cracking, etc., and hoses with bulges, soft spots, leaks, etc., should be replaced. The inspection should include heater hoses as well as radiator hoses.

Thermostatic Fan Clutch

Check with engine cool. Fan blade should rotate freely, with only a slight drag. If fan does not move, or drag is either excessive (or a rough grating is felt) or not present (fan revolves over 5 times when spun by hand), the clutch should be replaced.

WATER PUMP

Removal

1. Siphon or drain coolant from radiator. Remove fan shroud, if so equipped, and break loose fan pulley bolts.

WARNING
If coolant is siphoned from radiator, do not use mouth to start siphon. Coolant solution is poisonous and can cause death or serious illness if swallowed.

2. Disconnect heater hose, lower radiator hose and bypass hose (if so equipped) from water pump.

3. Remove Delcotron (alternator) upper brace (V6 and V8 engines), loosen swivel bolt, and remove fan belt. If so equipped, disconnect power steering and air conditioner belts and swivel power steering pump unit to one side.

4. Remove fan blade attaching bolts, fan, and pulley.

CAUTION
Bent or damaged fans should not be reused, as any distortion will affect fan balance and operation. Damaged fans cannot be properly repaired and should be replaced.

NOTE: *When a thermostatic fan clutch is removed from the car it should be supported so that clutch disc remains vertical to prevent silicone fluid leakage.*

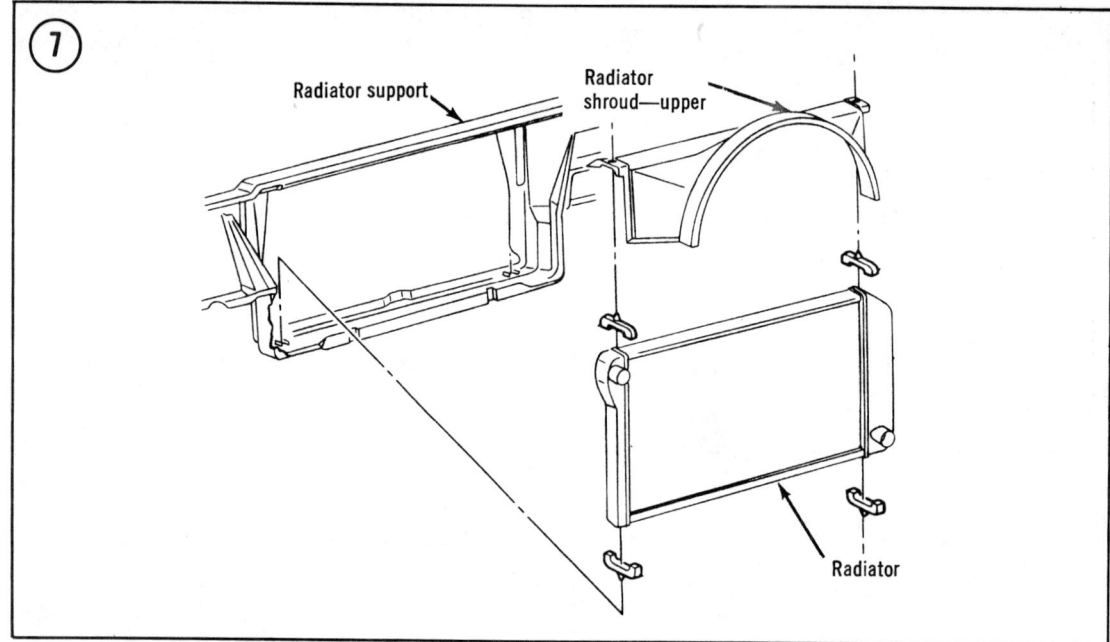

5. Remove pump-to-cylinder block and power steering-to-pump bolts. Remove water pump and old gasket from engine.

CAUTION
On inline 6-cylinder engines, pull pump straight forward out of block first to avoid impeller damage.

Installation

1. Install water pump assembly on engine block, using a new sealer-coated gasket. Tighten bolts 15 ft.-lb. for inline engines, or 30 ft.-lb. for V6's and V8's.

2. Install pump pulley and fan on pump hub and tighten to 15 ft.-lb.

NOTE: *For ease in aligning pulley and fan, install a $5/16$ in.-24 guide stud (bolt with head removed) in one hole of fan hub. Remove stud and install bolt after other 3 bolts have been started.*

3. If disconnected, install power steering and air conditioner belts.

4. Connect hoses and fill cooling system with solution of ethylene glycol antifreeze sufficient to withstand $-20°F$.

5. Install Delcotron upper brace (V6 and V8 only) and power steering pump bolt. Install fan

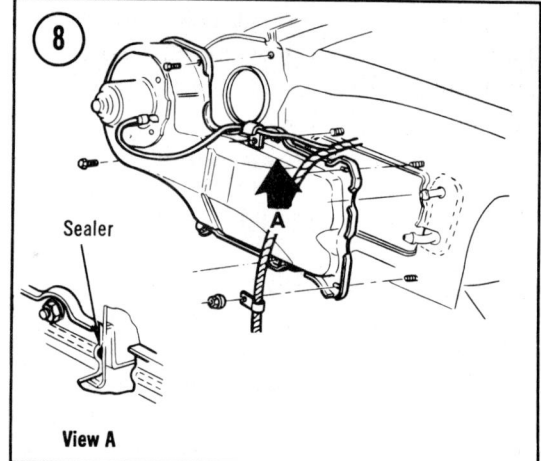

View A

belts and tension to 75 ±5 lb. (used belt) or 125 ±5 lb. (new belt).

6. Start engine and inspect installation for leaks.

RADIATOR

Removal

Refer to **Figure 7** for this procedure.

1. Disconnect the battery negative cable at the battery. Then drain the radiator, remove the fan shroud (if so equipped), and remove the fan.

COOLING, HEATING, AND AIR CONDITIONING SYSTEMS

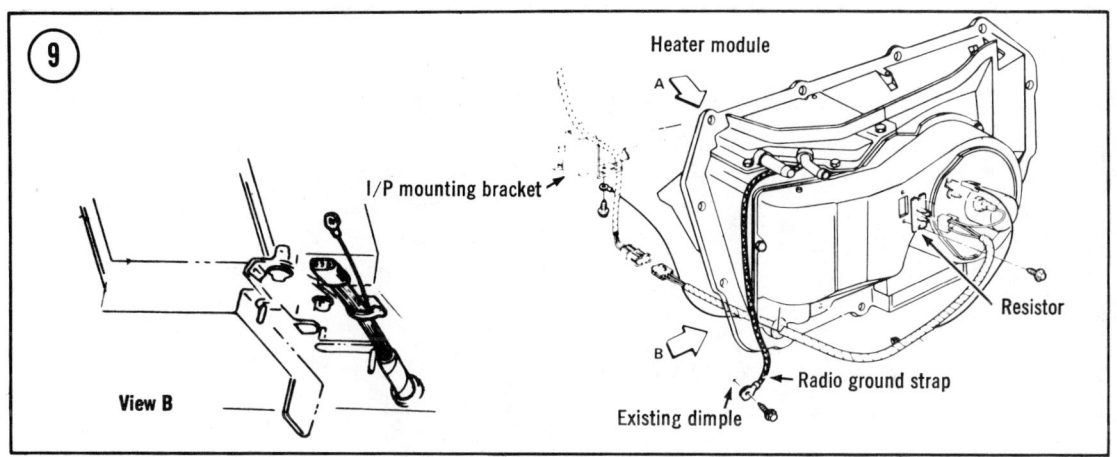

NOTE: *If the vehicle is equipped with a fan clutch, store the fan clutch in an upright position to prevent leakage of the clutch solution through the seal.*

2. Remove the upper and lower radiator hoses from the radiator.

3. If the vehicle has an automatic transmission, disconnect the transmission fluid cooler lines from the radiator. Plug the lines to prevent fluid loss and contamination.

4. Remove the radiator attaching bolts and remove the radiator and fan shroud assembly by lifting it straight up.

Installation

1. If a new radiator is being installed, remove the fan shroud and necessary fittings from the old radiator and install them on the new one.

2. Install the radiator and shroud assembly in the vehicle and securely tighten the attaching bolts.

3. Reconnect the automatic transmission fluid cooler lines to the radiator.

4. Connect the upper and lower radiator hoses to the radiator. If the hoses are more than 2 years old, or are in poor condition, it is a good idea to replace them at this time.

5. Install the fan and fan shroud and fill the radiator with the proper solution of ethylene glycol antifreeze and water. Follow the antifreeze manfacturer's instructions to arrive at the suitable solution for your climate.

6. Connect the battery negative cable and start the engine. Check the water hose and transmission cooler line connections for leaks. Also check the automatic transmission fluid level, if so equipped.

HEATER

Blower Motor Replacement (1970-1977)

Refer to **Figure 8**.

1. Disconnect the battery negative cable at the battery and then disconnect the electrical lead wire from the heater blower motor.

2. Remove the blower-to-case attaching screws and remove the blower assembly from the vehicle. Pry gently on the flange, if necessary, if the sealer does not allow easy removal.

NOTE: *On some early models, it may be necessary to shift the fender skirt to remove the blower assembly. In this case, remove all the skirt attaching screws except those holding the skirt to the radiator support. Then insert a 2 in. x 4 in. wood block between the skirt and the fender to provide the necessary clearance.*

3. Remove the blower wheel retaining nut and separate the wheel and motor.

4. Installation is the reverse of these steps. Make sure the open end of the blower wheel is away from the blower motor.

Blower Motor Replacement (1978 On)

Refer to **Figure 9**.

1. Disconnect the battery negative cable and then disconnect the electrical connector at the

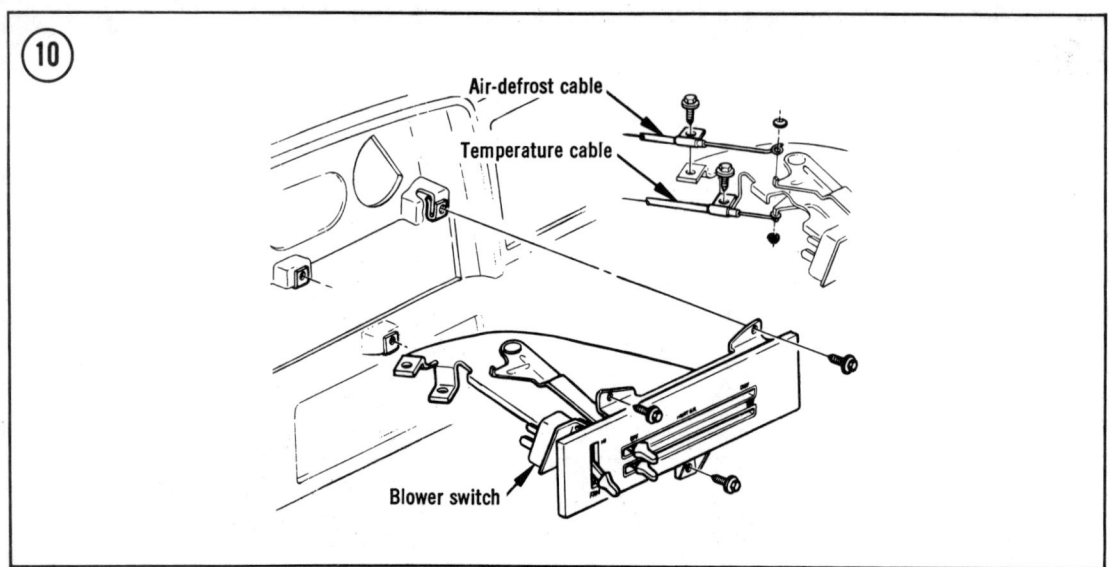

blower assembly. Disconnect the heater and radio ground straps.

2. Remove the blower-to-case attaching screws and remove the blower motor assembly.

3. Installation is the reverse of these steps. Make sure the ground straps are reinstalled in their original locations.

Control Head Assembly Replacement (All Models)

Refer to **Figure 10** (typical).

1. Disconnect the battery negative cable and remove the radio and control knobs.

2. Remove the instrument cluster bezel. See **Figure 11** (typical).

3. Remove the radio center speaker, if so equipped and if necessary for clearance.

4. Remove the 3 attaching screws from the control assembly and pull the assembly far enough to the rear to disconnect the bowden cables. Also disconnect the wiring harness from the blower switch. Then remove the control assembly. Take care not to kink the cables.

5. Installation is the reverse of these steps.

AIR CONDITIONING

Major service and repair to air conditioning systems requires specialized training and tools, and the difficulty of the work is compounded in the late heating/air conditioning systems. However, most air conditioning problems do not involve major repair; they are well within the ability of an experienced hobbyist mechanic, armed with an understanding of how the system works.

SYSTEM OPERATION

A typical air conditioning system is shown in **Figure 12**. (Actual component locations may differ, depending on model.)

Five basic components are common to all air conditioning systems:

1. Compressor
2. Condenser
3. Receiver/drier
4. Expansion valve
5. Evaporator

The components, connected with high-pressure hoses and tubes, form a closed loop. A refrigerant, dichlorodiflouromethane — more commonly referred to as R-12, circulates through the system under high pressure — as much as 300 psi. As a result, work on the air conditioning system is potentially hazardous if certain precautions are ignored. For safety's sake *read this entire section* before attempting any troubleshooting, checks, or work on the system.

A typical system is shown schematically in **Figure 13**. For practical purposes, the cycle

COOLING, HEATING, AND AIR CONDITIONING SYSTEMS

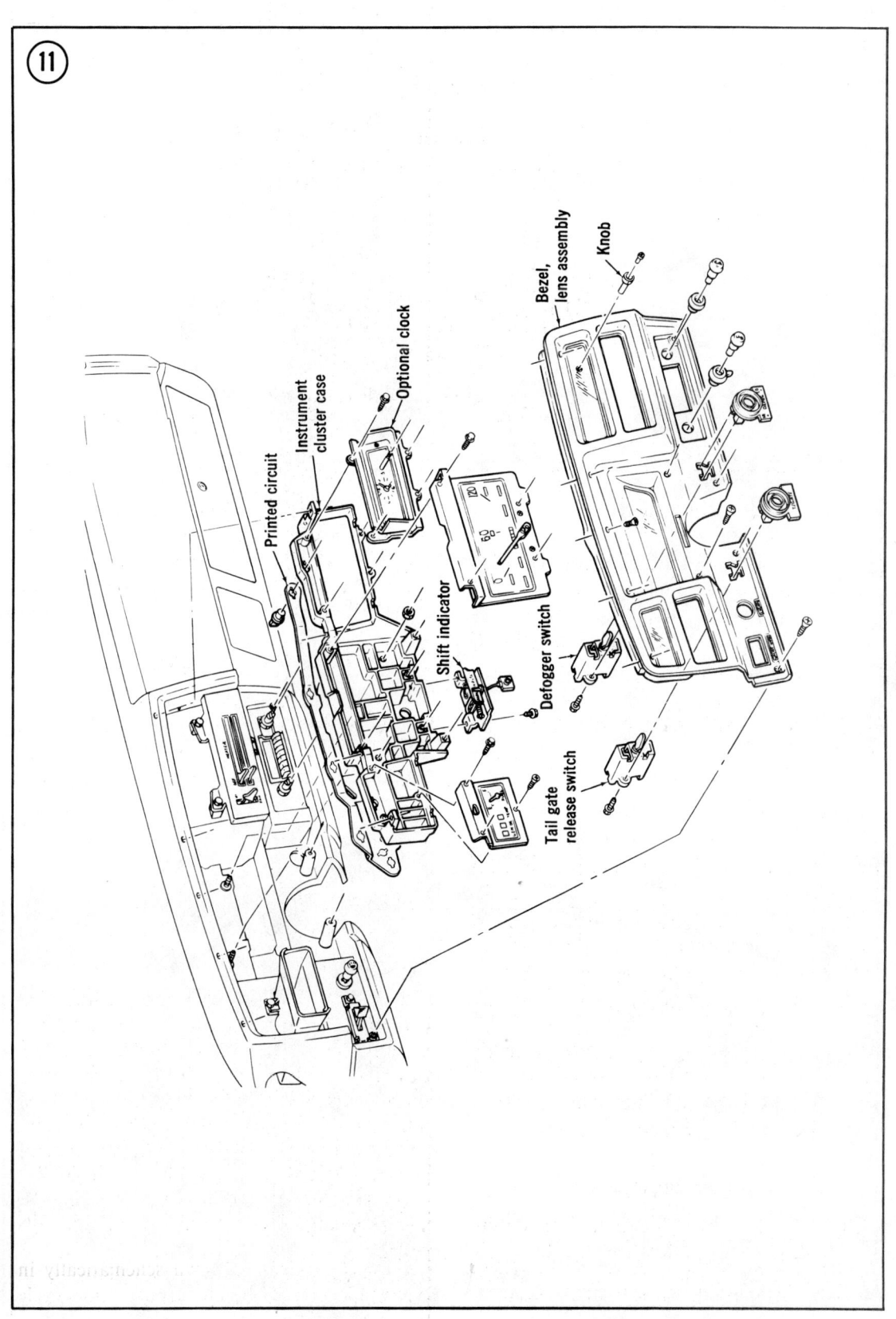

11

12

Engine
Radiator
Transmission

13

Expansion valve
Evaporator
(Liquid)
(Liquid)
Receiver-drier
Cap
(Vapor)
Check valve
(Vapor)
Compressor
Condenser
High pressure
Low pressure

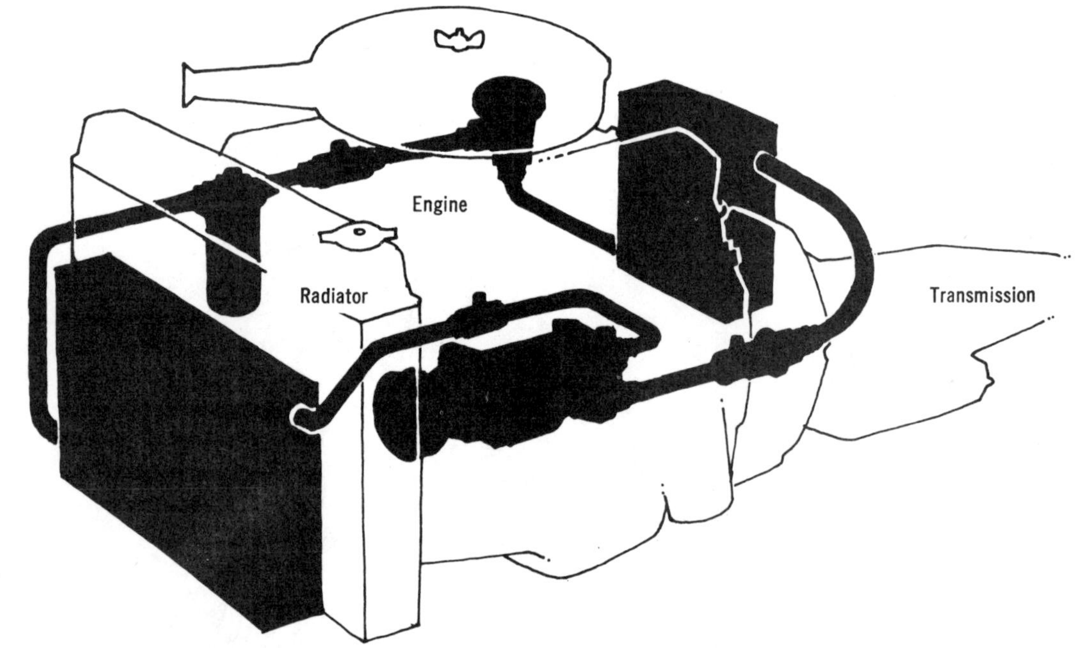

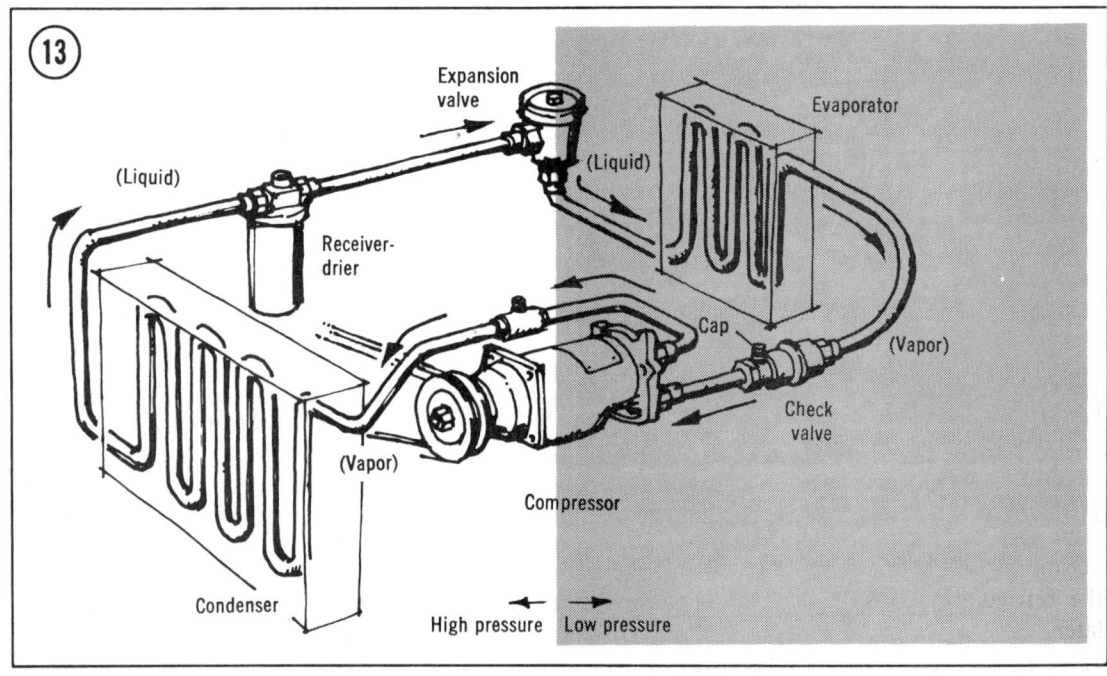

COOLING, HEATING, AND AIR CONDITIONING SYSTEMS

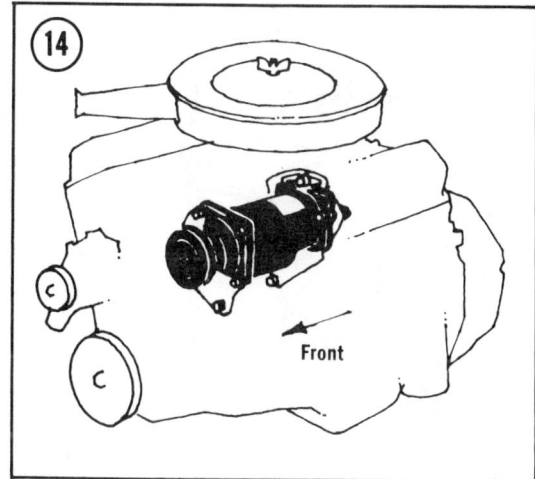

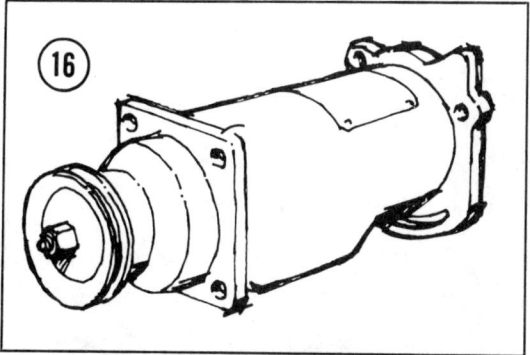

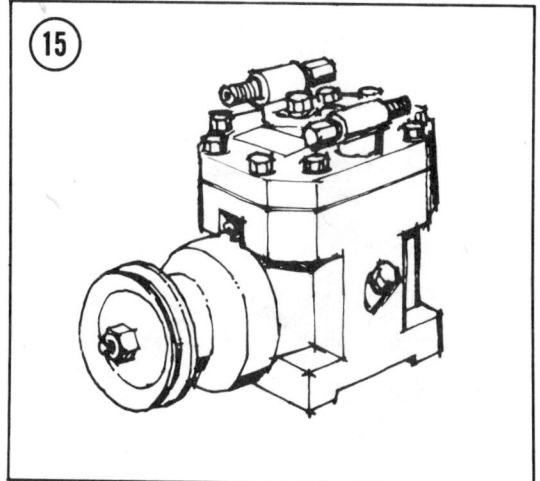

begins at the compressor. The refrigerant, in a warm, low-pressure vapor state, enters the low-pressure side of compressor. It is compressed to a high-pressure hot vapor and pumped out of the high-pressure side to the condenser.

Air flow through the condenser removes heat from the refrigerant and transfers the heat to the outside air. As the heat is removed, the refrigerant condenses to a warm, high-pressure liquid.

The refrigerant then flows to the receiver/drier where moisture is removed and impurities are filtered out. The refrigerant is stored in the receiver/drier until it is needed. Generally, the receiver/drier incorporates a sight glass that permits visual monitoring of the condition of the refrigerant as it flows. This is discussed later.

From the receiver/drier, the refrigerant flows to the expansion valve. The expansion valve is thermostatically controlled and meters refrigerant to the evaporator. As the refrigerant leaves the expansion valve it changes from a warm, high-pressure liquid to a cold, low-pressure liquid.

In the evaporator, the refrigerant removes heat from the cockpit air that is blown across the evaporator's fins and tubes. In the process, the refrigerant changes from a cold, low-pressure liquid to a warm, high-pressure vapor which flows back to the compressor where the refrigeration cycle began.

GET TO KNOW YOUR VEHICLE'S SYSTEM

With **Figure 12** as a guide, begin with the compressor and locate each of the following components in turn:

1. Compressor
2. Condenser
3. Receiver/drier
4. Expansion valve
5. Evaporator

Compressor

The compressor is located on the front of the engine, like an alternator, and is driven by one or two drive belts (**Figure 14**). The large pulley on the front contains an electromagnetic clutch that is activated and operates the compressor when the air conditioning controls are switched on. There are 2 compressor types — piston-and-crank (**Figure 15**), and swashplate (axial plate). See **Figure 16**.

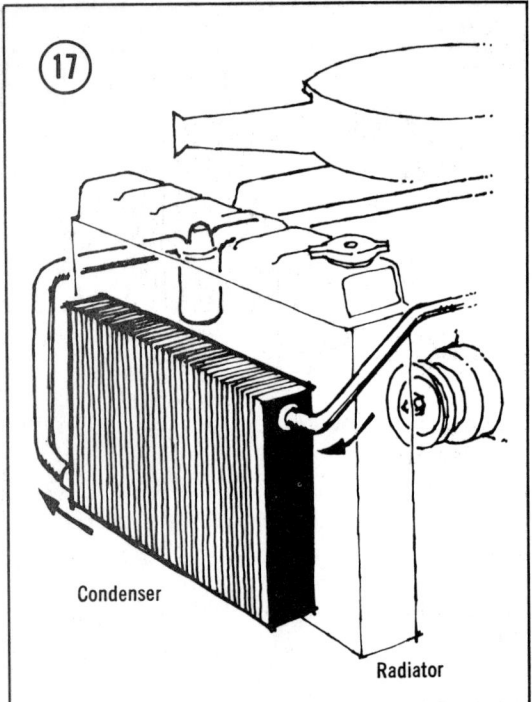

Condenser

In most cases, the condenser is mounted in front of the radiator (**Figure 17**). Air passing through the fins and tubes removes heat from the refrigerant in the same manner it removes heat from the engine coolant as it passes through the radiator.

Receiver/Drier

The receiver/drier is a small tank-like unit (**Figure 18**), usually found mounted to one of the wheel wells. Many receiver/driers incorporate a sight glass through which refrigerant flow can be seen when the system is operating (**Figure 19**). Some systems have an in-line sight glass (**Figure 20**). Some early systems do not have a sight glass but it's not essential to system operation — just handy to help diagnose air conditioning troubles.

Expansion Valve

The expansion valve (**Figure 21**) is located between the receiver/drier and the evaporator. It is usually mounted on or near the firewall, in the engine compartment. In some very late

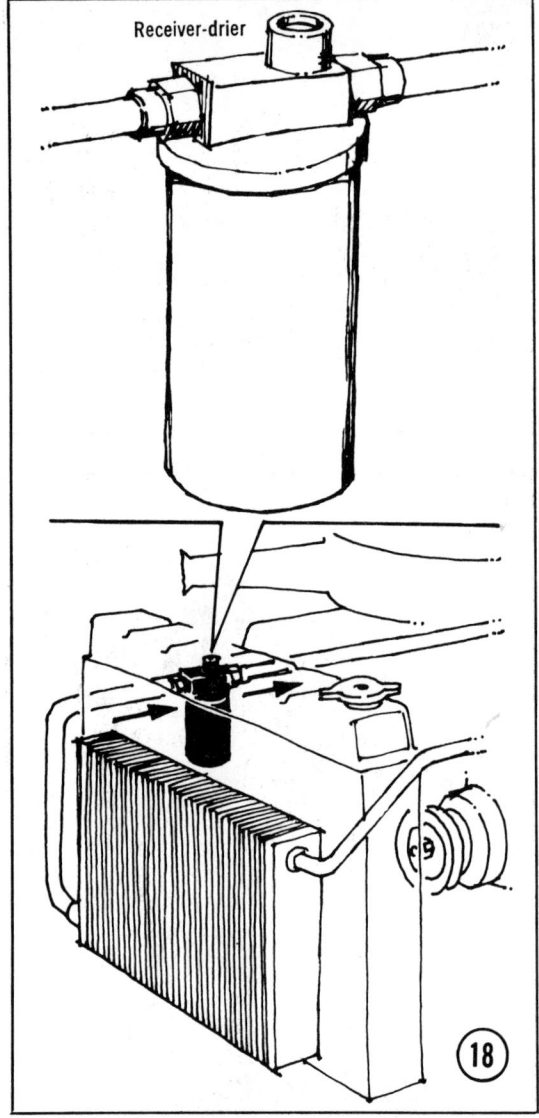

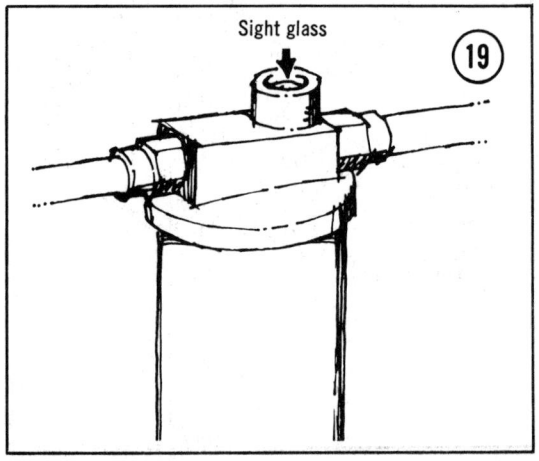

COOLING, HEATING, AND AIR CONDITIONING SYSTEMS

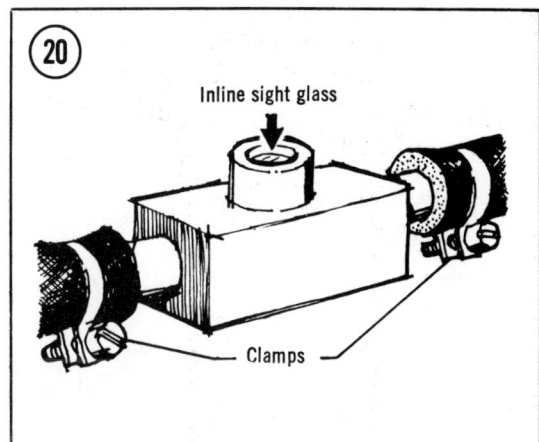

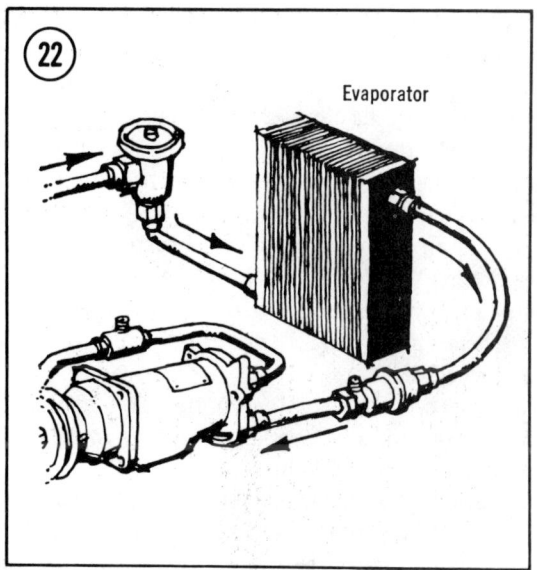

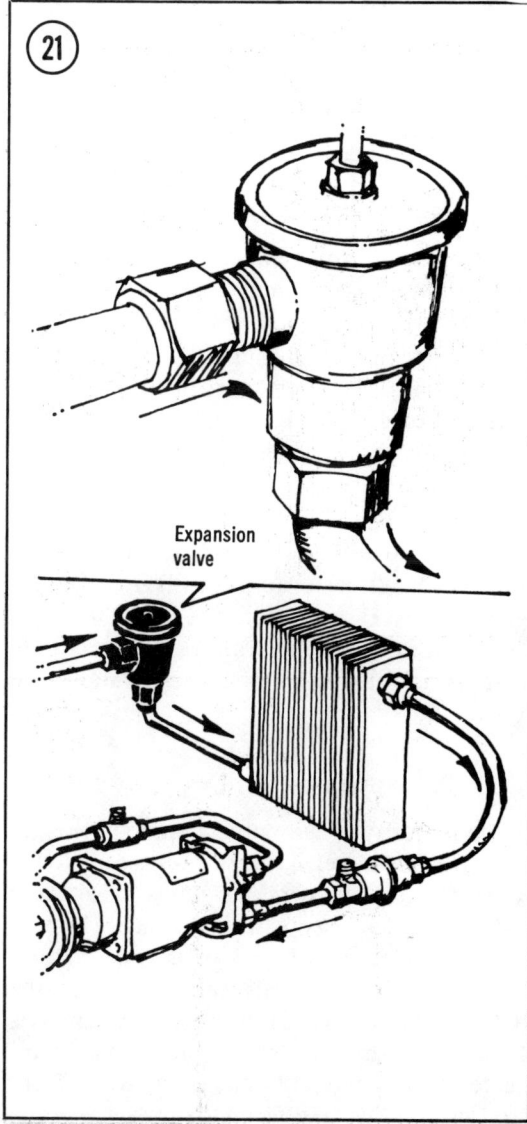

systems, the valve is concealed in a housing on the firewall.

Evaporator

The evaporator is located in the passenger compartment, beneath the dashboard, and is hidden from view by the fan shrouding and ducting (**Figure 22**). Warm air from the passenger compartment is blown across the fins and tubes in the evaporator where it is cooled and dried and then ducted back into the compartment through the air outlets.

ROUTINE MAINTENANCE

First echelon preventive maintenance for your air conditioning system couldn't be simpler; at least once a month, even in cold weather, start your engine and turn on the air conditioner and operate it at each of the switch and control settings. Allow it to operate for about 5 minutes. This will ensure that the compressor seal will not deform from sitting in the same position for a long period of time. If this occurs, the seal is likely to leak.

The efficiency of your air conditioning system depends in great part on the efficiency of your engine cooling system. Periodically check the coolant for level and cleanliness. If it is dirty, drain and flush the system and fill it with fresh coolant and water, following the coolant manufacturer's instructions for the coolant/water ratio. Have your radiator cap pressure tested and replace it if it will not maintain 13 psi pressure. If the system requires repeated topping up and the radiator cap is in good condition, it is likely that there is a leak in the system. Pressure test it as described earlier in this chapter.

With an air hose and a soft brush, clean the radiator fins and tubes to remove bugs, leaves and any other imbedded debris.

Check and correct drive belt tension as described earlier.

If the condition of the cooling system thermostat is in doubt, check it as described earlier and replace it if it is faulty.

When you are confident that the engine cooling system is working correctly, you are ready to inspect and test the air conditioning system.

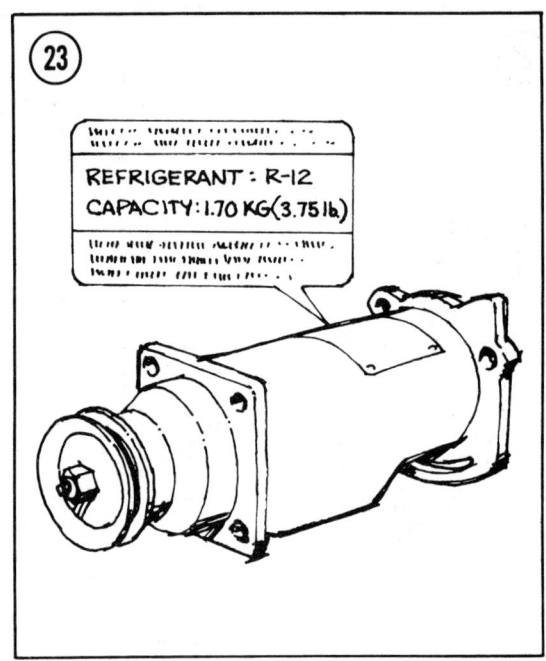

Inspection

1. Clean all lines, fittings, and system components with solvent and a clean rag. Pay particular attention to the fittings; oily dirt around connections almost certainly indicates a leak. Oil from the compressor will migrate through the system to the leak. Carefully tighten the connection, taking care not to overtighten and risk stripping the threads. If the leak persists it will soon be apparent once again as oily dirt accumulates. Clean the sight glass with a clean, dry cloth.

2. Clean the condenser fins and tubes with a soft brush and an air hose, or with a high-pressure stream of water from a garden hose. Remove bugs, leaves and other imbedded debris. Carefully straighten any bent fins with a screwdriver, taking care not to dent or puncture the tubes.

3. Check the condition and tension of the drive belts and replace or correct as necessary.

4. Start the engine and check the operation of the blower motor and the compressor clutch by turning the controls on and off. If either the blower or the clutch fails to operate, shut off the engine and check the condition of the fuses. If they are blown replace them. If not, remove them and clean the fuse holder contacts. Then, recheck to be sure the blower and clutch operate.

Testing

1. With the transmission in PARK (automatic) or NEUTRAL (manual) and the handbrake set, start the engine and run it at a fast idle.

2. Set the temperature control to its coldest setting and turn the blower to high. Allow the system to operate for 10 minutes with the doors and windows open. Then close them and set the blower on its lowest setting.

3. Place a thermometer in a cold-air outlet. Within a few minutes, the temperature should be 35-45°F. If it is not, it's likely that the refrigerant level in the system is low. Check the appearance of the refrigerant flow through the sight glass. If it is bubbly, refrigerant should be added.

REFRIGERANT

The majority of automotive air conditioning systems use a refrigerant designated R-12. However, a commercial grade, designated R-20, is used in heavy-duty systems. The two are not compatible. Look for an information sticker, usually mounted near the compressor to determine which refrigerant your system uses (**Figure 23**). Also, check the system capacity indicated on the sticker. Capacity can range from 2 to 5 pounds, depending on the system.

COOLING, HEATING, AND AIR CONDITIONING SYSTEMS

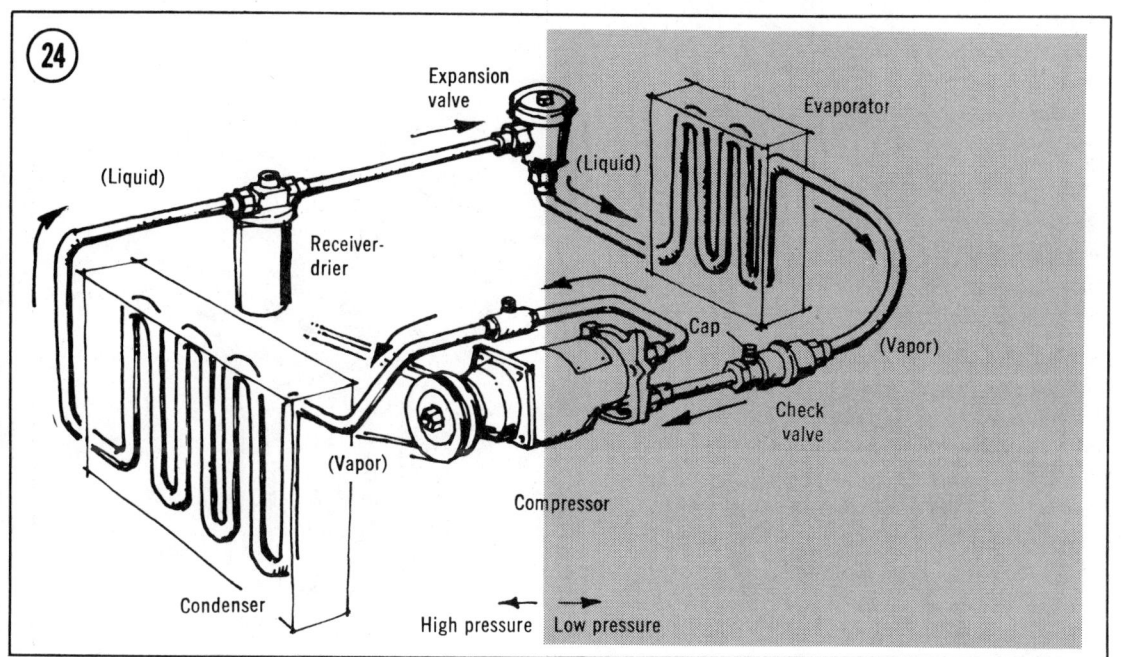

That harmless-looking little can of refrigerant is potentially hazardous. If hooked up to the high-performance side of the compressor, or hooked up without a gauge set, it becomes a hand grenade.

Charging

> **WARNING**
> *Do not attempt to add refrigerant to the system without using a gauge set; it's essential that the system pressure during charging not exceed 50 psi.*

1. Carefully read and understand the gauge manufacturer's instructions before charging the system.

2. Remove the cap from the Schrader valve on the low-pressure side of compressor (**Figure 24**). The low-pressure side is labelled SUCTION, SUCT., or SUC.

3. Connect the gauge set to the low-pressure Schrader valve. Connect the refrigerant can to the gauge set and hang the gauge set on the hood (**Figure 25**).

4. Start the engine and run it at a fast idle (about 1,000 rpm).

5. Set the temperature control at its coldest setting. Set the blower at its lowest setting.

6. Slowly open the refrigerant feed valve on the gauge set (**Figure 26**). Do not allow the refrigerant pressure to exceed 50 psi.

7. Watch the refrigerant as it flows through the sight glass (**Figure 27**). When it's free of bubbles, the system is charged. Shut off the refrigerant feed valve on the gauge set.

TROUBLESHOOTING

Preventive maintenance like that just described will help to ensure that your system is working efficiently. Still, trouble can develop and while most of it will invariably be simple and easy to correct, you must first locate it. The following sequence will help to diagnose system troubles when your air conditioning ceases to cool the passenger compartment.

1. First, stop the vehicle and look at the control settings. One of the most common sources of air conditioning trouble occurs when the temperature control is set for maximum cold and the blower is set on low. This arrangement promotes ice buildup on the fins and tubes of the evaporator, and particularly so in humid weather. Eventually, the evaporator will ice over completely, and restrict air flow. Turn the

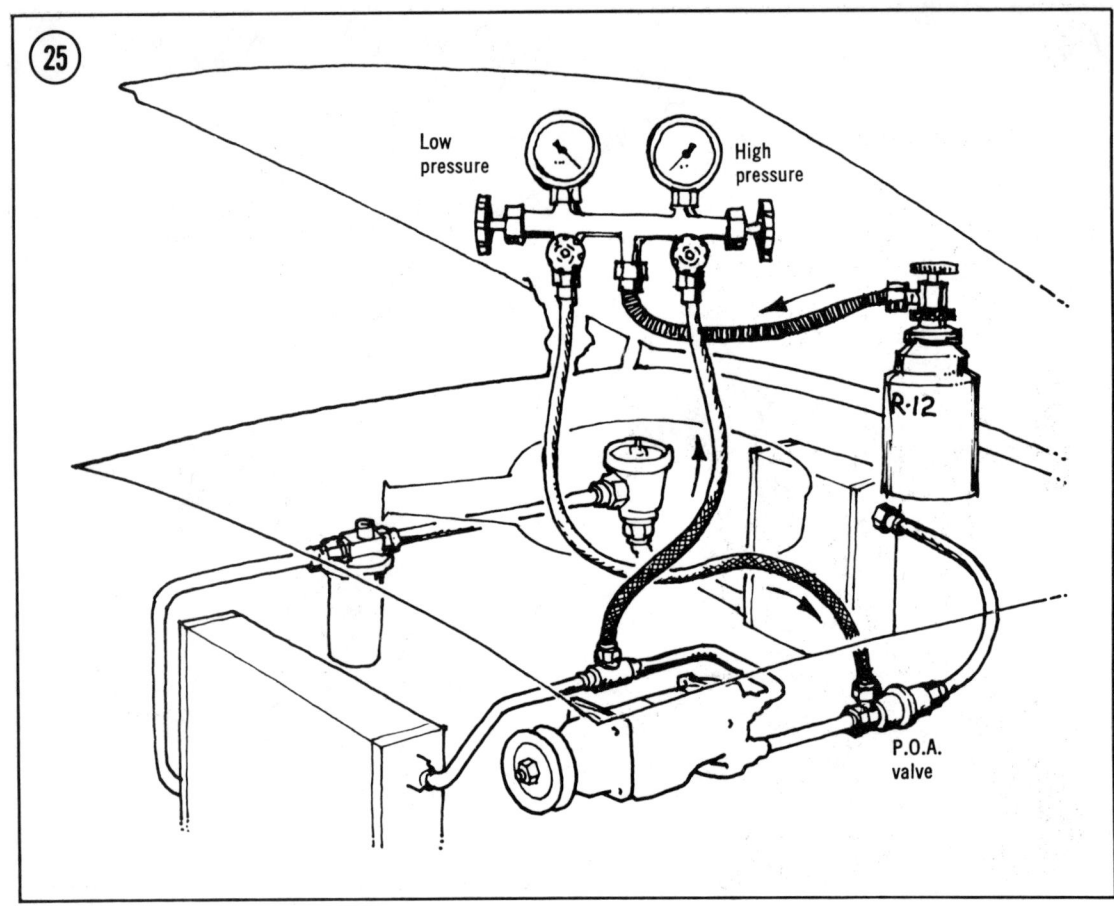

blower on high and place a hand over an air outlet. If the blower is running but there is little or no air flowing through the outlet, the evaporator is probably iced up. Leave the blower on high and turn off the temperature control or turn it down to its lowest setting — and wait; it will take 10 or 15 minutes before the ice begins to melt.

2. If the blower is not running, the motor may be burned out, there may be a loose connection, or the fuse may be blown. First check the fuse panel for a blown or incorrectly seated fuse. Then, check the wiring for loose connections.

3. Shut off the engine and check the condition and tension of the compressor drive belt. If it is loose or badly worn, tighten or replace it.

4. Start the engine and check the condition of the compressor clutch by turning the air conditioner on and off. If the clutch does not energize, it may be defective, its fuse may be blown, or the evaporator temperature-limiting switches may be defective. If the fuse is defective, replace it. If the clutch still does not energize, refer the problem to an air conditioning specialist.

5. If all components checked so far are OK, start the engine, turn on the air conditioner and watch the refrigerant through the sight glass; remember, if it's filled with bubbles after the system has been operating for a few seconds, the refrigerant level is low. If the sight glass is oily or cloudy, the system is contaminated and should be serviced by an expert as soon as possible. Corrosion and deterioration occur rapidly and if it's not taken care of at once it will result in a very expensive repair job.

6. If the system still appears to be operating satisfactorily but the air flow into the passenger compartment is not cold, check the condenser and cooling system radiator for debris that could block the air flow. Recheck the cooling system as described earlier under *Inspection*.

COOLING, HEATING, AND AIR CONDITIONING SYSTEMS

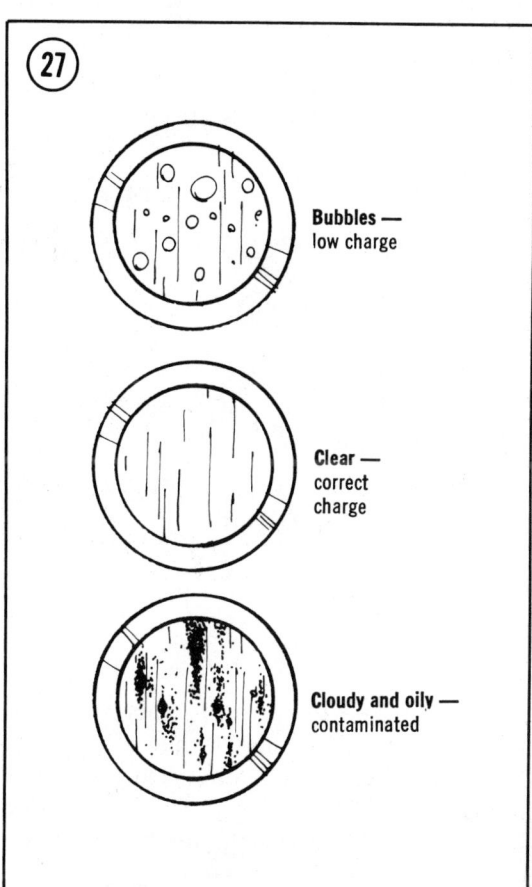

7. If the above steps do not uncover the difficulty, have the system checked and corrected by a specialist as soon as possible.

DISCHARGING THE SYSTEM

The pressure in the system must be relieved before any fittings are disconnected. To do this, connect the gauge set in the same manner as though the system were to be charged; however, do not attach a refrigerant can to the center hose. Slowly open both the high- and low-pressure valves on the gauge set. Then, slowly open the valve for the center hose.

WARNING
Route the center hose down through the engine compartment so it will discharge on the ground. Wear safety goggles and take care not to let the refrigerant touch your skin.

When the system pressure has been relieved, disconnect the gauge set. Slowly loosen any fittings that are involved until you are sure the system is not pressurized. *Wear safety goggles.*

Immediately plug any open fittings to keep moisture out of the system; corrosion will begin almost instantly if the system is left open to the atmosphere, and it will quickly make some very expensive components unserviceable.

When all of the system components and lines have been reconnected, charge the system as described earlier.

CHAPTER EIGHT

ELECTRICAL SYSTEM

Making repairs to electrical components such as the generator or starter motor is usually beyond the capability of both the inexperienced mechanic and his tool box. Such repairs are best left to the specialized mechanic who is equipped with specialized tools.

By using the troubleshooting procedures given in Chapter Two it is possible, however, to isolate problems in many cases to a specific component.

This chapter provides some additional tests and procedures for removing, inspecting (in some cases), and replacing electrical components.

In many cases it may be faster and more economical to obtain new or rebuilt components instead of making repairs. Make certain, however, that the new or rebuilt part is an exact replacement. Also, make sure that the cause of the failure has been isolated and corrected before installing a replacement. For instance, an uncorrected short in an alternator circuit will in all probability burn out a new alternator as quickly as it damaged the old one. If in doubt, always consult an expert.

Wiring diagrams are located at the end of the book.

BATTERY MAINTENANCE

NOTE: *Some late models are equipped with sealed "Freedom" batteries, which do not require routine maintenance.*

The level of the electrolyte in each cell of the battery should be checked regularly, especially in hot weather. The proper level is even with the split vent located at the bottom of the vent well in each cell. See **Figure 1**. Colorless, odorless drinking water should be added as required to

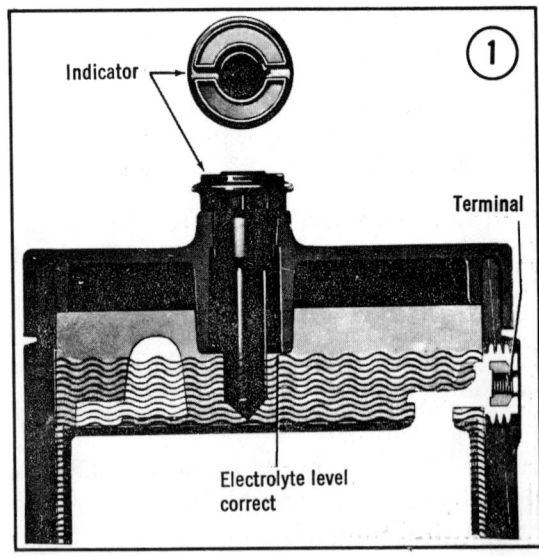

ELECTRICAL SYSTEM

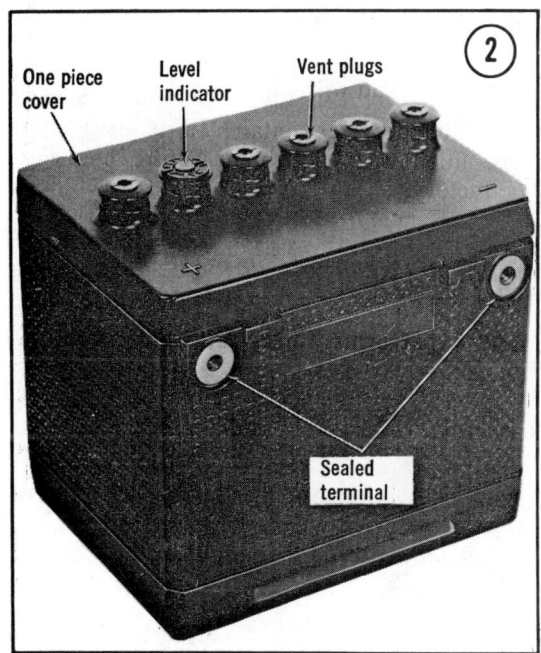

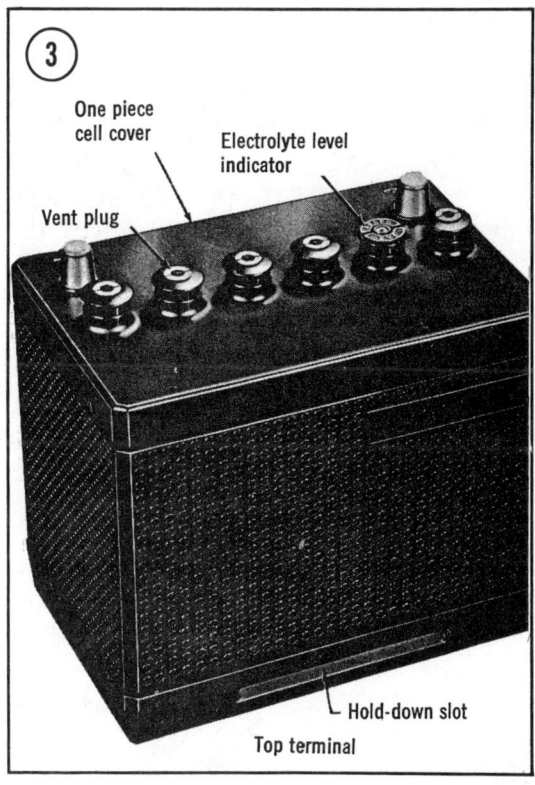

maintain this level in each cell. Never add electrolyte (acid).

If water must be added frequently, chances are that the battery is being overcharged (normal usage is one or two ounces per month per battery — not each cell). Common causes of overcharge are high battery operating temperature, voltage regulator set too high, and a poor ground connection from the regulator.

The battery terminals, top, carrier, and hold-down should be kept free of corrosion, oil, and dirt. The carrier should be in sound mechanical condition and the hold-down bolts should be tight to keep the battery level and prevent excessive shaking of the battery. However, the bolts should not be so tight that a severe strain is placed on the battery case.

Cleaning

To clean the battery, make certain the cell vent plugs are fully seated. Cover the vent holes, if accessible, with small pieces of tape. Wash the top and sides of the battery case and carrier, and the hold-down with a mild solution of ammonia or baking soda to neutralize any acid or corrosion present, then flush with clean water. Pay special attention to terminals and the areas around them. Be sure to remove the tape from the holes in the vent caps.

Most models use a battery with "sealed" terminals on the side. See **Figure 2**. This positions them out of the "wet" area on the top of the battery surrounding the vent holes. Normal spillage, spewage, and other sources of moisture are far less likely to collect around the terminal. These terminals, as a rule, require no maintenance.

Some older models use the "conventional" type battery with the terminals on the top (**Figure 3**). These terminals should be cleaned, as described above, whenever they become corroded. Following cleaning they should receive a light coat of petroleum jelly to help control future corrosion.

Common Cause of Battery Failure

All batteries eventually fail. Their life can be prolonged, however, with a good maintenance program. Some of the reasons for premature failure are listed below.

1. Vehicle accessories left on overnight or longer, causing a discharged condition.

2. Slow driving speeds on short trips, causing an undercharged condition.

3. Vehicle electrical load exceeding the generator capacity.

4. Charging system defects, such as high resistance, slipping generator belt, or faulty generator or regulator.

5. Abuse of the battery, including failure to keep battery top and terminals clean, failure to keep cable attaching bolts clean and tight, and failure to add water when needed, or habitually adding too much water.

BATTERY TESTING

Use of the following test procedures will provide a basis for deciding whether a battery is good and usable, requires recharging, or should be replaced. A complete analysis of battery condition requires a visual inspection, an instrument test, and, if required, a full charge hydrometer test.

Visual Inspection

This test is accomplished as follows.

1. Check outside of battery for broken or cracked case or cover. If any such damage is present, replace battery.

2. Check electrolyte level (except Freedom battery). Levels too high or too low may cause poor performance.

3. Check for loose and/or corroded cable connections. Correct as required before proceeding with instrument tests.

Instrument Tests

If an instrument with battery testing capabilities is available, follow the manufacturer's instructions. If such an instrument is not available, use a hydrometer to make a specific gravity cell comparison test (except Freedom battery) as follows:

1. Measure specific gravity of each cell.

2. Compare readings of all cells. If readings between highest and lowest cell show a difference of 0.050 (50 points) or more, the battery is defective and should be replaced.

Full Charge Hydrometer Test

This test should only be made on batteries which check out OK after the tests in Step 2 above but which subsequently fail in service.

1. Remove battery from vehicle and add colorless, odorless drinking water, if required, to adjust electrolyte to proper level.

2. Fully charge battery at the slow charging rate, using the procedure given below.

3. When battery is fully charged, use a hydrometer to measure the specific gravity of the electrolyte in each cell. Interpret results as follows:

   a. Full charge hydrometer reading of less than 1.230, corrected for temperature, indicates battery is defective and should be replaced.

   NOTE: *For every 10° above 80°F electrolyte temperature, add 0.004 to specific gravity reading. For every 10° below 80°F, subtract 4 points (0.004).*

   b. Hydrometer readings of above 1.310 corrected for electrolyte temperature indicate that the cells have been improperly filled (activation) or improperly serviced. Poor service and short battery life will result.

   c. Readings between 1.250 and 1.290 in all cells indicate that battery condition is good. The problem is probably in the charging system or battery cables.

CHARGING SYSTEM

The charging system consists of the battery (discussed above), the generator or alternator, the voltage regulator, the "tell-tale" lamp or ammeter and the wiring necessary to connect these components. Two basic types of alternators are used. The Series 1D Delcotron (see **Figure 4**), used on the older models, was an alternating current generator (alternator) requiring an external mechanical or transistorized voltage regulator. The Series 10-SI Delcotron (see **Figure 5**), also an alternating current generator, has an internal solid state voltage regulator which employs an integrated circuit.

ELECTRICAL SYSTEM

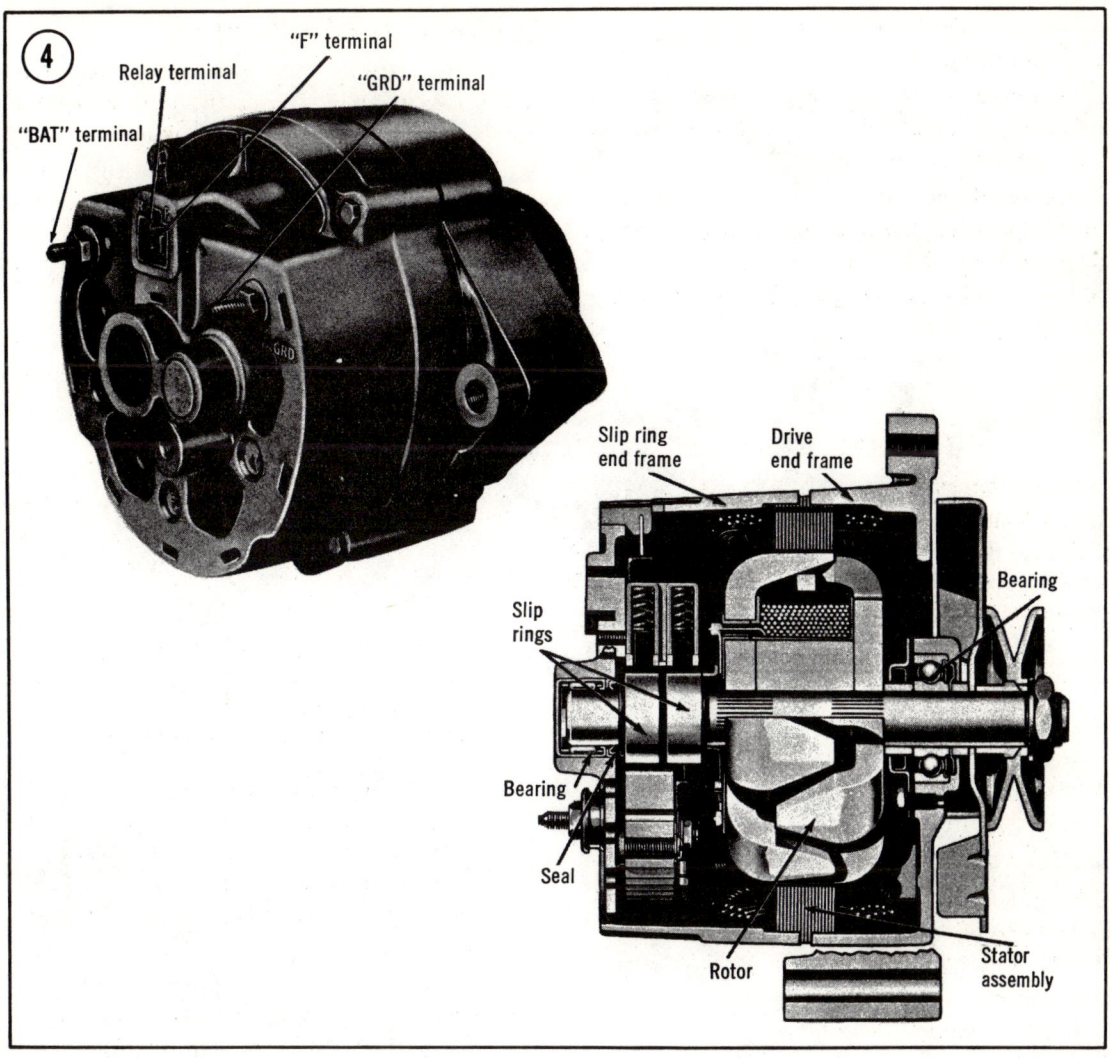

All models have an internal bridge which changes the stator AC voltages to DC voltages. The blocking action of the diodes in the bridge prevents battery discharge back through the Delcotron.

Neither type of Delcotron requires periodic maintenance, other than a check for loose mounting bolts and belt tension. Certain precautions should be observed, however, since the Delcotron and regulator are designed for use only on negative polarized systems. These are:

1. Do not attempt to "polarize" the alternator.
2. Do not short across or ground any of the terminals in the charging system except as specifically instructed in these procedures.
3. Never operate the alternator with the output terminal open-circuited.
4. Make sure the alternator, external voltage regulator (if so equipped), and battery are of the same ground polarity.
5. When connecting a charger or booster battery to the vehicle battery, connect negative terminal to negative terminal and positive terminal to positive terminal.
6. Whenever a lead is disconnected from the alternator, the battery ground cable should be disconnected.

Static Checks (All Series)

Before making any electrical checks on the charging system, visually inspect all connec-

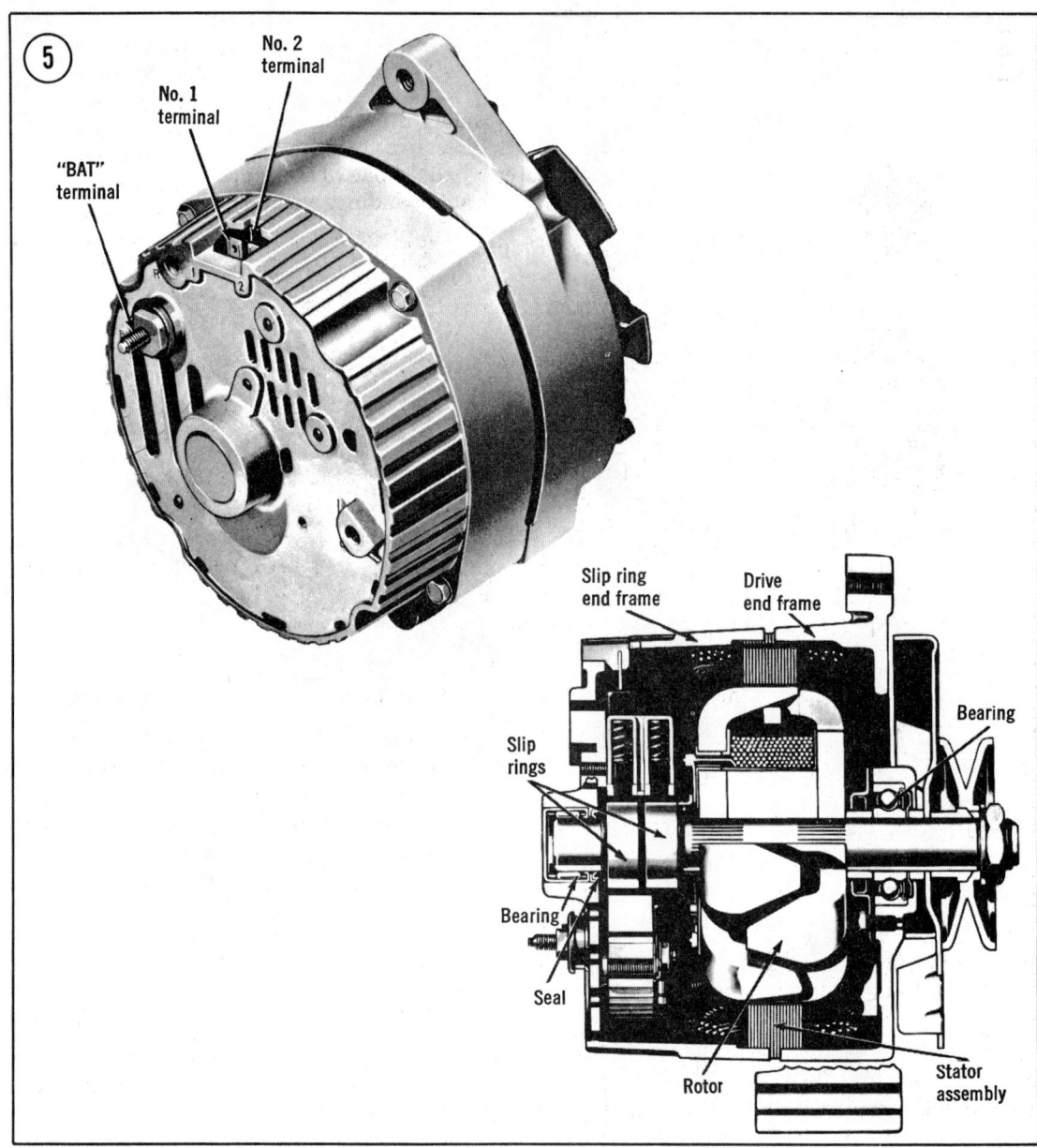

tions to make sure they are clean and tight. Verify that battery is serviceable and charged. Inspect wiring for frayed, broken, or cracked insulation. Check for loose mounting bolts and proper belt tension.

10-SI DELCOTRON TROUBLESHOOTING

NOTE: *If vehicle is equipped with ammeter instead of indicator lamp, omit indicator lamp circuit check.*

Indicator Lamp Circuit Check

Check indicator lamp for normal operation as follows:

Switch	Lamp	Engine
Off	Off	Stopped
On	On	Stopped
On	Off	Running

If indicator lamp operates normally, proceed to *Undercharged Battery* or *Overcharged Bat-*

ELECTRICAL SYSTEM

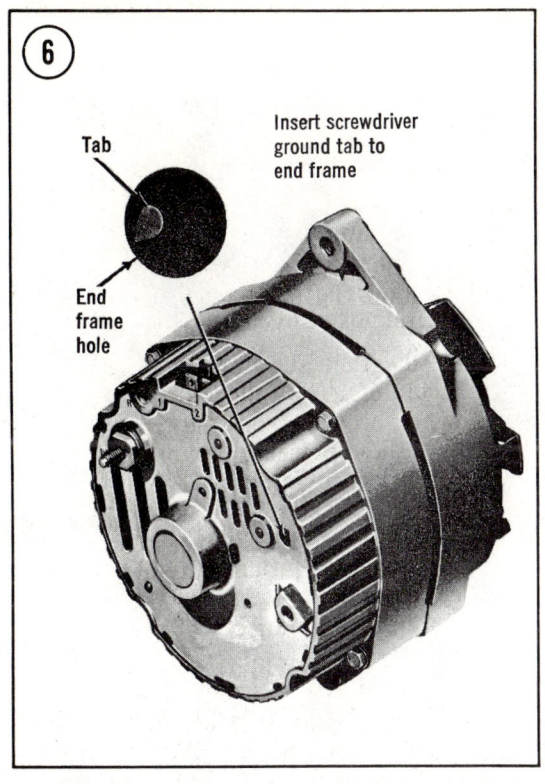

Figure 6
Tab — End frame hole — Insert screwdriver ground tab to end frame

CAUTION
Do not ground No. 2 lead. If lamp does not come on, check for blown fuse or fusible link, burned out lamp bulb, defective bulb socket, or open in No. 1 lead circuit between Delcotron and ignition switch. If lamp lights, remove ground at No. 1 terminal and reconnect No. 1 and No. 2 wires to Delcotron. Insert a screwdriver in test hole (see Figure 6) to ground wiring. If lamp does not go on, check connection between wiring harness and Delcotron No. 1 terminal. If wiring is OK, have Delcotron brushes, slip rings, and field winding checked. If lamp lights, repeat voltmeter check in Step (a). If reading is obtained, have the regulator replaced.

3. *Switch on, lamp on, engine running* — See *Undercharged Battery Condition Check*, below, for possible causes.

Undercharged Battery Condition Check

Symptoms of this condition are slow cranking and low specific gravity readings (battery cells). The condition can be caused by one or more of the following, even though the ammeter may be operating properly.

1. Accessories left on for extended periods.
2. Improper belt tension.
3. Defective or discharged battery (see battery tests above).
4. Wiring defects. Check all connectors for tightness and cleanliness, including connectors at Delcotron and firewall, cable clamps, and battery posts.
5. Open circuit between Delcotron and battery. Check by connecting voltmeter between BAT terminal on Delcotron and ground, No. 1 terminal and ground, and No. 2 terminal and ground. A zero reading indiates an open between the voltmeter connection point and the battery.

CAUTION
An open in the Delcotron No. 2 lead circuit will cause uncontrolled voltage, battery overcharge, and possible damage to the battery and accessories. The 10-SI series Delcotron has a built-in feature to prevent overcharge and ac-

tery. Otherwise, make the appropriate abnormal condition checks below:

1. *Switch off, lamp on* — Disconnect leads from Delcotron No. 1 and No. 2 terminals. If lamp remains on, there is a short circuit between the 2 leads. If the lamp goes out, the rectifier bridge is faulty and must be replaced, as this condition will result in an undercharged battery.

2. *Switch on, lamp off, engine stopped* — This defect can be caused by the defects listed in Step 1 above, or by reversal of the No. 1 and No. 2 leads at these 2 terminals, or by an open circuit. To determine where the open exists, proceed as follows:

  a. Connect voltmeter between Delcotron No. 2 terminal and ground. If reading is obtained, proceed to Step b. If reading is zero, repair open circuit between No. 2 terminal and battery. If lamp comes on, no further check is required.
  b. Disconnect leads from No. 1 and No. 2 terminals on Delcotron and turn ignition switch on. Momentarily ground No. 1 terminal lead.

cessory damage by preventing the Delcotron from turning on if there is an open in the No. 2 lead circuit. Such an open could occur between terminals, at the crimp between the harness wire or terminal, or in the wire itself.

Overcharged Battery Condition Check

1. Verify that battery is in serviceable condition and fully charged.

2. Connect a voltmeter from Delcotron No. 2 terminal to ground. If reading is zero, an open exists in No. 2 lead circuit.

3. If battery and No. 2 lead circuit are OK, but an obvious overcharged condition, (such as excessive battery water usage, exists), have the following items checked:

   a. Field winding (for shorts)
   b. Brushes and brush leads (for grounding or defective regulator)

1D DELCOTRON TROUBLESHOOTING

Charging System Condition Test

This test may be used to indicate overall charging system condition and to isolate a malfunctioning component (if present).

1. Turn ignition off and make static tests outlined above. If tests fail to isolate trouble, set handbrake and place transmission in NEUTRAL.

2. Connect voltmeter from junction block on horn relay to ground at voltage regulator base.

> CAUTION
> Make certain voltmeter lead does not touch a resistor or terminal extension under the regulator, as this could cause damage.

3. Connect a tachometer to engine, using the tachometer manufacturer's instructions.

4a. *On models equipped with indicator lamp in charging circuit* — Turn ignition switch ON. If indicator fails to light, perform *Indicator Lamp Circuit Tests* below, and make corrections as required before proceeding.

4b. *On ammeter-equipped models* — Turn ignition switch to ACC. If ammeter fails to register

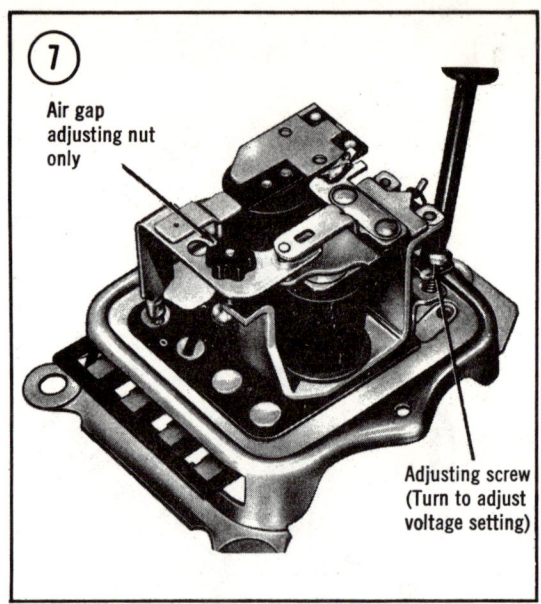

discharge, perform *Field Circuit Resistance Check* below before proceeding.

5a. *On "indicator lamp" models* — If lamp lights, start engine and run at 1,500 rpm or above. If lamp fails to go out, perform *Indicator Lamp Circuit Tests* and make corrections before proceeding.

5b. *On "ammeter" models* — If ammeter shows discharge, start engine and run at 1,500 rpm or above. If meter fails to move toward charge, perform *Field Circuit Tests* below before proceeding.

6. Start engine and run at 1,500 rpm a few minutes to develop battery surface charge. Turn on headlights and heater blower motor (high speed), operate at 1,500 rpm and read voltmeter. If reading is 12.5 volts or more, turn off headlights and heater blower and stop engine. Have voltage regulator adjusted (see procedure below). If reading was below 12.5 volts, check Delcotron output using the procedure given below.

   a. *Delcotron tests bad* — Have Delcotron repaired or replaced.
   b. *Delcotron tests OK* — Disconnect regulator connector, remove cover, and reconnect connector. Repeat load test at beginning of Step 6. Turn adjusting screw (see **Figure 7**) to raise voltmeter reading to 12.5 volts. On transistorized regulator,

ELECTRICAL SYSTEM

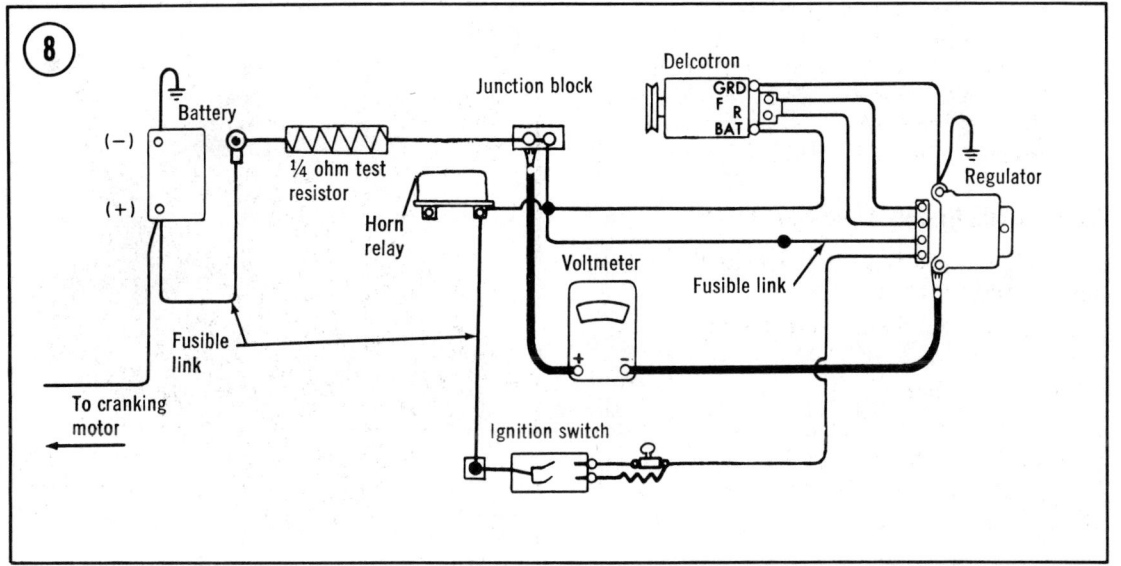

remove pipe plug, insert screwdriver into slot and turn clockwise one or two turns to increase voltage. Turning counter-clockwise decreases voltage. If 12.5 volts cannot be obtained, install new regulator and repeat Step 6. Turn off loads and stop engine.

Regulator Voltage Adjustment

1. Connect a ¼ ohm, 25 watt resistor into charging circuit at horn relay junction block as shown in **Figure 8** (between both leads and terminal).

2. Operate engine at 1,500 rpm for at least 15 minutes (longer in cold weather), then cycle regulator voltage control (by disconnecting and reconnecting regulator connector) and read voltage. If reading is between 13.5-15.2, regulator is OK. If not within these limits, leave engine running at 1,500 rpm and perform the following:

  a. Disconnect the regulator connector and remove regulator cover. Reconnect connector and set voltage to 14.2-14.6 volts. See Step 6 under *Charging System Condition Test* and **Figure 7**.

  b. Disconnect connector, reinstall cover, and reconnect connector.

  c. Allow engine to operate another 5-10 minutes at 1,500 rpm to reestablish regulator internal operating temperature.

  d. Cycle regulator voltage (by disconnecting and reconnecting connector), then read voltage. Readings between 13.5 and 15.2 volts indicate regulator is OK. Stop engine, remove voltmeter and resistor, and reconnect leads to horn relay junction block.

Generator Output Test (Voltmeter Method)

1. Disconnect 2-terminal connector from Delcotron "F" and "R" terminals.

2. Connect a jumper between BAT terminal to F terminal to provide field excitation.

3. Connect voltmeter between BAT terminal and GRD terminal.

4. Start engine and turn on high beam headlights and high speed heater blower. Slowly increase engine speed to 1,500 rpm (with 2D 6.2 in. Delcotron run at 600 rpm) and read voltage. A reading of 12.5 volts or higher indicates output is OK. Stop engine, remove voltmeter and reconnect wiring. If reading was less than 12.5 volts, have Delcotron repaired or replaced.

Indicator Lamp Initial Field Circuit Tests

The indicator lamp circuit (on models so equipped) provides initial field excitation, causing the indicator lamp to glow. The light is cancelled by closing the field relay which applies battery voltage to both sides of the bulb (bulb goes out). Thus the indicator lamp should go on when the ignition switch is turned on but should go out almost immediately when the engine is started. Ammeter-equipped models use the same initial field excitation and control circuits, except that the indicator lamp is not used. Continuity tests on both circuits can be made as follows.

1. If lamp fails to light or ammeter fails to function, probable causes are faulty bulb or socket, faulty ammeter, an open circuit in the wiring, regulator, or field, or a shorted positive diode (which also may cause lamp to light when switch is off). Test as follows. See **Figure 9**.

   a. Disconnect connector from regulator and turn ignition switch to ON position. Connect a continuity test lamp from connector terminal 4 to ground as shown in Step I of **Figure 9**. If test lamp does not light, check for faulty bulb, socket, or open circuit between switch and regulator connector. Repair as required. If test light goes on, failure is regulator, Delcotron, or wire between regulator F terminal and Delcotron. In this case, proceed to Step b.

   b. Disconnect test lamp lead from ground and connect between connector F and "4" terminals as shown in Step II of **Figure 9**. If test lamp lights, problem is open circuit in regulator or relay contacts are stuck closed. Have regulator repaired or replaced. If test lamp fails to light, trouble is in wire between connector F terminals on Delcotron and regulator or in field windings. In this case, proceed to Step c.

   c. Connect test lamp between F terminal on Delcotron and "4" terminal in regulator connector, as shown in Step III of **Figure 9**. If lamp lights, there is an open circuit in wire between Delcotron and regulator F terminals. Correct as required. If lamp fails to light, an open exists in Delcotron field windings. Have repaired or replaced.

2. If indicator lamp fails to go out, or ammeter shows discharge while engine is running, possi-

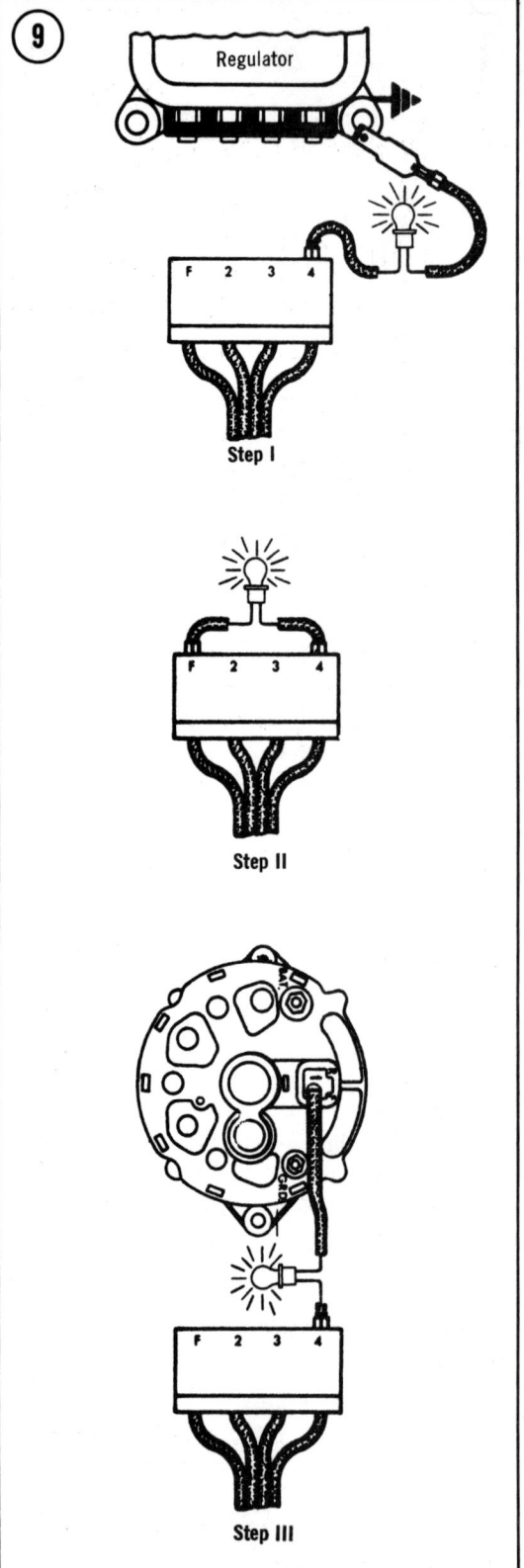

ELECTRICAL SYSTEM

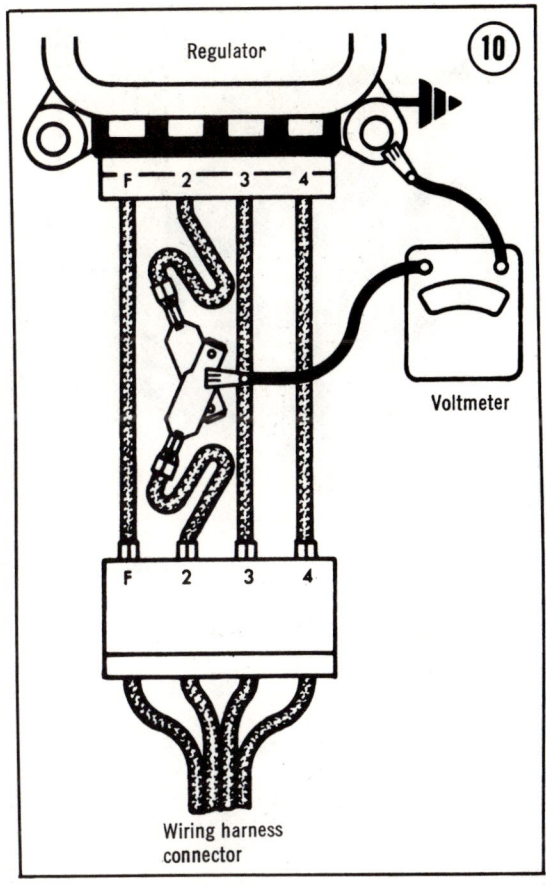

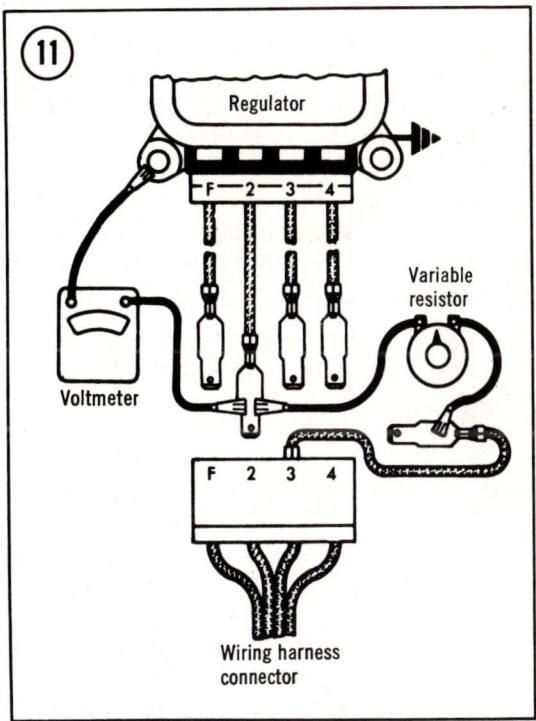

ble causes are loose drive belt (adjust as required), faulty field relay (see *Field Relay Checks and Adjustments* below), or defective Delcotron (see *Output Test* procedure above). Other causes could be (at normal idle) parallel resistance wire open (see *Field Circuit Resistance Wire Check* below), or, on ammeter models, the initial field excitation wire to ACC terminal is open. Correct as required. If lamp fails to go out after switch is turned off, positive diode is shorted (see *Diode Test* below).

Field Circuit Resistance Wire Checks

The resistance wire is a part of the wiring harness. It cannot be soldered, however, All splices must be made with crimp-type connectors. The wire is rated at 10 ohms, 6.25 watts minimum. To check for open resistor or field excitation wire (connected to ignition switch ACC terminal), proceed as follows.

1. Connect test lamp between regulator connector terminal "4" and ground as shown in Step I of **Figure 9**.
2. Turn ignition switch ON. If test lamp lights, resistance wire is OK. If lamp fails to light, resistor wire is open circuited and must be replaced. Note that dash indicator lamp does not light in this test because series resistance of the 2 bulbs causes amperage to be too low.

Field Relay Checks and Adjustment

To check for a faulty relay, proceed as follows.

1. Connect voltmeter between regulator connector terminal "2" and ground. See **Figure 10**.
2. Operate engine at fast idle (1,500-2,000 rpm). If voltmeter shows zero reading, check voltage between No. 2 terminal on regulator and R terminal on Delcotron. If voltage at regulator exceeds closing voltage (1.5-3.5 volts), regulator field relay is faulty. Check and adjust, using the procedure given below.

Closing Voltage Adjustment

1. Make connections as shown in **Figure 11**, using a 50 ohm variable resistor.

2. Turn resistor to OPEN position.
3. Turn ignition switch OFF.
4. Slowly decrease resistance and note closing voltage of relay. Adjust (1.5-3.5 volts) by bending heel iron as shown in **Figure 12**, as required.

DELCOTRON REMOVAL/INSTALLATION (ALL MODELS)

1. Disconnect battery ground cable at battery to prevent damage to diodes.
2. Disconnect wiring leads at Delcotron.
3. Remove Delcotron brace bolt (if equipped with power steering), loosen pump brace and mount nuts, and then remove drive belt(s).

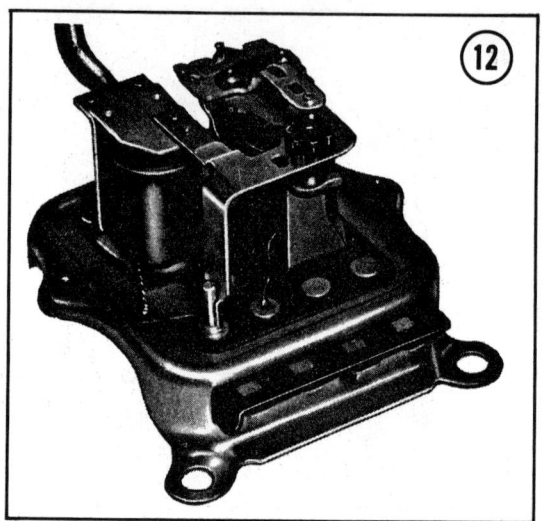

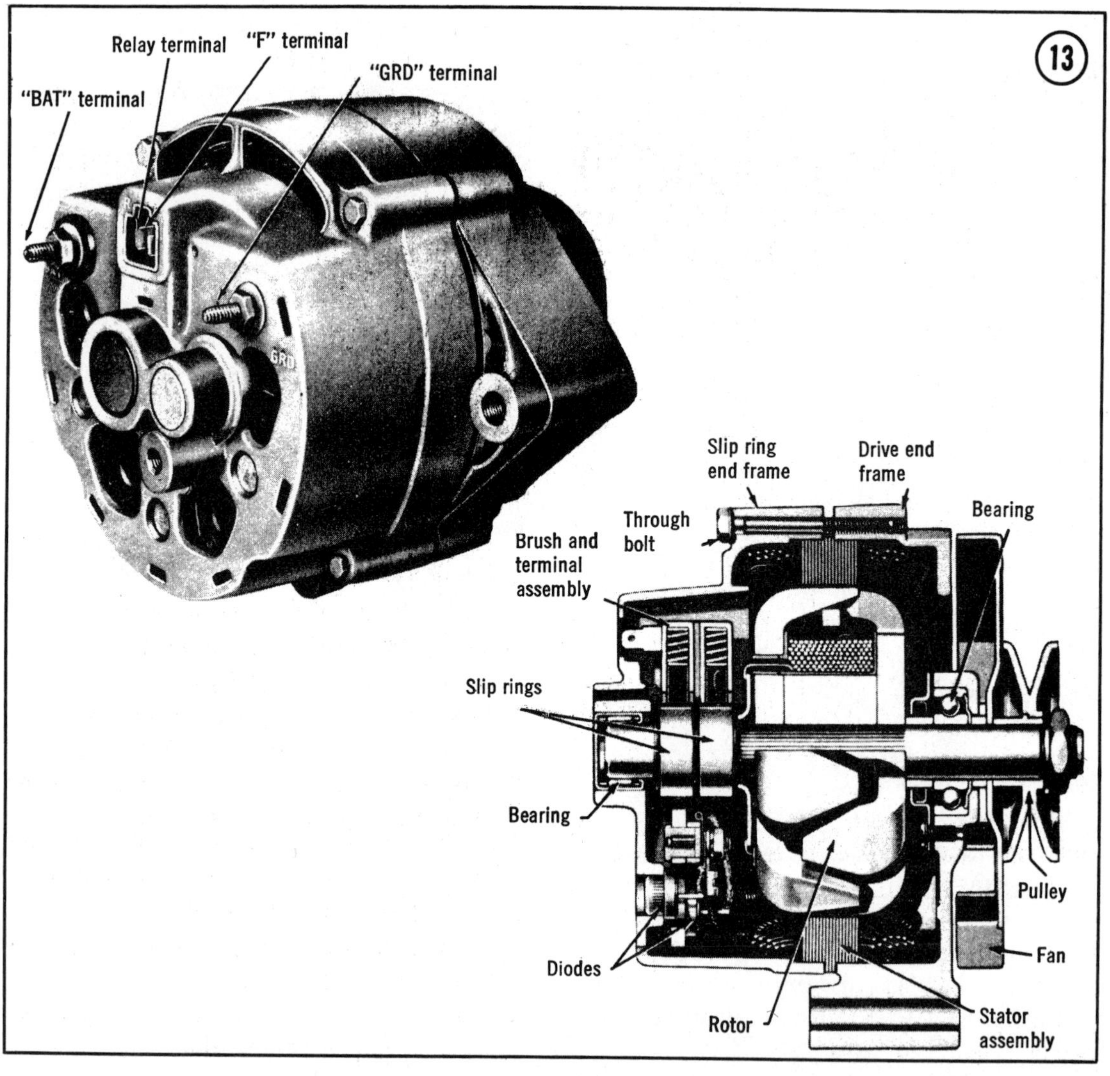

ELECTRICAL SYSTEM

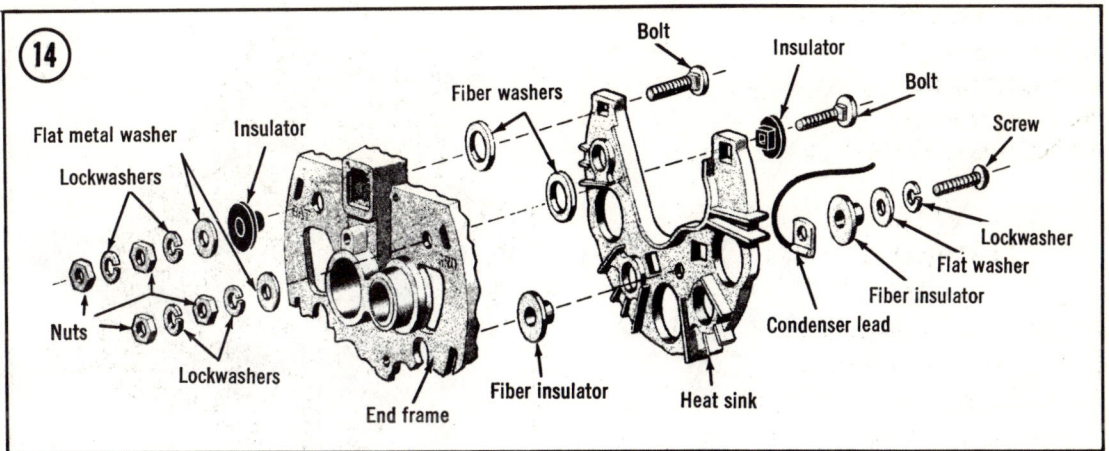

4. Support the Delcotron and remove bolt(s). Remove Delcotron from vehicle.

5. Reverse the procedure to install Delcotron, and then adjust drive belt(s) tension.

DELCOTRON 1D ALTERNATOR SERVICE

Disassembly

Refer to **Figure 13** for this procedure.

1. Hold alternator in a vise with soft jaws.
2. Make a scribe mark on frame halves to aid in reassembly.
3. Remove 4 through bolts and separate slip ring end frame/stator assembly from drive end/rotor assembly.
4. Place masking tape over slip ring and bearing to prevent entry of dirt. Also tape rotor at slip ring end.
5. Remove stator lead nuts and separate stator from frame.

6. Remove brushes and brush holder.
7. Remove BAT and GRD terminals, and one mounting screw from heat sink, then remove heat sink. See **Figure 14**.
8. If grease in the slip ring end frame bearing is exhausted, replace bearing as described in procedure below. Do not regrease a dry bearing.
9. Hold shaft with $\frac{5}{16}$ in. Allen wrench and remove nut with $1\frac{5}{16}$ wrench (**Figure 15**).
10. Remove pulley and fan from alternator.
11. Remove rotor and spacers from drive end frame.
12. Remove drive end bearing retaining plate, gasket, bearing, and slinger.

Cleaning and Mechanical Inspection

1. Wash all metal parts except rotor and stator assemblies in solvent.
2. Clean drive end bearing in solvent. Do not clean slip end bearing in solvent; instead, wipe outside with clean, lint-free cloth.
3. Inspect both bearings for pitting, discoloration, and roughness. Replace if necessary.
4. Check slip rings for wear, pitting, and out-of-roundness. Rings may be trued to within 0.001 in. or less on a lathe. Finish with 400 grain polishing cloth. If excessive damage is present, rotor must be replaced.
5. Inspect brushes. Replace if chipped, worn more than halfway, or oily.
6. Check brush springs for distortion or weakening. Replace if doubtful.

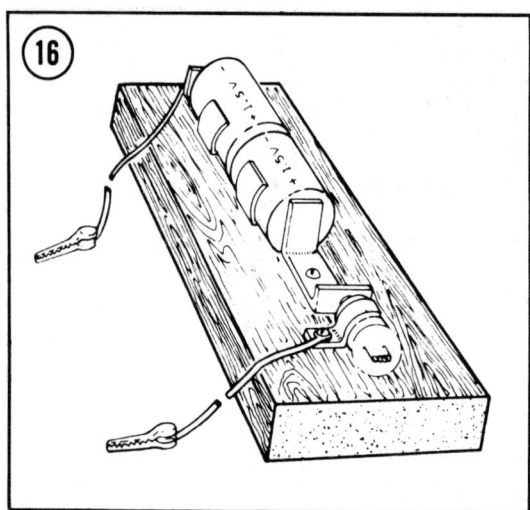

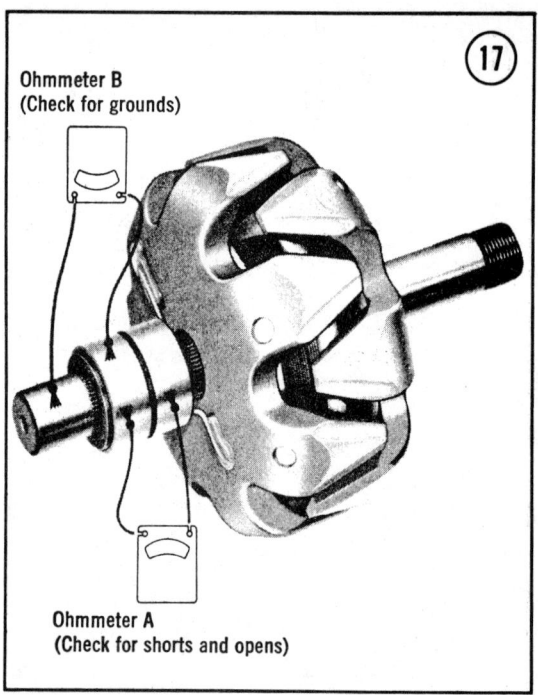

Ohmmeter B
(Check for grounds)

Ohmmeter A
(Check for shorts and opens)

Electrical Testing

For the following tests, use an ohmmeter containing a 1½-volt battery and set to RX1, or use a continuity tester, shown in **Figure 16**.

Rotor Check

1. Connect one tester lead to each slip ring (see ohmmeter A, **Figure 17**). If lamp fails to light, or ohmmeter reads infinity, rotor windings are open. Replace rotor.

2. Connect one tester lead to either slip ring (see ohmmeter B, **Figure 17**). If lamp lights or ohmmeter reads low, field windings are grounded. Replace rotor.

3. Connect 12-volt battery and 0-10 amp ammeter in series with slip ring. Compare current reading with field current specifications (**Table 1**). Replace rotor if reading is significantly higher than the higher specification limit.

Stator Checks

1. Connect tester to ohmmeter A position in **Figure 18**. If lamp lights or ohmmeter reads low, windings are grounded to stator frame. Replace stator.

2. Connect tester to ohmmeter B and C positions in **Figure 18**. If lamp fails to light or ohmmeter reads infinity, at either position, windings are open. Replace stator.

3. Check stator windings for discoloration indicating a shorted winding (which are usually difficult to locate without special test equipment). If all other electrical checks are normal and alternator still fails to supply rated output, have stator checked at an auto electrical shop or replace it.

Diode Checks

1. Using a self-powered test lamp (see **Figure 16**) of not more than 12 volts, and with the stator removed and disconnected, connect one tester lead to the heat sink and the other to the lead of the diode being tested, and note if the light comes on. Then reverse the tester leads and again note whether or not the light comes on. If the light comes on in one direction, but not the other, the diode being tested is OK. See **Figure 19**.

2. Repeat Step 1 for the remaining diodes in the heat sink.

3. Repeat Steps 1 and 2 for the diodes in the end frame.

4. Replace all defective diodes.

Diode Replacement

Diodes are very fragile and must be carefully pressed in and out of the heat sink. If a defective diode is discovered, take the heat sink to a

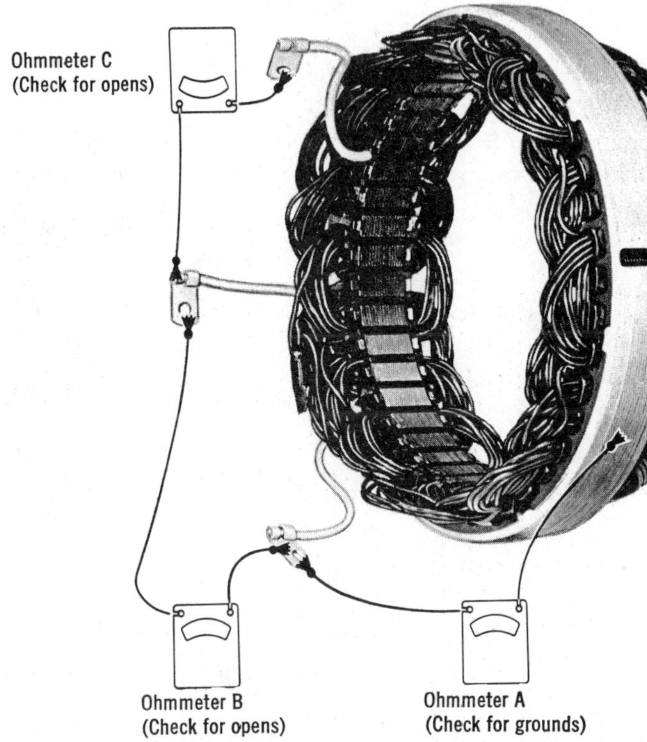

ELECTRICAL SYSTEM

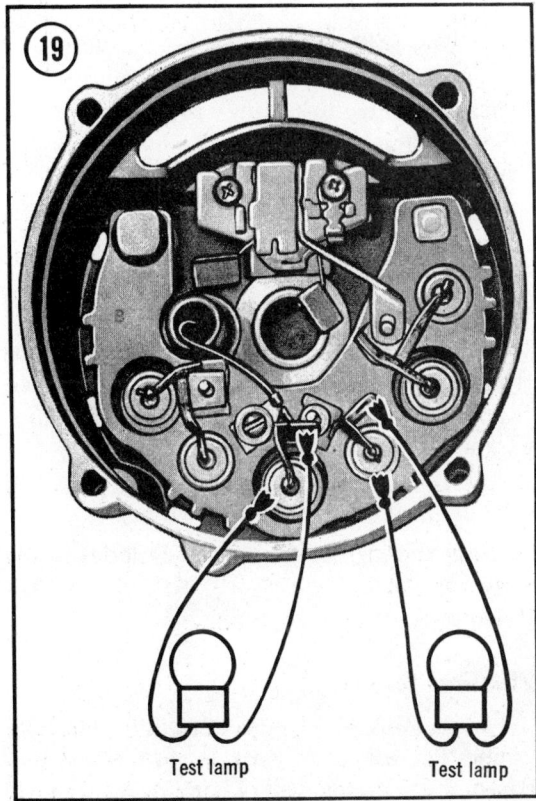

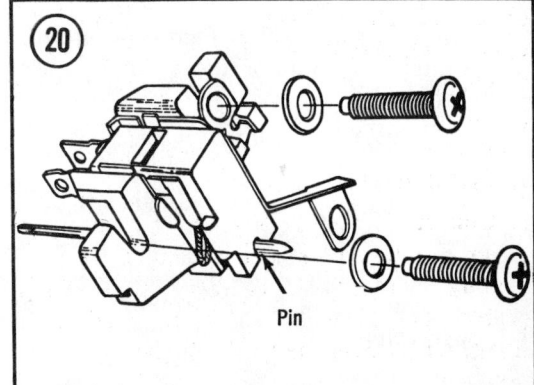

dealer or an alternator shop to have the diode replaced.

NOTE: *Chevrolet recommends replacing all 3 diodes on a heat sink if one is defective.*

Brush Replacement

Refer to **Figure 20** for this procedure.

Perform Steps 1 through 6 of *Disassembly* procedure above. Reverse Steps 1 through 6 of *Disassembly* procedure above.

Slip Ring End Bearing Replacement

Replace bearing if grease supply is exhausted or bearing is damaged.

1. Press out old bearing from outside of frame, using a suitable tool over outer bearing race.
2. Fit new bearing over opening at outside of frame.
3. Place a flat plate over bearing and support bottom of bore. Press bearing in until bearing is flush with outside of frame.
4. Saturate felt seal with SAE 20 oil. Install seal and steel retainer.

Assembly

Refer to **Figure 13** for this procedure.

1. Install drive end bearing.
2. Install rotor in drive end frame. Attach spacer, fan, pulley washer, and nut.
3. Hold frame in vise with soft jaws. Hold shaft with $5/16$ in. Allen wrench and tighten $15/16$ in. shaft nut to 40-50 ft.-lb.
4. Install brush holder.
5. Install heat sink. Follow **Figure 14** carefully for location of all components. Do not omit or interchange any parts.
6. Install stator and connect leads to bridge with nuts.
7. Remove masking tape from the bearing and rotor shaft (if still present).
8. Note scribe marks made during disassembly and join end frames. Secure frames with the 4 through bolts.

DELCOTRON 10-SI ALTERNATOR SERVICE

Disassembly

Refer to **Figure 5** for this procedure.

1. Hold alternator in a vise with soft jaws.
2. Make a scribe mark on frame halves to aid in reassembly.
3. Remove 4 through bolts and separate slip ring end frame/stator assembly from drive end/rotor assembly.
4. Place masking tape over slip ring end bearing to prevent entry of dirt. Also tape rotor at slip ring end.
5. Remove stator lead nuts and separate stator from frame.
6. Remove diode trio mounting screw and remove trio from frame.

NOTE: *Electrical checks may be made without further disassembly. After performing electrical checks, return to this procedure for further disassembly, if necessary.*

7. Remove rectifier bridge screw and BAT screw. Disconnect capacitor lead. Remove rectifier bridge.
8. Remove the brush holder and the regulator assemblies.

CAUTION
The 2 screws holding the brush clips have insulating washers and sleeves. If the third screw does not have an insulating washer and sleeve, do not interchange it with the other 2 screws.

9. Remove capacitor.
10. If grease in slip ring end frame is exhausted, replace bearing as described in separate procedure below. Do not regrease a dry bearing.
11. Hold shaft with $5/16$ in. Allen wrench and remove shaft nut with $15/16$ wrench, as shown in **Figure 15**.
12. Remove pulley and fan from alternator.
13. Remove rotor and spacers from drive end frame.
14. Remove drive end bearing retaining plate, gasket, bearing, and slinger.

Cleaning and Mechanical Inspection

Clean and inspect Series 10-SI Delcotrons exactly as described above for Series 1D Delcotrons.

Electrical Testing

For the following tests, use ohmmeter containing 1½-volt battery and set to RX1, or use continuity tester shown in **Figure 16**.

ELECTRICAL SYSTEM

Table 1 ALTERNATOR SPECIFICATIONS

	1100934 1100566 1100837 1100836 1100834	1100567 1100841 1100839	1100896 1100843	1100917 1100846	1100497[1]
Voltage (volts)	14	14	14	14	14
Output current (hot) (Amps)	37	42	61	63	37
Field current (Amps)	2.2-2.6	2.2-2.6	2.2-2.6	2.8-3.2	4.4-4.9
	1102394 1102491 1102483 1100497[2] 1100934	1100573 1102346 1100573	1102486 1102480 1100597 1102347	1100542 1102354	
Voltage (volts)	14	14	14	14	
Output current (hot) (Amps)	37	42	61	63	
Field current (Amps)	4-4.5	4-4.5	4-4.5	4-4.5	

1. 1973
2. 1974-1975

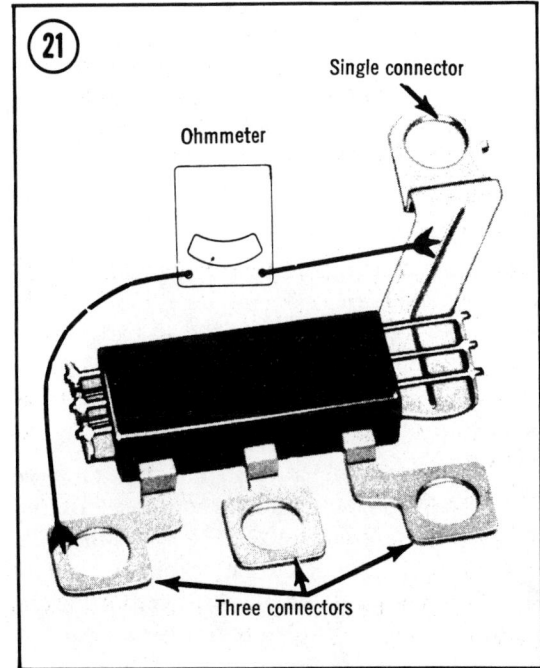

Rotor Checks

1. Connect one tester lead to each step ring (see ohmmeter A, **Figure 17**). If lamp fails to light or ohmmeter reads infinity, rotor windings are open. Replace rotor.

2. Connect one tester lead to either slip ring and the other to rotor shaft (see ohmmeter B, **Figure 17**). If lamp lights or ohmmeter reads low, field windings are grounded. Replace rotor.

3. Connect a 12-volt battery and 0-10 amps ammeter in series with slip rings. Compare current reading with field current specifications **(Table 1)**. Replace rotor if reading is significantly higher than the specification.

Stator Checks

1. Connect tester to position A shown in **Figure 18**. If lamp lights or ohmmeter reads low, windings are grounded to stator frame and stator must be replaced.

2. Connect tester to positions B and C of **Figure 18**. If lamp fails to light or ohmmeter reads infinity in either position, windings are open and stator must be replaced.

3. Check stator windings for discoloration indicating a shorted winding (which are difficult to locate without special test equipment). If all electrical checks are normal and alternator still fails to supply rated output, have stator checked by an alternator shop or replace it.

Diode Trio Checks

1. With diode trio removed completely, connect one ohmmeter lead to single connector and the other lead to any one of 3 connectors (see **Figure 21**). Observe reading. Reverse ohmmeter

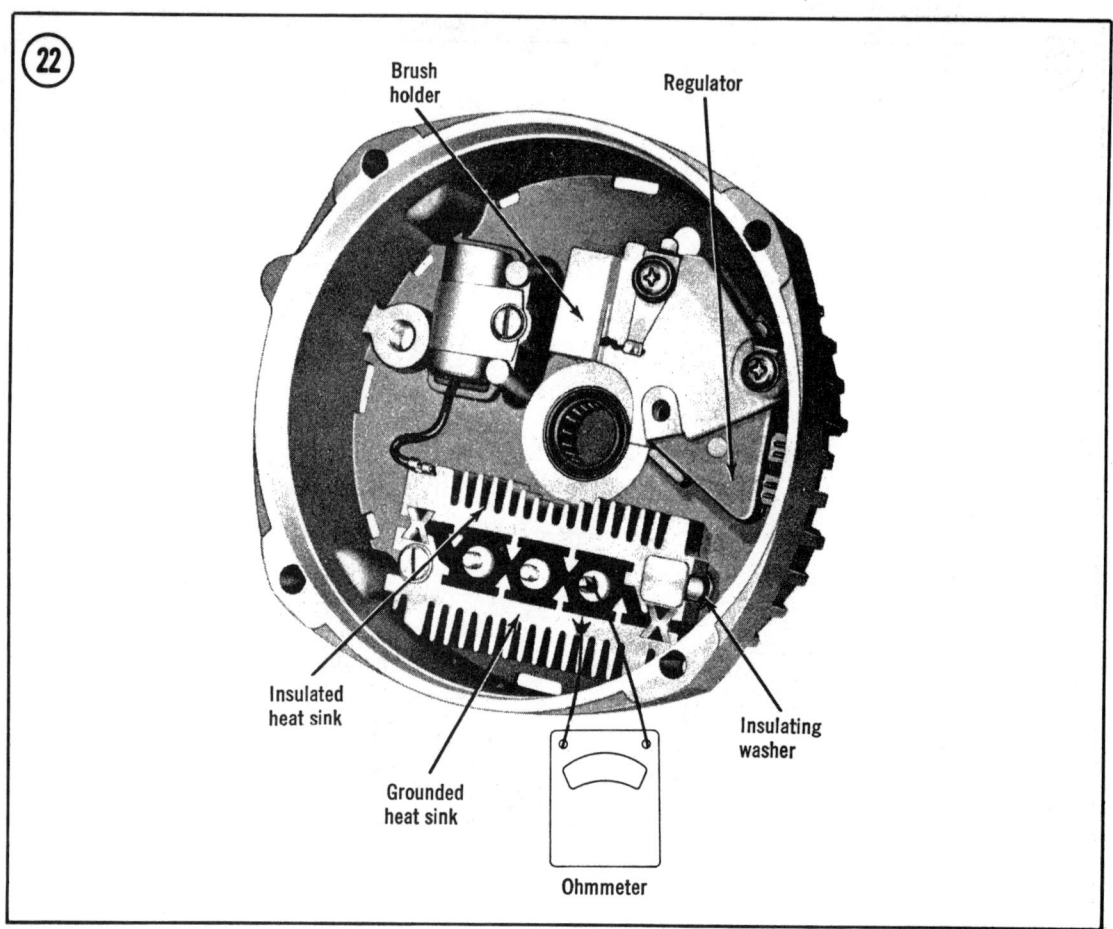

leads to same connectors and again observe reading. A good diode reads high in one direction and low in the other. If both readings are the same, either high or low, replace diode trio.

2. Repeat Step 1 for each of the 3 connectors.

> NOTE: *Two diode trios, differing only in appearance, have been used in these alternators. They are completely interchangeable.*

Rectifier Bridge Checks

1. Connect one ohmmeter lead to the grounded heat sink and the other lead to one of the 3 bridge terminals (see **Figure 22**). Observe reading. Reverse leads to same terminal and again observe reading. If readings are the same, replace bridge. If readings differ (one high, one low), repeat procedure on the other 2 terminals.

2. Repeat Step 1 on the insulated heat sink instead of the grounded heat sink.

Brush Lead Clip Check

1. Assemble alternator by following Steps 1 through 8 of the *Assembly* procedure below.

2. Connect ohmmeter from brush lead clip to frame (see ohmmeter 1, **Figure 23**). Observe reading. Reverse leads and observe reading again. If both readings are zero, either the brush lead is grounded or the regulator is defective. Check insulated washer and insulated sleeve on screw. If washer and screw are present and undamaged, replace regulator unit.

Slip Ring End Bearing Replacement

Replace bearing if grease supply is exhausted or bearing is damaged.

1. Press old bearing out from outside frame, using suitable tool over outer race of bearing.

2. Fit new bearing over the opening at outside of frame.

ELECTRICAL SYSTEM

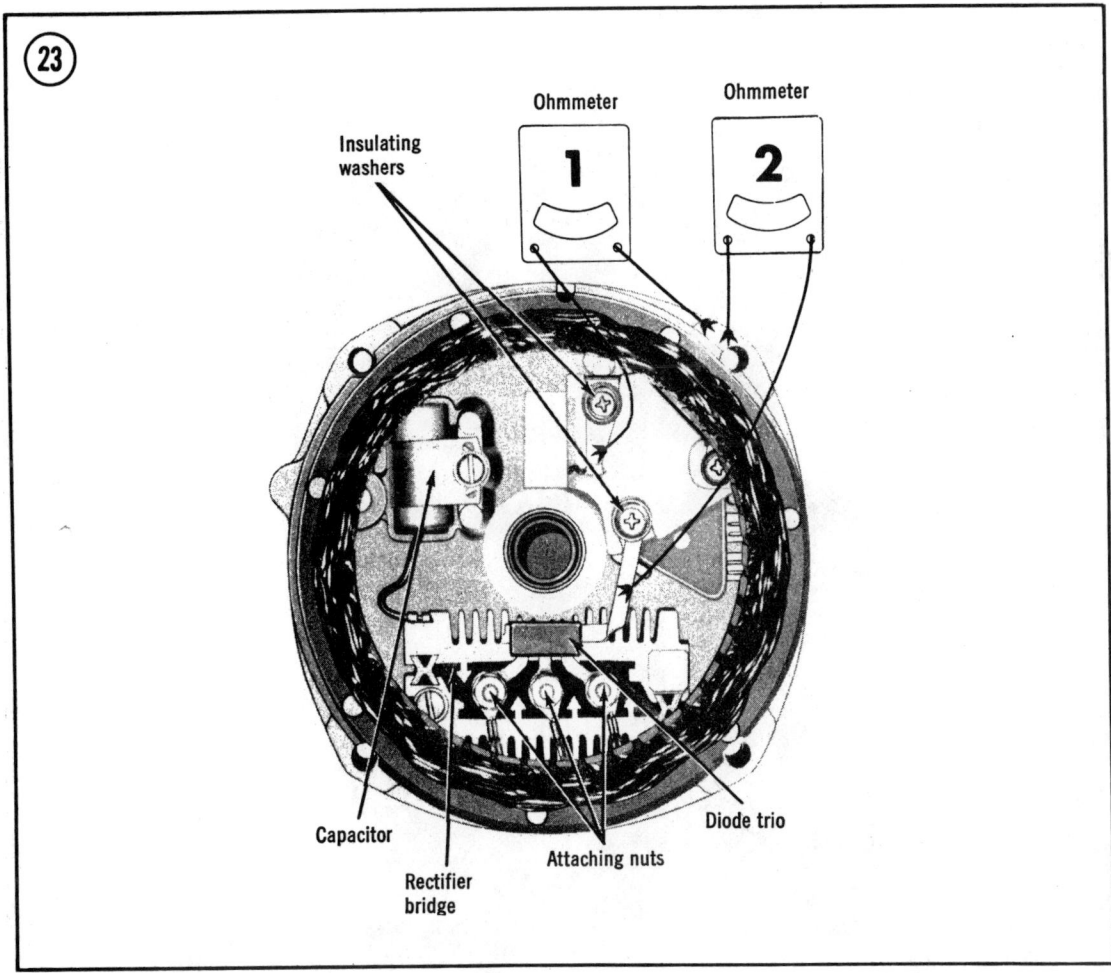

3. Place a flat plate over bearing and support bottom of bore. Press bearing in from outside until bearing is flush with outside of frame.

4. Install new seal. Lightly oil seal lip and press it in with seal lip toward inside of frame.

Assembly

1. Install drive end bearing.
2. Install rotor in drive end frame. Attach spacer, fan, pulley, washer and nut.
3. Hold frame in vise with soft jaws. Hold shaft with 5/16 in. Allen wrench and tighten 15/16 in. shaft nut to 40-50 ft.-lb.
4. Install capacitor.
5. Install brush holder and regulator units.

CAUTION
The 2 screws holding the brush clips have insulating sleeves and washers. If the third screw does not have an insulating sleeve and washer, do not interchange it with the other 2.

STARTER MOTOR

The procedure below is general in nature and applies, with slight variations, to all starters used on all vehicles. **Figure 24** is a cross-sectional view of a typical starter motor/solenoid assembly. **Figure 25** shows typical starter installations on various engine models. **Table 2** shows starter specifications.

Periodic lubrication or other maintenance is not required for either the starter motor or the solenoid.

Diagnostic procedures for troubleshooting the starter motor are given in Chapter Two. If these procedures are used and it is determined that the starter motor and/or solenoid are

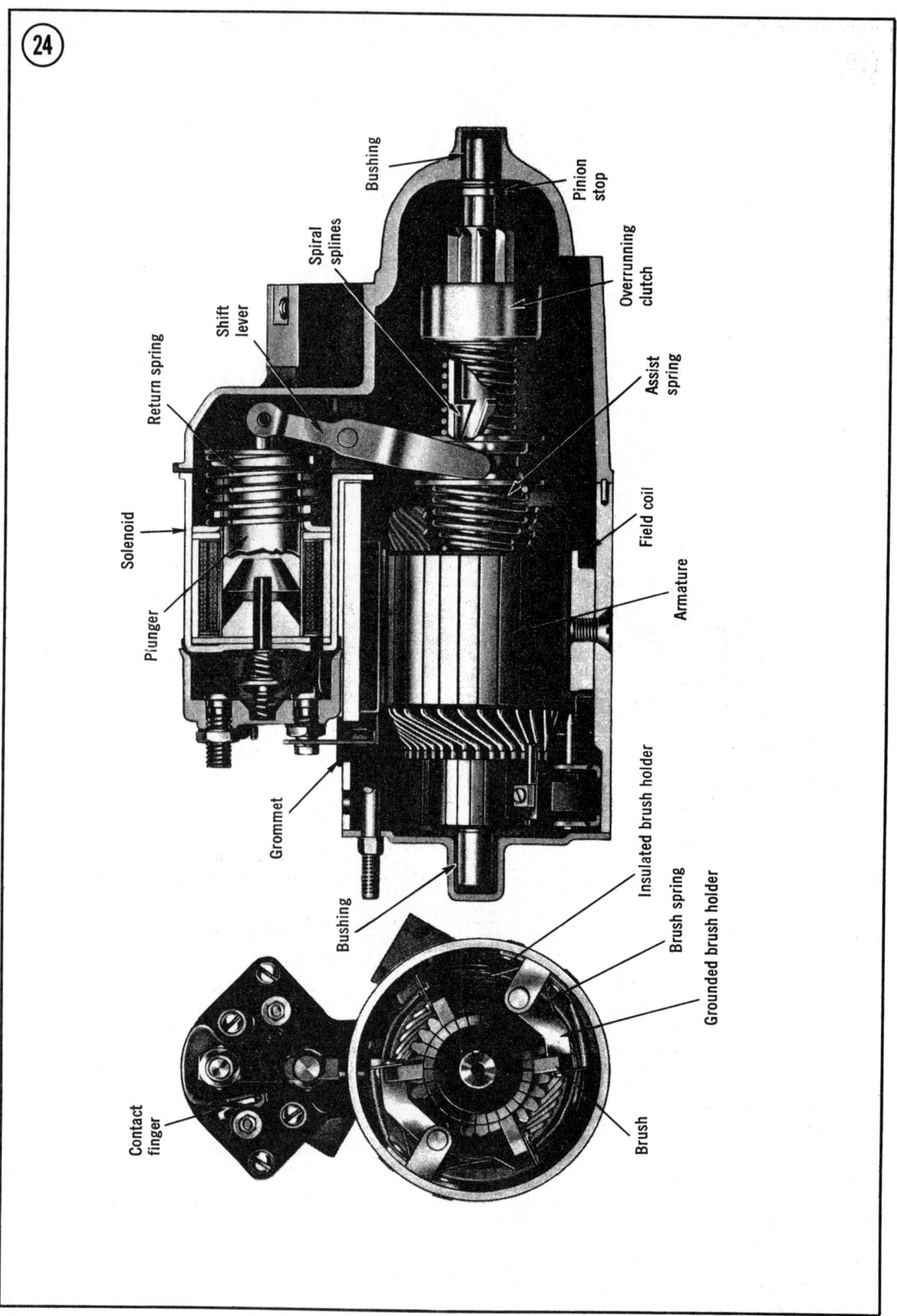

ELECTRICAL SYSTEM

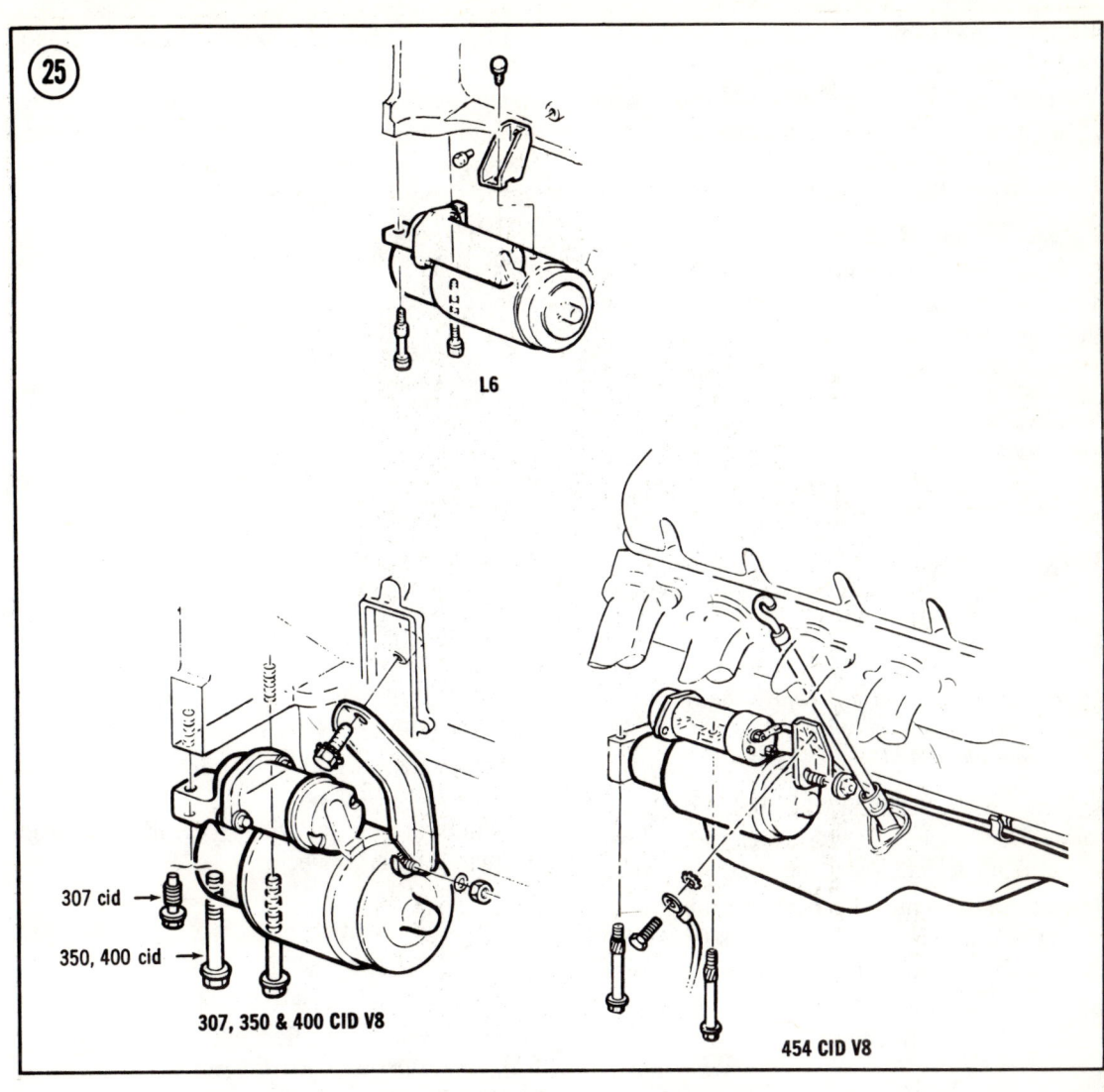

Table 2 STARTER SPECIFICATIONS

			(No Load)			
	1108799 1108365 1108774 1108512 1108367 1109056	1109059 1108418 1108775 1108430 1108776 1109052	1108427 1108338	1108361	1107368 1107320 1107388	1107399
Voltage	9[1]	9[1]	9	9	10.6	10.6
Current	50-80[2]	65-95	50-80	55-85	65-100	49-87
RPM	5,500[3]- 10,500	7,500- 10,500	3,500- 6,000	3,100- 4,900	3,600- 5,100	6,200- 10,700

1. 10.6 volts on 1979 and later models.
2. 60-88 A on 1979 and later Model No. 1108774.
3. 6,500-10,100 rpm on 1979 and later Model No. 1108774; 7,500-11,400 rpm on 1979 and later Model No. 1109056.

defective, the starter/solenoid assembly should be removed from the engine and repaired or replaced. (Starter repairs are beyond the scope of this book. See a qualified mechanic.)

Starter Motor Replacement

Refer to **Figure 25**.

1. Disconnect battery ground cable at the battery and, if possible, raise vehicle to a good working height.
2. Disconnect all wires at solenoid terminals.

> NOTE: *Reinstall nut as each wire is removed. Thread sizes are different and stripped threads could result if nuts become mixed.*

3. Loosen starter front bracket (nut on V6 and V8 engines, bolt on 6-cylinder engines), and then remove 2 mount bolts.

> NOTE: *On V6 and V8 engines using a solenoid heat shield, remove the front bracket upper bolt and detach bracket from starter motor.*

4. Remove the front bracket bolt or nut and rotate bracket clear of work area. Then lower starter from vehicle, front end first.
5. Reverse the removal procedure to install the starter motor. Tighten mount bolts first (25-35 ft.-lb.), then tighten bracket nut or bolt. Make sure all electrical connections are tight.
6. Check starter operation.

BODY ELECTRICAL SYSTEM

Instructions and pointers for troubleshooting the electrical wiring system are given in Chapter Two. Service which the owner/mechanic can be expected to perform includes isolation and repair of short and open circuits, wire, connector, and fuse replacement, and sealed beam and other bulb replacement.

Maintenance of the lighting and wiring systems consists of an occasional check to see that wiring connections are tight and clean, that lighting units are tightly mounted and that headlights are properly adjusted. The latter task should be performed by a qualified mechanic having access to the specialized equipment required.

Loose or corroded connectors can result in a discharged battery, dim lights, and possible damage to the alternator or voltage regulator. Should insulation become burned, cracked, abraded, etc., the affected wire or wiring harness should be replaced.

Rosin core solder — never use acid core solder on electrical connections — must always be used when splicing wires. Splices should be covered with insulating tape.

Replacement wires must be of the same gauge as the replaced wire — never use a smaller gauge. All harnesses and wires should be held in place with clips, cable ties, or other holding devices so that chafing and abrasion can be avoided.

Headlight Replacement

> NOTE: *On models so equipped, open headlamp doors.*

1. Remove headlamp bezel retaining screws and then remove bezel.
2. Disengage spring from retaining ring and remove 2 attaching screws.
3. Remove retaining ring. Disconnect sealed beam unit from connector.
4. Install replacement sealed beam unit in connector and position in place with number molded into lens face at top of unit.
5. Position and install retaining ring, using screws and spring (as applicable).
6. Check operation of the sealed beam unit, then install bezel.

Parking and Rear Lamp Replacement

Because of the large number of models, it would be impossible to provide individual procedures covering all types of lamp installations. Close examination will usually reveal the approach to be taken for bulb replacement. Many installations use twist-out sockets which are removed from the rear of the lamp housings. Other installations require the removal of the lamp lens (held in place by 2 or more screws through the lens face) for access to the bulb and socket. In other installations, nuts must be removed from the rear of the lamp unit to release the lens. See **Table 3** for bulb usage.

ELECTRICAL SYSTEM

FUSES/CIRCUIT BREAKERS

Each electrical circuit is protected from overload by either a circuit breaker, a fuse, or a fusible link.

Circuit Breaker

The circuit breaker is a protective device that causes an open circuit when the current load exceeds the rated breaker capacity. The circuit breaker will remain open until the problem causing the overload is found and corrected. There is one circuit breaker, which is located on the headlight switch and protects the headlamp circuit. The headlamp switch circuit breaker will cause a flickering condition whenever an overload is present in the circuit.

Fuses

Most major circuits are protected by fuses **(Table 4)**. The fuse contains an element that melts when the protected circuit is overloaded. When this occurs, the fuse must be replaced. The cause of the overload should be located and corrected before replacing a burned out fuse. If this does not happen, chances are that

Table 3 LAMP USAGE (MONTE CARLO — MALIBU CLASSIC)

Circuit	1970	1971	1972	1973	1974	1975
Headlamp	6012A	6014	6014	6014	6014	6014
Front park/turn	1157NA	1157NA	1157NA	1157NA	1157NA	1157NA
Tail stop/turn	1157	1157	1157	1157	1157	1157
Turn signal ind.	194	194	194	194	168	168
High beam ind.	194	194	194	194	168	168
Instrument lamps	194	194	194	168	168	168
Courtesy lamp	211	211	211	211	211	211
License plate lamp	67	67	194	67	168	168
Radio dial lamp	293/1893	1816	1816	1816	1816	1816
Brake alarm	194	194	194	168	168	168
Back-up lamp	1156	1156	1156	1156	1156	1156
Glove box lamp	1893	1893	1893	1893	1893	1891
Heater cont. panel	1445	1445	1445	1445	1445	1445
A.C. control panel	1445	1445	1445	1445	1445	1445

Circuit	1976	1977	1978	1979	1980
Headlamp, upper[1]	4652	4652	4652	4652	N.A.
Headlamp, lower[2]	4651	4651	4651	4651	
Front park/turn	1157NA	1156NA	1156NA	1156NA	
Tail stop/turn	1157	1157	1157	1157	
Turn signal ind.	168	168	168	168	
High beam ind.	168	168	168	168	
Instrument lamps	168	168	168	168	
Courtesy lamp (dome)	211-2	211-2	211-2	211-2	
License plate lamp	168	194	194	168	
Radio dial lamp	1816	216/1893	216/1893	216/1893	
Brake alarm	168	168	168	168	
Back-up lamp	1156	1156	1156	1156	
Glove box lamp	1893	1893	1893	1893	
Heater cont. panel	1445	194	194	194	
A.C. control panel	1445	194	194	194	

1. Or outboard.   2. Or inboard.

Table 3 LAMP USAGE (CHEVELLE-MALIBU) (continued)

Circuit	1970	1971	1972	1973	1974	1975
Headlamp, outer	4002	N/A	N/A	N/A	N/A	N/A
Headlamp, inner	4001	6014	6014	6014	6014	6014
Front park/turn	1157	1157NA	1157NA	1157	1157NA	1157NA
Tail stop/turn	1157	1157	1157	1157	1157	1157
Turn signal ind.	194	194	194	194	168	168
High beam ind.	194	194	194	194	168	168
Instrument lamps	194	194	194	168	168	168
Courtesy lamp (dome)	211	211	211	211	211	211
License plate lamp	67	67	194	67	168	168
Radio dial lamp	293/1893	1816	1816	1816	1816	1816
Brake alarm	194	194	194	168	168	168
Back-up lamp	1156	1156	1156	1156	1156	1156
Glove box lamp	1893	1893	1893	1893	1893	1891
Heater cont. panel	1445	1445	1445	1445	1445	1445
A.C. control panel	1445	1445	1445	1445	1445	1445

Circuit	1976	1977	1978	1979	1980
Headlamp	6014	6014	6014	6014	N/A
Front park/turn	1157NA	1156	1156NA	1156NA	
Tail stop/turn	1157	1157	1157	1157	
Turn signal ind.	168	168	168	168	
High beam ind.	168	168	168	168	
Instrument lamps	168	168	168	168	
Courtesy lamp (dome)	211-2	211-2	211-2	211-2	
License plate lamp	194	194	194	194	
Radio dial lamp	1816	216/1893	216/1893	216/1893	
Brake alarm	168	168	168	168	
Back-up lamp	1156	1156	1156	1156	
Glove box lamp	1893	1893	1893	1893	
Heater cont. panel	1445	194	194	194	
A.C. control panel	1445	194	194	194	

the replacement fuse will burn out too. Replacement fuses must be of the same amperage capacity as the fuses they are replacing. Fuses of higher ratings could cause damage to wiring and insulation, or even cause fires. A fuse with a lower than specified rating could cause unnecessary interruption of the circuit at normal loading.

Fusible Links

Fusible links provide protection to circuits which are not normally fused, such as the ignition system circuit. The links are usually four gauges smaller than the wiring in the circuit they are protecting, and are covered with heavy red insulation. The gauge size is plainly marked on the insulation. If a link burns out, the cause must be isolated and corrected. A new fusible link must then be spliced into the circuit. The following procedure should be used to replace fusible links.

1. Disconnect battery ground cable at battery.

2. Disconnect old fusible link from components to which it is attached.

3. Cut harness behind connector (**Figure 26**) to remove damaged fusible link.

4. Strip harness wire insulation approximately to ½ inch.

ELECTRICAL SYSTEM

Table 4 FUSES

Circuit	Fuse Rating (amps)					
	1970	1971	1972	1973	1974	1975
Wiper	25	25	25	25	25	25
Back-up lamp and gauges	25	20	20	20	20	20
Heater/AC	25	25	25	25	25	25
Radio	10	10	10	10	10	10
Instrument lamps	4	2	3	3	4	4
Stop and taillamp	20	20	20	20	20	20
Clock, lighter and courtesy	20	25	20	20	20	20
In-line AC	30	30	30	30	30	30
Gauges and telltales	10	10	10	10	10	10
Tail, license, luggage & park			20	20	20	20

Circuit	Fuse Rating (amps)				
	1976	1977	1978	1979	1980
Wiper	25	25	25	25	N.A.
Back-up lamp and gauges	20	20	20	20	
Heater/AC	25	25	25	25	
Radio	10	10	10	10	
Instrument lamps	4	4	4	5	
Stop and taillamp	20	20	20	20	
Clock, lighter and courtesy	20	20	20	20	
In-line AC	30	30	30	—	
Gauges and telltales	10	10	10	10	
Tail, license, luggage & park	20	20	20	20	

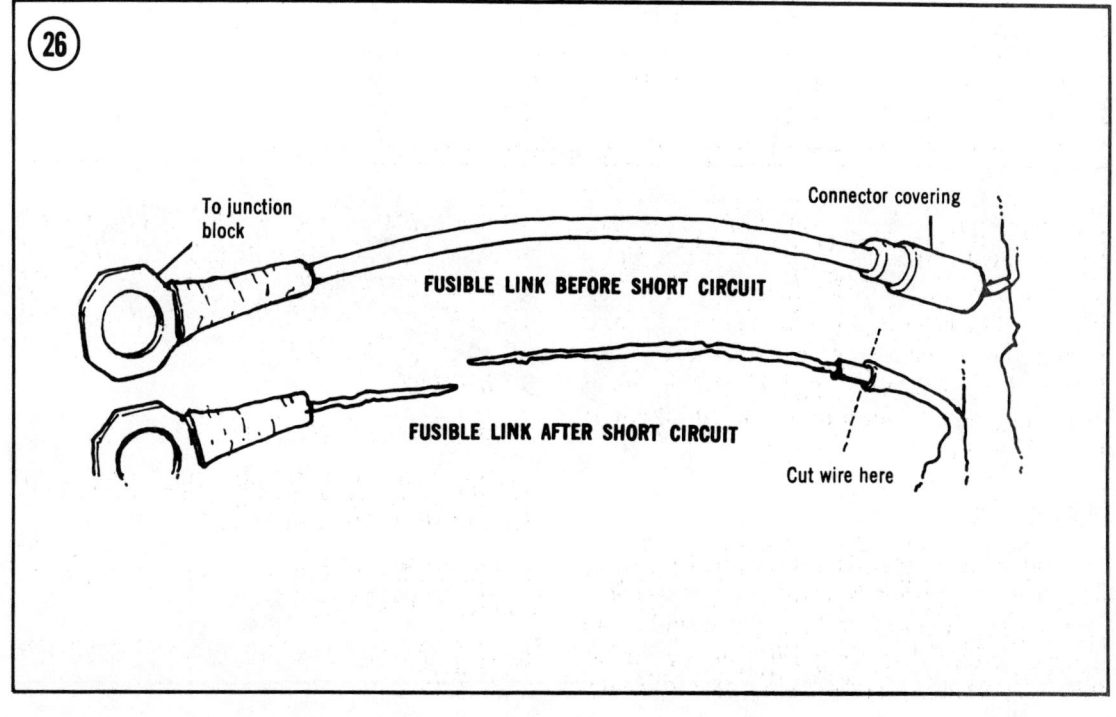

27

Fusible link
Connector
SERVICE PACKAGE
Thoroughly tape all exposed wires
SERVICE PACKAGE INSTALLED
Connector crimped and soldered

28

Ground strap

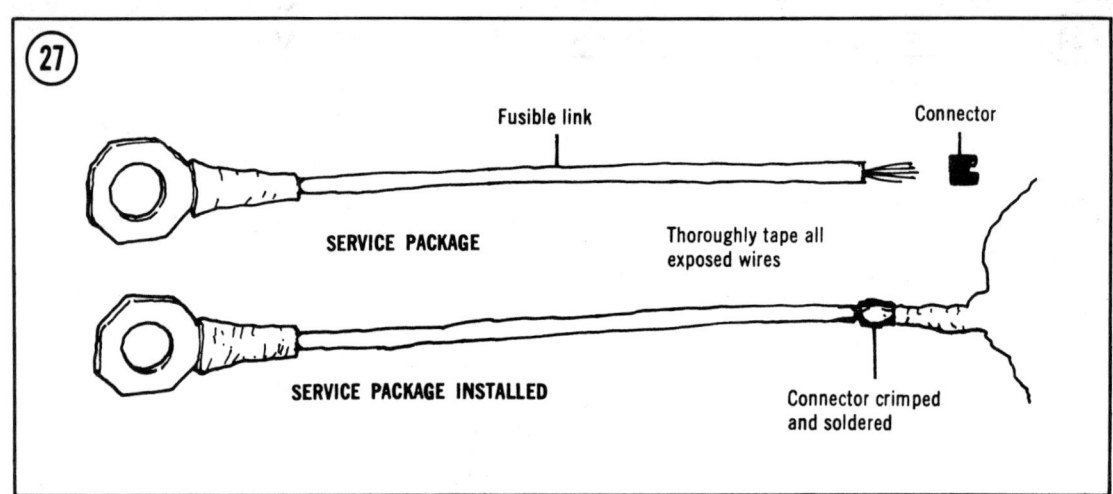

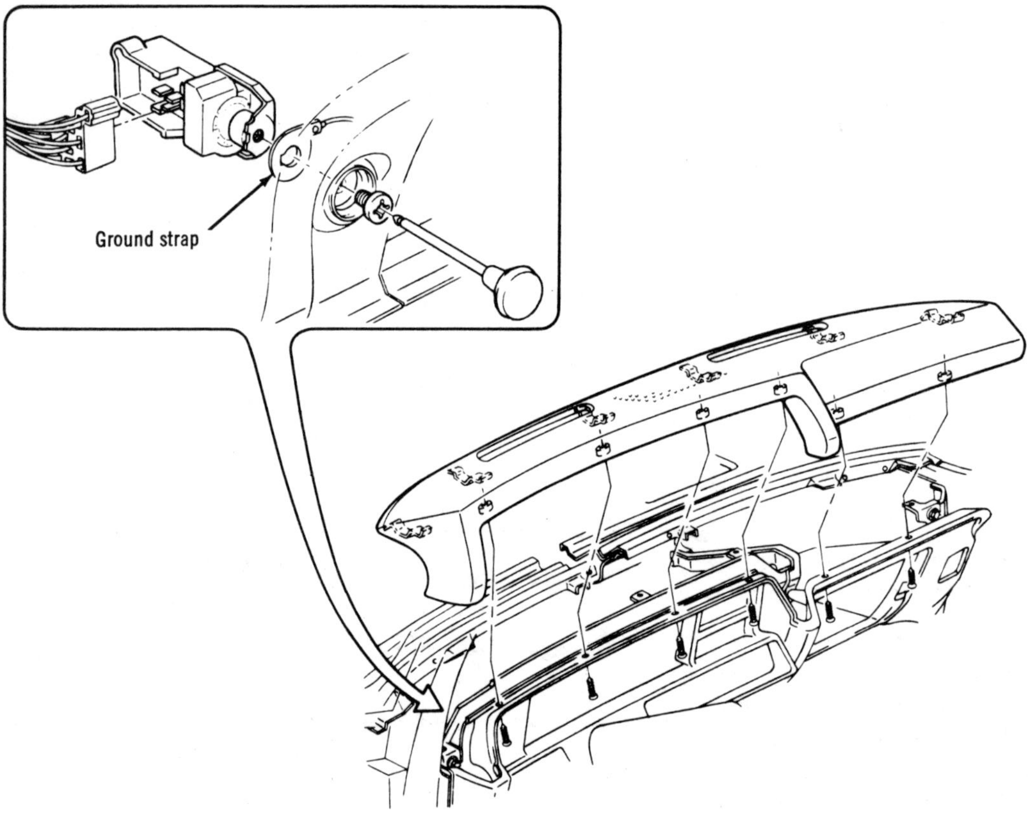

ELECTRICAL SYSTEM

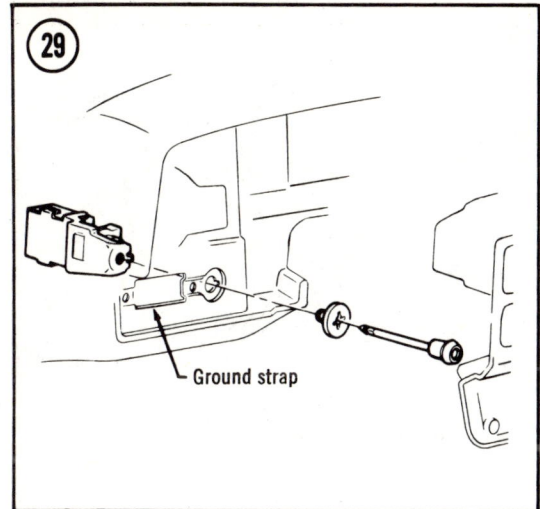

Ground strap

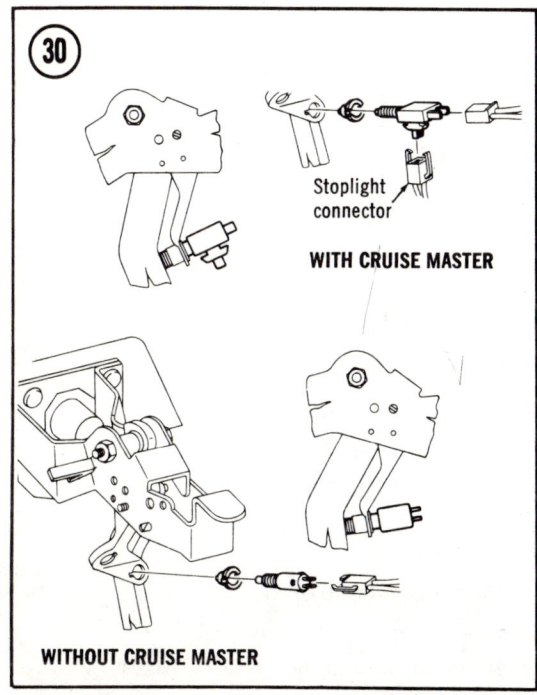

Stoplight connector

WITH CRUISE MASTER

WITHOUT CRUISE MASTER

5. Position clip around ends of new fusible link (**Figure 27**) and harness wire and crimp so that both wires are securely fastened.

6. Solder the connection, using rosin core solder. Use sufficient heat to obtain a good solder joint, but do not overheat.

7. Tape all exposed wires with insulating tape.

8. Connect fusible link to component from which it was removed.

9. Reconnect ground cable to battery.

LIGHTING SWITCH REPLACEMENT

Refer to **Figure 28** (1970-1972) or **Figure 29** (1973-on).

1. Disconnect the battery negative cable.

2. On 1970-1972 models, remove steering column lower cover. On 1973 and later models, remove instrument panel bezel.

3. Pull headlight switch knob to ON position.

4. On 1978 and later models, remove the screws (3) attaching the switch mounting plate to the cluster and pull the assembly rearward.

5. If necessary for clearance, move the radio speaker by removing attaching screws.

6. Reach under panel and depress switch shaft retaining button and pull out shaft and knob.

7. Remove nut attaching switch to plate or panel and remove switch.

8. Disconnect wiring connector from switch.

9. Installation is the reverse of these steps.

DIMMER SWITCH REPLACEMENT

1. Fold back floor mat and disconnect connector for dimmer switch.

2. Remove dimmer switch by removing attaching screws.

3. Install new switch by connecting it to connector and checking operation. Then attach switch to floor pan with attaching screws and replace floor mat.

STOPLIGHT SWITCH REPLACEMENT

See **Figure 30** (typical) for this procedure.

1. Disconnect wiring harness connector(s) from switch.

2. Remove retainer (nut, clip, screw, etc.) and unscrew the switch from the bracket.

3. Press brake pedal in and install new switch in bracket until shoulders bottom out. Plug in connector.

4. Check switch position to see that electrical contact is made (observe stoplights) when brake pedal is depressed $\frac{3}{8}$-$\frac{5}{8}$ in.

BACK-UP LIGHT SWITCH REPLACEMENT

Mast Jacket Mounted (Manual Transmission)

See **Figure 31** (typical) for this procedure.

1. Disconnect wiring connector(s) from switch terminal(s).
2. Remove switch unit from mast jacket by removing retaining screws.
3. Position gearshift in REVERSE.
4. Place switch in position (all models). Make sure that the actuating tang is in the slot in the shift tube, and the right side of the tang is contacting the right-hand surface of the slot.
5. Install the attaching screws while holding the switch in the position described in Step 4 and tighten securely.
6. Reconnect the wiring connector(s) to connector(s) on switch.

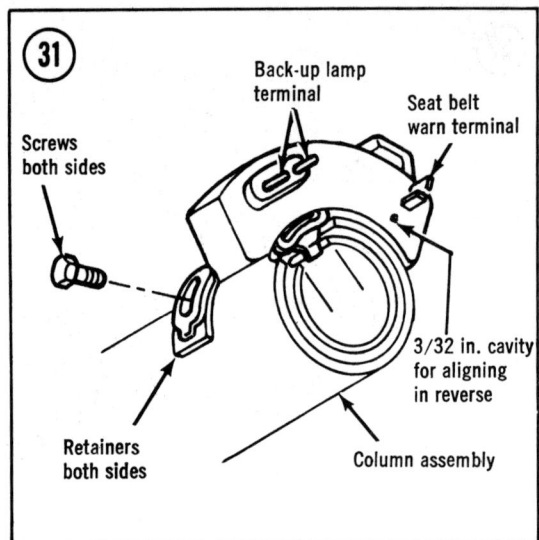

NEUTRAL SAFETY SWITCH REPLACEMENT (AUTOMATIC TRANSMISSION)

NOTE: *On 1973 and later models, this component also includes backing lamp and seat belt warning switches.*

Column Shift

1. Disconnect the wiring connector(s) from the switch(es). See **Figure 32** (typical).
2. Remove the switch by removing the retaining screws.
3. On 1973 and later models:
   a. Place shift lever against NEUTRAL gate (rotate lower shift tube counterclockwise as viewed from driver seat).
   b. Insert switch actuating tang into shifter tube slot and secure switch to column with attaching screws. Tighten screws securely.
   c. Connect wiring harness to switches.
   d. Move the selector out of NEUTRAL to shear the pin.

   NOTE: *If pin has been sheared, align 3/32 in. hole in switch assembly with actuating tang and insert pin 1/4 in. to hold tang. Perform Steps a through c above, then remove pin before shifting out of* NEUTRAL.

4. On 1972 and earlier models:
   a. Position shift lever in DRIVE. Position actuating tang against selector plate.
   b. Align slot in contact support with 3/32 in. hole in switch and insert pin in hole. Switch now is in DRIVE position.
   c. Locate contact support drive slot over shifter tube drive tang and tighten screws. Remove clamp and pin.
   d. Connect wiring harness to switch and check switch operation.

Floor Shift

1. Disconnect control rod from shift lever arm.
2. Remove knob from shift lever.
3. Remove retaining screws from control assembly and remove control assembly from seal.
4. Remove switch from control assembly.
5. Position gearshift in PARK and assemble switch assembly to lever and bracket by inserting pin into actuating tang of switch.
6. Install and tighten attaching screws.
7. Connect wiring harness and check switch operation.
8. Install trim plate and knob, and reconnect shift lever arm to control rod.

ELECTRICAL SYSTEM

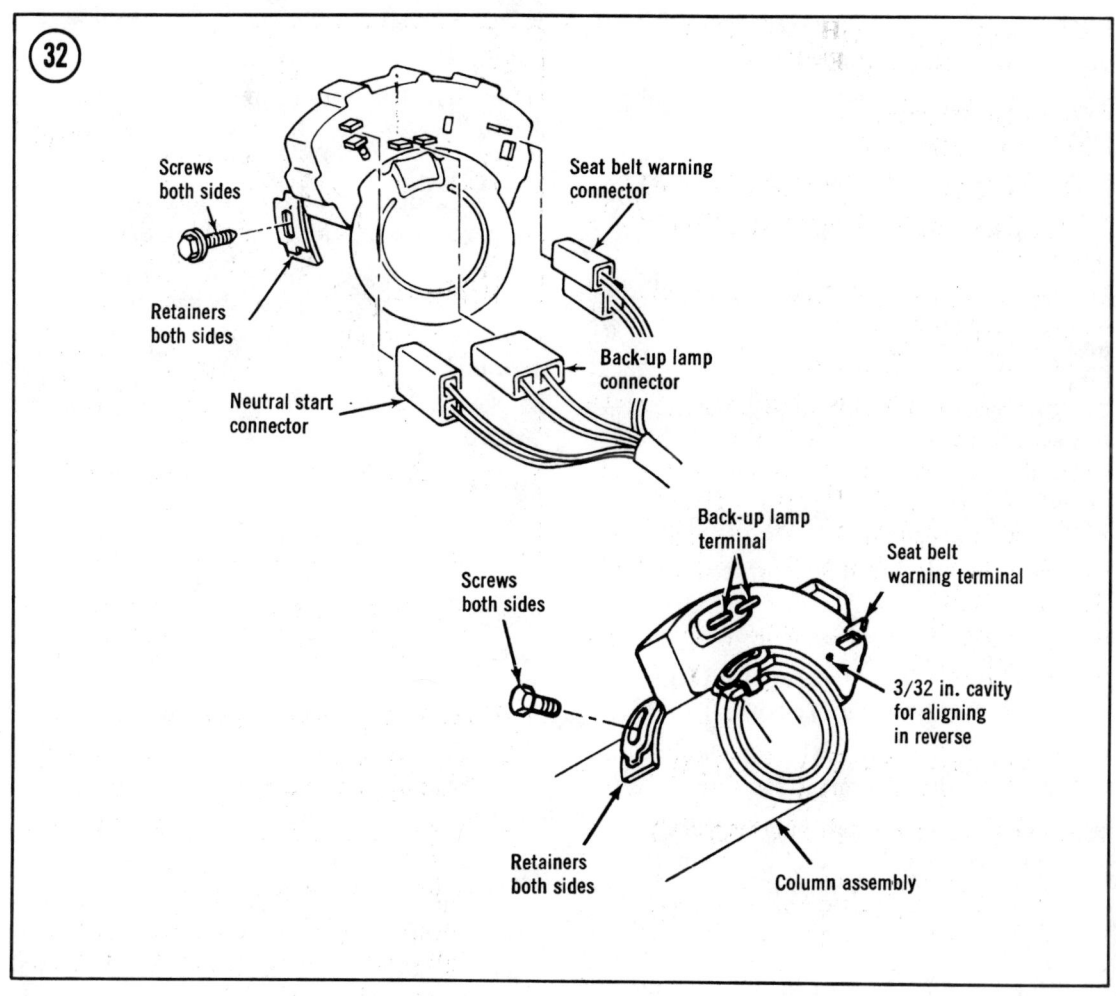

DISTRIBUTOR REMOVAL

1. Remove distributor cap and position it out of the way.

   NOTE: *If necessary, remove secondary wires (spark plug and coil high tension wires) from distributor cap, after first marking cap tower for No. 1 cylinder spark plug wire.*

2. Disconnect distributor primary wire from coil terminal.

3. Crank engine until No.1 cylinder is in the firing position and the timing mark on the harmonic balancer is aligned with the proper mark on the timing tab on the engine front cover.

4. Make alignment marks on the distributor housing and the engine block in line with the rotor pointer.

5. Disconnect vacuum line to distributor.

6. Remove the distributor hold-down bolt and clamp. Remove distributor from engine. Note that rotor moved during removal, and mark new position of rotor pointer on distributor housing.

   NOTE: *If possible, avoid any further cranking of engine until distributor has been replaced.*

DISTRIBUTOR INSTALLATION (ENGINE NOT DISTURBED)

1. Turn rotor pointer to align with second mark made on the distributor housing (about $\frac{1}{8}$ turn clockwise from first mark) during the removal procedure.

2. Insert distributor into its opening in engine block and align first mark on housing with

mark made on block during removal. Push down into position. Rotor should now be aligned with first mark on distributor housing and mark on engine block.

> NOTE: *During installation it may be necessary to move rotor slightly so drive gear will mesh with camshaft gear.*

3. Install distributor hold-down clamp and screw. Connect vacuum line to vaccum advance mechanism. Connect primary lead wire to coil terminal and install distributor cap. Replace spark plug wires and coil high tension lead wire, if removed.

> NOTE: *Verify that spark plug wires are connected to proper spark plugs.*

4. Time ignition, using the procedure given above.

DISTRIBUTOR INSTALLATION (ENGINE DISTURBED)

1. Locate No. 1 piston in firing position, using either of the following methods:
   a. Remove No. 1 spark plug (the one closest to front of engine) and place finger over spark plug hole. Crank engine until compression is felt, then continue cranking until timing mark on crankshaft pulley lines up with proper mark on timing tab attached to front cover of engine.
   b. Remove rocker cover (left side on V6 and V8 engines) and crank engine until No. 1 intake valve closes. Continue to slowly crank until timing mark on pulley lines up with mark on timing tab.

2. Place distributor in opening in block in normal installed position. Hold distributor in this position and position rotor to point toward front of engine.

3. Turn rotor approximately ⅛ turn more toward left of engine and push distributor down to mesh drive gear with camshaft gear. It may be necessary to turn rotor slightly to engage the gears.

4. Press down firmly on distributor housing and crank engine a few times to verify that oil pump shaft is engaged. Install hold-down clamp and bolt.

5. Turn distributor body slightly until points just open, then tighten hold-down clamp bolt.

6. Hold distributor cap in place and verify that rotor lines up with terminal for No. 1 spark plug.

7. Install distributor cap. If spark plug wires were removed, replace them in their proper towers.

8. Connect vacuum hose to vacuum advance unit, and distributor primary lead wire to coil terminal.

9. Set dwell angle and time ignition using the procedures given above.

HIGH ENERGY IGNITION SYSTEM

All 1975 and later Chevrolet engines are equipped with High Energy Ignition (HEI) systems. All 8-cylinder and V6 engines are equipped with a distributor that contains all ignition components in one unit. The system used on 6-cylinder inline engines is identical in operating theory and uses similar system components, but has an externally mounted ignition coil.

The HEI is a pulse-triggered, transistor-controlled, inductive discharge system. Conventional breaker points are not used. Principal system elements are the ignition coil, electronic module, pick-up assembly, and the centrifugal and vacuum advance mechanisms.

Ignition Coil

The HEI coil operates in basically the same way as a conventional coil, but is smaller in size and generates higher secondary voltage when the primary circuit is broken. The coil is built into the cap of the 8-cylinder system, and is externally mounted in the 6-cylinder system.

Electronic Module

Circuits within the electronic module perform 5 functions. These are spark triggering, switching, current limiting, dwell control, and distributor pickup.

ELECTRICAL SYSTEM

Pick-up Assembly

The pick-up assembly consists of a rotating timer core with external teeth (turned by the distributor shaft), a stationary pole piece with internal teeth, and a pick-up coil and magnet located between the pole piece and a bottom plate.

Centrifugal and Vacuum Advance

The centrifugal and vacuum advance mechanisms are basically identical to units used in breaker point ignition systems.

Operation

As the distributor shaft turns the timer core teeth out of alignment with the pole piece teeth, a voltage is created in the magnetic field of the pick-up coil. The pick-up coil sends this voltage to the electronic module, which determines from the rotational speed of the distributor shaft when to start building current in the ignition coil primary windings. When the timer core teeth are again aligned with the pole piece teeth, the magnetic field is changed, creating a different voltage. This signal is sent to the electronic module by the pick-up coil and causes the module to shut off the ignition coil primary circuit. This collapses the coil magnetic field and induces a high secondary voltage to fire one spark plug.

The electronic module limits the 12-volt current to the ignition coil to 5-6 amperes. The module also triggers the opening and closing of the coil primary circuit with zero energy loss.

Corrective Maintenance

If a part in the HEI systems fails, it must be replaced. See *Troubleshooting*, this chapter, for diagnostic procedures. If the replacement of a part is necessary (other than the ignition coil), it may be necessary or more convenient to remove the distributor from the engine.

In a conventional breaker point ignition system some energy can be lost due to point arcing and/or capacitor charging-time lag. (Although a capacitor is present in the HEI system, it is used only as a radio noise suppressor.) The efficiency of the triggering system allows up to approximately 35,000 volts to be delivered through the secondary wiring system to the spark plugs.

The module circuit controlling dwell angle causes the angle to increase as engine speed increases.

Troubleshooting

The following procedures are for the diagnosis of problems in High Energy Ignition (HEI) system distributors.

1. *Engine cranks but will not start* — Turn ignition on and place automatic transmission selector in PARK or manual transmission in NEUTRAL. Connect a test light to the BAT lead terminal on side of distributor. Connect other test light lead to ground. If light goes on, remove a spark plug and hold it ¼ inch from engine block and crank engine. If spark is present, problem is not in the ignition system. Check spark plugs, fuel system, or for flooded condition. If test light did not go on, check ignition switch and repair or replace as necessary. If engine still does not start, perform spark test described above. If no spark resulted from spark test, connect test light to distributor B+ terminal by inserting test lead into red B+ wire. If test light does not go on, repair or replace B+ wire or connector. If test light goes on, or if engine fails to start after repairs, electronic module should be checked. Replace module if necessary.

> NOTE: *Checking of the electronic module requires special equipment. This task should be referred to your Chevrolet dealer or a qualified mechanic who has the necessary equipment.*

If the electronic module is OK, remove the distributor cap assembly, disconnect the 3-wire connector (HEI system), and inspect cap and distributor for evidence of moisture, dust, cracks, burns, etc., and replace or repair as required. If no defects were noted, remove green and white leads from module and connect an ohmmeter from either lead to ground. If any reading less than infinity is obtained (on the X1000 scale), pickup coil should be replaced.

If the engine still does not start, connect an ohmmeter between the 2 outside terminals in the distributor cap side connector. If a reading above one ohm (on the X1 scale) is obtained, replace the ignition coil.

If the reading from the green or white module lead to ground was infinite, connect the ohmmeter between the 2 leads (green and white). If the reading is between 500 and 1500 ohms, repeat the reading while moving the vacuum advance with a screwdriver. If the reading is still within the 500-1500 ohm limits, check the ignition coil as outlined in the paragraph above.

If a reading of less than 500 or more than 1500 ohms was obtained, replace the pick-up coil. If engine still does not start, check the ignition coil as outlined above.

If the ohmmeter reads less than one ohm (on the X1 scale) when the ignition coil is checked, connect the ohmmeter between the center terminal in the distributor cap side connector and the ignition coil case. If the reading is above 30,000 ohms or less than 6000 ohms, replace the ignition coil. If the reading is between 6000 and 30,000 ohms, repeat the procedure above until the problem is isolated.

2. *Engine runs rough or cuts out* — First check spark plugs and spark plug wires and repair or replace as required. If engine still runs rough, have the electronic module checked and replace if indicated. If this is not the problem, remove the distributor cap and inspect cap and distributor for evidence of moisture, dust, cracks, burns, etc. and repair or replace as necessary. If electronic module was OK, remove green and white leads from module and connect an ohmmeter between either and ground. If the reading is less than infinity (on the X1000 scale), replace the pick-up coil. If the reading was infinity, connect the ohmemter between the green and white leads. If the reading is between 500 and 1500 ohms, retake the reading while moving the vacuum advance with a screwdriver. If the reading is still within the 500-1500 ohm limits, check the ignition coil as outlined below. If the reading was not within the 500-1500 ohm limits, replace the pick-up coil. If engine still does not function properly, check the ignition coil as outlined below.

To check ignition coil, connect an ohmmeter between the center terminal in the distributor cap side connector and the ignition coil case. If the reading is more than 30,000 ohms or less than 6000 ohms, replace the ignition coil. If the reading is between 6000 and 30,000 ohms, repeat the procedure above until the problem is isolated.

Distributor Removal

1. Unplug wiring harness connectors at side of distributor cap.
2. Remove distributor cap and position to one side out of the way.
3. Disconnect hose from the vacuum advance mechanism.
4. Note the alignment of the rotor and make marks on the engine and the distributor housing so alignment can be duplicated when distributor is reinstalled.
5. Remove distributor hold-down clamp.
6. Lift distributor straight up and away from engine. Note and mark housing with final position of rotor. Also note position of vacuum advance unit in relation to engine.

**Distributor Installation
(Engine Undisturbed)**

If engine has not been disturbed since distributor was removed, proceed as follows:

1. Position the rotor to align with the last mark made on distributor body (Step 6 above).
2. Insert the distributor shaft into its hole and push down into position in the engine block. When fully inserted, the rotor should be aligned with the marks made in Step 4 above.

> NOTE: *It may be necessary to turn rotor slightly to mesh distributor gear with camshaft gear. However, rotor should line up with mark when fully inserted. If not, remove and repeat Steps 1 and 2.*

3. Install hold-down clamp and nut. Tighten nut snugly after making certain distributor body and rotor are in same relative position to engine as noted during removal.
4. Reinstall distributor cap with 4 latches after verifying that tab in base of cap is aligned with notch in distributor housing.

ELECTRICAL SYSTEM

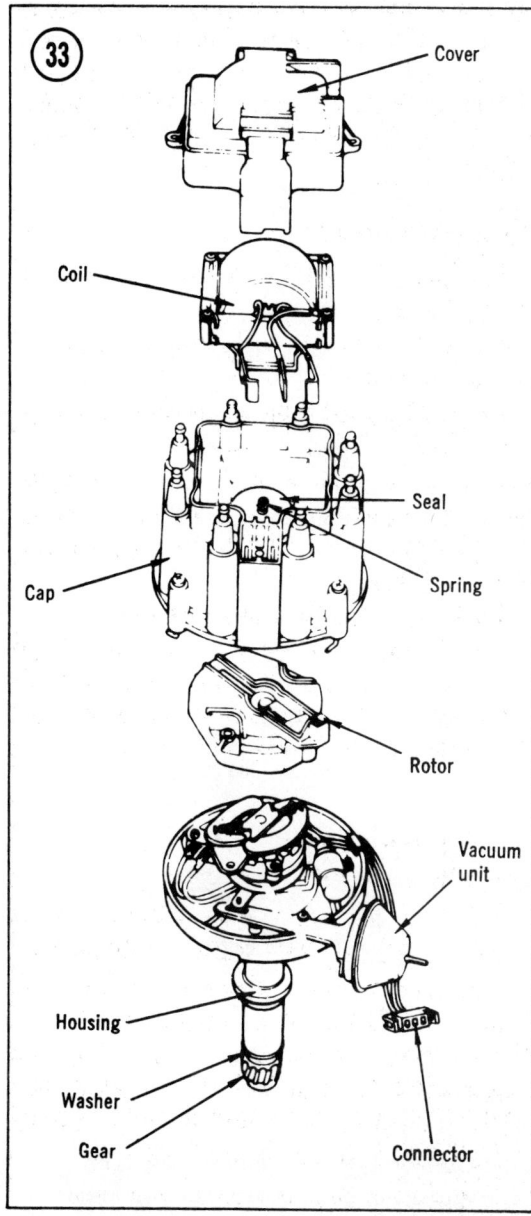

NOTE: *The No. 1 cylinder, piston, or spark plug is the one closest to the front of the vehicle. In V6 and V8 engines, it is located on the left side (note that the left bank of cylinders is offset slightly further forward than the right bank).*

a. Remove spark plug from No. 1 cylinder and place finger over plug hole. Crank engine until compression is felt. Continue cranking until timing mark on crankshaft pulley lines up with proper timing mark on tab attached to engine front cover.

b. Remove rocker cover (left cover on V8 engines) and crank engine until intake valve for No. 1 cylinder closes. Continue cranking until timing marks on pulley and timing tab are aligned.

2. Place the distributor in its hole in the normally installed position, as noted in the removal procedure.

3. While holding the distributor in the position described in Step 2, turn the rotor to point to the front of the engine, then turn rotor about $\frac{1}{8}$ turn counterclockwise. Push down on distributor to engage gear with camshaft gear, rotating the rotor slightly to assist in engagement, if necessary.

4. Verify that rotor lines up with No. 1 spark plug terminal on distributor cap when cap is in proper position.

5. Install distributor hold-down clamp and nut.

6. Install distributor cap and secure with 4 latches, making certain tab in base of cap is aligned with notch in distributor housing.

7. Plug wiring harness connectors into terminals on side of distributor cap.

8. Time engine per the procedure given earlier in this chapter.

5. Plug in wiring harness connector to terminal on side of distributor cap.

6. Time engine in accordance with procedure given earlier in this chapter.

**Distributor Installation
(Engine Disturbed)**

If the engine has been disturbed, reinstall the distributor as follows:

1. Position the No. 1 piston in firing position by one of the following methods.

**Ignition Coil Removal
(V8 Engines)**

Refer to **Figure 33** for this procedure.

1. Disconnect battery wire and harness connector from distributor cap.

2. Remove 3 coil cover attaching screws and remove coil cover.

3. Remove 4 ignition coil attaching screws.

CHAPTER EIGHT

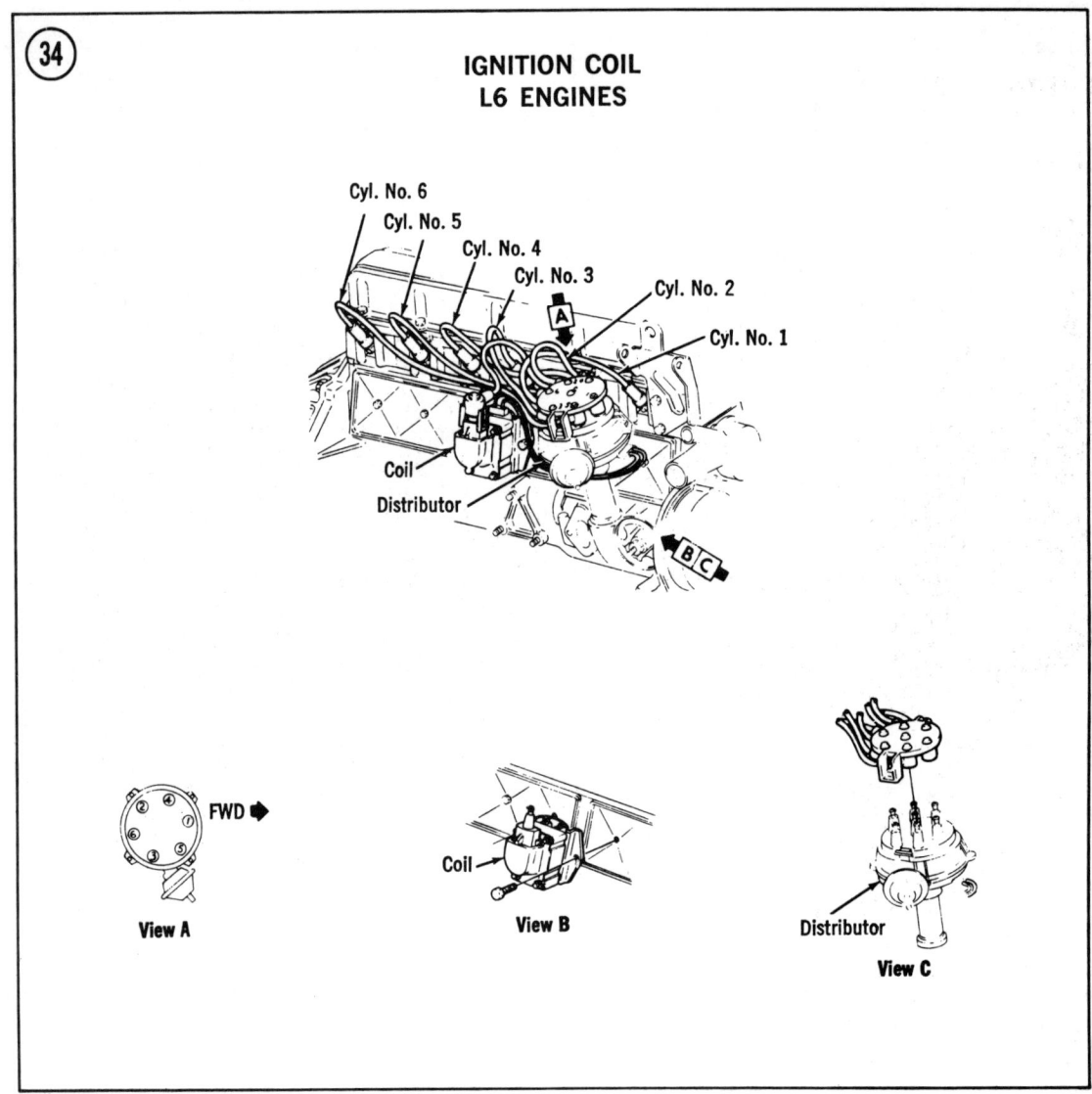

4. Remove ground wire from coil.

5. Remove coil leads from connectors, then remove coil.

Ignition Coil Installation
(V8 Engines)

1. Place coil into distributor cap with terminals over connector at side of cap.

2. Push coil lead wires into connector on side of cap as follows: brown wire next to vacuum advance unit, black wire in center, and pink wire opposite vacuum advance unit.

3. Secure coil to distributor cap with 4 attaching screws, placing ground wire under one of the screws.

4. Install coil cover and secure with 3 attaching screws.

Ignition Coil Removal/Installation
(L6 Engines)

Refer to **Figure 34**.

1. Disconnect ignition switch and distributor lead wires from coil.

2. Remove 4 attaching screws, then remove coil from side of engine.

3. Installation is the reverse of these steps.

CHAPTER NINE

CLUTCH AND TRANSMISSIONS

This chapter provides procedures for replacement of manual and automatic transmissions and for clutch removal, replacement, and adjustment.

A number of manual transmissions were offered during the period covered by this book. Disassembly and assembly procedures are provided for 3- and 4-speed Saginaw transmissions (offered in all covered years). Before attempting these procedures, read them carefully, make sure you understand what is required, and have the necessary special tools or suitable substitutes on hand.

Repairs requiring disassembly are not recommended for home mechanics or garage mechanics without special skills and a large assortment of special tools. In fact, the cost of the necessary tools far exceeds the price of a professionally rebuilt transmission.

Considerable money can be saved by removing the transmission and installing a new or rebuilt one yourself. See Chapter Two, *Troubleshooting*, for procedures which will help isolate the fault.

Table 1, at the end of the chapter, gives tightening torques.

CLUTCH

There is one linkage adjustment (clutch fork pushrod or pedal pushrod) to compensate for all normal clutch wear. The clutch should have a specified amount of free travel before the throwout bearing engages clutch diaphragm spring levers (see individual procedures below for amounts specified). Lash is required to prevent clutch slippage which would occur if the bearing was held against the fingers or to prevent the bearing from running continually until failure.

A clutch that was slipping before adjustment may still slip after adjustment because of prior heat damage. Allow clutch to cool for at least 12 hours, then check for slippage as follows.

1. Drive in high gear at 20-25 mph.
2. Depress clutch pedal to floor and accelerate engine speed to about 3,000 rpm.
3. Snap foot off clutch pedal and at the same time press accelerator to floor board. Engine speed should drop and then accelerate. If clutch is bad, engine will increase immediately (before vehicle accelerates).

> NOTE: *Clutch will overheat if this test is repeated before allowing clutch to cool for at least 12 hours.*

Clutch Linkage Adjustment

Refer to **Figure 1**.

1. Disconnect return spring (at clutch fork).

1
Lever and
shaft assembly
Gage hole
Nut
Rod
Fork rod
Return spring
Clutch
fork
V-8 ENGINE
Lever and
shaft assembly
Spacer
Nut
Gage hole
Return spring
Swivel
Fork rod
Clutch
fork
L6 ENGINE

Flywheel
Driven plate assembly
Clutch housing
cover
Pressure plate
and cover assembly
Clutch release bearing
Clutch fork
Clutch housing
Clutch fork
ball stud
2

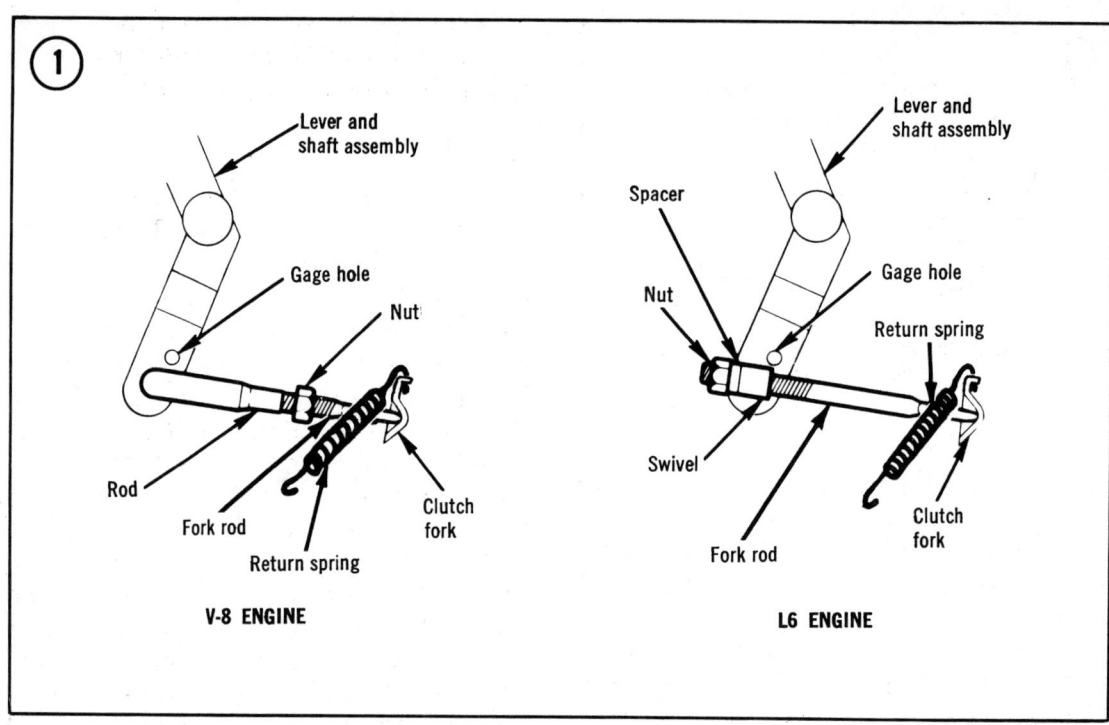

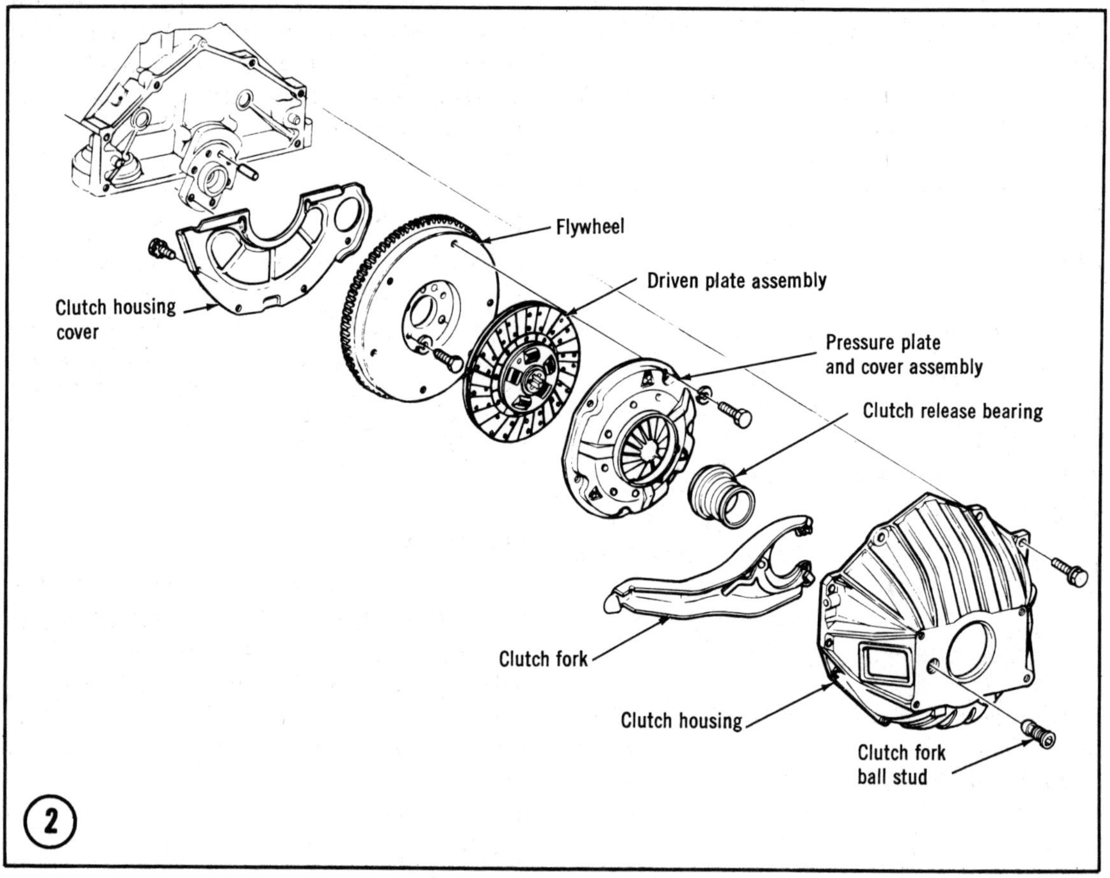

CLUTCH AND TRANSMISSIONS

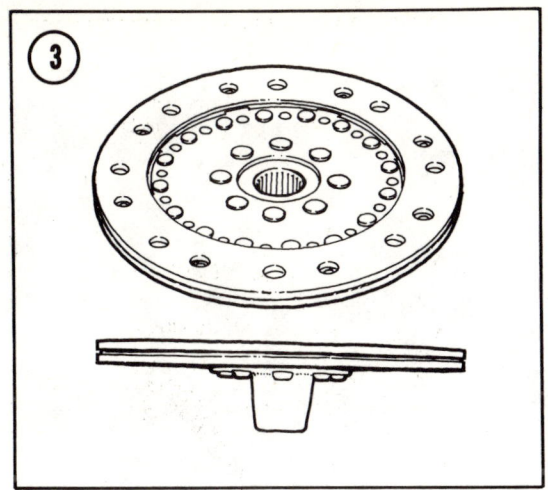

2. Raise clutch lever and shaft assembly until clutch pedal rests against rubber bumper and dash brace.

3. Move outer end of clutch fork rearward until throwout bearing just contacts pressure plate fingers.

4. Install pushrod in gauge hole and adjust length until all lash is removed from linkage system.

5. Remove rod from gauge hole and reinstall in lower hole in lever. Install retainer and tighten locknut. Be careful not to alter length of rod.

6. Reinstall return spring and check free travel of pedal. Travel should be $1\frac{1}{8}$-$1\frac{3}{4}$ in. (1970-1972), or $\frac{3}{4}$-$1\frac{5}{16}$ in. (1973-on).

Clutch Mechanism Removal

Refer to **Figure 2**.

1. Support engine and remove transmission as described later.

2. Disconnect clutch fork pushrod and spring.

3. Remove flywheel housing.

4. Slide clutch fork from ball stud and remove fork from dust boot.

> NOTE: *Ball stud is threaded into clutch housing and is easily replaced.*

5. Install tool J-5824 or equivalent clutch pilot to support clutch.

6. Look for "X" mark on flywheel and clutch cover. If not visible, make small punch marks on these parts to aid reassembly.

7. Loosen clutch-to-flywheel bolts evenly, one turn at a time, until spring pressure is released.

8. Remove bolts and clutch assembly.

Inspection

Never replace clutch parts without giving thought to the reason for failure. To do so only invites repeated troubles.

1. Clean the flywheel face and pressure plate assembly in a non-petroleum base cleaner.

2. Check the friction surface of the flywheel for cracks and grooves. Attach a dial indicator and check runout. Compare with specifications for your engine. If necessary, have the flywheel reground; replace it in cases of severe damage.

3. Check the pressure plate for cracked or broken springs, evidence of heat, cracked or scored friction surface and looseness. Check release lever ends for wear. On diaphragm spring clutches, check the spring fingers for wear. If there is any damage, replace with a professionally rebuilt pressure assembly.

4. Check the clutch disc (drive plate) lining for wear, cracks, oil, and burns. The assembled thickness of the disc should be at least 0.36 in. See **Figure 3**. Check for loose rivets and cracks in the spring leaves or carrier plate. Ensure that the disc slides freely on the transmission spline without excessive radial play. If the disc is defective, replace it with a new one.

5. Check the release bearing for wear to determine if it caused the original trouble. Never reuse a release bearing unless necessary. When other clutch parts are worn, the bearing is probably worn. If it is necessary to reinstall the old bearing, do not wash it in solvent; wipe it with a clean cloth.

Installation

1. Wash your hands *clean* before proceeding.

2. Sand the friction surface of the flywheel and pressure plate with a medium-fine emery cloth. Sand lightly across the surfaces (not around) until they are covered with fine scratches. This breaks the glaze and aids seating a new clutch disc.

3. Clean the flywheel and pressure plate with non-petroleum solvent.

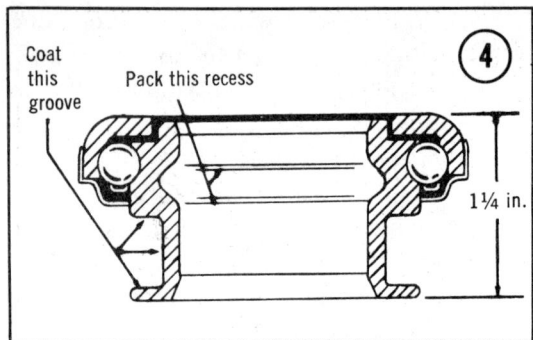

4. Position clutch disc and pressure plate on engine and support with tool J-5824 or equivalent pilot tool. Clutch disc damper springs face the pressure plate.

> NOTE: *An excellent pilot tool can be made by cutting off about one foot from the forward end of an old transmission main shaft. Other tools made from wooden dowellings are available from most auto parts suppliers.*

5. Align "X" marks or punch marks on clutch cover and flywheel. Install bolts finger-tight.

6. Tighten diagonally opposite bolts a few turns at a time until all are tight. Torque to 11 ft.-lb.

7. Remove pilot tool.

8. Unhook clutch fork. Lubricate ball socket with high melting point grease, e.g., graphite, and reinstall fork on ball stud.

9. Lubricate recess on inside of throwout bearing with molybdenum grease. See **Figure 4**.

10. Install clutch fork and dust boot into clutch housing.

11. Install throwout bearing to fork.

12. Install flywheel housing.

13. Install transmission.

14. Connect fork pushrod and spring.

15. Adjust clutch pedal travel and free play.

Clutch Pilot Bearing Removal/Installation

The clutch pilot bearing should be removed for inspection whenever the clutch is removed.

1. Pull bearing from crankshaft with tool J-1448 or equivalent puller. See **Figures 5 and 6**.

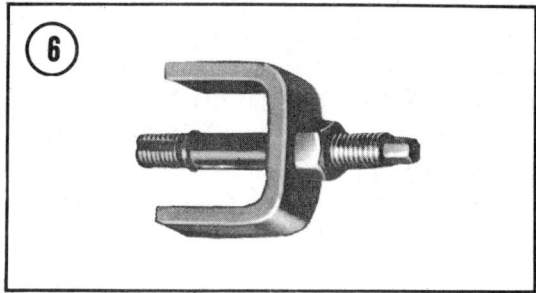

2. Clean bearing with a clean cloth dipped in solvent.

3. Check bearing for excessive wear or other damage. Replace if necessary.

4. Drive bearing into the crankshaft with tool J-1522 or equivalent driver. See **Figure 7**.

TRANSMISSION REPLACEMENT

> NOTE: *This procedure is general and covers all transmissions, manual or automatic.*

1. On manual transmissions, remove shifter knob and on 4-speed models, remove spring and T-handle.

2. Raise vehicle.

3. Disconnect speedometer cable and all electrical leads from transmission.

CLUTCH AND TRANSMISSIONS

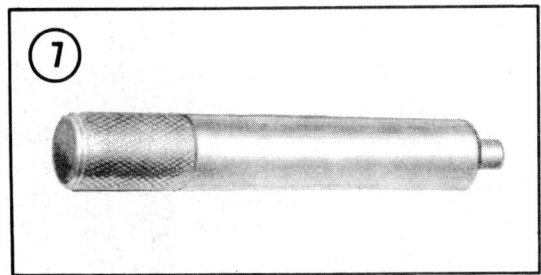

4. On automatic transmissions, disconnect the shift control linkage, oil cooler pipes (if present), and the vacuum line modulator (if so equipped).

5. Disconnect and remove the drive shaft (see Chapter Twelve).

6A. Remove manual transmission:

  a. Remove the transmission mount-to-crossmember bolts and crossmember-to-frame bolts. Raise and support engine and remove crossmember.

  b. Remove shift levers from transmission and disconnect back drive rod at bell housing (floor shift models).

  c. Remove shift control assembly-to-transmission support attaching bolts (floor shift models) and pull unit down until shift lever clears boot. Remove assembly.

  d. Remove upper bolts holding transmission to clutch housing and install guide pins in holes. Then remove lower bolts.

  e. Support transmission while sliding it to the rear to remove it from vehicle.

6B. Remove automatic transmission:

  a. Support transmission with a suitable jack.

  b. Disconnect the rear mount from crossmember. Then remove attaching bolts and remove crossmember from frame.

  c. Remove converter under pan.

  d. Remove converter-to-flywheel bolts.

  e. Lower transmission until jack is barely supporting it and then remove transmission-to-engine mounting bolts. Remove oil filter tube at transmission.

  f. Raise transmission to normal position, support engine with a jack, and remove transmission by sliding it rearward. Lower it away from vehicle.

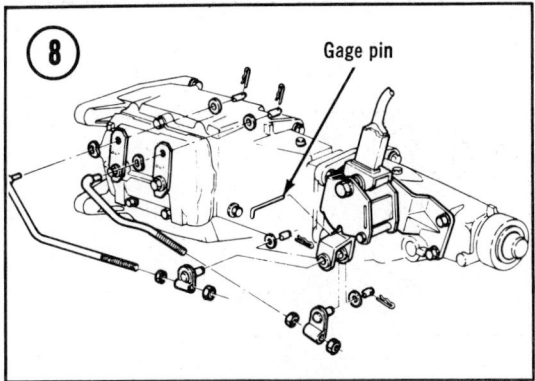

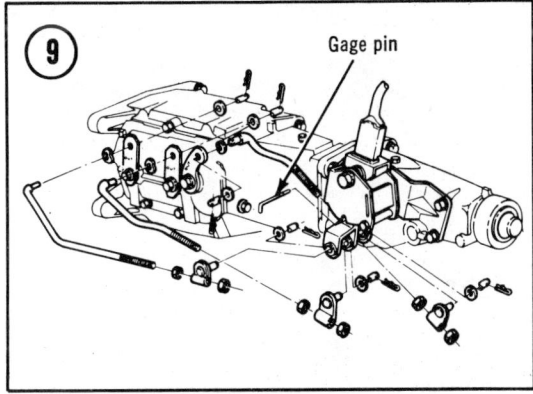

7. Reverse the above procedures to install transmission.

NOTE: *Tighten all nuts and bolts to specifications (see Table 1, end of chapter).*

NOTE: *On automatic transmissions, before installing flexplate-to-converter bolts, make certain weld nuts on converter are flush with flexplate and the converter freely rotates in this position. Then hand start all bolts and tighten finger-tight before torquing to specifications. This will ensure proper alignment of converter.*

SHIFT LINKAGE ADJUSTMENT

3- and 4-Speed Transmission Floor Shift

Refer to **Figures 8 and 9** (typical).

1. Raise vehicle. Ignition should be OFF.
2. Loosen locknuts on shift rod swivels so rods move freely through swivels.

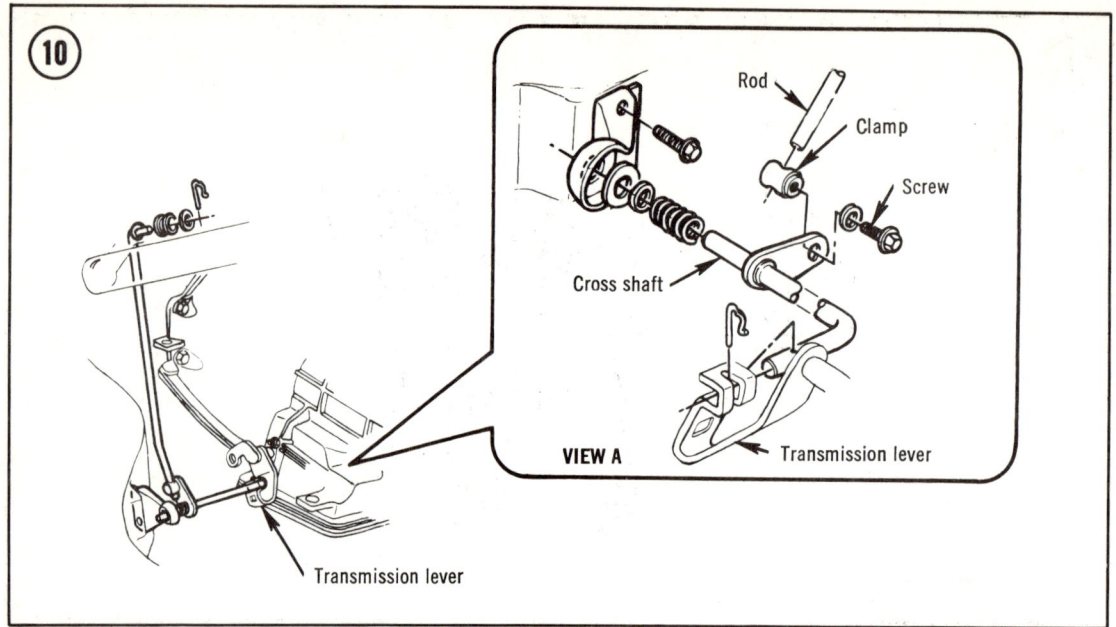

3. Place all transmission levers in NEUTRAL.

4. Move shift control lever to neutral detent and install locating gauge (¼ in. diameter) into lever alignment slot.

5. Tighten the shift rod swivel locknuts and remove gauge.

6. Shift control lever into reverse and place ignition switch in LOCK position. Loosen back drive control rod swivel locknut. Pull down slightly on rod to remove all slack and tighten jam nut.

7. Verify that ignition key moves freely to and from LOCK position. Readjust if required.

8. Check shift operation through all gears. Readjust rods if necessary. Lower vehicle.

Powerglide Column Shift Linkage Adjustment

Refer to **Figure 10**.

1. Verify that shift tube and lever assembly move freely in steering column jacket. If required, align steering column using the procedure given in Chapter Eleven.

2. Lift shift lever toward steering wheel and place transmission in drive (D) detent.

> NOTE: *Do not use pointer to position selector, as pointer is adjusted last in this procedure.*

3. Release lever, then try to move to low range without lifting lever. The lever should not go into low range unless it is lifted toward the steering.

4. Lift lever and place transmission in NEUTRAL (N) detent.

5. Release lever, then try to move to REVERSE (R) without lifting lever. The lever should not go into REVERSE unless it is lifted toward steering wheel.

6. If adjustment is required, place lever in (D) detent and loosen the adjustment swivel or clamp at the cross shaft. Rotate transmission lever until it contacts the drive stop in the steering column. Tighten swivel or clamp and recheck the adjustment.

7. Readjust indicator pointer to agree with detent positions. Loosen screw on pointer shaft clamp, adjust as required, and tighten screw. See **Figure 11**.

8. If required, readjust neutral safety switch using the procedure given in Chapter Eight.

Powerglide Floor Shift Linkage Adjustment

Refer to **Figure 12**.

1. Set shift lever in PARK position, turn ignition switch to lock position, and raise vehicle.

11

12

Shift lever
Selector lever
Clip
Nut
Cross shaft
Rod
Transmission lever
Cable
Retainer
Retainer
5.5 in.
Clip
Pin
Cable
Bracket
VIEW A

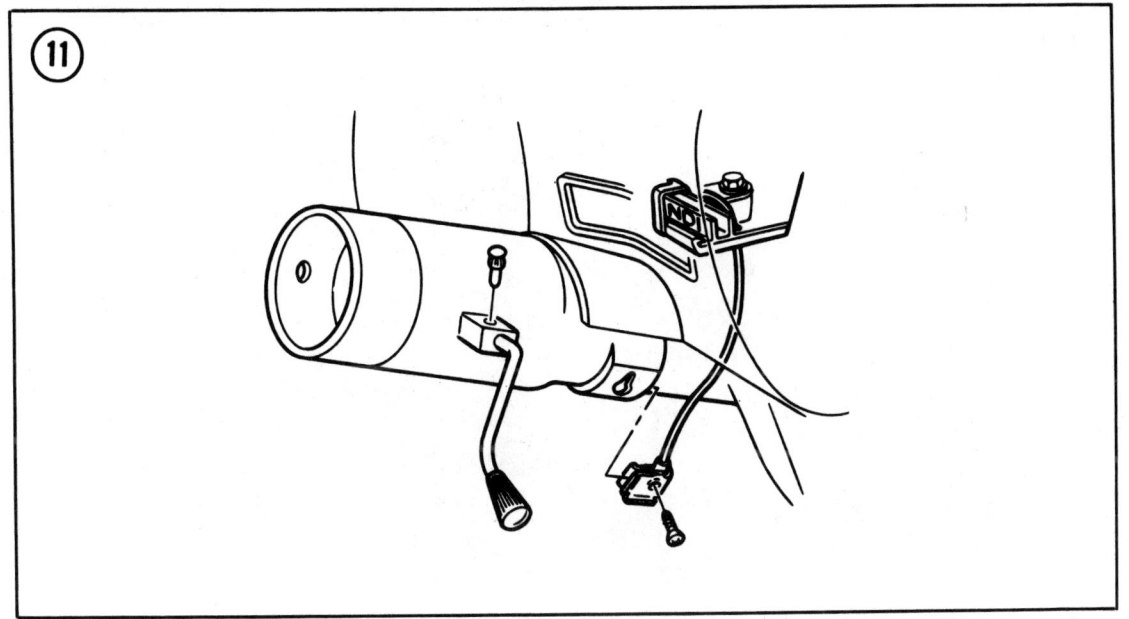

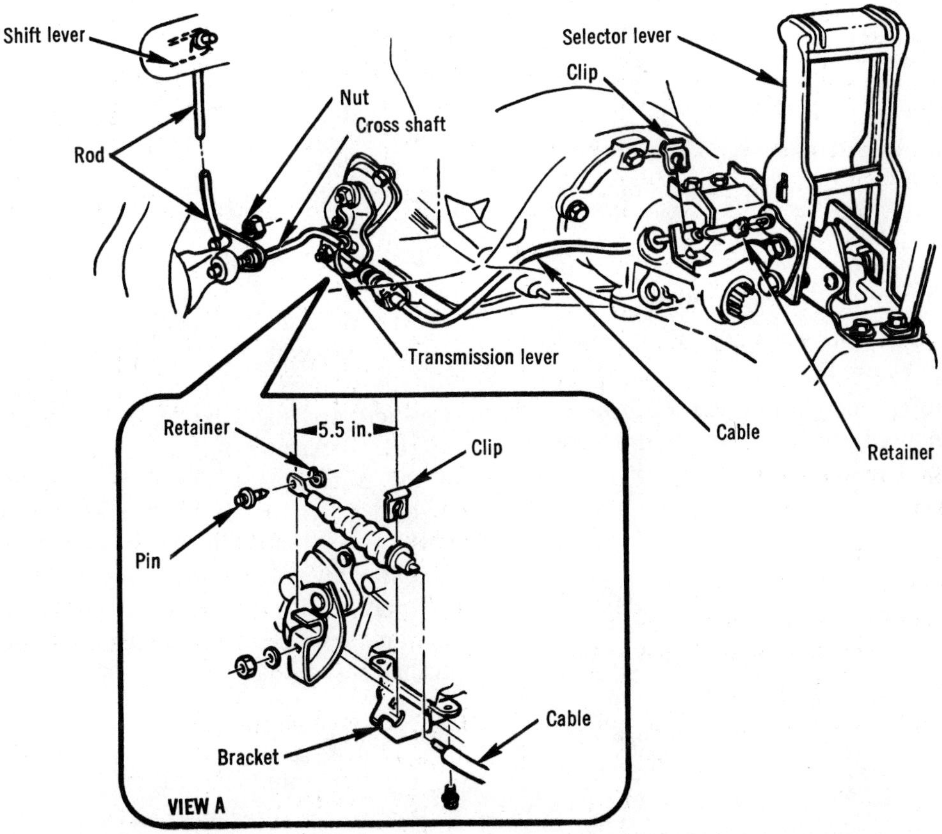

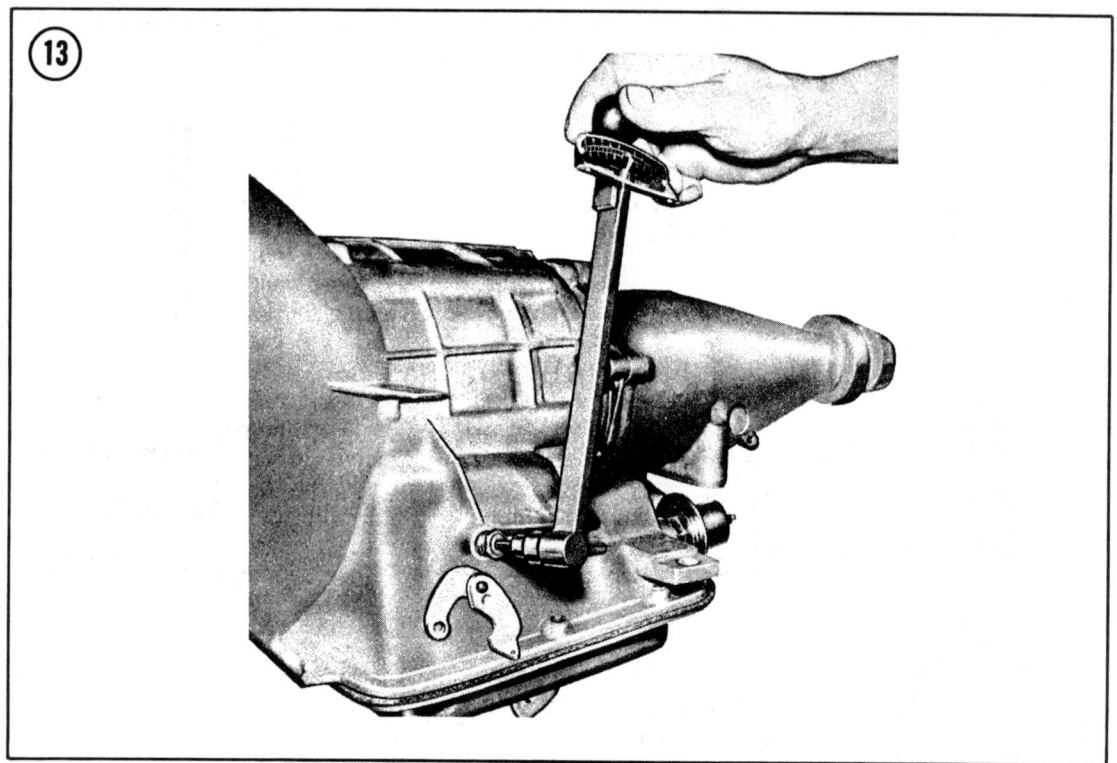

2. Loosen rod retaining screw and remove all lash from linkage by rotating shift lever downward. Tighten screw and lower vehicle.

3. Check shift linkage for proper operation.

Powerglide Low Band Adjustment

1. Tighten the low servo adjusting screw to 40 in.-lb. See **Figure 13**.

2. Back off 4 complete turns for a band which has been used for 1,000 miles or more, or 3 turns for a new band.

3. Tighten the locknut to 15 ft.-lb.

Turbo-Hydraulic Shift Linkage Adjustment

Refer to **Figure 12** for this procedure.

1. Disconnect shift cable at both ends.
2. Place transmission control lever in DRIVE position.
3. Place selector lever in DRIVE position.
4. Install cable as shown in **Figure 12**.

SAGINAW 3-SPEED MANUAL TRANSMISSION

The special tools required to perform these procedures are shown in **Figure 14**. If you do not have these tools or suitable substitutes, and/or if you are not absolutely sure of your mechanical ability, you should not try to perform the procedures.

These procedures require that the transmission be removed from the car as described elsewhere in this chapter and placed on a suitable workstand or workbench. The outside of the transmission should be cleaned thoroughly before starting disassembly, and the work area should be clean and dust free. Cleanliness is especially important during assembly of the transmission, as dirt or other contamination that finds its way into the mechanism could lead to early failure.

Disassembly

Refer to **Figure 15** for this procedure.

1. Remove the attaching bolts and remove the transmission side (shift) cover. See (62) **Figure 15**.

CLUTCH AND TRANSMISSIONS

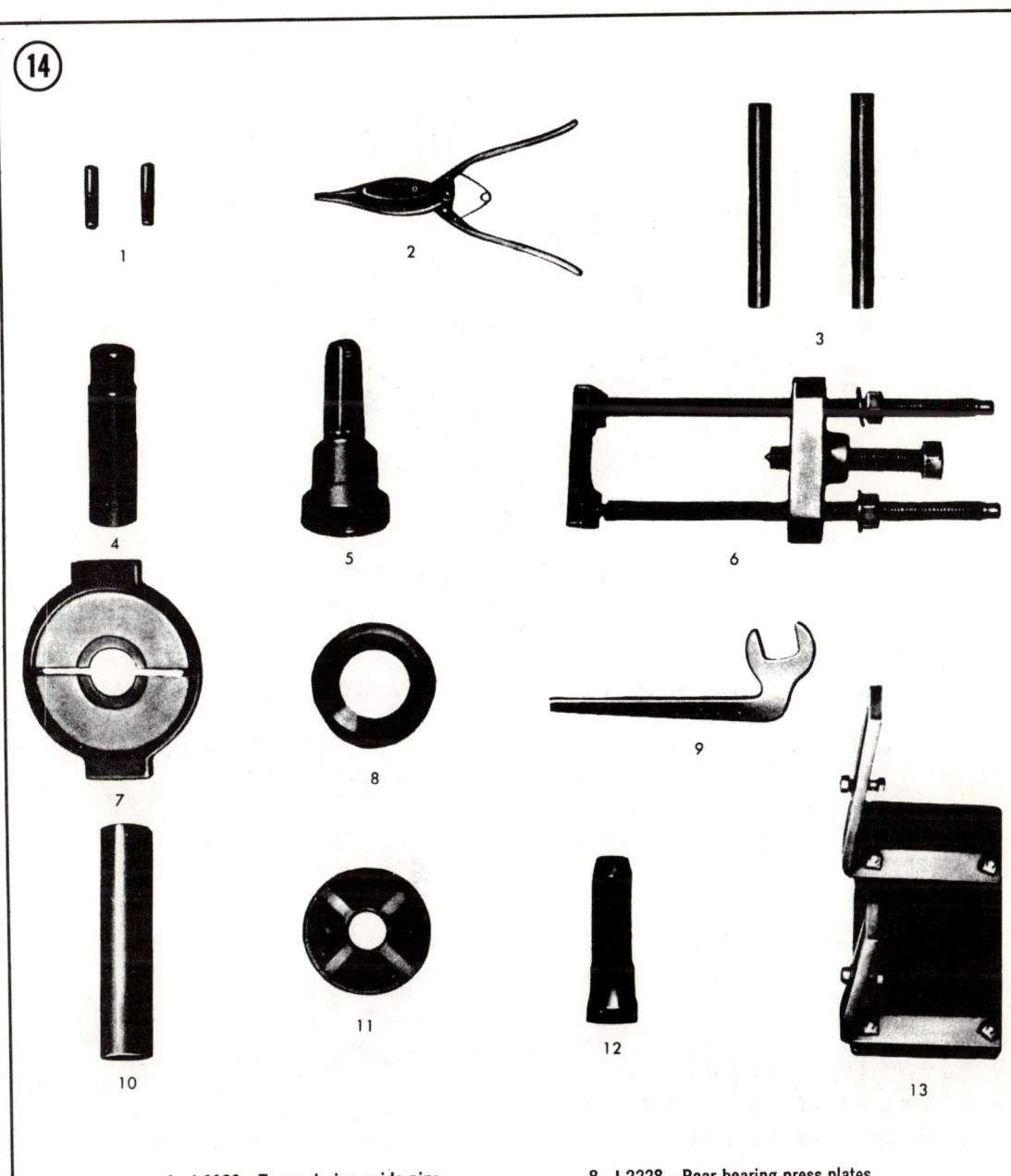

1. J-1126—Transmission guide pins
2. J-8059—Retainer snap ring pliers
3. J-22246/J-22379—Countergear loading tool
4. J-5778—Extension bushing remover and installer
5. J-5154—Extension seal installer
6. J-5814-01—Speedometer drive gear remover and adapter J-5814-15
7. J-1453-01—Speedometer drive gear press plates and press plate holder J-358-1
8. J-2228—Rear bearing press plates
9. J-933—Main drive gear wrench
10. J-5590—Clutch gear bearing installer
11. J-9772—Clutch gear bearing installer
12. J-23096—Clutch gear retainer seal installer
13. J-5752—Transmission holding fixture

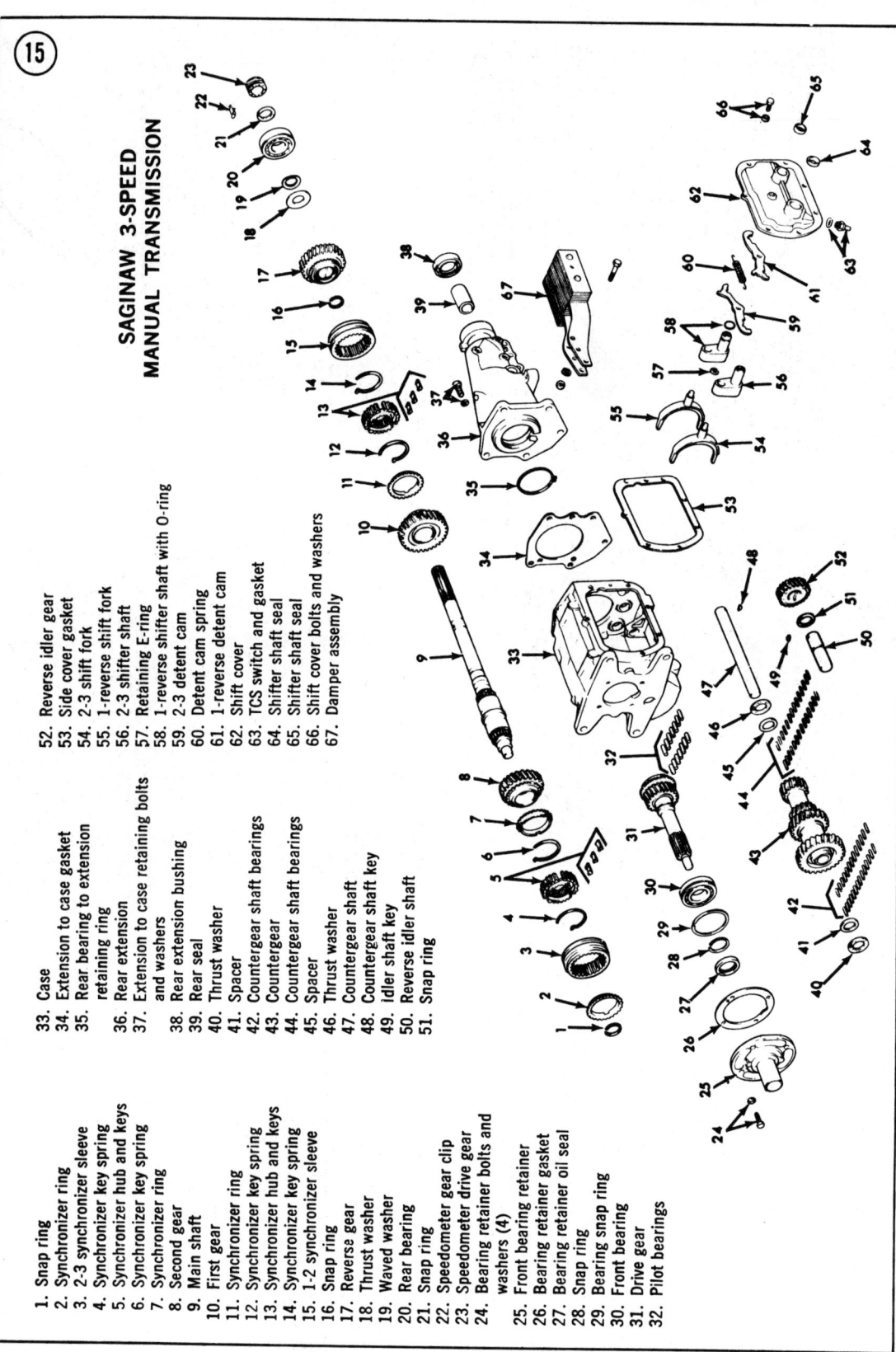

SAGINAW 3-SPEED MANUAL TRANSMISSION

1. Snap ring
2. Synchronizer ring
3. 2-3 synchronizer sleeve
4. Synchronizer key spring
5. Synchronizer hub and keys
6. Synchronizer key spring
7. Synchronizer ring
8. Second gear
9. Main shaft
10. First gear
11. Synchronizer ring
12. Synchronizer key spring
13. Synchronizer hub and keys
14. Synchronizer key spring
15. 1-2 synchronizer sleeve
16. Snap ring
17. Reverse gear
18. Thrust washer
19. Waved washer
20. Rear bearing
21. Snap ring
22. Speedometer gear clip
23. Speedometer drive gear
24. Bearing retainer bolts and washers (4)
25. Front bearing retainer
26. Bearing retainer gasket
27. Bearing retainer oil seal
28. Snap ring
29. Bearing snap ring
30. Front bearing
31. Drive gear
32. Pilot bearings
33. Case
34. Extension to case gasket
35. Rear bearing to extension retaining ring
36. Rear extension
37. Extension to case retaining bolts and washers
38. Rear extension bushing
39. Rear seal
40. Thrust washer
41. Spacer
42. Countergear shaft bearings
43. Countergear
44. Countergear shaft bearings
45. Spacer
46. Thrust washer
47. Countergear shaft
48. Countergear shaft key
49. Idler shaft key
50. Reverse idler shaft
51. Snap ring
52. Reverse idler gear
53. Side cover gasket
54. 2-3 shift fork
55. 1-reverse shift fork
56. 2-3 shifter shaft
57. Retaining E-ring
58. 1-reverse shifter shaft with O-ring
59. 2-3 detent cam
60. Detent cam spring
61. 1-reverse detent cam
62. Shift cover
63. TCS switch and gasket
64. Shifter shaft seal
65. Shifter shaft seal
66. Shift cover bolts and washers
67. Damper assembly

CLUTCH AND TRANSMISSIONS

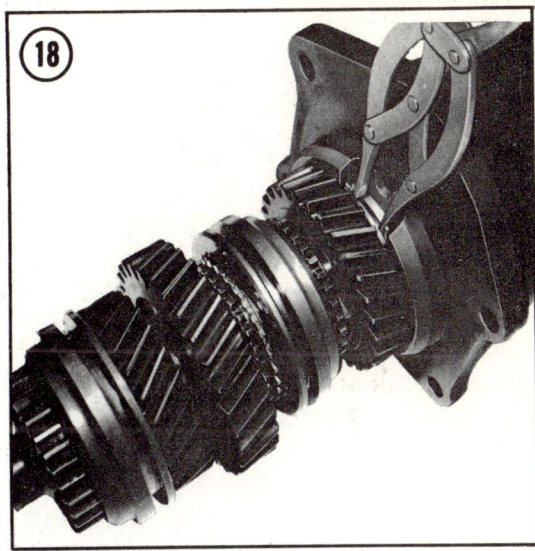

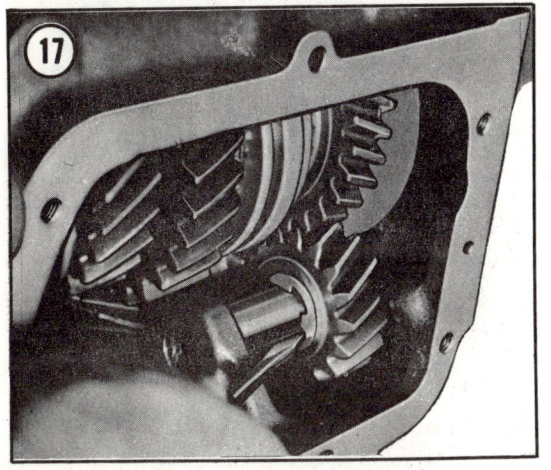

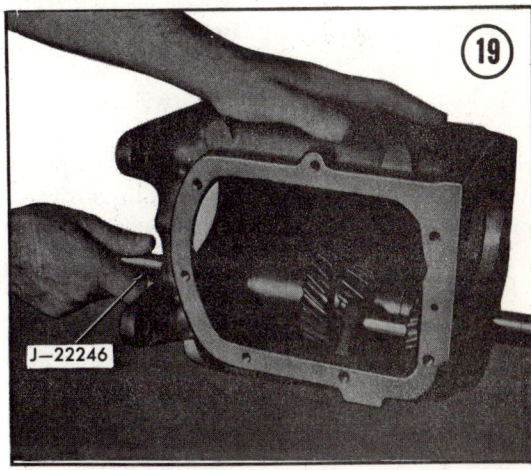

2. Remove the attaching bolts (24) and remove the drive gear bearing retainer (25) and gasket (26).

3. Remove the drive gear bearing-to-gear stem snap ring (28). Pull outward on the drive gear (31) until a screwdriver or similar tool can be inserted to remove the larger snap ring (29). See **Figure 16**.

4. Remove the drive gear bearing (30, **Figure 15**), which is a slip fit on the gear shaft and in the transmission case bore.

5. Remove the speedometer driven gear assembly from the transmission extension.

6. Remove the extension-to-transmission case attaching bolts.

7. Remove the snap ring (51) from the reverse idler shaft (50). See **Figure 17**.

8. Remove the drive gear, main shaft, and extension assembly as an assembly from the rear of the transmission case. Then separate the drive gear and synchronizer ring (31, **Figure 15**) from the main shaft (9) and remove the needle bearings (32). If the bearings are to be reused, make sure all 14 are accounted for and store them in a safe place.

9. Remove the snap ring (35) with snap ring pliers as shown in **Figure 18** and remove the transmission extension from the main shaft.

10. Using special tool J-22246 or equivalent, drive the countershaft (47, **Figure 15**) and its Woodruff key (48) out through the rear of the case. See **Figure 19**.

NOTE: *The special tool will support the front end rear bearings, the counter gear, and the thrust washers (40 through 46, Figure 15) while the countershaft is being driven out. If bearings are to be reused, make sure all 27 front bearings (42) and all 27 rear bearings (44) are accounted for and stored separately in a safe place.*

11. Remove the front and rear bearings, the counter gear, and the thrust washers and spacers from the transmission housing.

12. Use a long drift bar to drive out the reverse idler shaft (50) and Woodruff key (49) through the rear of the transmission case. See **Figure 20**.

13. Use snap ring pliers as shown in **Figure 21** to remove the snap ring (1, **Figure 15**) and then remove the 2nd-3rd speed clutch assembly and synchronizer rings (2 through 7) as an assembly. Then remove the 2nd gear (8).

14. Remove the speedometer drive gear retaining clip (22) from the main shaft and slide or tap off the speedometer drive gear (23). Special tool J-5814-01 or equivalent can be used for this purpose.

15. Remove the snap ring (21) from the main shaft groove as shown in **Figure 22**.

16. Support the reverse gear with press plates (special tool J-1453-01 and holder J-358-1 or equivalent) and press on the main shaft to remove the reverse gear (17, **Figure 15**), thrust washer (18), waved washer (19), and rear bearing (20) from the rear of the main shaft. See **Figure 23**.

17. Remove the snap ring (16, **Figure 15**) and remove the 1st-2nd speed gear synchronizer assembly (12 through 15) as an assembly, the synchronizer ring (11) and 1st gear (10) from the rear of the main shaft. If the synchronizer assembly and gear do not slide off readily, use a press to remove them.

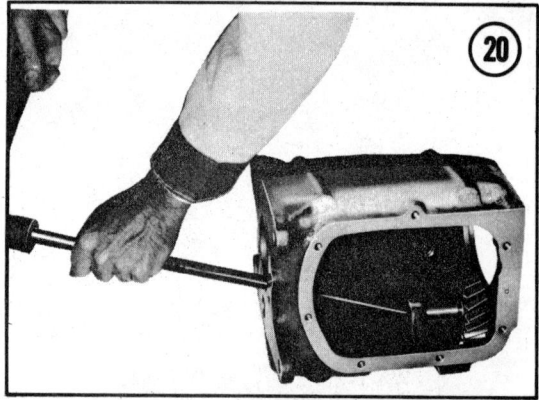

Inspection and Repair

1. Thoroughly clean the transmission case in solvent and check the case for cracks. If the case is damaged, replace it.

2. Check the front and rear mating surfaces of the case for burrs and other damage. Small burrs can be removed with a carefully-applied fine mill file. If damage to the faces is present that would prevent sealing, replace the case.

3. Carefully examine the bearing bores in the case for damage that would prevent proper seating of the bearings. If such damage is present, replace the case.

4. Clean the front and rear ball bearings in solvent and blow them dry with compressed air, taking care not to let the bearings spin.

5. Lightly lubricate the bearings with engine oil and turn them by hand. If roughness can be felt while the bearings are being turned, replace them.

NOTE: *If there is any doubt at all as to the condition of the bearings, replace them as a matter of insurance.*

6. Clean and dry all clutch gear and counter gear roller bearings and check them for condi-

CLUTCH AND TRANSMISSIONS

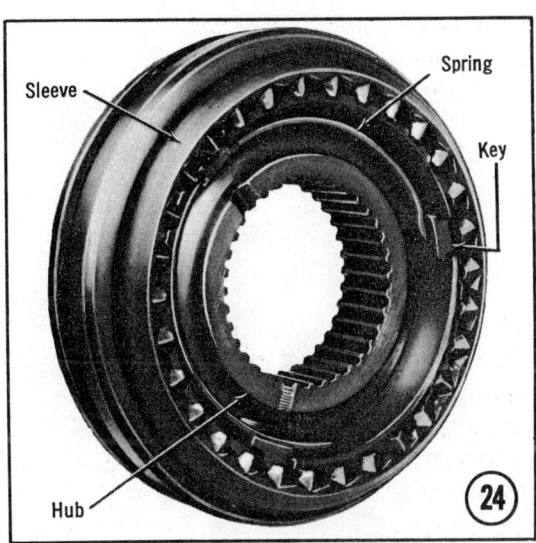

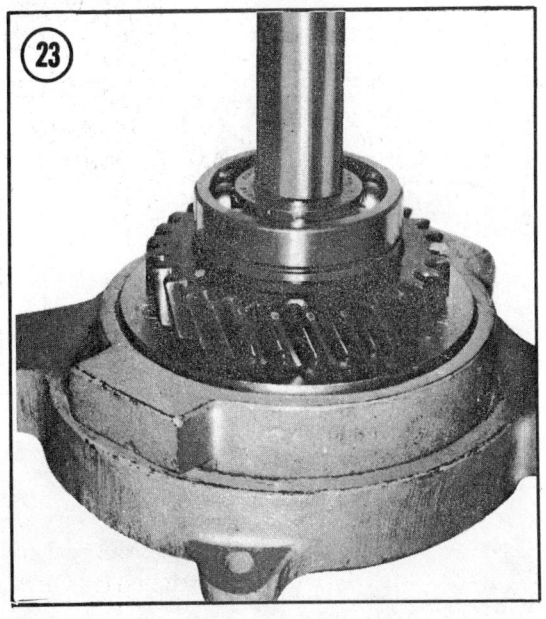

tion. Replace them if they show signs of wear. At the same time, check the countershaft and the reverse idler shaft for wear and replace them if necessary.

7. Clean and check all thrust washers and replace those that show signs of wear.

8. Clean and dry all of the gears and check them for excessive wear, chips, cracks, and broken teeth. Replace all gears that are worn or damaged.

9. Check the reverse gear and reverse idler bushings for wear and damage. Replace the entire gears if the bushings are not serviceable (the bushing alone cannot be replaced in either gear).

10. Clean the synchronizer clutches in solvent and dry them with compressed air. Check to make sure the clutch sleeves slide freely on their hubs.

> NOTE: *The clutch sleeves and hubs are manufactured as a matched set. Springs and keys that are damaged can be replaced, but if either the hub or the sleeve is damaged, the entire assembly should be replaced.*

11. To replace synchronizer springs and/or keys, mark the sleeve and hub so they can be reassembled to their original positions and remove the hub from the sleeve. See **Figure 24**. The keys and springs now can be easily removed. To reassemble, install the keys and springs on the hub with one spring on each side of the hub and with all 3 keys engaged by both springs. The tangs on the ends of the springs should engage the cavities in the keys on both sides. Slide the hub assembly into the sleeve, making sure the marks made before disassembly are aligned.

12. Clean and dry the bushing in the rear end of the transmission extension. Check the

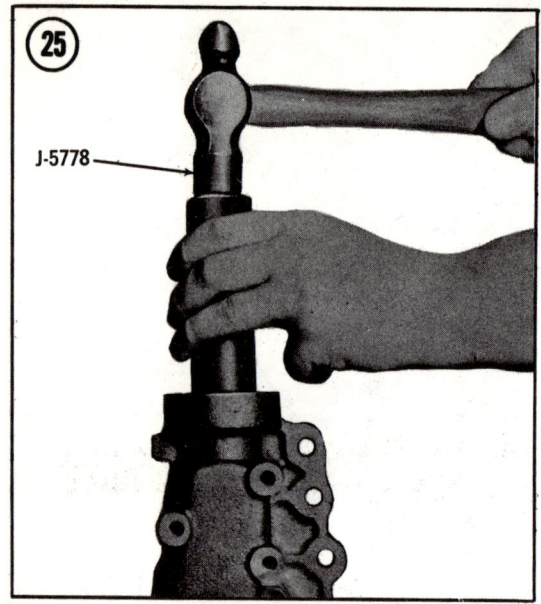

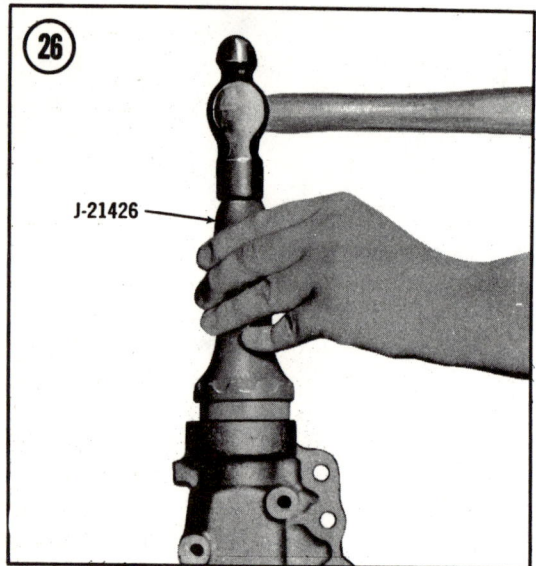
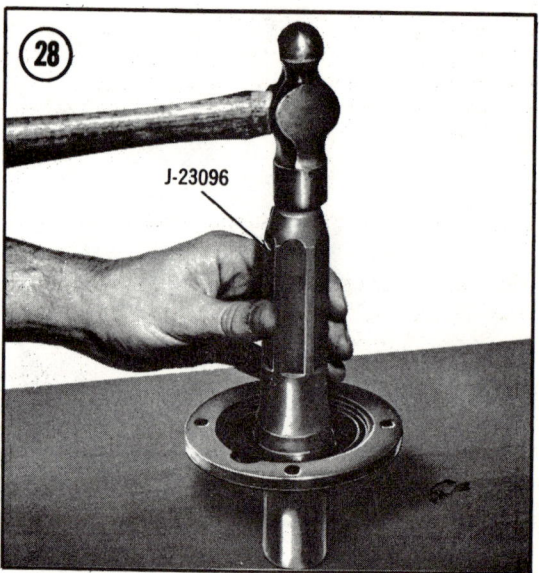

bushing for signs of wear and damage. If replacement is required, remove the oil seal and use a special tool (J-5778 or equivalent) to drive out the bushing as shown in **Figure 25**. Use the same tool to drive in a new bushing. Coat the inside diameter of the bushing and the seal with transmission oil and install the seal in the extension, using a special tool (J-21426 or equivalent) as shown in **Figure 26**.

13. Use a screwdriver to pry out the seal in the drive bearing retainer as shown in **Figure 27**. Use special tool J-23096 or equivalent, as shown in **Figure 28**, to seat the new seal in the bore.

Assembly

Refer to **Figure 29** for the relative positions of components on drive (clutch) gear and the main shaft.

1. Support the main shaft (20, **Figure 29**) with the front end upward and install the 2nd speed gear (8) on the shaft. The clutching teeth on the

CLUTCH AND TRANSMISSIONS

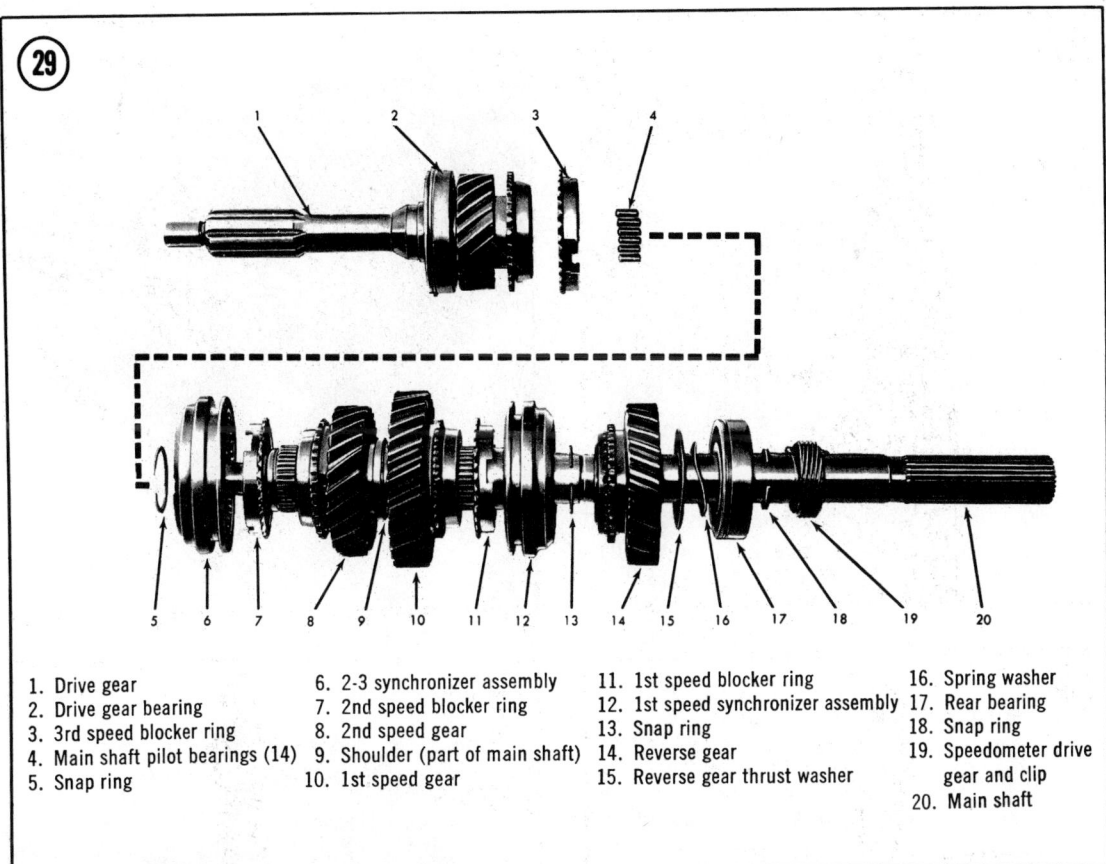

1. Drive gear
2. Drive gear bearing
3. 3rd speed blocker ring
4. Main shaft pilot bearings (14)
5. Snap ring
6. 2-3 synchronizer assembly
7. 2nd speed blocker ring
8. 2nd speed gear
9. Shoulder (part of main shaft)
10. 1st speed gear
11. 1st speed blocker ring
12. 1st speed synchronizer assembly
13. Snap ring
14. Reverse gear
15. Reverse gear thrust washer
16. Spring washer
17. Rear bearing
18. Snap ring
19. Speedometer drive gear and clip
20. Main shaft

side of the gear must face upward, and the rear face of the gear must rest against the flange on the main shaft.

2. Install a synchronizer (blocking) ring (7) on the main shaft with the clutching teeth facing downward over the synchronizing surface of the 2nd speed gear.

NOTE: *The 3 synchronizer rings used in this transmission are identical and interchangeable.*

3. Install the 2nd/3rd speed synchronizer clutch assembly (6) on the main shaft with the fork slot facing downward. Guide the assembly onto the main shaft splines until it bottoms out. Make sure the notches in the synchronizer ring engage the keys of the synchronizer clutch assembly.

NOTE: *The notches on the outside diameter of the synchronizer clutch hub must face toward the front end of the main shaft.*

4. Install the snap ring (5) to retain the synchronizer clutch assembly on the main shaft.

5. Turn the main shaft so the rear end is facing upward and support it. Install the 1st speed gear (10) on the shaft with the clutching teeth facing upward and the other face of the gear resting against the flange on the main shaft.

6. Install a synchronizer ring (11) with the clutching teeth facing downward over the synchronizing surface of the 1st speed gear.

7. Install the 1st/REV speed synchronizer clutch assembly (12) on the main shaft with the fork slot facing downward. Engage the assembly with the shaft splines, making sure it bottoms out.

8. Install the snap ring (13) to secure the synchronizer clutch assembly to the main shaft, making sure the notches of the synchronizer ring are engaged with the clutch assembly keys.

9. Install the reverse gear (14) on the main shaft with the clutching teeth facing downward.

10. Install the reverse gear thrust washer (15). This is a steel washer.

11. Install the spring washer (16) on the main shaft.

12. Install the rear ball bearing (17) on the main shaft with the snap ring slot facing downward and then press the bearing onto the shaft. Make sure the bearing inner race is supported.

13. Install the snap ring (18) to retain the bearing to the main shaft.

14. Install the speedometer drive gear and retaining clip (19).

15. Using special tool J-22246 or equivalent, load the 27 roller bearings (42 and 44, **Figure 15**), and a spacer (41 and 45) into each end of the counter gear (43). Use heavy grease to hold the parts in position. See **Figure 30**.

16. Install the counter gear assembly in the transmission case through the rear opening, at the same time installing a tanged thrust washer (40 and 46) at each end of the gear.

17. Install the counter gear shaft in the transmission case, making sure it picks up the tanged thrust washers and that the thrust washer tangs are aligned with the notches in the case. Insert the shaft and its Woodruff key (47 and 48) from the rear of the case through the bore of the counter gear, pushing the loading tool out of the case.

18. Install the reverse idler gear (52) in the case and install its shaft and Woodruff key (50 and 49) from the rear of the case. Do not install the snap ring (51) at this time.

19. Expand the extension snap ring with snap ring pliers (see **Figure 18**) and install the extension over the rear of the main shaft. Make sure the bearing seats in the extension bore and the snap ring seats in its groove.

20. Install the 14 roller bearings (4, **Figure 29**) into the drive (clutch) gear (1) cavity, using heavy grease to retain them in place (**Figure 31**) and install the 3rd gear synchronizer (blocking) ring (3) on the drive gear clutching surface with its teeth toward the gear.

21. Install drive gear assembly (as assembled in Step 20) on the front end of the main shaft, making sure that none of the 14 roller pilot bearings are lost. Make sure the notches of the

Main drive gear

3rd gear synchronizer ring are aligned with the 2nd/3rd gear synchronizer clutch keys.

22. Install the case-to-extension gasket (use a new gasket) on the rear face of the transmission case and hold it in place with grease.

23. Install the assembled drive gear, main shaft assembly, and extension in the transmission case by inserting it from the rear. Install the

CLUTCH AND TRANSMISSIONS

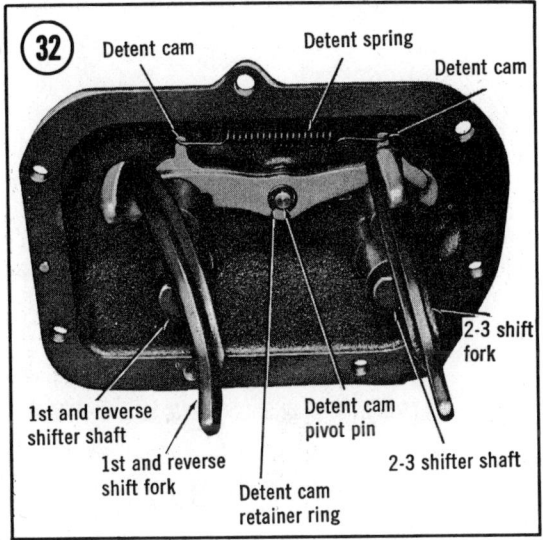

32 Detent cam, Detent spring, Detent cam, 2-3 shift fork, 1st and reverse shifter shaft, 1st and reverse shift fork, Detent cam pivot pin, Detent cam retainer ring, 2-3 shifter shaft

extension-to-case attaching bolts and tighten them to 45 ft.-lb.

24. Install the snap ring (29, **Figure 15**) in the front ball bearing (30) and position the bearing over the drive gear shaft and into the bore in the case.

25. Install the snap ring (28) on the drive gear shaft. Then install the gasket (26) and the front bearing retainer (25) with the attaching bolts (24). Make sure the retainer oil return hole is at the bottom and tighten the bolts to 15 ft.-lb.

26. Install the snap ring (51) on the reverse idler shaft.

27. Make sure the synchronizer sleeves are in the NEUTRAL position and install the gasket (use a new gasket), cover and fork assembly (53 through 66), making sure the shift forks enter the clutch sleeve grooves. Tighten the attaching bolts to 15 ft.-lb.

28. Install the speedometer driven gear in the transmission extension.

29. Rotate the drive gear shaft and shift the transmission to check freedom of movement in all gears.

Shift Cover Disassembly/Assembly

Refer to **Figure 15** and **Figure 32** for this procedure.

1. Remove the TCS switch (63, **Figure 15**) from the shift cover.
2. Remove the shift forks (54 and 55) from the shifter shaft assemblies (56 and 58). Remove both shifter shafts from the cover.
3. If replacement is planned, pry out the shifter shaft seals (64 and 65) from the outside of the cover.
4. Remove the detent cam spring (60) and remove the retaining ring (57). Remove the detent cams (59 and 61).
5. Clean and inspect all parts and replace those that are damaged.
6. Install the 1st/REV detent cam (61) on the pivot pin with the tang facing up over the opening for the 2nd/3rd shifter shaft in the cover. Then install the 2nd/3rd detent cam (59) with the tang facing up in the 1st/REV shifter shaft opening. Install the retaining ring (57) and then install spring in detent cam tang notches. See **Figure 32**.
7. Install the shifter shaft seals (if removed) and then install the shifter shafts, taking care not to damage the seals.
8. Install the shift forks in the shifter shafts. Lift up on the detent cams to allow the forks to fully seat.
9. Install the TCS switch.

SAGINAW 4-SPEED MANUAL TRANSMISSION

The special tools shown in **Figure 14** are required to perform this procedure. If you do not have these tools or suitable substitutes, and/or if you are not absolutely sure of your mechanical abilities, you should not try to perform the procedures.

These procedures require that the transmission be removed from the car as described elsewhere in this chapter and placed on a suitable workstand or workbench. The outside of the transmission should be cleaned thoroughly before starting disassembly, and the work area should be clean and dust free. Cleanliness is especially important during assembly of the transmission, as dirt or other contamination that finds its way into the transmission could lead to early failure.

Disassembly

Refer to **Figures 33 and 34** for this procedure.

1. Remove the attaching bolts and then remove

CHAPTER NINE

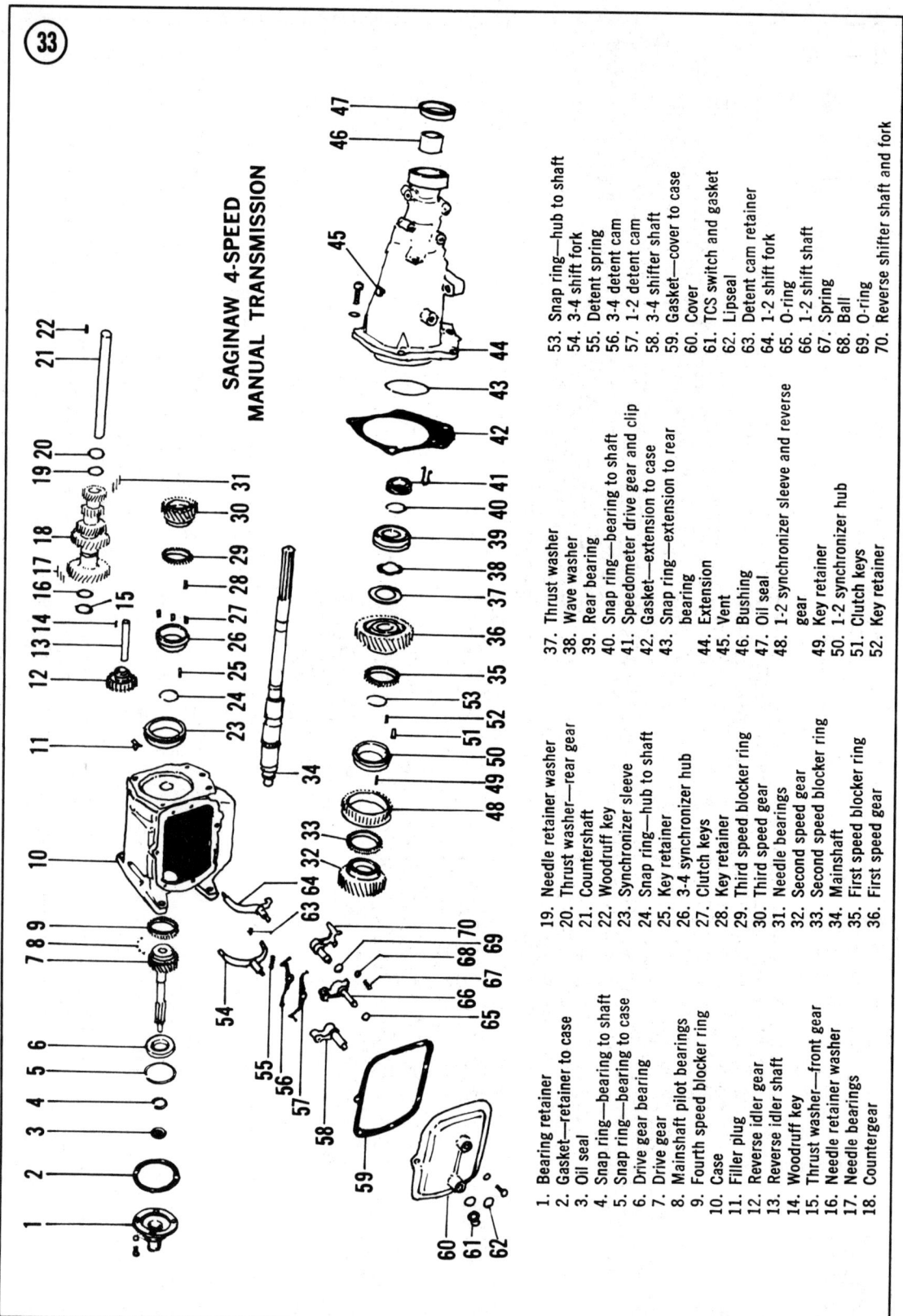

SAGINAW 4-SPEED MANUAL TRANSMISSION

1. Bearing retainer
2. Gasket—retainer to case
3. Oil seal
4. Snap ring—bearing to shaft
5. Snap ring—bearing to case
6. Drive gear bearing
7. Drive gear
8. Mainshaft pilot bearings
9. Fourth speed blocker ring
10. Case
11. Filler plug
12. Reverse idler gear
13. Reverse idler shaft
14. Woodruff key
15. Thrust washer—front gear
16. Needle retainer washer
17. Needle bearings
18. Countergear
19. Needle retainer washer
20. Thrust washer—rear gear
21. Countershaft
22. Woodruff key
23. Synchronizer sleeve
24. Snap ring—hub to shaft
25. Key retainer
26. 3-4 synchronizer hub
27. Clutch keys
28. Key retainer
29. Third speed blocker ring
30. Third speed gear
31. Needle bearings
32. Second speed gear
33. Second speed blocker ring
34. Mainshaft
35. First speed blocker ring
36. First speed gear
37. Thrust washer
38. Wave washer
39. Rear bearing
40. Snap ring—bearing to shaft
41. Speedometer drive gear and clip
42. Gasket—extension to case
43. Snap ring—extension to rear bearing
44. Extension
45. Vent
46. Bushing
47. Oil seal
48. 1-2 synchronizer sleeve and reverse gear
49. Key retainer
50. 1-2 synchronizer hub
51. Clutch keys
52. Key retainer
53. Snap ring—hub to shaft
54. 3-4 shift fork
55. Detent spring
56. 3-4 detent cam
57. 1-2 detent cam
58. 3-4 shifter shaft
59. Gasket—cover to case
60. Cover
61. TCS switch and gasket
62. Lipseal
63. Detent cam retainer
64. 1-2 shift fork
65. O-ring
66. 1-2 shift shaft
67. Spring
68. Ball
69. O-ring
70. Reverse shifter shaft and fork

CLUTCH AND TRANSMISSIONS

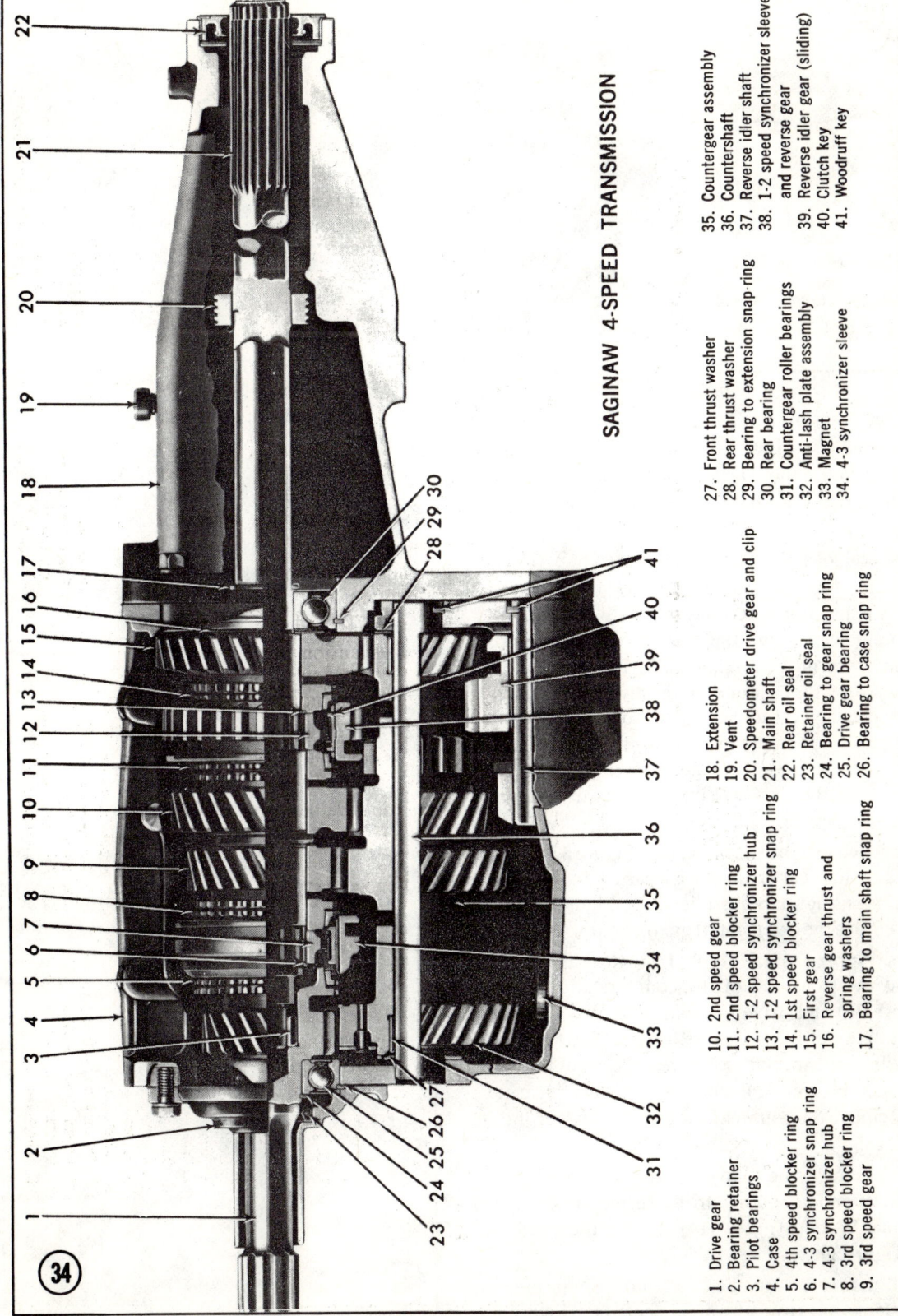

SAGINAW 4-SPEED TRANSMISSION

1. Drive gear
2. Bearing retainer
3. Pilot bearings
4. Case
5. 4th speed blocker ring
6. 4-3 synchronizer snap ring
7. 4-3 synchronizer hub
8. 3rd speed blocker ring
9. 3rd speed gear
10. 2nd speed gear
11. 2nd speed blocker ring
12. 1-2 speed synchronizer hub
13. 1-2 speed synchronizer snap ring
14. 1st speed blocker ring
15. First gear
16. Reverse gear thrust and spring washers
17. Bearing to main shaft snap ring
18. Extension
19. Vent
20. Speedometer drive gear and clip
21. Main shaft
22. Rear oil seal
23. Retainer oil seal
24. Bearing to gear snap ring
25. Drive gear bearing
26. Bearing to case snap ring
27. Front thrust washer
28. Rear thrust washer
29. Bearing to extension snap ring
30. Rear bearing
31. Countergear roller bearings
32. Anti-lash plate assembly
33. Magnet
34. 4-3 synchronizer sleeve
35. Countergear assembly
36. Countershaft
37. Reverse idler shaft
38. 1-2 speed synchronizer sleeve and reverse gear
39. Reverse idler gear (sliding)
40. Clutch key
41. Woodruff key

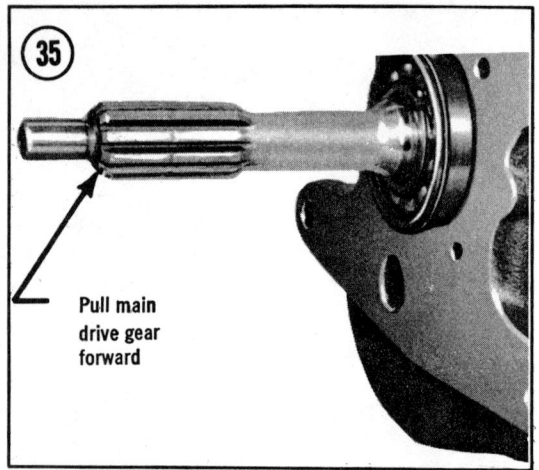
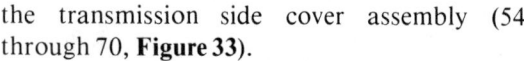

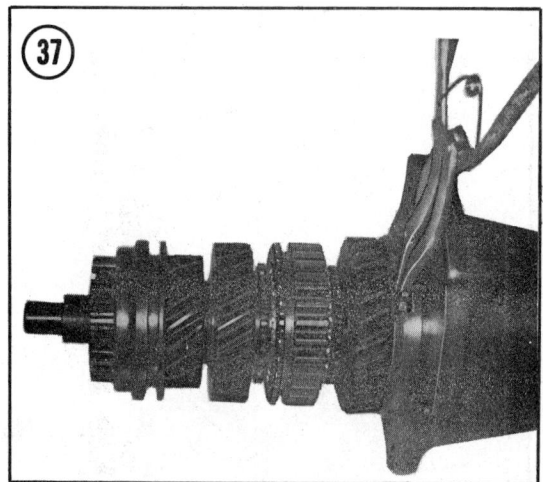

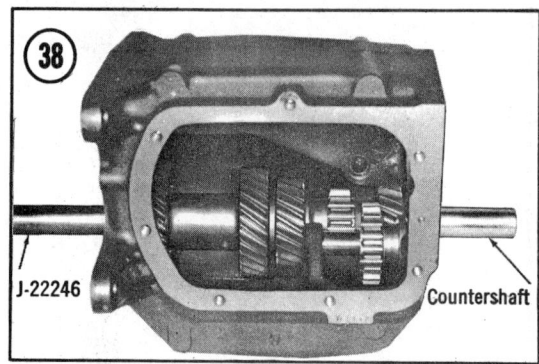

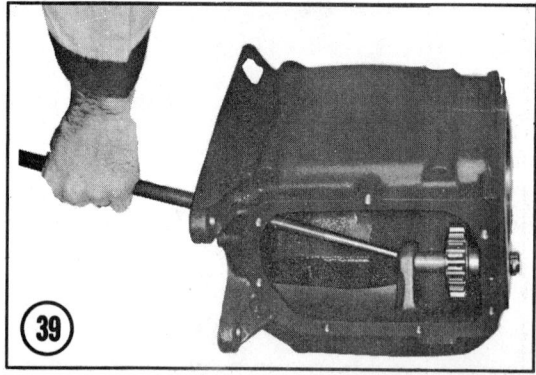

the transmission side cover assembly (54 through 70, **Figure 33**).

2. Remove the attaching bolts and then remove the drive gear bearing retainer and gasket (1 and 2).

3. Remove the bearing-to-shaft snap ring (4) from the drive gear shaft. Pull out on the drive gear and use a screwdriver between the large snap ring (5) and the case to remove the drive gear bearing. See **Figure 35**. This bearing is a slip fit in the case and on the drive gear shaft.

4. Remove the extension-to-case attaching bolts and then remove the extension, main shaft assembly, and drive gear assembly from the rear of the transmission case as an assembly. See **Figure 36**.

5. Remove the snap ring (43, **Figure 33**) that retains the rear main shaft bearing in the extension, using snap ring pliers. See **Figure 37**. Then remove the extension from the main shaft.

6. Apply a special tool (J-22246 or equivalent) to the front end of the countershaft (21, **Figure 33**) and drive the shaft and Woodruff key (22) out of the transmission case. See **Figure 38**. Allow the special tool to retain the counter gear (18, **Figure 33**) and the needle bearings. Remove the counter gear and bearings from the case.

7. Remove the snap ring from the reverse idler shaft and use a long drift to drive the shaft (13) and Woodruff key (14) out of the case. See **Figure 39**.

8. Use snap ring pliers to remove snap ring (14, **Figure 40**) from the main shaft (see **Figure 41**)

CLUTCH AND TRANSMISSIONS

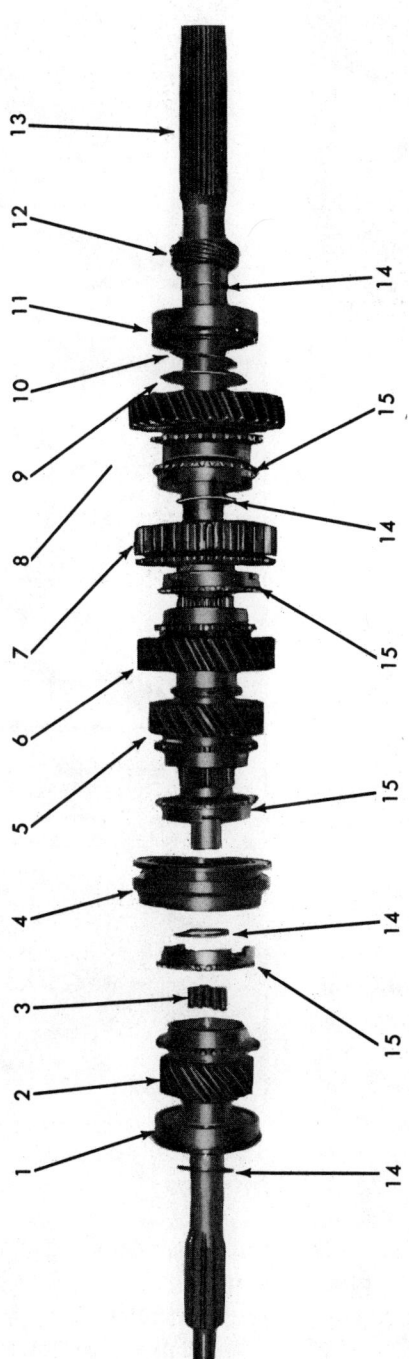

MAIN SHAFT

1. Drive gear bearing
2. Drive gear
3. Main shaft pilot bearings
4. 3-4 synchronizer assembly
5. Third speed gear
6. Second speed gear
7. 1-2 synchronizer and reverse gear assembly
8. First speed gear
9. Thrust washer
10. Spring washer
11. Rear bearing
12. Speedometer drive gear
13. Main shaft
14. Snap ring
15. Synchronizing "blocker" ring

and remove the 3rd/4th gear synchronizer clutch (4) and the synchronizing blocker ring (15). Then remove the 3rd speed gear (5) from the main shaft.

9. Depress the retaining clip and then slide the speedometer drive gear (12) off the main shaft.

10. Remove the rear bearing snap ring (40, **Figure 33**) from the main shaft, using snap ring pliers. See **Figure 42**.

11. Support the 1st speed gear with press plates (see **Figure 43**) and press on the rear end of the main shaft to press off the 1st speed gear (8, **Figure 40**), thrust washer (9), spring washer (10), and rear bearing (11).

12. Remove the snap ring (see **Figure 44**) and then remove the 1st/2nd gear synchronizer clutch and reverse gear (7, **Figure 40**), blocker rings (15), and 2nd speed gear (6) from the rear of the main shaft.

Inspection and Repair

1. Clean the inside and outside of the transmission in solvent and dry it with compressed air, if available. Check the case for cracks and replace it if cracks are present.

2. Check the front and rear mating surfaces of the case for burrs and other damage. Small burrs can be removed with a carefully-applied fine mill file. If damage to the faces is present that would prevent sealing, replace the case.

3. Carefully check the bearing bores in the case for damage that would prevent proper seating of the bearings. If such damage is present, replace the case.

4. Clean the front and rear ball bearings in solvent and blow them dry with compressed air, taking care not to let the bearings spin.

5. Lightly lubricate the bearings with engine oil and turn them by hand. If roughness is felt while the bearings are being turned, replace them.

CLUTCH AND TRANSMISSIONS

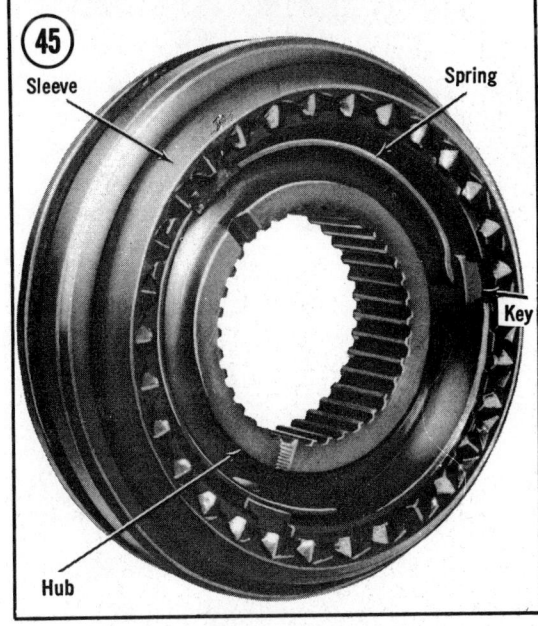

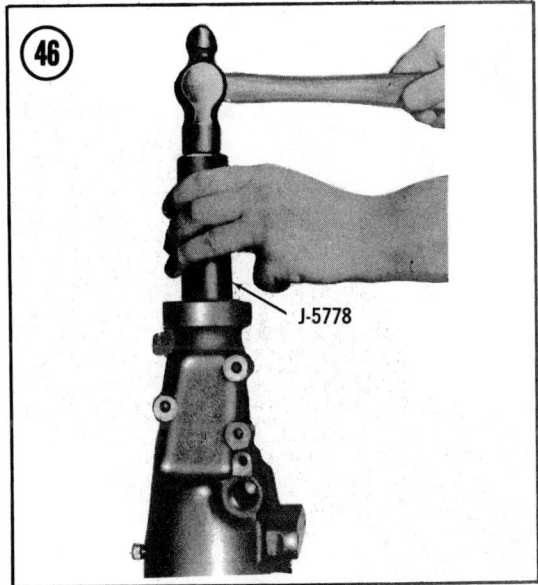

NOTE: *If there is any doubt as to the condition of the bearings, replace them as a matter of insurance.*

6. Clean and dry all clutch gear and counter gear needle bearings and check them for condition. Replace them if they show signs of wear. At the same time, check the countershaft and the reverse idler shaft for wear and replace them if necessary.

7. Clean and check all thrust washers and replace those that show any signs of wear.

8. Clean and dry all of the gears and check them for excesive wear, chips, cracks, and broken teeth. Replace all gears that are worn or damaged.

9. Check the bushing in the reverse idler gear for wear and damage. Replace the entire gear if the bushing is not serviceable (bushings alone cannot be replaced).

10. Clean the synchronizer clutches in solvent and dry them with compressed air. Check to make sure the sleeves slide freely on their hubs.

NOTE: *The clutch sleeves and hubs are manufactured as a matched set. Springs and keys can be replaced if damaged, but if either the hub or the sleeve is damaged, the entire assembly should be replaced.*

11. To replace synchronizer springs and/or keys, mark the sleeve and the hub with matching marks so they can be reassembled to their original positions and remove the hub from the sleeve. See **Figure 45**. The keys and springs now can be easily removed. To reassemble, install the keys and springs on the hub with one spring on each side of the hub and all 3 keys engaged by both springs. The tangs on the end of the springs should engage the cavities in the keys on both sides. Slide the hub into the sleeve, making sure the marks made before disassembly are aligned.

12. Clean and dry the bushing in the end of the transmission extension. Check the bushing for signs of wear and damage. If replacement is required, remove the oil seal and use a special tool (J-577-8 or equivalent) to drive out the bushing as shown in **Figure 46**. Use the same

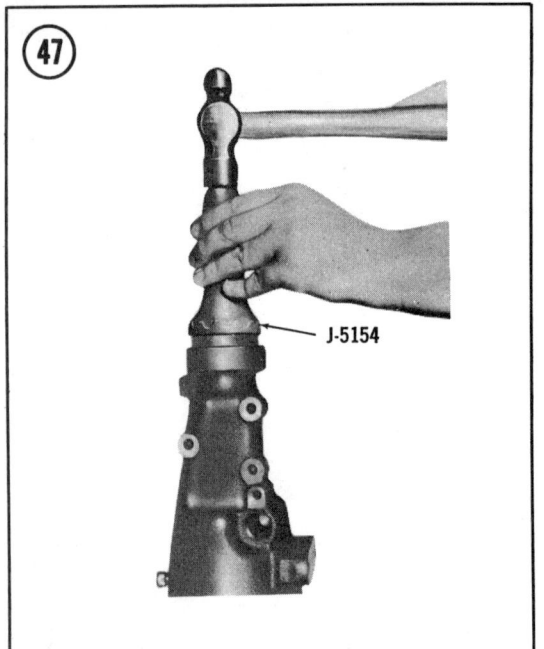

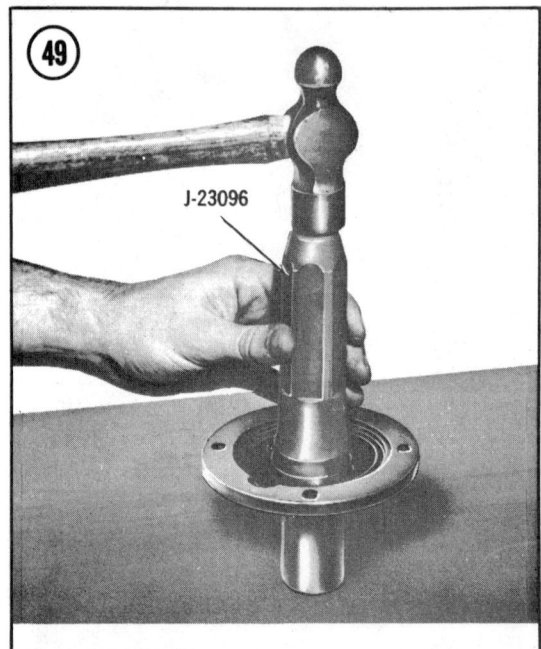

tool to drive in a new bushing. Coat the inside diameter of the bushing and the seal with transmission oil and install the seal in the extension, using special tool J-5154 or equivalent as shown in **Figure 47**.

13. Use a screwdriver to pry out the oil seal in the drive gear bearing retainer as shown in **Figure 48**. Use a special tool (J-23096 or equivalent) to install a new seal in the bore. See **Figure 49**. Lip of the seal must face to the rear of the bearing retainer.

CLUTCH AND TRANSMISSIONS

Assembly

1. Support the main shaft with the front end upward and install the 3rd speed gear (5, **Figure 40**) on the shaft with the clutching teeth facing upward. The rear face of the gear must rest against the main shaft flange.

2. Install a blocker ring (15) with the clutching teeth facing down over the synchronizing surface of the 3rd speed gear (5).

> NOTE: *The 4 synchronizing blocker rings used in this transmission are identical and interchangeable.*

3. Install the 3rd/4th speed synchronizer clutch (4) on the main shaft with the fork slot down. Guide the clutch onto the main shaft splines until it bottoms out. Notches in the blocker ring must mesh with the clutch keys.

4. Install the snap ring (14) to retain the clutch assembly on the main shaft. See **Figure 41**.

5. Turn the main shaft so the rear end is facing up and support it in this position. Install the 2nd speed gear (6, **Figure 40**) with the clutching teeth facing upward and the other face of the gear resting on the main shaft flange.

6. Install a blocker ring (15) on the main shaft with the clutching teeth facing down over the synchronizing surface of the 2nd speed gear (6).

7. Install the 1st/2nd speed synchronizer clutch (17) on the main shaft with the fork slot down. Engage the assembly with the main shaft splines, making sure it bottoms out.

8. Install a snap ring (14) to retain the synchronizer clutch assembly to the main shaft. See **Figure 44**.

9. Install a blocker ring (15, **Figure 40**) on the main shaft with the notches down so they mesh with the synchronizer clutch keys. Then install the 1st speed gear (8) with the clutching teeth downward.

10. Install the thrust washer (9) and spring washer (10) on the main shaft.

11. Place the rear ball bearing on the main shaft with the snap ring slot facing down (toward front of shaft). Support the bearing as shown in **Figure 50** and press on the end of the main shaft as shown to install the bearing.

12. Install the rear bearing-to-main shaft snap ring. See **Figure 42**.

13. Install the speedometer drive gear and clip (12, **Figure 40**) on the main shaft.

14. Install a special tool (J-22246 or equivalent) in the bore of the counter gear and install 27 needle bearings in each end of the gear. Hold the bearings in place with grease. See **Figure 51**. Install a thrust washer (16 and 19, **Figure 33**) at each end of the gear.

15. Install the counter gear into the transmission case through the rear opening. Position the gear in the case with a tanged thrust washer (15 and 20) at each end. Tangs must face away from the gear and fit into notches in the case.

16. Insert the counter gear shaft (21) and Woodruff key (22) through the hole in the rear of the case and use the shaft to push out the special tool, at the same time picking up the counter gear, bearings, and thrust washers on the shaft.

17. Install the reverse idler gear (12), shaft (13), and Woodruff key (14) in the transmission case.

18. Install the extension on the main shaft. Use snap ring pliers (**Figure 37**) to expand the snap

ring and make sure the snap ring seats in the rear bearing groove.

19. Load the 14 needle bearings into the drive gear cavity as shown in **Figure 52**, using heavy grease to hold the bearings in place.

20. Install the drive gear and bearings (2 and 3, **Figure 40**) and the 4th speed blocker ring (15) on the main shaft. Make sure that none of the bearings fall out and that the clutching teeth of the blocker ring face the gear. See **Figure 52**. Also make sure the notches in the blocker ring align with the keys in the 3rd/4th speed synchronizer clutch assembly.

21. Install a new extension-to-case gasket on the rear opening of the transmission case and hold it in place with heavy grease. Insert the drive gear main shaft-extension assembly into the rear of case and install the extension-to-case attaching bolts. Tighten the bolts to 45 ft.-lb. Use a sealing cement on the bottom bolt only.

22. Install the front bearing outer snap ring (5, **Figure 33**) on the bearing (6) and slide the bearing over the drive gear shaft into the case bore.

23. Install the bearing-to-shaft snap ring (4) on the drive gear shaft to retain the bearing in place.

24. Install the bearing retainer (1) and a new gasket (2) over the drive gear shaft and install the retaining bolts. Make sure the oil return hole in the bearing retainer is at the bottom, and tighten the retaining bolts to 15 ft.-lb.

25. Make sure the synchronizer clutches are in the NEUTRAL position and install the side cover. Make sure the shift forks are aligned with the grooves in the clutch sleeve. Tighten the attaching bolts to 15 ft.-lb.

26. Install the speedometer driven gear assembly in the extension.

27. Rotate the drive gear and shift the transmission to check freedom of movement in all gears.

Shift Cover Disassembly/Assembly

Refer to **Figures 33 and 53** for this procedure.

1. Remove the outer shift levers and the TCS switch (61, **Figure 33**) from the cover.

2. Remove the shift forks (54 and 64) from the shift shaft assemblies (58 and 66).

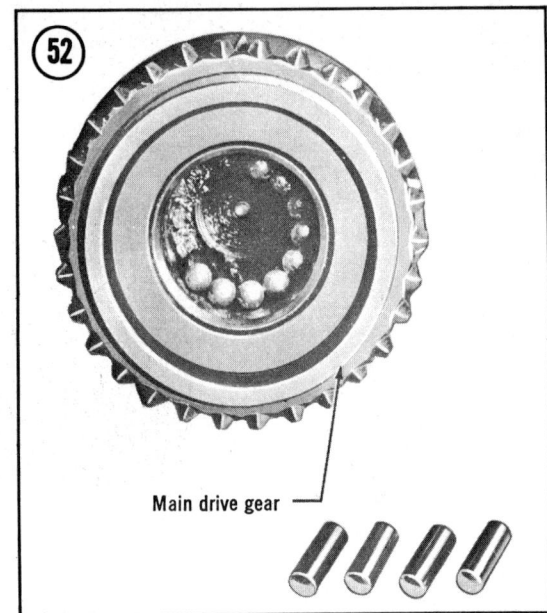

Main drive gear

3. Remove the 3 shifter shaft assemblies (58, 66, and 70) from the cover. If replacement is planned, remove the lip seal (62) and the O-rings (65 and 69). Remove the reverse shifter detent ball (68) and spring (67).

4. Remove the detent cam spring (55) and the pivot retainer (63) from the cover. See **Figure 53**.

5. Mark the detent cams so they can be returned to their original positions and then remove them from the cover.

6. Clean all parts in solvent and check them for wear and/or damage. Replace parts as required.

7. Install the detent cams on the pivot pin in their original positions with the spring tangs facing up.

8. Install the retainer on the pivot pin and hook the spring in the notches on the detent cam tangs. Install lip seal and O-rings, if removed.

9. Install the 1st/2nd speed and 3rd/4th speed shifter shafts in the cover, taking care not to damage seals. Install the shift forks in the shifter shafts, lifting up on the detent cams to allow the forks to fully seat.

10. Install the detent spring and ball in the cover and then install the reverse shifter shaft in the cover.

11. Install the TCS switch.

CLUTCH AND TRANSMISSIONS

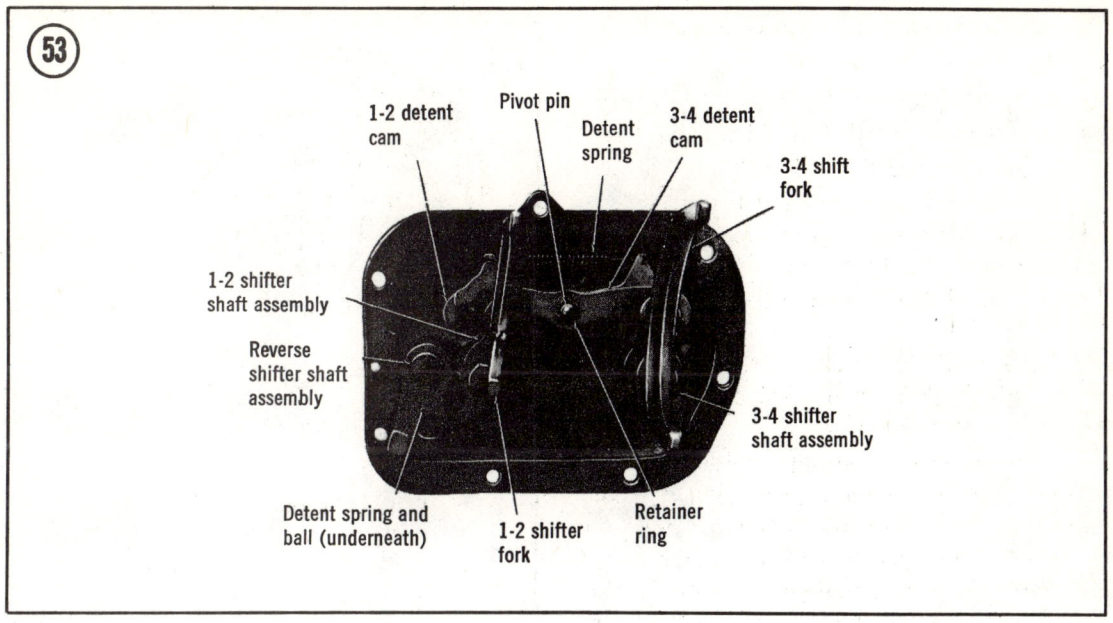

Table 1 is on the following pages.

Table 1 TRANSMISSION TIGHTENING TORQUES

3-speed Saginaw

Clutch gear retainer to case bolts	15 ft.-lb.
Side cover to case bolts	15 ft.-lb.
Extension to case bolts	45 ft.-lb.
Shift lever to shifter shaft bolts	25 ft.-lb.
Lubrication filler plug	18 ft.-lb.
Transmission case to clutch housing bolts	75 ft.-lb.
Cross member to frame nuts	25 ft.-lb.
Crossmember to mount bolts	40 ft.-lb.
2-3 cross over shaft bracket retaining nut	18 ft.-lb.
1-revolution swivel attaching bolt	20 ft.-lb.
Mount to transmission bolt	50 ft.-lb.

4-speed Muncie

Rear bearing retainer	18 ft.-lb.
Cover bolts	25 ft.-lb.
Filler plug	35 ft.-lb.
Drain plug	35 ft.-lb.
Clutch gear bearing retainer bolts	18 ft.-lb.
Universal joint front flange nut	95 ft.-lb.
Power take off cover bolts	18 ft.-lb.
Parking brake	22 ft.-lb.
Countergear front cover screws	25 ft.-lb.
Rear mainshaft locknut (4 wheel drive models)	95 ft.-lb.
Transmission to clutch housing bolts	75 ft.-lb.
Crossmember to mount	40 ft.-lb.
Mount to transmission	50 ft.-lb.

3-speed Muncie

Clutch gear retainer to case bolts	15 ft.-lb.
Side cover to case bolts	15 ft.-lb.
Extension to case bolts	45 ft.-lb.
Shift lever to shifter shaft bolts	25 ft.-lb.
Lubrication filler plug	18 ft.-lb.
Transmission case to clutch housing bolts	75 ft.-lb.
Cross member to frame nuts	25 ft.-lb.
Crossmember to mount bolts	40 ft.-lb.
Transmission drain plug	30 ft.-lb.
2-3 cross over shaft bracket retaining nut	18 ft.-lb.
1-revolution swivel attaching bolt	20 ft.-lb.
Mount to transmission bolt	50 ft.-lb.

Powerglide

Transmission case to engine	35 ft.-lb.
Oil pan bolts	8 ft.-lb.
Low band adjustment locknut	15 ft.-lb.
Converter to engine bolts	35 ft.-lb.
Oil pan drain plug	20 ft.-lb.

Turbo-Hydramatic

Transmission to engine bolts	35 ft.-lb.
Rear mount to transmission bolts	40 ft.-lb.
Rear mount to crossmember bolt	40 ft.-lb.
Crossmember mounting bolts	25 ft.-lb.
Strainer retainer bolt	10 ft.-lb.
Converter to flywheel bolts	35 ft.-lb.

Table 1 TRANSMISSION TIGHTENING TORQUES (continued)

4-speed Saginaw		4-speed Warner	
Clutch gear retainer to case bolts	15 ft.-lb.	Clutch gear retainer to case bolts	18 ft.-lb.
Side cover to case bolts	15 ft.-lb.	Side cover to case bolts	18 ft.-lb.
Extension to case bolts	45 ft.-lb.	Extension to case bolts	40 ft.-lb.
Shift lever to shifter shaft bolts	25 ft.-lb.	Shift lever to shifter shaft bolts	20 ft.-lb.
Lubrication filler plug	13 ft.-lb.	Lubrication filler plug	15 ft.-lb.
Transmission case to clutch housing bolts	75 ft.-lb.	Transmission case to clutch housing bolts	52 ft.-lb.
Crossmember to frame nuts	25 ft.-lb.	Crossmember to mount and mount to extension bolts	25 ft.-lb.
Crossmember to mount and mount to extension bolts	40 ft.-lb.	Rear bearing retainer to case bolts	25 ft.-lb.
Mount to transmission bolts	32 ft.-lb.	Extension to rear bearing retainer bolts (short)	25 ft.-lb.
		Retainer to case bolts	35 ft.-lb.
		Transmission drain plug	20 ft.-lb.

CHAPTER TEN

BRAKES

Front disc brakes are standard on all 1975 and later vehicles and have been optional equipment from 1970 to 1974. Rear drum brakes are standard on all models. All covered models use a split brake system, with a dual master hydraulic cylinder providing essentially 2 separate braking systems, one for the front wheels and one for the rear wheels. When a failure is encountered in either system, the other can be used to stop the car. Failure of either system will cause a dash warning lamp to light. All brakes are self-adjusting (except parking brake).

BRAKE INSPECTION

At the intervals recommended in Chapter Three, brake linings and pads and other internal brake components should be inspected (drums, rotors, wheel cylinders, springs, hydraulic lines, etc.). More frequent checks should be made if brakes are used frequently. Parking brake should also be checked and adjusted, if required, when linings are checked. Linings and pads should be replaced when worn to within $\frac{1}{32}$ in. of shoe table or rivet head, whichever applies. Brake shoes and pads should be replaced in axle sets (right and left sides). Disc brakes are checked by inspecting the outboard shoe, or pad, at both ends of the caliper (inboard shoe should also be checked in case premature wear has occurred).

BLEEDING

The hydraulic brake system must be bled when brakes are overhauled, when any hydraulic line has been disconnected, or whenever air enters the hydraulic system in any way.

CAUTION
Do not attempt to bleed the system while any brake drum is off the car or any caliper is removed from a disc, as damage to wheel cylinders and linings could occur.

NOTE: *Bleed one valve at a time to prevent master cylinder fluid level from becoming low enough to introduce more air into system. Proper sequence is to bleed valve nearest master cylinder (either front or back) first, then bleed the opposite valve.*

NOTE: *Do not mix brake fluid types. Use only Delco Supreme 11 fluid or other DOT-3 type.*

BRAKES

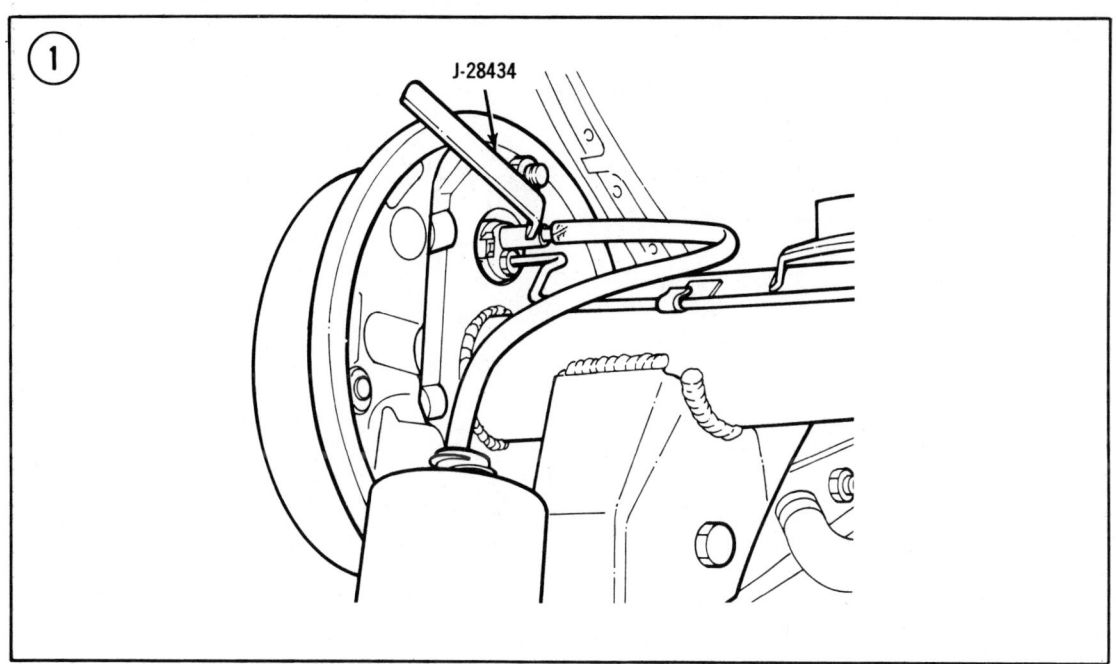

1. Clean all dirt, etc., from top of master cylinder and remove cover and rubber diaphragm.

2. If fluid level is low, fill and reinstall diaphragm and cover.

3. Install bleeder wrench (GM Part No. J-21472 or equivalent) on bleeder valve nearest master cylinder. See **Figure 1**. Install a bleeder hose on the valve. Place other end of hose in a transparent container with enough brake fluid to ensure end of hose will remain submerged.

NOTE: *Carefully monitor fluid level in master cylinder and add fluid as required during bleeding to prevent draining of reservoirs.*

4. Open bleeder valve ¾ turn and have an assistant depress brake pedal. As pedal nears the end of its travel, close bleeder valve tightly and allow brake pedal to return slowly to released position. Repeat this step until expelled brake fluid is completely free of air bubbles, then close bleeder valve tightly.

CAUTION
Do not re-use brake fluid which is bled from brakes. To do so might introduce contamination into the hydraulic system and result in damage.

5. Remove bleeder hose and wrench.

6. Repeat Steps 1 through 5 for opposite bleeder valve, then for valves on other end of car, starting with valve closest to master cylinder.

7. Fill master cylinder to level shown in **Figure 2**, then reinstall diaphragm and cover, as shown in **Figure 3**.

BRAKE LININGS

Refer to **Figure 4** (front) and **Figure 5** (rear), typical.

It may be necessary to retract adjusting screw if brake drums are severely worn. If drum does not slide off easily, either knock out lanced area on side of brake drum or remove access slot plug in backing plate (depending upon model). See **Figures 6 and 7**.

CAUTION
If drum lanced area is knocked out, make certain all metal is removed from drum and brake mechanism. After adjusting screw is released and drum is removed, plug hole with a metal cover (obtainable from dealer).

Removal

1. Mark drum position so it can be reinstalled in original position.

2
¼ in.
¼ in.
LEVEL CYLINDER
¼ in.
¼ in.
TILTED CYLINDER

3
Reservoir cover
Reservoir diaphragm
Reservoir
Reservoir grommet
Master cylinder body
Spring
Spring retainer
Primary seal
Secondary piston
Secondary seal
Primary piston assembly
Lock ring

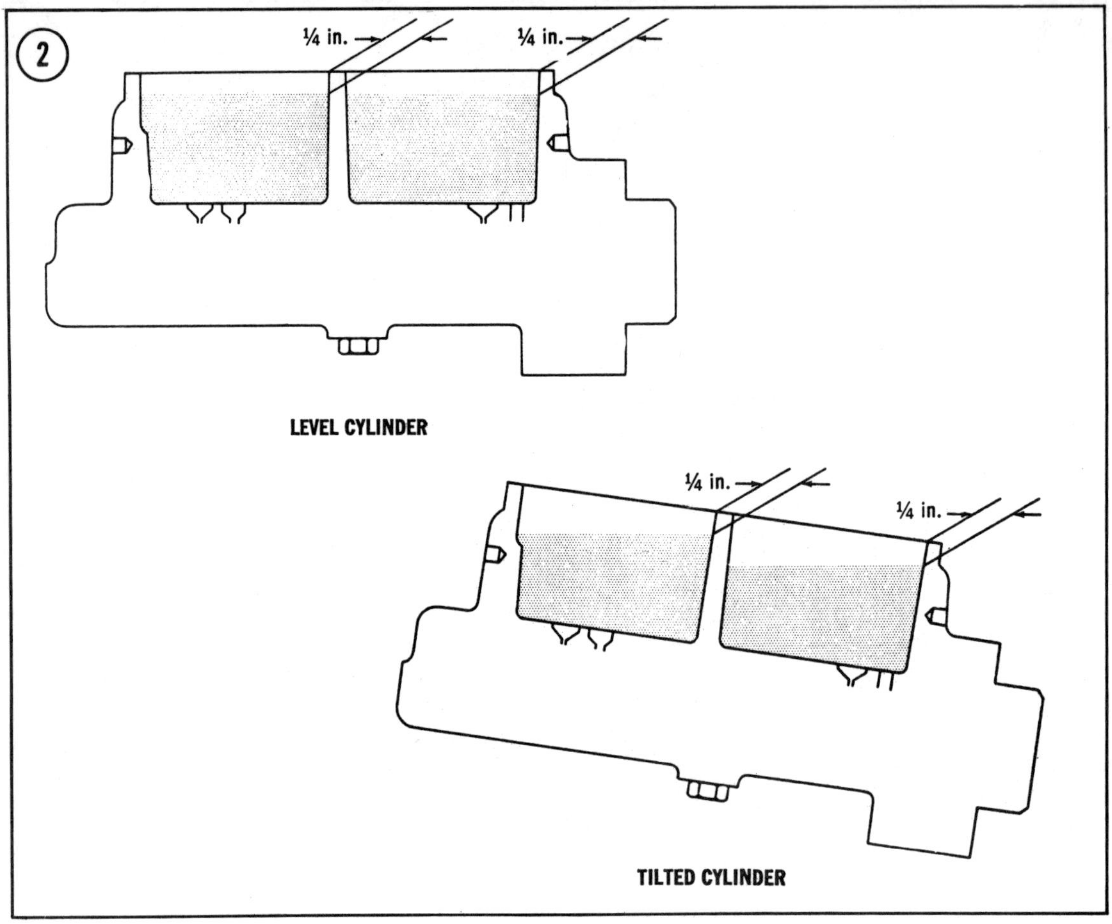

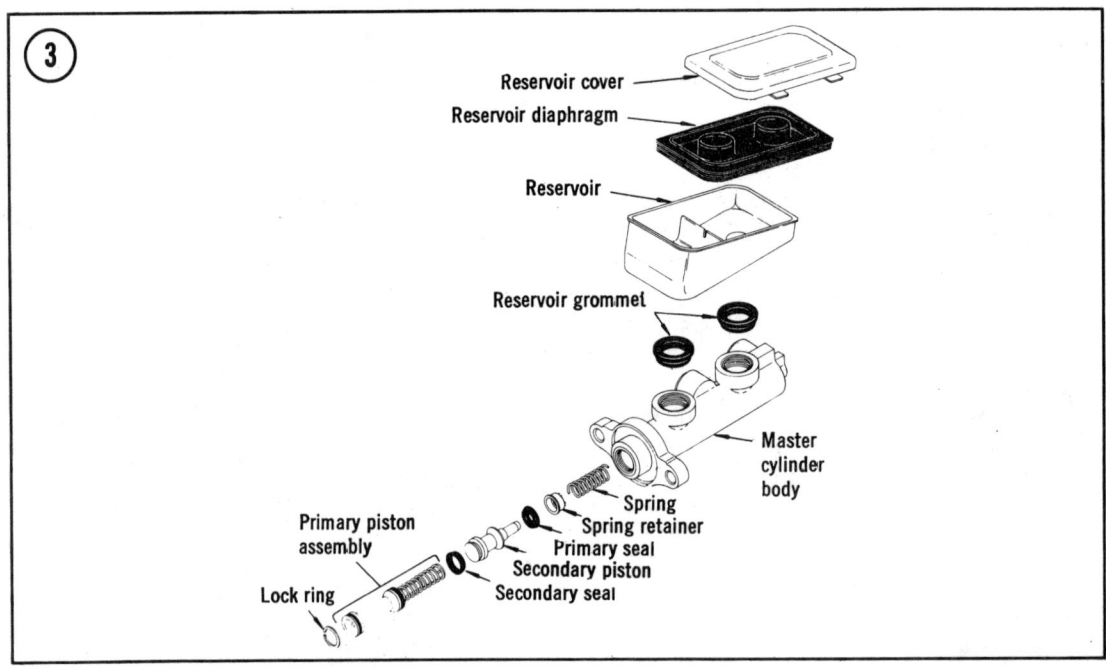

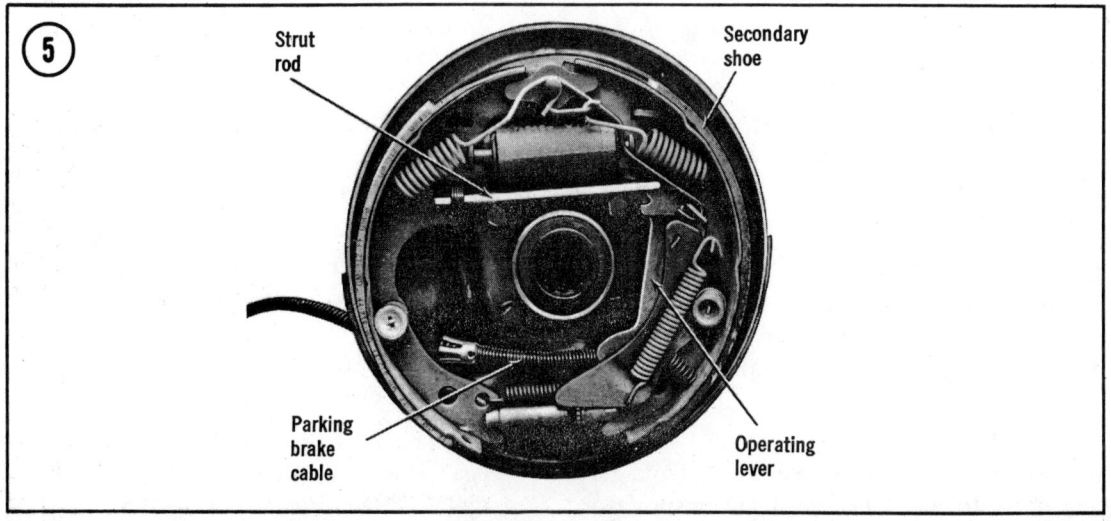

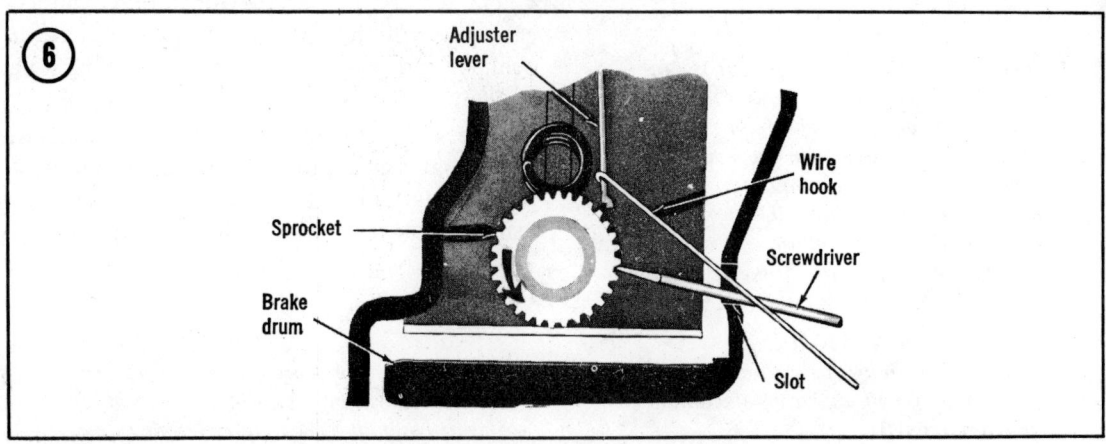

2. Loosen check nuts at parking brake equalizer enough to remove all tension from cables.

3. Remove brake drums.

CAUTION
Do not depress brake pedal while drums are removed from brakes.

4. Unhook brake shoe pull back springs. See **Figure 8**.

NOTE: *A spring removal tool (GM J-8049 or equivalent) will save time and possibly help prevent damage to springs and brake parts.*

5. Remove actuator return spring and link.

6. Remove the brake shoe hold-down pins and springs (**Figure 9**) and remove actuator assembly (**Figures 4 and 5**).

NOTE: *Mark positions of brake shoes if they are to be replaced.*

NOTE: *Actuator pivot and override spring are an assembly and should not be disassembled unless it is necessary to replace a broken part.*

7. Separate brake shoes by removing spring and adjusting screw.

8. Remove parking brake lever from secondary brake shoe (**Figure 5**) on rear brakes only.

Inspection

1. Clean all dirt out of brake drum (avoid getting dirt into wheel bearings) and inspect for roughness, scoring, or out-of-round condition.

NOTE: *If measuring equipment is not available, take all drums to an automotive machine shop or brake repair shop and have them checked, and turned on a lathe, if required. Brake drums that are more than 0.006 out-of-round must be reconditioned or replaced to ensure effective brake action.*

CAUTION
Brake drums must be refinished not more than 0.060 in. over maximum standard diameter.

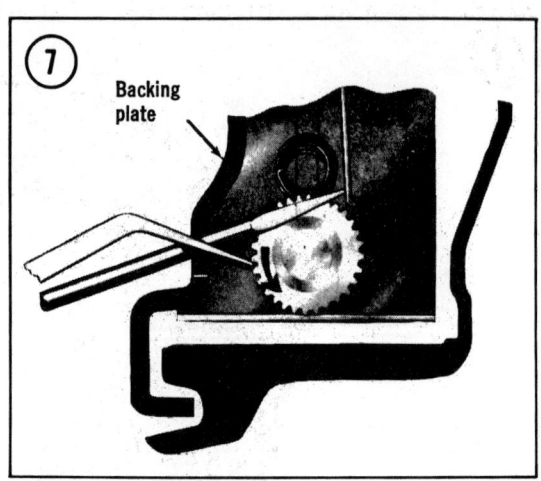

WARNING
Under no circumstances should a cracked brake drum be used. Do not attempt to weld a cracked drum, as welding heat will cause distortion. Use of a cracked or deformed brake drum could result in brake failure, which could result in injury or loss of life.

2. Remove and inspect wheel bearings and grease seals and replace parts as required. Repack and replace bearings, and replace seals.

3. On wheel cylinders with external rubber boots, carefully pull lower edges of boot away from cylinders and check to see if interior is wet with brake fluid. If excessive fluid is noted, cylinder must be overhauled or replaced.

NOTE: *A slight amount of fluid is almost always present and is needed to lubricate the piston.*

4. On wheel cylinders having internal boots, carefully pull a small part of boot out of cylinder and check as in Step 3, observing note above.

5. Check all flange plate attaching bolts to verify that they are tight. Clean all dirt and rust from contact faces on flange plate (**Figure 10**), using fine emery or crocus cloth, if required.

Installation

CAUTION
Make certain only exact replacement shoe and lining assemblies are used. If old assemblies are to be reinstalled, they

BRAKES

291

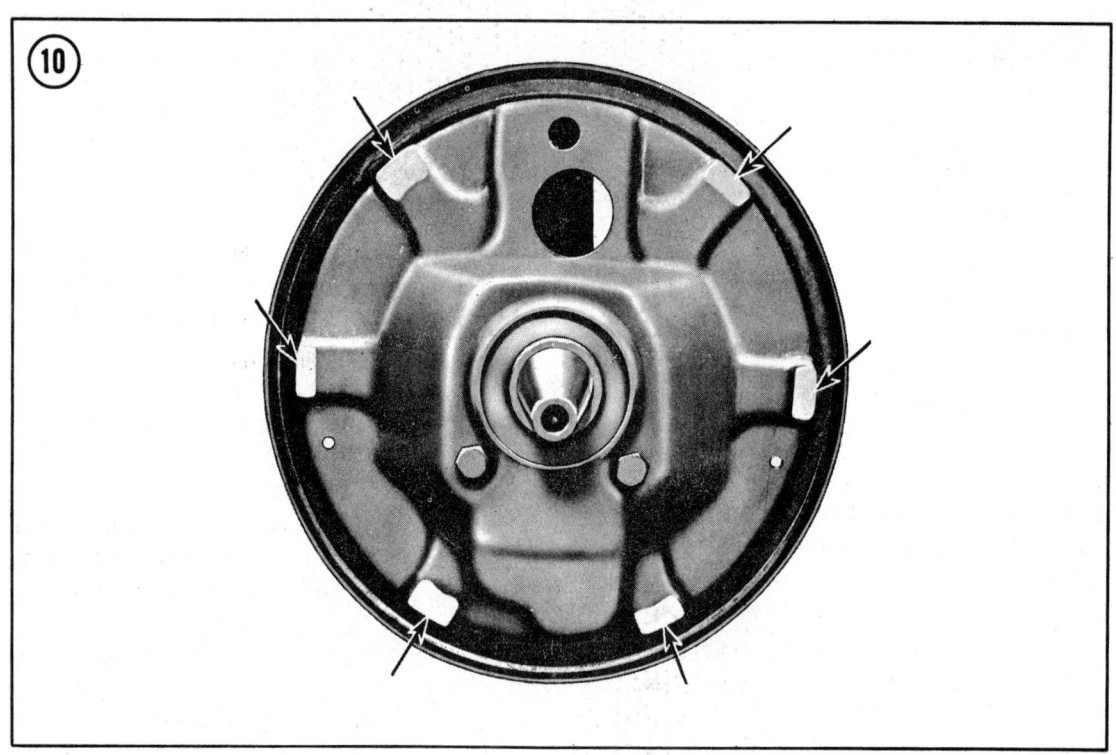

BRAKES

must be installed in original positions (as marked when removed).

NOTE: *Use only Delco Brake Lubricant (#5450032) or equivalent in the following procedure when lubrication is required.*

1. Inspect new shoe and lining assemblies and remove all nicks and burrs on bonding material, shoe edges, or on any contact surface.

CAUTION
Do not permit oil or grease to come into contact with linings. Keep hands clean.

2. Lubricate parking brake cable (if working on rear brakes). Also lubricate fulcrum end of parking brake lever, then attach lever to secondary shoe. Verify that lever moves freely.

3. Lightly lubricate pads on flange plate and threads of adjusting screw (make sure adjusting screw is clean before lubricating).

4. Connect primary and secondary brake shoes together with adjusting screw spring. Then position adjusting screw, socket and nut in position.

CAUTION
Verify that proper adjusting screw is being used (marked "L" for left side, "R" for right side of vehicle). Star wheel should be installed closer to secondary shoe. Adjusting screw spring must not interfere with star wheel operation. Verify that star wheel lines up with adjusting slot in flange plate.

5. On rear wheels, connect parking brake cable to lever.

6. Attach primary brake shoe (shorter lining faces forward) using hold-down spring and pin. Make certain shoes engage with wheel cylinder connecting links.

7. Install and secure actuator assembly and secondary brake shoe with hold-down spring and pin. Position parking brake strut and strut spring (rear wheels).

8. Install guide pin over anchor pin and install wire link.

CAUTION
To avoid damage, fasten wire link to actuator assembly first and place other end over anchor pin stud by hand (do not use tool) while holding adjuster assembly in full down position.

9. Install actuator return spring by easing it into place with a screwdriver or other suitable flat tool. Do not pry on actuator lever to install it.

NOTE: *If old brake return (pull back) springs are nicked, distorted or if strength or condition is doubtful, new springs should be installed.*

10. Hook brake return in shoes and install other ends of primary spring over anchor pin. Then hook other end of secondary shoe spring over the wire link end (see **Figure 11**).

11. After installation is complete, make certain actuator lever works easily by hand operating the self-adjusting feature (**Figure 11**).

12. Repeat the above procedure for all wheels equipped with drum brakes.

13. If wheel cylinders were removed and replaced, bleed hydraulic lines using procedure given above.

14. Adjust service and parking brakes, using procedures given below, and install drums and wheels.

15. Lower vehicle to floor.

SERVICE BRAKE ADJUSTMENT

1. Disengage actuator from star wheel and rotate star wheel by hand.

2. Using brake drum as a fixture, turn star wheel until drum slides over brake linings with slight drag.

3. Use star wheel to retract lining 1¼ turns.

4. Install drum and wheel, making certain drum is in some position as when removed.

5. Make final adjustment after all drums and wheels are replaced, using Steps 1 through 4. This is accomplished by making numerous forward and reverse stops, using firm pressure on brake pedal, until satisfactory pedal height and braking action are achieved.

PARKING BRAKE ADJUSTMENT

1. Raise vehicle (if not already raised).
2. Apply parking brake 2 notches from fully released position.
3. Loosen the brake equalizer check nut until a light to moderate drag is felt when rear wheels are rotated frontward.
4. Hold front nut while tightening jam nut securely.
5. Release parking brake and rotate rear wheels. No drag should be felt.
6. Lower vehicle to floor.

WHEEL CYLINDER SERVICE (DRUM BRAKES)

Removal

1. Raise vehicle and remove wheel and brake drum, and brake shoe pull back springs, using procedure given above.
2. Carefully clean all dirt away from line to wheel cylinder connection. Disconnect hydraulic line from cylinder and cover end of line with clean, lint-free cloth to prevent contamination from entering hydraulic system.
3. On 1977 and earlier models, remove screws securing cylinder to flange plate. Disengage cylinder pushrods from brake shoes and remove cylinder.
4. On 1978 and later models, insert small ($\frac{1}{8}$ in. diameter or less) awls or similar tools into access slots between cylinder pilot and retainer locking tabs (see **Figure 12**). Bend both tabs at the same time to release cylinder from backing plate. Discard retaining clip.

Disassembly

Refer to **Figure 13** (1970-1977) or **Figure 14** (1978 and later).

1. Remove boots from cylinder ends and discard.
2. Remove and discard pistons and caps.

Cleaning and Inspection

NOTE: *Most brake fluids are colored to assist in detecting hydraulic leaks. Staining from this coloring should not be confused with corrosion, which can be identified by pits or roughness in the cylinder bore.*

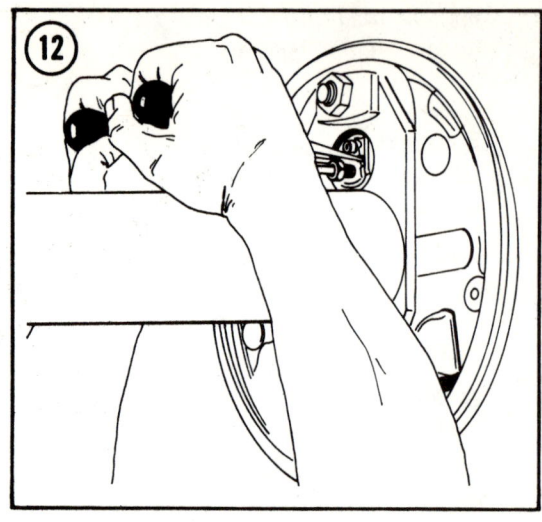

1. Check cylinder bore for staining and corrosion. Discard if corroded (see note above) and obtain new cylinder.
2. Remove stains by polishing cylinder bore with crocus cloth by revolving the cylinder while supporting the cloth in the bore with a finger. Do not polish by sliding the cloth lengthwise in the bore under pressure.
3. Clean hands thoroughly with soap and water (do not use gasoline or any other petroleum based solvent) and then clean the cylinder and all metal parts in clean brake fluid.
4. Shake excess fluid from cylinder. Never use a rag to dry cylinder bore as even the smallest fragment of lint could cause problems.

Assembly

1. Lubricate cylinder bore and counterbore with clean brake fluid and insert spring- expander assembly.
2. Install new cups with flat surfaces toward ends of cylinder. Do not lubricate cups prior to assembly.
3. Install new pistons with flat surfaces toward center of cylinder. Do not lubricate pistons before assembly.
4. Press new boots into cylinder counterbore by hand. Do not lubricate boots prior to installation.

BRAKES

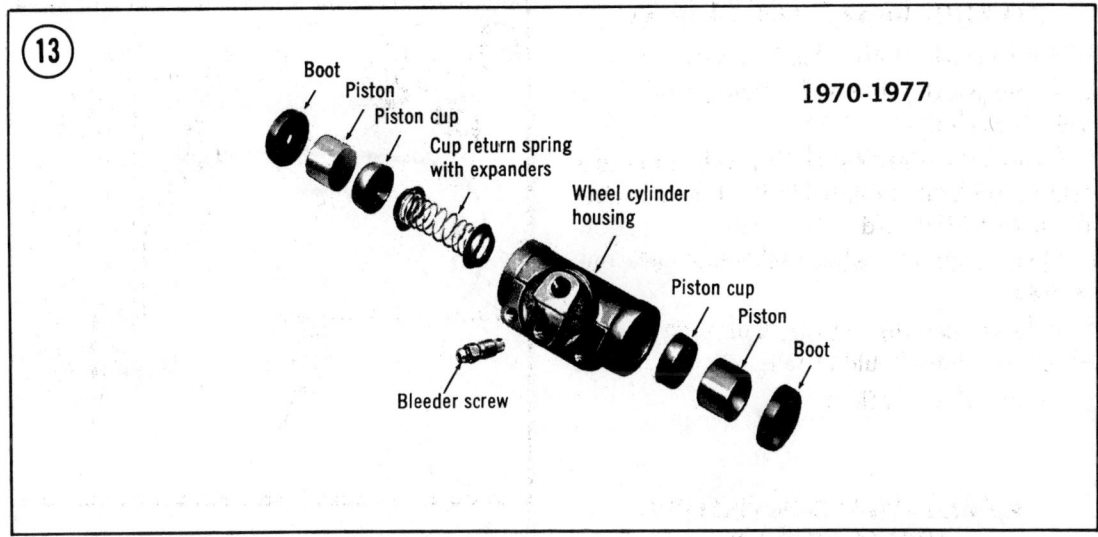

1970-1977

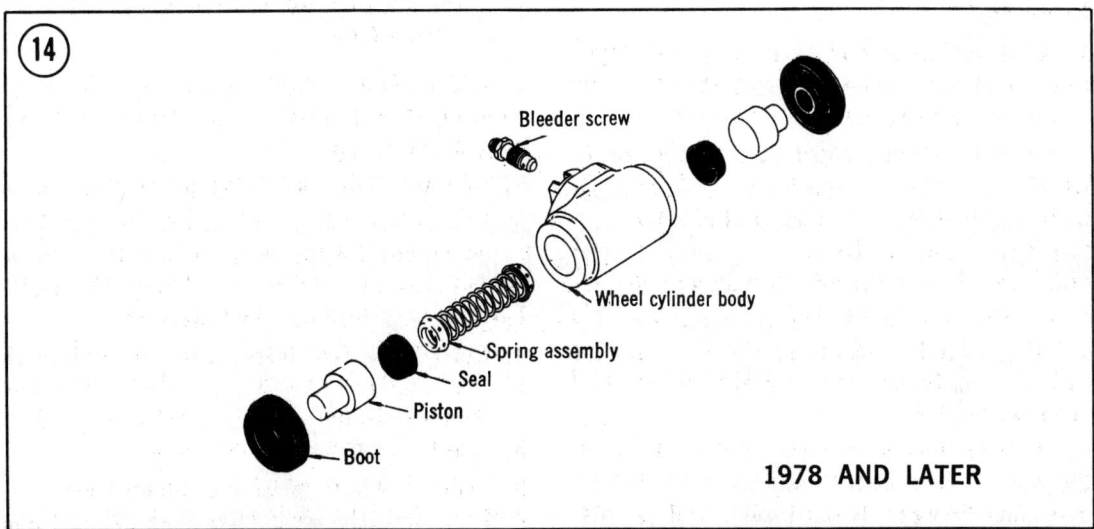

1978 AND LATER

Installation

1. On 1970-1977 models, position wheel cylinder to flange plate and install and securely tighten screws.

2. On 1978 and later models, hold cylinder against backing plate with a block installed between the cylinder and the wheel flange. Then install new retaining spring over cylinder, lining tabs up with the cylinder grooves. Drive retainer into position with a 1⅛ in. socket and a 10 in. extension. Make sure both retainer tabs are properly engaged.

3. Replace all pushrods and pull back screws.

4. Reconnect hydraulic line to wheel cylinder.

5. Install brake drum and wheel.

6. Bleed hydraulic system using the procedure given above.

7. Lower vehicle to floor.

PAD REPLACEMENT — FRONT WHEEL DISC BRAKES

Removal

1. If master cylinder reservoir is more than ⅓ full, siphon out enough fluid to reduce level to ⅓ full (to avoid overflow when caliper piston is pushed back into bore). Discard fluid removed as brake fluid should never be reused.

2. Raise vehicle and remove front wheels.

3. Using a C-clamp as shown in **Figure 15**, push caliper piston back into its bore.

4. Remove the 2 mounting bolts attaching caliper to support (see **Figure 16**).

5. Lift caliper from disc.

6. Remove inboard shoe and dislodge outboard shoe (see **Figure 17**). Position caliper on front suspension arm so brake hose does not support caliper weight.

> NOTE: *Mark shoe positions if shoes are to be reinstalled.*

7. Remove shoe support spring from piston.

8. Remove 2 sleeves from inboard ears of caliper; then remove 4 rubber bushings from the grooves in caliper ears.

Cleaning and Inspection

> NOTE: *If shoes have $\frac{1}{32}$ in. or less lining left above rivet heads they should be replaced. Always replace shoes in axle (2 wheel) sets.*

1. Thoroughly clean mounting bolts and holes and bushing grooves in caliper ears.

> CAUTION
> *If bolts are damaged or corroded they should be replaced to prevent later brake failure.*

2. Examine inside of calipers for leakage evidence. If present, caliper should be overhauled (see procedure below).

3. Clean inside of calipers, including exterior of dust boot. Check boot for damage.

> CAUTION
> *Do not use compressed air for cleaning as it could unseat boot.*

4. If disc is badly worn or scored it should be refinished or replaced. Remove hub and take it to an automotive machine shop or brake repair shop for checking and refinishing.

Installation

> CAUTION
> *If old shoes are to be reinstalled they must be installed in original positions.*

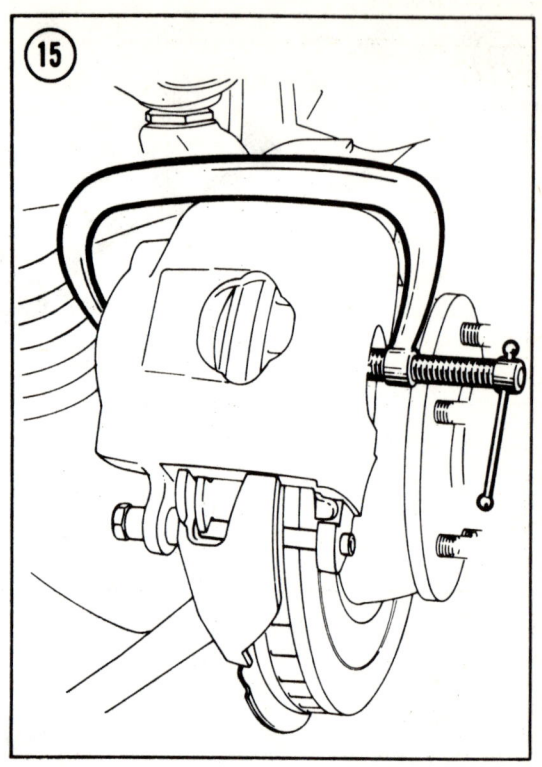

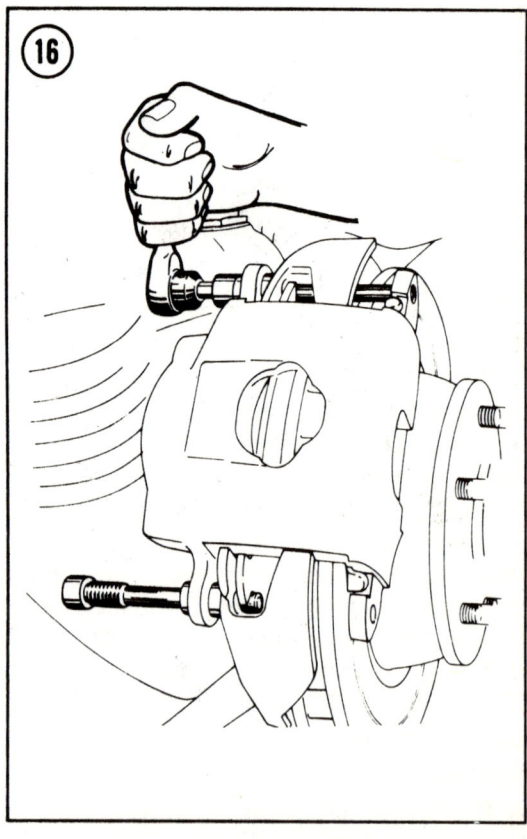

BRAKES

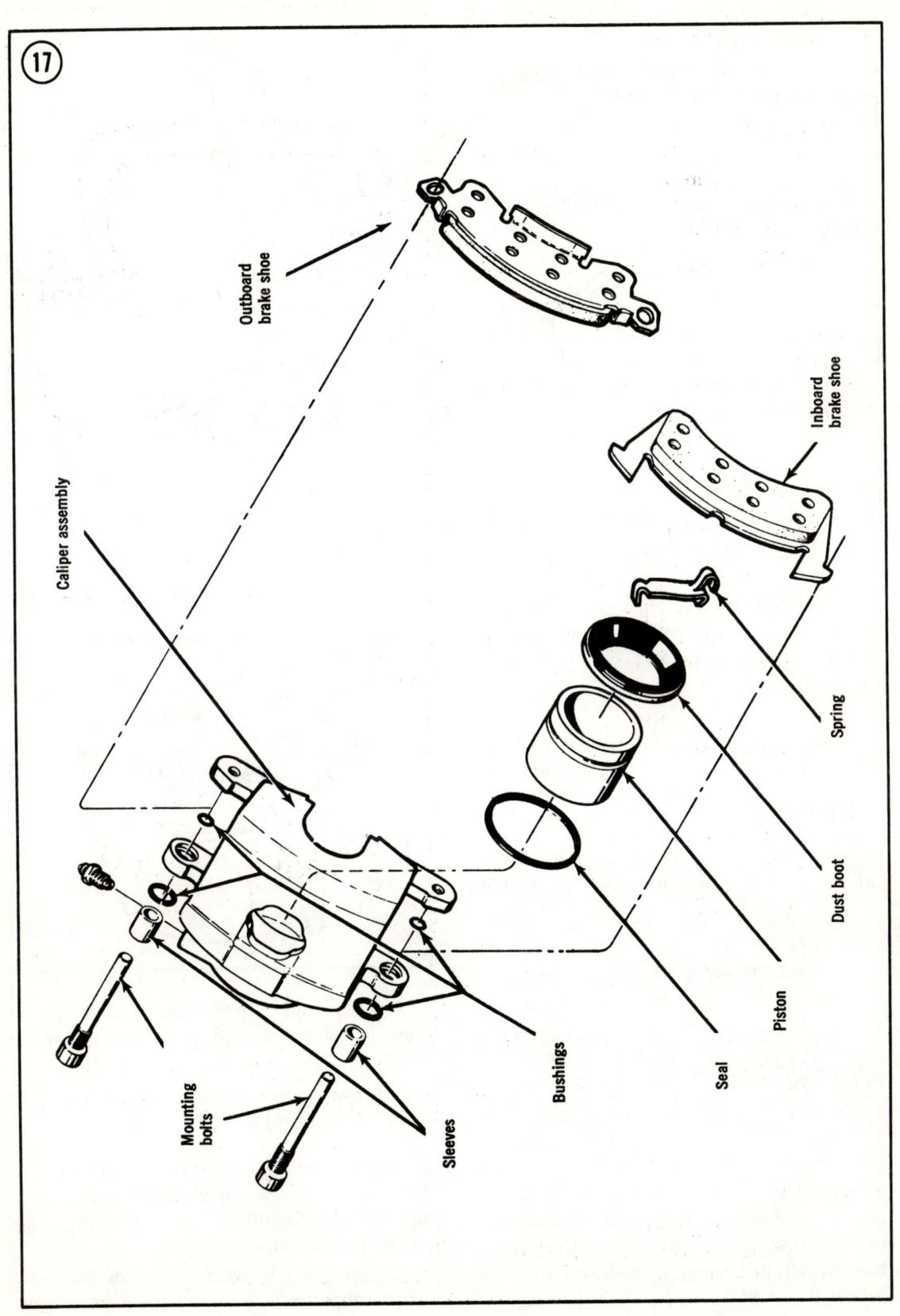

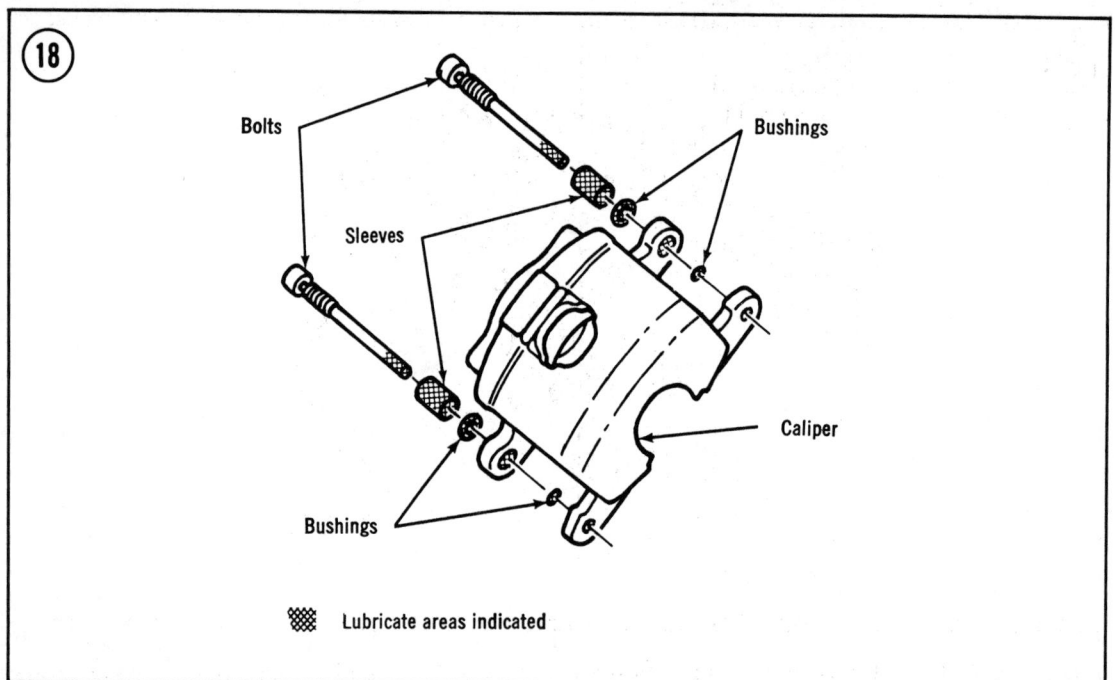

1. Lubricate new sleeves, new rubber bushings, bushing grooves and ends of mounting bolts with Delco Silicone Lube No. 5459912 or equivalent (see **Figure 18**).

> **CAUTION**
> *New sleeves and rubber bushings must be used and lubrication instructions followed to ensure that calipers function properly.*

2. Install new rubber bushings in grooves in caliper ears and install new sleeves to inboard ears.

> NOTE: *Position sleeves with ends toward shoe and lining assemblies flush with machined surfaces of ears.*

3. Install shoe support spring and inboard shoe in center of piston cavity as shown in **Figure 19**. Push down until shoe lays flat against caliper.

4. Position the outboard shoe in the caliper with ears at top of shoe over caliper ears and tab at bottom of shoe engaged in caliper cutout.

5. Lift caliper (with shoes installed) up and rest bottom edge of outboard lining on outer edge of brake disc. Verify that there is no clearance between tab at bottom of outboard shoe and caliper abutment.

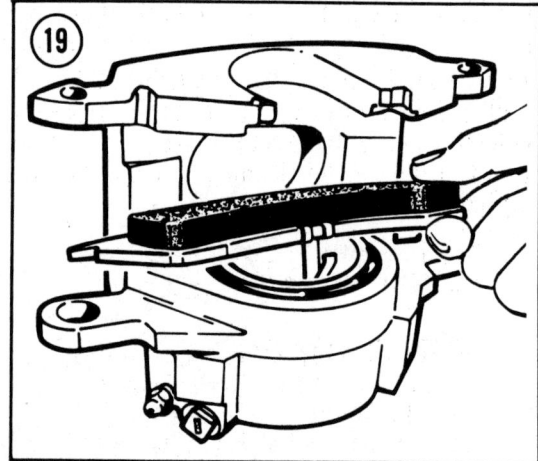

6. Position caliper over brake disc, lining up holes in caliper ears with holes in mounting bracket.

> NOTE: *Verify that brake hose is not kinked or twisted.*

7. Insert mounting bolts through sleeves in inboard caliper ears and mounting bracket, making sure that ends of bolts pass under retaining ears on inboard shoe.

8. Engage holes in outboard shoes and outboard caliper ears with mounting bolts, and

BRAKES

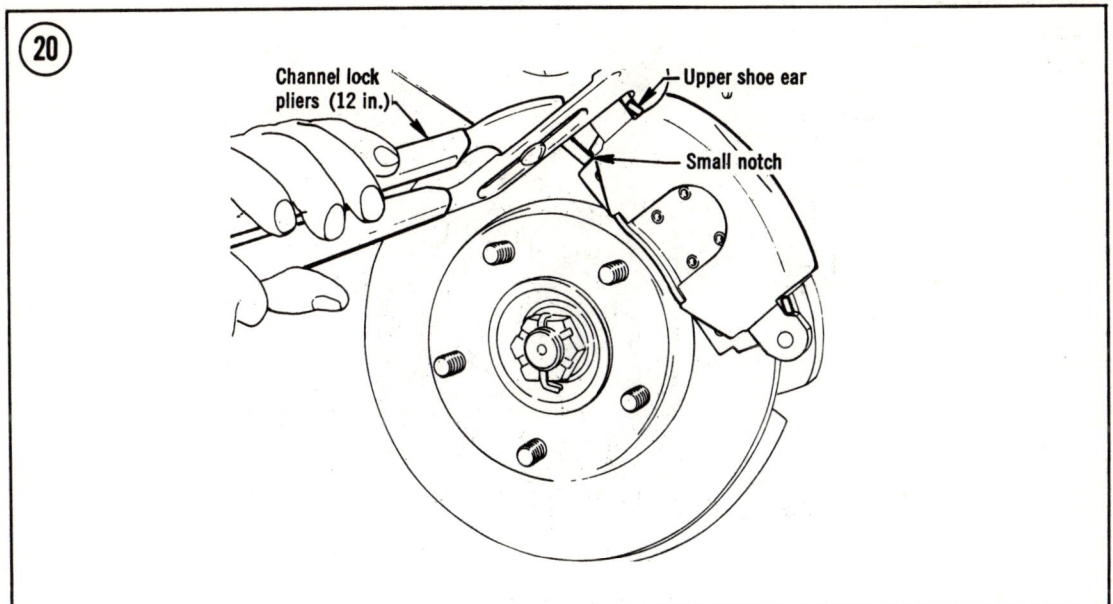

thread bolts into mounting bracket. Torque to 35 ft.-lb.

9. Pump brake pedal to seat linings against the discs.

10. Using arc-jointed pliers, bend upper ears of outboard shoe until no radial clearance exists between shoe and caliper housing (see **Figure 20**). Locate pliers on small notch of caliper housing during bending procedure.

> NOTE: *Repeat Step 10 if radial clearance still exists after initial clinching.*

> NOTE: *Outboard shoes (with formed ears) are designed for original installation only and are fitted to calipers. The shoes should never be relined or reconditioned for reinstallation.*

11. Reinstall front wheel and lower vehicle.

12. Add brake fluid to master cylinder reservoir to bring level to within ¼ in. of top.

13. Before moving vehicle, pump brake pedal several times to make certain it is firm. Recheck master cylinder after firm pedal is achieved.

CALIPER

Removal/Installation

1. Remove caliper assembly from disc, using procedure given above.

2. Remove hose to caliper bolt and cap or tape open connections to prevent introduction of contamination.

3. Remove caliper assembly from vehicle; then remove brake shoes from caliper.

4. Installation is the reverse of these steps.

5. Connect hydraulic lines to calipers, using new copper gaskets.

> CAUTION
> *Hose must be positioned in caliper locating gate (between locating beads) to assure proper positioning to caliper.*

6. Bleed calipers, using procedure given elsewhere in this chapter.

Disassembly

1. Clean caliper exterior with clean brake fluid and place caliper on a clean working surface.

2. Drain brake fluid from caliper.

> WARNING
> *Follow Step 3 and do not attempt to catch or protect piston with fingers, as injury could result.*

3. Pad interior of caliper with clean shop towels and remove piston by application of compressed air as shown in **Figure 21**.

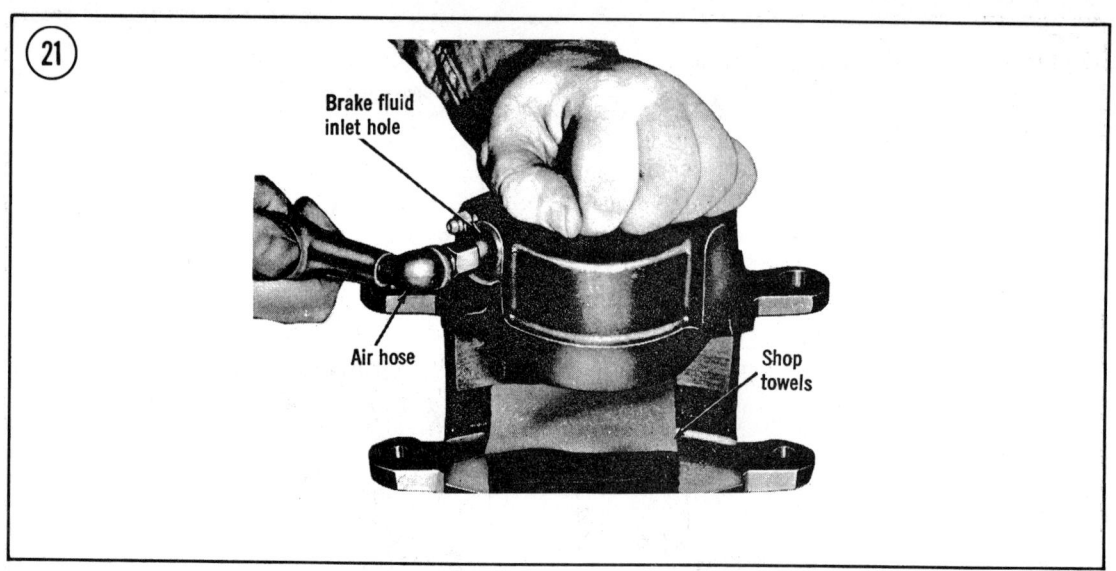

CAUTION
Do not blow piston out of bore. Use only enough air to ease the piston out.

NOTE: *If compressed air is not available, remove piston by applying brake pedal after caliper has been removed from disc, but before disconnecting hydraulic line.*

4. Pry dust boot (see **Figure 17**) out of caliper piston bore, exercising care not to scratch or damage piston bore.

5. Use a small piece of wood or plastic to pry piston seal from its groove in piston bore.

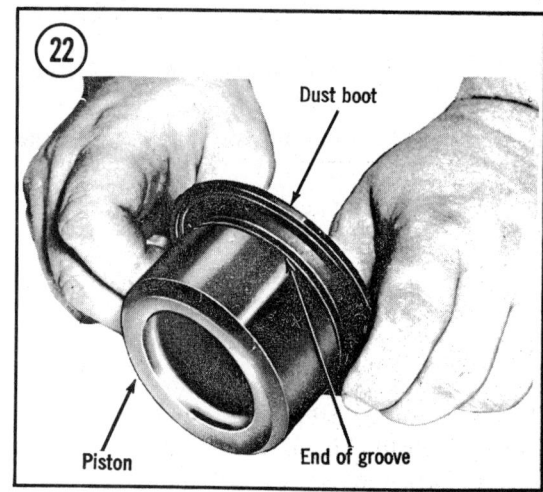

CAUTION
Do not use metal tool of any kind, as damage to bore may result.

6. Remove bleeder valve from caliper.
7. Remove and discard sleeves and bushings from caliper ears.

Cleaning and Inspection

CAUTION
Dust boot, piston seal, rubber bushings, and sleeves must be discarded when the caliper is overhauled. Replace them with new parts.

1. Clean all parts (except those discarded) in clean brake fluid. Use only dry, filtered, compressed air to blow out all passages in caliper and bleeder valve.

CAUTION
Use of unfiltered shop air may leave a film of mineral oil on internal passages. Mineral oil may damage rubber parts. If filtered air is not available, allow parts to drain completely dry before attempting reassembly.

2. Examine all mounting bolts and replace any that appear corroded or damaged.

3. Carefully examine piston for scoring, nicks, corrosion and worn or damaged chrome plating. If any surface defects are noted, replace piston.

BRAKES

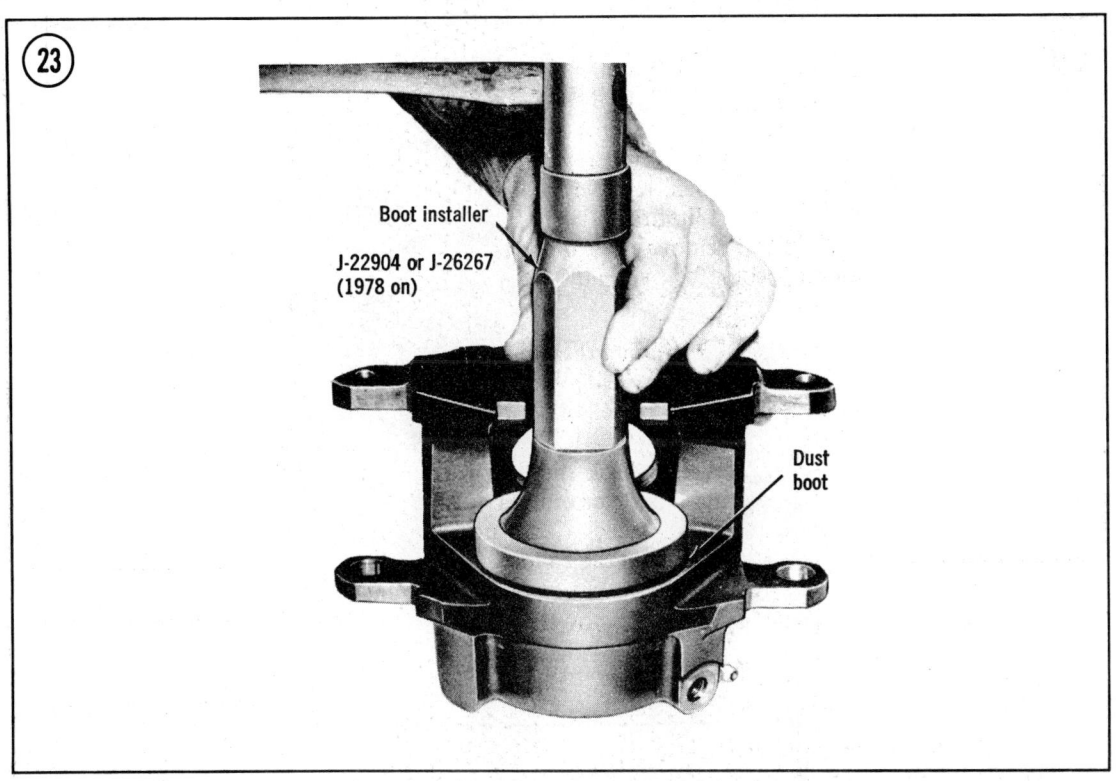

CAUTION
Piston surfaces and plating are manufactured to close tolerances. Do not attempt to resurface the piston by any means. The use of any abrasive, including crocus cloth, is not acceptable.

4. Check bore of caliper for same defects listed in Step 3. Piston bore is not plated and stains or minor corrosion can be polished out with a crocus cloth. A thorough clean-up is required after the use of crocus cloth.

CAUTION
Do not use emery cloth or any other type of abrasive. If crocus cloth does not remove the stain or corrosion, the caliper must be replaced.

Assembly

1. Lubricate caliper piston bore and new piston seal with clean brake fluid and place seal in caliper bore groove.
2. Lubricate piston with clean brake fluid and assemble new dust boot into groove in piston. Fold must face open end of piston as shown in **Figure 22**.

3. Insert piston into caliper bore, using care not to disturb the seal seating. Force piston to bottom of bore (50-100 lb. of force required).
4. Position dust boot in caliper counterbore and seat, using a boot installer tool (GM No. J-22904 or J-26267 for 1978-on as shown in **Figure 23**).

CAUTION
Check boot installation carefully to ensure that retaining ring molded into boot is not bent and that boot is installed below caliper face and evenly all around.

NOTE: *If boot installer tool is not available, the inventive mechanic can improvise a substitute. Care should be taken, however, not to damage the boot, piston, or caliper. A better alternative is to take the assembly to a brake shop and have the boot installed there.*

5. Install bleeder valve and torque to 65 in.-lb. (if valve will not seal at 65 in.-lb., torque to 100 in.-lb. maximum to obtain seal. If valve will not seal at 100 in.-lb., replace valve).

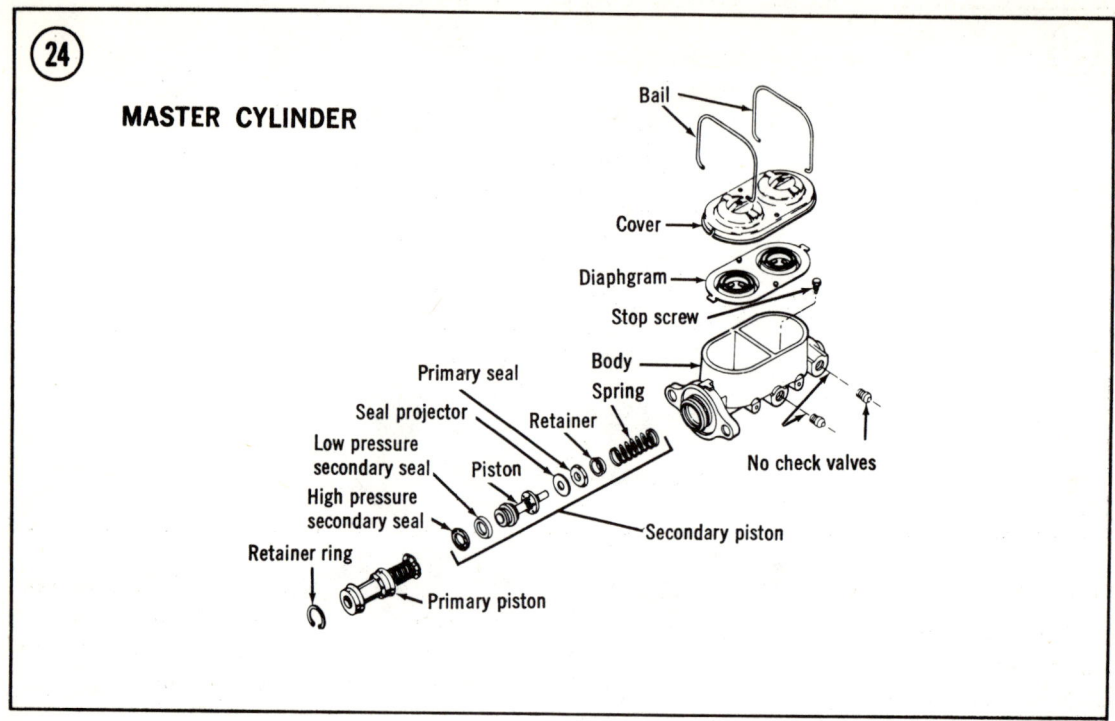

㉔ MASTER CYLINDER

POWER BRAKE UNIT

Inspection

1. Check vacuum line and connections, including check valve, for vacuum loss.
2. Inspect hydraulic lines and connections for leaks.
3. Check for scored brake drums or worn or contaminated linings.
4. Check and fill fluid reservoirs, as required.
5. Inspect for loose master cylinder and power unit mounting bolts.
6. Check power piston air filter and replace if indicated.
7. Check for misalignment and/or binding of brake pedal linkage.

Bleeding

Bleeding of brake hydraulic system should be done in accordance with procedure given elsewhere in this chapter.

CAUTION
Do not use power assist when bleeding brake lines. Engine must be off. Pump brake pedal several times to expend vacuum reserve before starting.

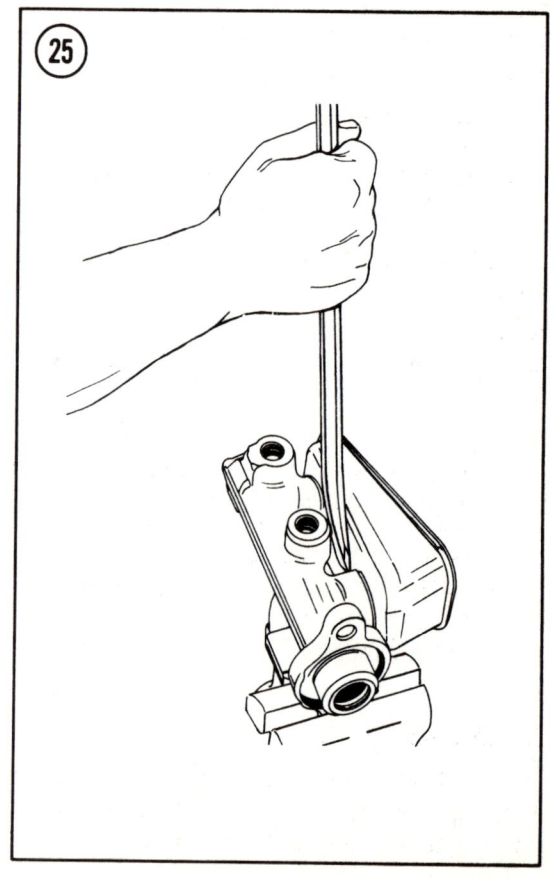

㉕

BRAKES

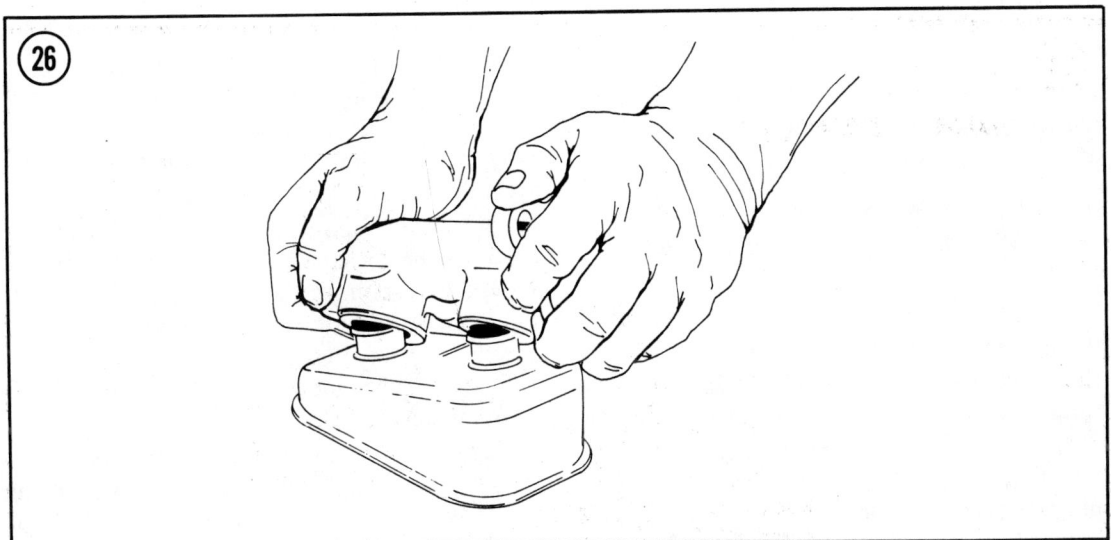

Removal

1. Disconnect vacuum line from power unit check valve.

2. Remove nuts holding power unit to master cylinder (in models so attached). Pull forward on master cylinder until cylinder clears power unit mounting studs. Carefully move cylinder to one side with all hydraulic lines attached.

> NOTE: *On older models, the power unit and master cylinder must be removed as a unit. In this case, disconnect hydraulic lines from master cylinder.*

3. Remove nuts holding power cylinder to dash.

4. Remove retainer holding pushrod to brake pedal. Remove power unit from engine compartment.

> NOTE: *Repair of the power unit requires special tools and knowledge. Instead of attempting to make repairs, the home mechanic should take the unit to a dealer or competent power brake mechanic for repairs or replacement. Always replace vacuum check valve and grommet when the power unit is overhauled.*

Installation

1. Install all components in reverse order of the removal procedure. See **Table 1** at the end of the chapter for torque specifications.

2. Check brake operation. If hydraulic lines were removed, the system must be bled to remove all air from lines and components. Check stoplight switch adjustment.

> CAUTION
> *Start engine and allow time for vacuum to build up in power unit before operating brake pedal.*

MASTER CYLINDER

Removal

1. Remove the brake fluid from the master cylinder reservoir. A suction-type turkey baster makes an excellent tool for fluid removal.

2. Disconnect the brake lines from the master cylinder. Plug the lines or place tape over their ends to prevent contamination from entering the system.

3. On non-power brake systems, disconnect the brake pedal from the master cylinder pushrod.

4. Remove the nuts attaching the master cylinder to the fire wall or the power brake booster. Remove the master cylinder from the car, taking care not to drip brake fluid on the paint.

Installation

1. Position the master cylinder in the engine compartment and install the nuts to attach it to

the fire wall or power brake booster. Tighten nuts to 24 ft.-lb.

2. On non-power brake models, connect the master cylinder pushrod to the brake pedal pin and install the retainer.

> NOTE: *It may be easier to connect the pushrod before installing the attaching nuts (as described in Step 1).*

3. Install plugs in the brake line outlets and fill the master cylinder reservoir with fresh brake fluid. Bleed the master cylinder by placing a suitable container under it, then opening the plugs one at a time. With the plug open, push the brake pedal in and hold it. Then tighten the plug and release the pedal. Continue until all air is expelled, and then repeat the procedure with the other plug.

4. Remove the plugs one at a time and install the brake lines. Then remove the container that was placed under the master cylinder in Step 3.

5. Replenish the brake fluid and then bleed the entire hydraulic system as described in *Bleeding*, this chapter.

6. Road test the vehicle to ensure that the brake system is working properly.

Disassembly

> NOTE: *Although master cylinders can be rebuilt, and kits containing all of the necessary replacement parts are available, it is recommended that a defective master cylinder be replaced with a new or professionally rebuilt unit. If you are sure of your mechanical skills and you decide to rebuild the cylinder yourself, bear in mind that a number of master cylinder configurations were used in the models covered by this book, and that the following disassembly, inspection, and assembly procedures are general in nature. During disassembly, carefully note the arrangement of parts so the cylinder can be correctly reassembled.*

Refer to **Figures 3 and 24** for typical master cylinder configurations.

1. Thoroughly clean the outside of the master cylinder. Then remove the cover and pour out any residual brake fluid.

2. On non-power brake models, pull the boot away from the master cylinder and pry up on the retainer tab to release the pushrod retainer.

3. Depress the piston and remove the secondary piston stop screw from the front master cylinder reservoir.

4. Install the master cylinder in a vise and remove the lock ring from the groove in the bore. Then remove the primary piston assembly. On 1978 and later models, use a pry bar or similar tool (see **Figure 25**) to separate the reservoir from the cylinder. Remove and discard the grommets.

5. Remove the secondary piston, spring retainer, and spring. This is best done by applying compressed air through the stop screw hole. If air is not available, the parts can be removed by hooking a bent piece of wire under the edge of the secondary piston and pulling it out. Take care not to damage the cylinder surface.

6. If replacement of the brake line outlet inserts is required, use a $13/64$ in. drill to enlarge the holes in the inserts. Then install a heavy washer over an outlet and thread a $1/4$-20 x $3/4$ in. screw into the enlarged insert hole. Turn the screw as required to remove the insert, then repeat for the other insert.

7. Remove the primary and secondary seals from the secondary piston.

8. Clean all metal parts in clean brake fluid. Blow the parts dry with compressed air and store them in a clean place on clean paper or a clean lint-free cloth.

Inspection

1. Check the cylinder bore for scoring and/or corrosion (pits or excessive roughness). If damage is present, replace the cylinder.

2. Remove stains or discoloration from the cylinder bore with crocus cloth. Do not use emery cloth or other abrasive material that might remove metal or leave harmful residue.

3. Clean the master cylinder bore thoroughly with fresh brake fluid. Do not use any other solvent. Shake excessive fluid from the cylinder and blow dry with compressed air. If air is not available, allow the cylinder to drain thoroughly.

BRAKES

> CAUTION
> *Do not dry the cylinder with a cloth, as even a small amount of lint could lead to brake failure.*

4. Make sure the compensating port in the cylinder is clear.

Assembly

1. Install new inserts in the brake line outlets and then seat them by installing a spare brake line nut in the outlet and tightening it as far as it will go. Remove the nut and repeat for the other outlet.

2. Install a new secondary seal in the groove in the end of the secondary piston. Then install a new primary seal over the secondary piston so that the flat side of the seal seats against the piston flange. See **Figure 19**.

3. Install a second new secondary seal in the groove in the pushrod end of the secondary piston.

4. Lubricate the master cylinder bore and the secondary and primary seals on the secondary piston with clean brake fluid.

5. Install the spring retainer into the secondary piston spring and then install the retainer and spring assembly onto the end of the secondary piston. The retainer should be located inside the lip of the primary seal.

6. Install the secondary piston and spring assembly in the master cylinder bore so that the spring seats in the closed end of the bore.

7. Install the master cylinder in a vise with the open end up and lubricate the seals of the primary piston with clean brake fluid. Install the primary piston assembly, spring end first, into the bore. Hold the primary piston down and install the lock ring.

8. Continue holding the primary piston down and install the secondary piston stop screw. On 1978 and later models, lubricate new grommets in clean brake fluid and install them in the cylinder casting. After grommets are properly seated, remove the casting from the vise and install the reservoir as shown in **Figure 26**, using a rocking motion. The shoulder of the grommet should be in continuous contact with both the casting and the reservoir.

9. Install a new diaphragm in the cover, if required, and install the cover on the master cylinder.

10. On non-power brake models, install the pushrod through the pushrod retainer (if disassembled) and push the retainer over the end of the master cylinder. Install a new boot over the pushrod and the retainer. Thread the thread nut down to the shoulder on the pushrod and thread the clevis down to the jam nut. Tighten the nut against the clevis to 14 ft.-lb.

Table 1 TORQUE SPECIFICATIONS

Component	Torque
Master cylinder to dash	24 ft.-lb.
Master cylinder to power unit	24 ft.-lb.
Power unit to dash	24 ft.-lb.
Brake line nuts (all)	150 in.-lb.
Brake bleeder valves	65-100 in.-lb.
Brake shoe to anchor pin	120 ft.-lb.
Wheel cylinder to flange plate	50 in.-lb. (except 1978-on)
Caliper mounting bolt	35 ft.-lb.
Flex hose-to-caliper bolt	32 ft.-lb.
Pedal mounting pivot bolt (nut)	25 ft.-lb.
Combination valve mounting	15 ft.-lb.

CHAPTER ELEVEN

FRONT SUSPENSION AND STEERING

The independent front suspension is conventional. It uses unequal upper and lower control arms and coil springs. The shock absorbers mount inside the coil springs. A stablizer bar attaches to the lower control arms and the frame. **Figure 1** shows the complete front suspension.

Figure 2 shows the steering system. Pitman arm motion is transmitted to a center relay rod supported at one side by an idler arm. Center relay rod movement in turn moves the steering knuckles through a tie rod. A steering damper incorporated in the steering linkage dampens road shock.

The power steering system uses the same recirculating ball steering gear, and linkage, with an added hydraulic power assist. An engine driven vane-type pump delivers hydraulic pressure to a control valve. The control valve senses the requirement for power assistance and supplies the power cylinder which operates the linkage.

WHEEL ALIGNMENT

Several front suspension dimensions affect running and steering of the front wheels. These variables must be properly aligned to maintain directional stability, ease of steering, and proper tire wear.

The dimensions involved define:
  a. Caster
  b. Camber
  c. Toe-in
  d. Steering axis inclination
  e. Front axle height

All except steering axis inclination are adjustable. Since these adjustments are critical, they must be done by a competent front end alignment shop or your dealer.

Pre-alignment Check

Several factors influence suspension angles, or steering. Before any adjustments are attempted, perform the 10 following checks.

1. Check tire pressure and wear.
2. Check play in front wheel bearings. Adjust, if necessary.
3. Check play in ball-joints.
4. Check for broken springs.
5. Remove any excessive load.
6. Check shock absorbers.
7. Check steering gear adjustments.
8. Check play in pitman arm and tie rod parts.
9. Check wheel balance.
10. Check *rear* suspension for looseness.

FRONT SUSPENSION AND STEERING

1

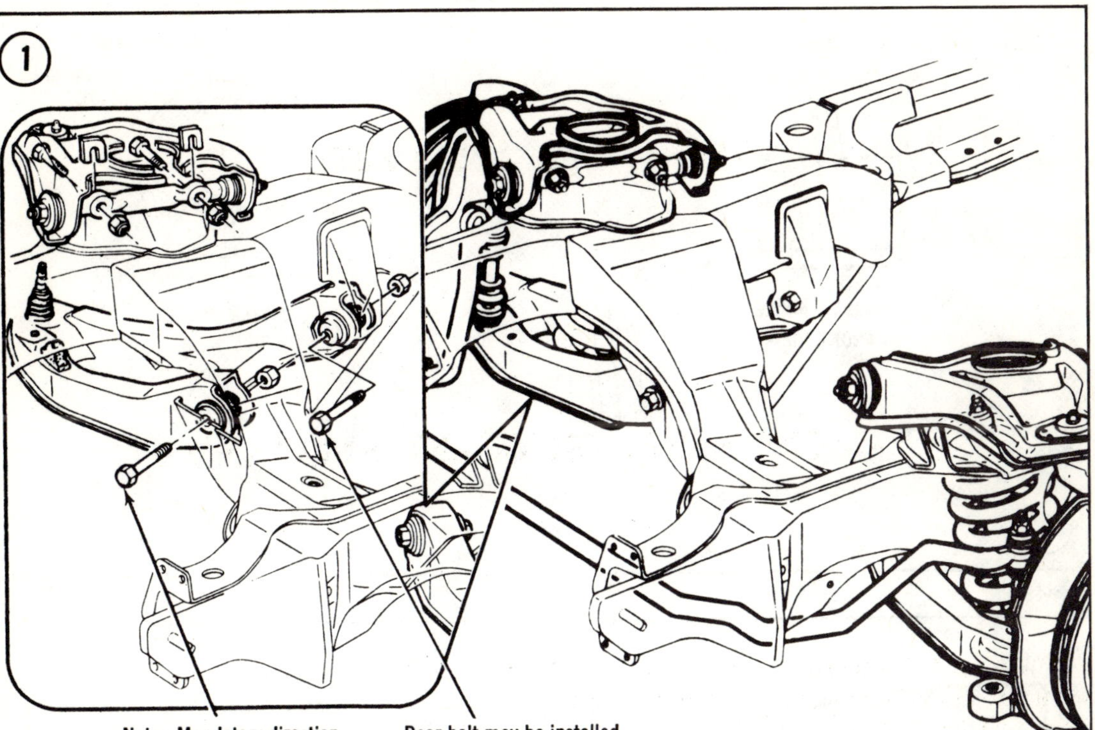

Note: Mandatory direction of front bolt installation

Rear bolt may be installed in either direction

2

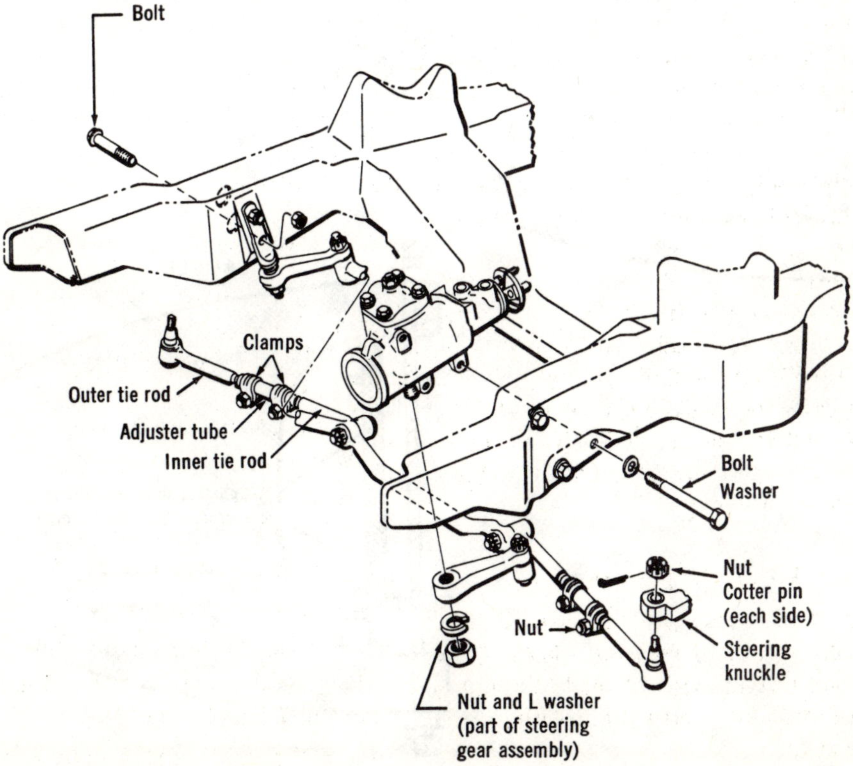

- Bolt
- Clamps
- Outer tie rod
- Adjuster tube
- Inner tie rod
- Bolt
- Washer
- Nut
- Cotter pin (each side)
- Steering knuckle
- Nut
- Nut and L washer (part of steering gear assembly)

Centerline of vehicle
Centerline of wheel
90°
Positive direction
CAMBER ANGLE
FRONT VIEW
Centerline of wheel
Centerline of ball-joint axis
FWD
90°
Positive direction
CASTER ANGLE
SIDE VIEW
FWD
WHEEL TOE-IN
TOP VIEW

3

CASTER
Subtract shims here to increase "positive" caster
Frame
Add shims here to increase "positive" caster
FORWARD
CAMBER
Frame
Subtract shims equally to increase "positive" camber

4

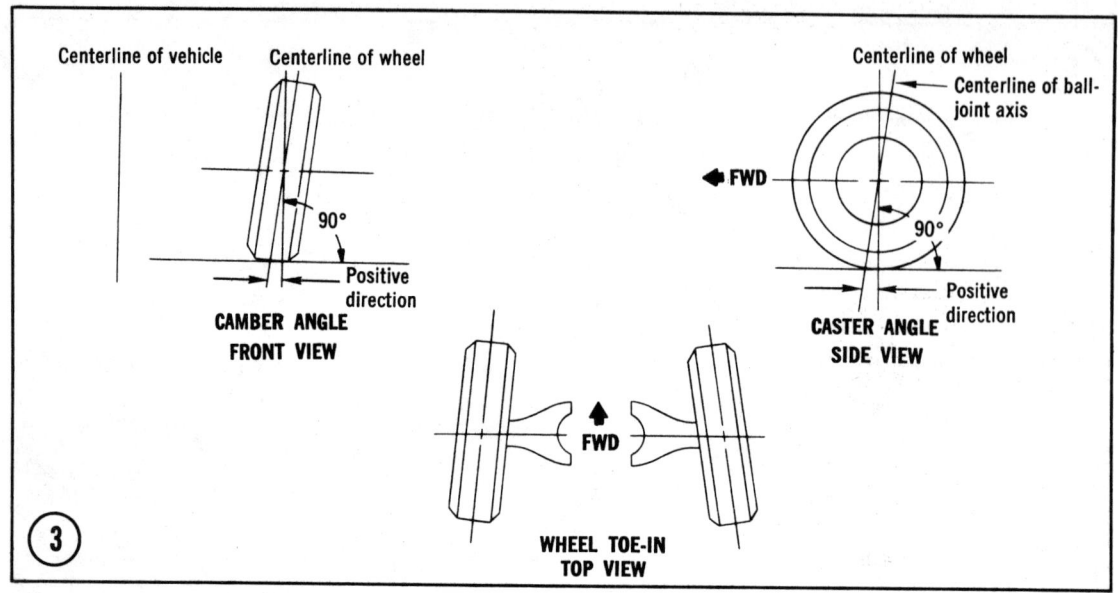

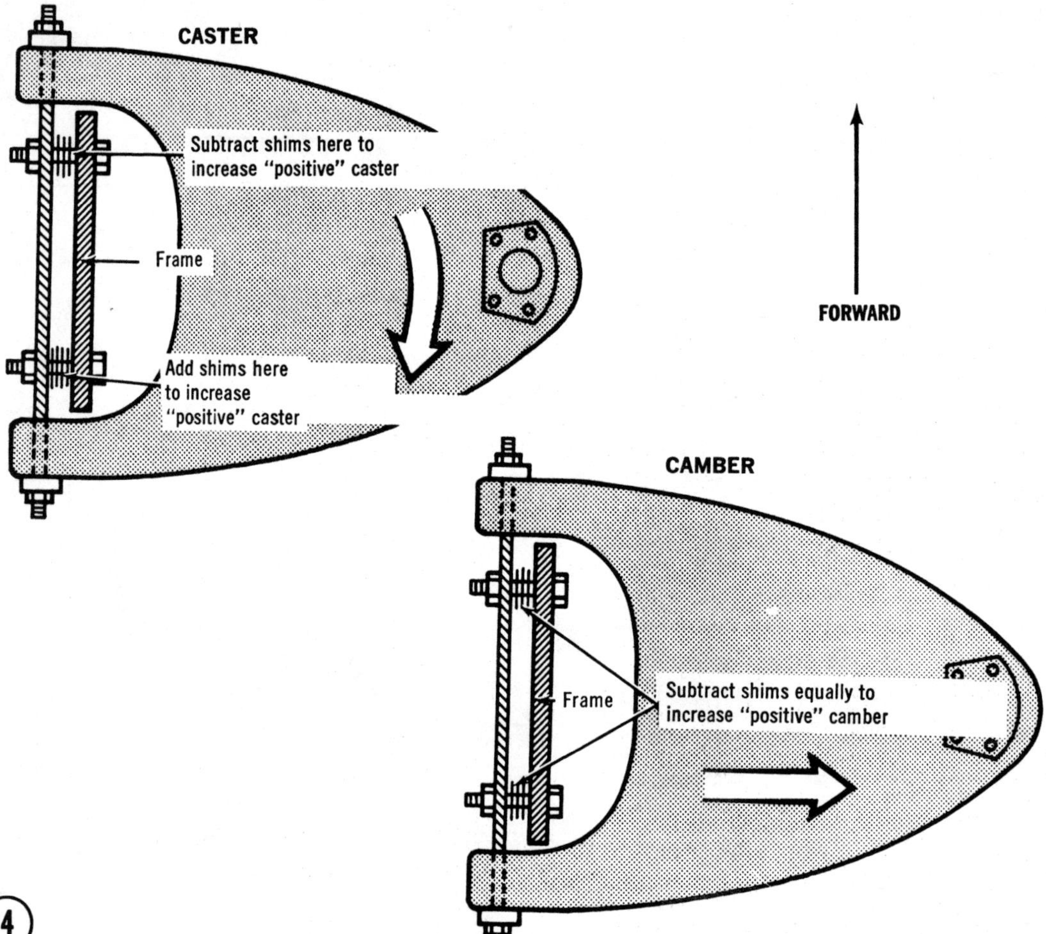

FRONT SUSPENSION AND STEERING

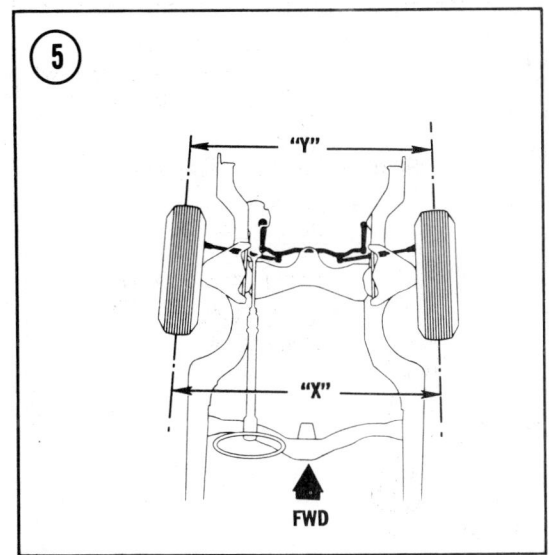

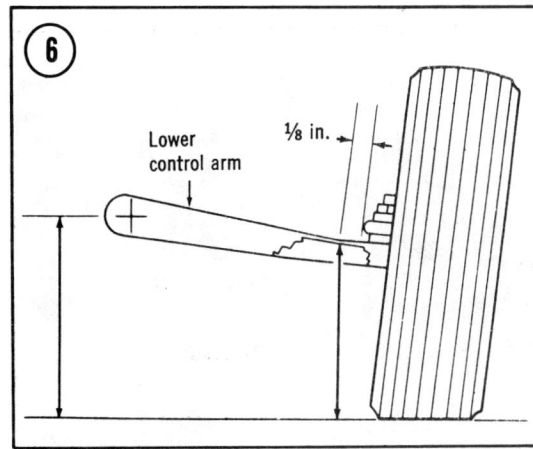

A proper inspection of front tire wear can point to several alignment problems. Tires worn primarily on one side show problems with toe-in. If toe-in is incorrect on one wheel, the car probably pulls to one side or the other. If toe-in is incorrect on both wheels, the car probably is hard to steer in either direction.

Incorrect camber may also cause tire wear on one side. Tire cupping (scalloped wear pattern) can result from worn shock absorbers, one wheel out of alignment, a bent spindle, or a combination of all. Tires worn in the middle, but not on the edges, or worn nearly even on both edges, but not in the middle, are probably over-inflated or under-inflated, respectively. These conditions are not caused by suspension misalignment.

Camber

Camber is the inclination of the wheel from vertical as shown in **Figure 3**. Note that camber angle is positive camber, i.e., the top of the tire inclines outward more than the bottom.

Camber is adjusted by adding or removing shims at both front and rear bolts of upper control arm shaft. See **Figure 4**.

Caster

Caster is the fore and aft inclination of the steering knuckle centerline from vertical. See **Figure 3**. The car has positive caster, i.e., the bottom of the wheel is shifted forward. Caster causes the wheel to return to a position straight ahead after a turn. It also prevents the car from wandering due to wind, potholes, or uneven road surfaces.

Caster is adjusted by adding or subtracting shims to the front bolt or the rear bolt of the upper control arm shaft. See **Figure 4**.

Steering Axis Inclination

Steering axis inclination, shown in **Figure 3**, is the inward inclination of the steering knuckle centerline from vertical. This angle is not adjustable, but can be checked with proper front end racks to find bent suspension parts.

Toe-in

Camber and rolling resistance tend to force the front wheels outward at their forward edge. To compensate for this tendency, the front edges are turned slightly inward when the car is at rest. This is toe-in. See **Figure 3**.

Toe-in is adjusted by lengthening or shortening the tie rods. Each tie rod is threaded so that the center section can be rotated to make the adjustment. See **Figure 5**.

Ride Height

Before making any other suspension adjustments, check ride height as follows.

1. Park car on smooth, level floor.
2. Bounce car several times to permit it to settle to normal height.
3. Measure and record distance from floor to center of the front inner pivot of lower arm. See **Figure 6**.

4. Measure and record distance from floor to lower face of lower steering knuckle boss. See **Figure 6**.

5. Record the difference in the two measurements.

6. Measure other side of vehicle in the same manner (repeat Steps 2 through 5).

7. The differences between the 2 sides must not exceed ½ in.

SHOCK ABSORBERS

Replacement

Refer to **Figure 7**.

1. Raise front of vehicle on hoist or jackstands.
2. Hold upper stem of shock absorber with a ¼ in. wrench. Remove upper nut, retainer, and grommet.
3. Disconnect lower shock mount from control arm. Pull shock out of arm.
4. Installation is the reverse of these steps. Tighten upper nut to 15-25 ft.-lb. Tighten lower bolts to 8-11 ft.-lb.

FRONT SPRINGS

WARNING
Due to the danger of working with a compressed spring, special tools must be used. Do not improvise.

Removal

1. Raise car on jackstands.
2. Remove wheel and tire.
3. Remove stabilizer bar and shock absorber.
4. Loosen the lower ball-joint to the steering knuckle nut.
5. Loosen the lower control arm cross shaft bushing bolts.

> NOTE: *As a safety precaution, a chain can be installed through spring and lower control arm as shown in **Figure 8**.*

6. Install tool J-23028 as shown in **Figure 9**.

> NOTE: *Place tool under lower control arm bushings so that control arm bushings seat in tool grooves.*

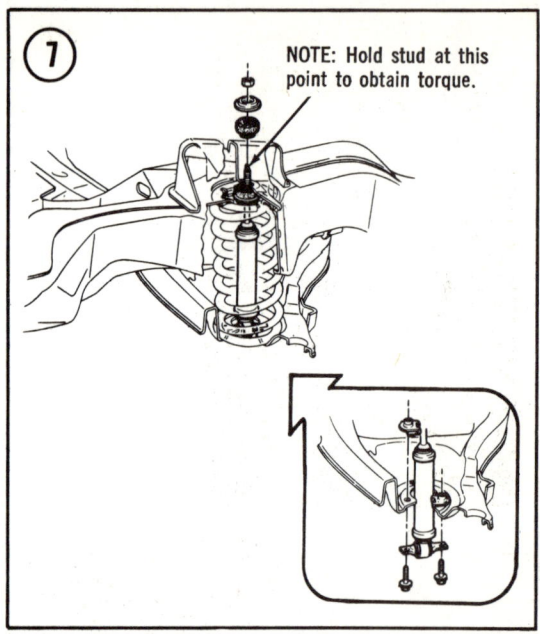

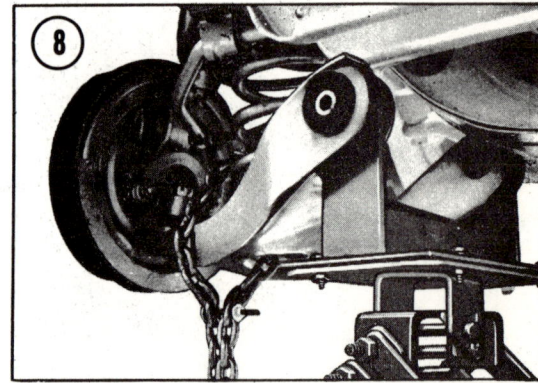

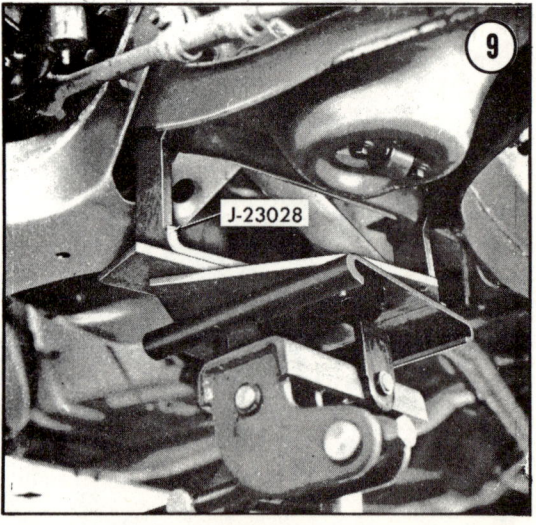

FRONT SUSPENSION AND STEERING

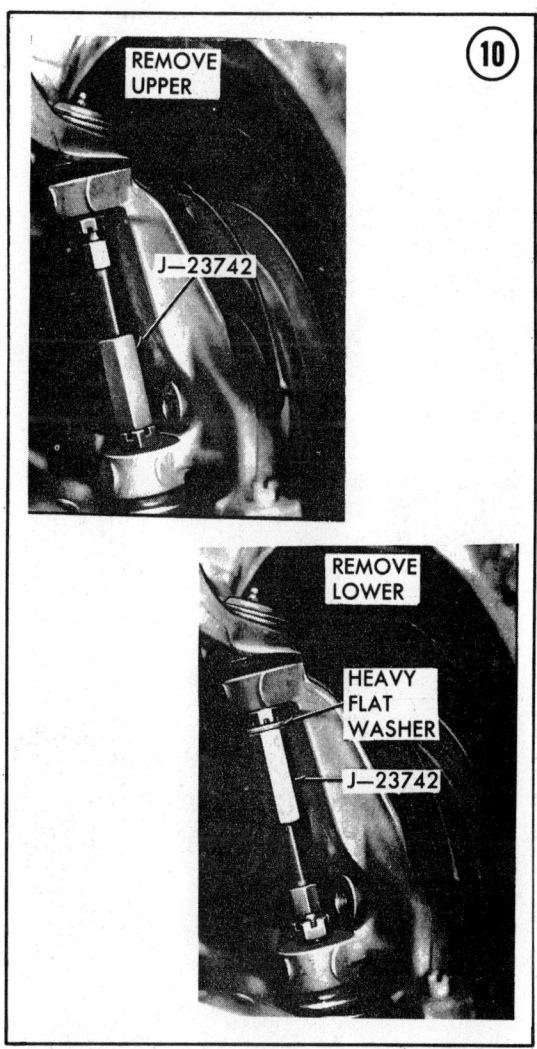

7. Raise jack to remove tension on lower control arm pivot bolts, then remove bolts.

8. Carefully lower control arm by slowly releasing jack. Remove chain, if installed, and remove spring.

Installation

1. Place spring on control arm. Lift control arm with special tool J-23028 and jack.
2. Position control arm to crossmember and install pivot bolts. Lower jack and remove tool.
3. Replace stabilizer bar link, shock absorber, and wheel and tire.
4. Lower vehicle and install upper shock absorber stem retainer nut, retainer, and grommet.

BALL-JOINTS

Replacement

1. Raise vehicle on jackstands and remove the wheels.
2. Remove cotter pin and loosen ball-joint stud nut about one turn.
3. Place jack under lower control arm. Jack the arm up just enough to support control arm.
4. Install tool J-23742-1 or equivalent between ball studs as shown in **Figure 10**. Turn threaded end of tool until stud is free of steering knuckle.
5. Remove ball-joint stud nut and disconnect steering knuckle from ball-joint.
6. Grind or chisel rivets off ball-joint assembly and remove ball-joint.
7. Check ball-joint mounting area on control arm for cracks or metal fatigue.
8. Drill rivet holes in control arm to diameter specified in ball-joint service kit.
9. Install new ball-joint with special bolts supplied in service kit. Insert bolts from bottom with nut on top. Torque bolts to 15-25 ft.-lb.
10. Turn ball stud cotter pin hole so it faces fore and aft.
11. Raise control arm and insert upper ball stud into steering knuckle. Install ball stud nut and torque to 50 ft.-lb. Secure with new cotter pin. If cotter pin hole does not line up, *tighten* nut additionally; do not *loosen* nut.
12. Connect stabilizer link to the control arm bracket. Torque to 9-12 ft.-lb.

STABILIZER BAR

Removal

Refer to **Figure 11** (typical) for these procedures.

1. Raise car on jackstands.
2. Disconnect both stabilizer links from control arms. Remove link bolt bushings and retainers.
3. Remove bolts securing stabilizer supports to frame. Bolt heads are accessible through holes in frames.
4. Remove stabilizer bar.

CHAPTER ELEVEN

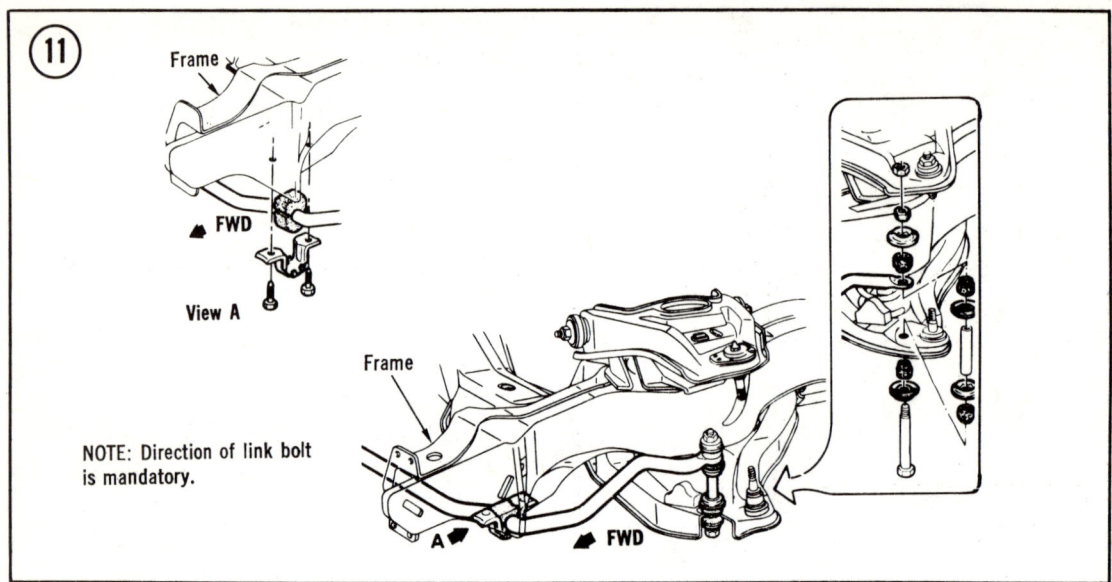

NOTE: Direction of link bolt is mandatory.

Installation

1. Secure stabilizer bar with brackets.
2. Assemble link bolts, bushings, and retainers. See **Figure 11**. Secure to control arms.
3. Lower car. Bounce front several times to settle rubber bushings.
4. Retighten all bolts.

STEERING KNUCKLE

Removal

1. Raise vehicle on jackstands. Support lower control arms with jacks so that spring stays compressed as if car were on floor.
2. Remove front wheels.
3a. On models with front drum brakes, remove brake drum, wheel hub, and brake shoes. Disconnect backing plate and steering knuckle. See Chapter Ten.
3b. On models with disc brakes, remove brake caliper and disc. See Chapter Ten.
4. Remove upper and lower ball-joint nuts.
5. Loosen steering knuckle from ball-joints by tapping knuckle bosses with hammer.
6. Installation is the reverse of these steps. Tighten ball-joint nuts to 75 ft.-lb. Use new cotter pins.
7. Adjust wheel bearings as described earlier.
8. Check wheel alignment.

FRONT WHEEL HUB (DRUM BRAKES)

Removal/Installation

1. Perform Steps 1 and 2 of *Wheel Bearing Replacement* procedure to remove hub.
2. Remove, clean, and repack bearings as described in Steps 3-5, 11, and 12, *Wheel Bearing Replacement*.
3. Reinstall wheel hub by following Steps 13-18, *Wheel Bearing Replacement*.

WHEEL BEARINGS

Adjustment

1. Raise front of vehicle until tires clear the ground.
2. Remove hub cap and dust cap. Remove cotter pin from spindle.
3. Tighten spindle nut to 15 ft.-lb. while rotating wheel.
4. Back off nut one flat ($\frac{1}{6}$ turn) and insert cotter pin. If slot and hole do not line up, back off an additional $\frac{1}{2}$ flat or less, as necessary.
5. Spin wheel to be sure it turns freely.
6. Install dust cap and hub cap.
7. Lower vehicle.

Replacement

1a. On models so equipped, remove brake drums as described in Chapter Eleven. Remove

FRONT SUSPENSION AND STEERING

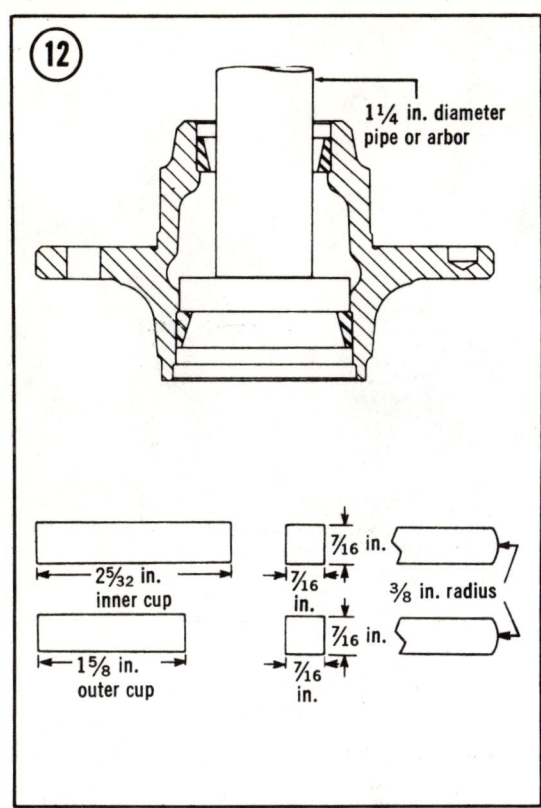

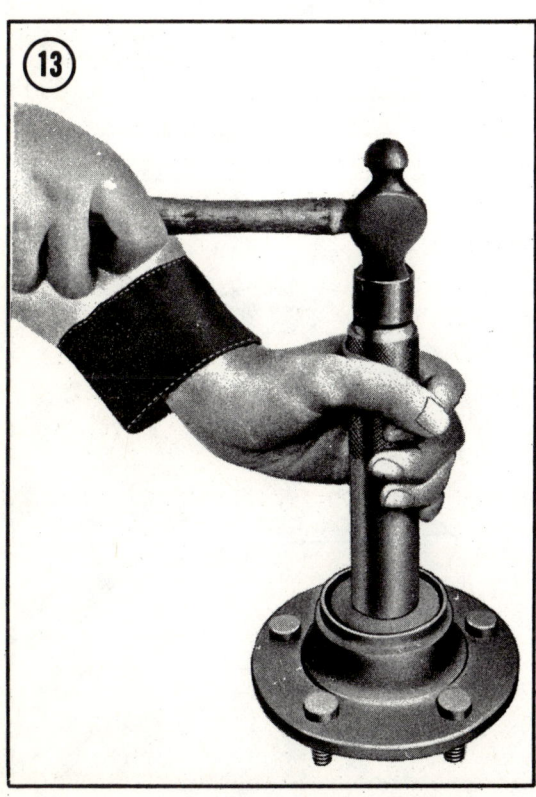

hub dust cap, cotter pin, spindle nut, and washer. Pull hub off.

1b. On disc brake models, remove brake caliper and disc.

2. Remove outer bearing from hub on the brake disc.

3. Pry out inner grease seal from hub on brake disc and remove inner bearing.

4. Clean wheel bearings in hub or brake disc in solvent. Blow dry. Check bearing rollers for scores, wear, and evidence of overheating (bluish tint). Check bearing races in hub or brake disc. Do not mix bearings up if they are good; they must be replaced on the same wheel assembly.

CAUTION
If a bearing race is damaged, the bearing and race must be replaced.

5. If bearings and races are good, reinstall by following Steps 11-17. Otherwise, proceed to Step 6.

6. Make 2 bearing race removers out of $7/16$ in. square steel stock as shown in **Figure 12**. A brass drift can be used, if care is taken.

7. Lay remover across race in hub (**Figure 12**). Ends fit in slots in hub shoulder.

8. Press race from hub with a pipe or rod which bears on the remover.

9. Press new races into hub using tool J-8849 (outer race), tool J-8850 (inner race), or equivalents. See **Figure 13**. Be sure that the races seat fully against hub shoulder and do not cock in seats.

10. Pack inner and outer bearings with high temperature wheel bearing grease. Do not pack cavity in hub with grease.

11. Fit inner bearing in hub or brake disc. Install new grease seal with flange facing bearing race.

12a. On drum brake models, install wheel hub over steering spindle.

12b. On disc brake models, install brake disc.

13. Install outer bearing. Press it in firmly by hand.

14. Install spindle washer and adjusting nut.

15. Adjust wheel bearings as described earlier.

16a. On drum brake models, install brake drum and wheel.

16b. On disc brake models, install brake caliper and wheel.

17. Lower vehicle.

STEERING ADJUSTMENTS

Before attempting to correct steering deficiencies with these adjustments, carefully check:

- a. Front end alignment
- b. Shock absorber condition
- c. Wheel balance
- d. Tire pressure

Manual Steering Gear Adjustment

1. Remove pitman arm nut. Mark relation of pitman arm on sector shaft.

2. Remove pitman arm as shown in **Figure 14**. Use tool J-6632, J-5504, or suitable puller.

3. Loosen lash adjuster locknut (**Figure 15**) and turn adjuster a few turns counterclockwise.

4. Turn steering wheel very gently and slowly in one direction until stopped by gear. Back away about one turn.

> CAUTION
> *Do not turn wheel hard against stops when relay rod is disconnected or ball guides in steering gear will be damaged.*

5. Using a torque wrench, measure pull required to keep wheel in motion. See **Figure 16**. Scale centerline must remain at right angles to wheel spoke. From 5-8 in.-lb. should be required; adjust worm bearings if force is higher or lower.

> NOTE: *Use a torque wrench with a maximum reading of 50 in.-lb.*

6. To adjust worm bearings, loosen worm adjuster locknut. See **Figure 15**. Turn adjuster until there is no appreciable end play in worm. Tighten locknut.

7. Repeat Steps 5 and 6 as necessary.

8. If gear feels rough after adjustment, bearings are probably damaged.

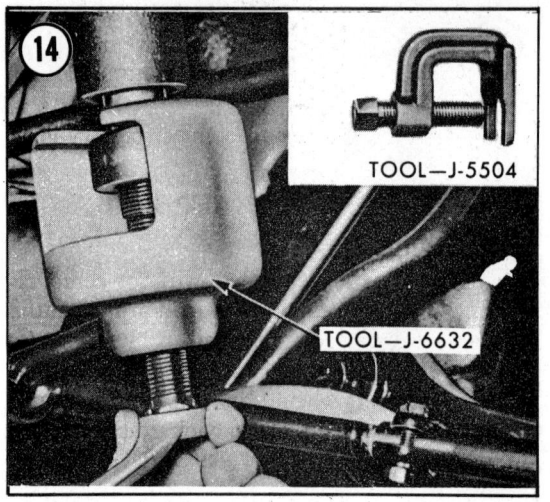

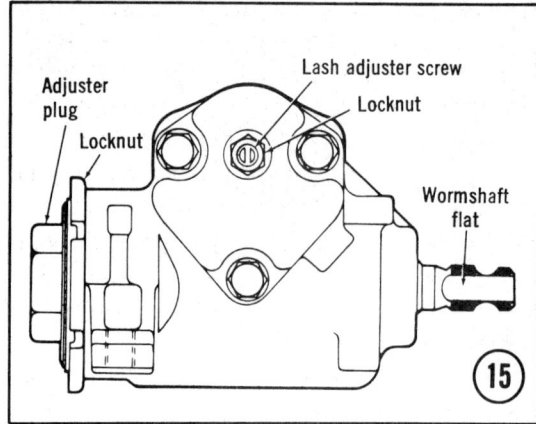

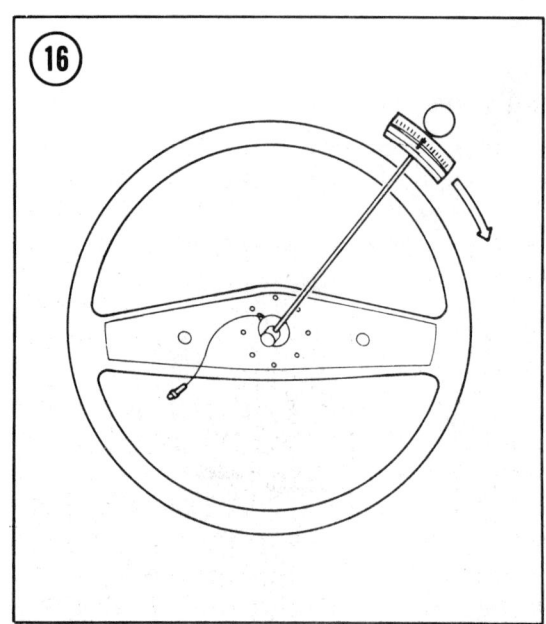

FRONT SUSPENSION AND STEERING

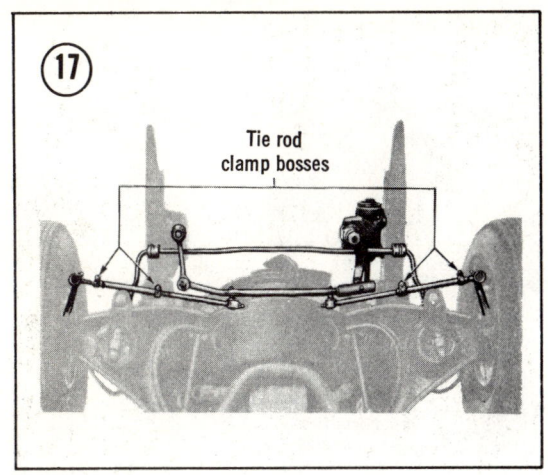

9. Count number of turns required to turn steering wheel from one stop to the other. Turn wheel back from stop exactly ½ the number of turns. Mark on top of worm shaft just below clamp should be at top of shaft and in line with the sawcut at the lower coupling clamp.

10. Turn lash adjuster screw clockwise to take out all lash in gear teeth. Tighten locknut.

11. Check force required to pull wheel through center position with spring scale. From ⅞-1½ lb. force should be required. Repeat Steps 9-11 as necessary to obtain correct force.

12. Reassemble pitman arm to sector shaft. Line up marks made during disassembly. Torque nut to 180-185 ft.-lb.

Steering Wheel Alignment

1. Drive the car onto any level smooth surface such as a driveway or parking lot. The front wheels must point straight ahead. One method to ensure this is to take advantage of wheel caster. Drive the car straight forward without touching the steering wheel. Stop the car with the handbrake. The wheels should stop straight ahead.

2. Mark on top of worm shaft just below clamp should be at top of shaft and in line with sawcut on lower coupling clamp.

3. If mark on worm shaft is not at top of shaft when wheels are straight ahead, loosen both clamps on left and right tie rods. See **Figure 17**. Turn each tie rod sleeve an equal amount in the *same direction* to bring mark to the top. Tighten clamps.

> NOTE: *If tie rod sleeves are turned unequally, toe-in will be changed and should be adjusted.*

4. If the steering wheel is not centered, remove it and reinstall so that it is centered.

Steering Wheel Removal/Installation

1. Disconnect the column harness at wiring connector.
2. Disconnect battery ground cable.
3. Pry off horn button cap.
4. Remove screws securing horn contact to spacer and hub.
5. Remove screws securing lock screw to lock knob and remove lock screw, lock knob and spacer.
6. Remove screws securing steering wheel to hub and remove wheel.
7. Installation is the reverse of these steps.

STEERING GEAR

Removal/Installation

1. Remove lower clamp bolt on steering coupling to worm shaft.
2. Raise vehicle on hoist or jackstands.
3. Remove pitman arm with suitable puller. See **Figure 14**.
4. Remove steering gear mounting bolts and remove steering gear.
5. Installation is the reverse of these steps. Be sure that wheels are straight ahead, steering wheel is straight, and steering gear is set to the middle of its travel.

Disassembly

Refer to **Figure 18** for this procedure.

1. Clean exterior thoroughly in solvent to remove road dirt and grease.
2. Cover *clean* work surface with clean paper or rags.
3. Hold gear in vise by lower mounting ear.

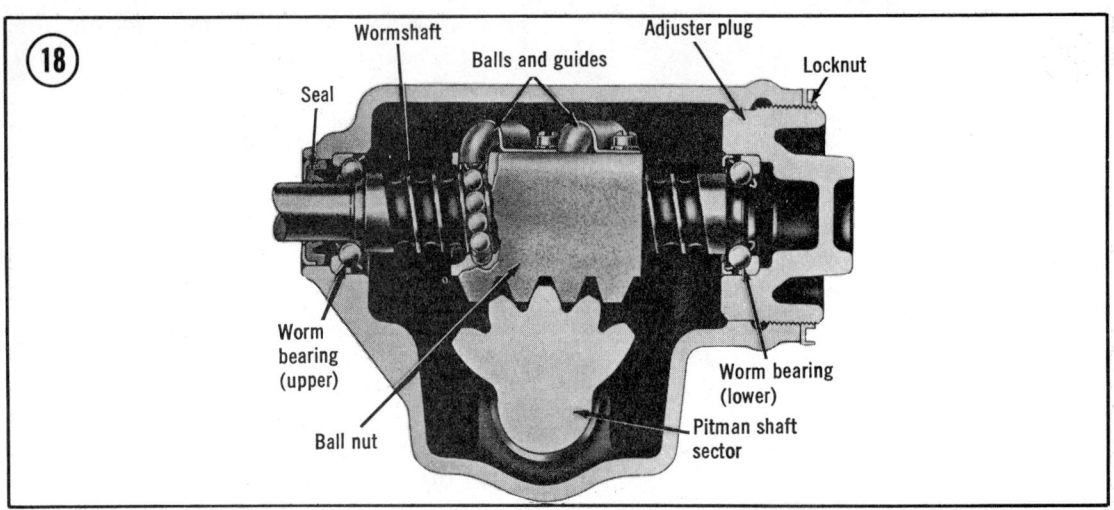

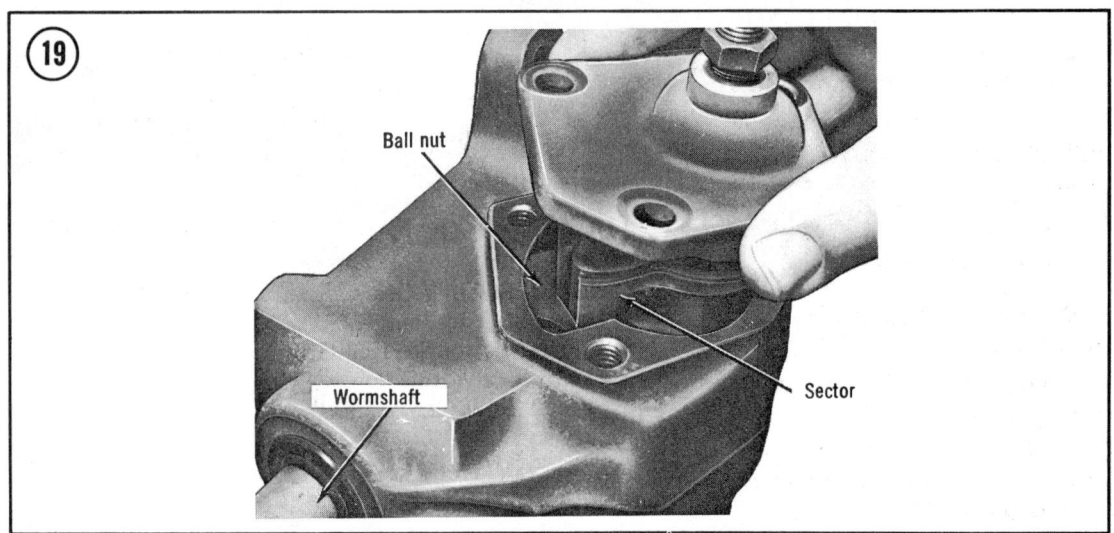

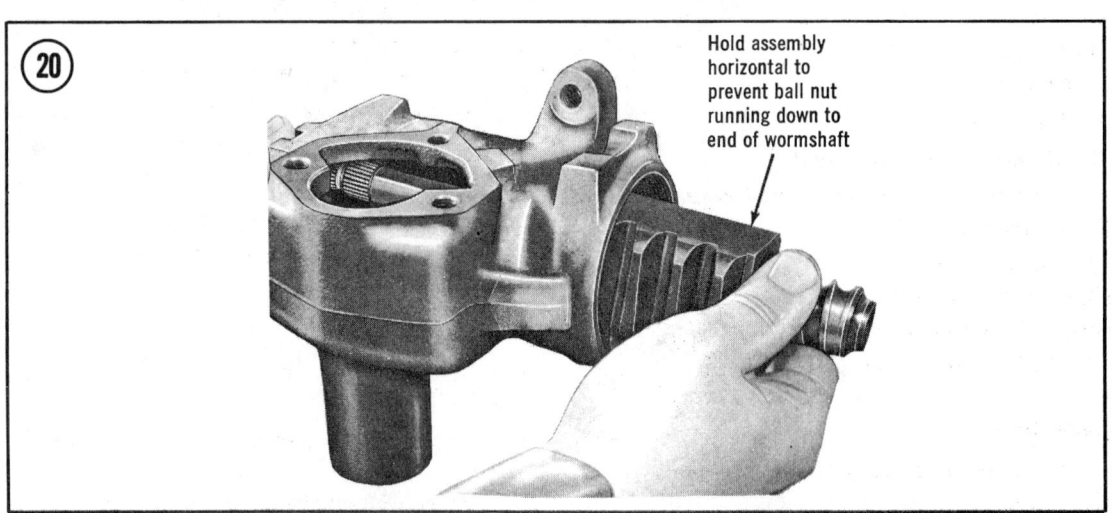

FRONT SUSPENSION AND STEERING

㉑

4. Loosen lash adjuster locknut and turn lash adjuster several turns counterclockwise.

5. Remove 3 cap screws holding side cover. Pull side cover/sector shaft assembly out. See **Figure 19**.

> NOTE: *Turn worm shaft if necessary until sector clears opening.*

6. Remove worm bearing adjuster and locknut from housing.

7. Remove worm shaft and ball nut assembly from housing. See **Figure 20**. Remove upper ball bearing from worm shaft. Remove lower bearing retainer and bearing from adjuster plug housing (**Figure 21**).

8. Remove locknut from lash adjuster and unscrew adjuster clockwise from side cover. Slide adjuster and shim out of slot in end of sector shaft.

9. Remove sector shaft seal.

10. Remove worm shaft seal.

11. Remove clamp retaining ball guides in nut. Draw guides out of nut.

12. Turn nut upside down and rotate worm shaft back and forth until all the balls have dropped out into a clean pan.

13. Pull nut off worm.

Inspection

1. Wash all parts in solvent. Dry them with clean rags.

2. Check ball bearings, races, worm and nut grooves for signs of wear and scoring.

3. Inspect sector shaft and bushings for wear and check the fit of the shaft in the housing bushings.

4. Check ball guides for damaged ends. Replace if necessary.

5. Check worm shaft assembly for damage.

Rebuilding and Assembly

Refer to **Figure 22**.

> NOTE: *Bushings require replacement only if worn or damaged. Consider dealer replacement with proper tools.*

1. Press the pitman shaft bushing out of housing. See **Figure 23**.

2. Coat new bushing with steering gear lube.

3. Press new pitman shaft bushing into end of bore so that bushing is flush with, or slightly lower, than end of bore.

> NOTE: *Worm shaft seal and sector shaft seal must be replaced anytime gear is assembled.*

4. Pry out old seals.

5. Coat new seals with steering gear lube inside and out.

6. Position new worm shaft seal over center hole in worm bearing adjuster and press in using socket of suitable outside diameter. Seal

318   CHAPTER ELEVEN

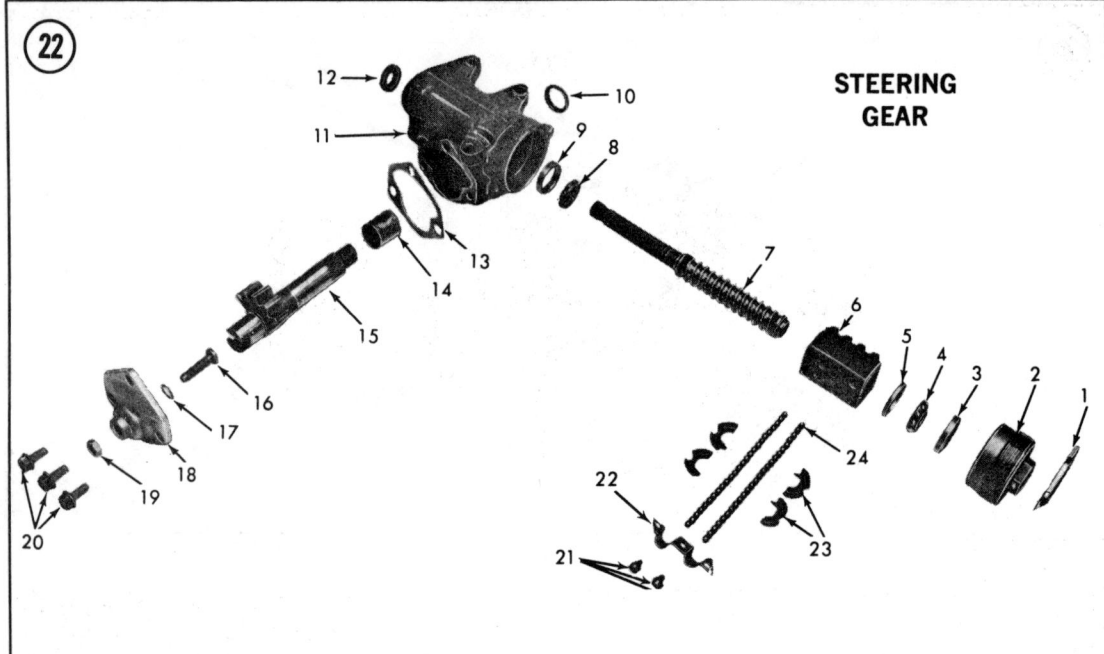

1. Worm bearing adjuster locknut
2. Worm bearing adjuster
3. Lower worm bearing race
4. Lower ball bearing
5. Lower bearing retainer
6. Ball nut
7. Wormshaft
8. Upper ball bearing
9. Upper wormshaft bearing race
10. Pitman shaft seal
11. Housing
12. Wormshaft seal
13. Side cover gasket
14. Pitman shaft bushing
15. Pitman shaft
16. Lash adjuster
17. Lash adjuster shim
18. Housing side cover and bushing assembly
19. Lash adjuster locknut
20. Side cover bolts
21. Ball guide clamp screws
22. Ball guide clamp
23. Ball guides
24. Balls

edge should be flush with inner edge of small bore adjuster.

7. Position new pitman shaft seal over pitman shaft bore. Press in with the same method used for worm shaft seal.

> NOTE: *Worm shaft bearing races require replacement only if worm is damaged. Consider dealer replacement with proper tools.*

8. Place steering gear housing or worm adjuster in vise with bore horizontal.

9. Pull race out with suitable slide hammer and accessories. See **Figure 24**.

10. Press new bearing races in with tool J-5755. See **Figure 25**.

> NOTE: *Steps 11 through 20 assemble ball nut on worm shaft.*

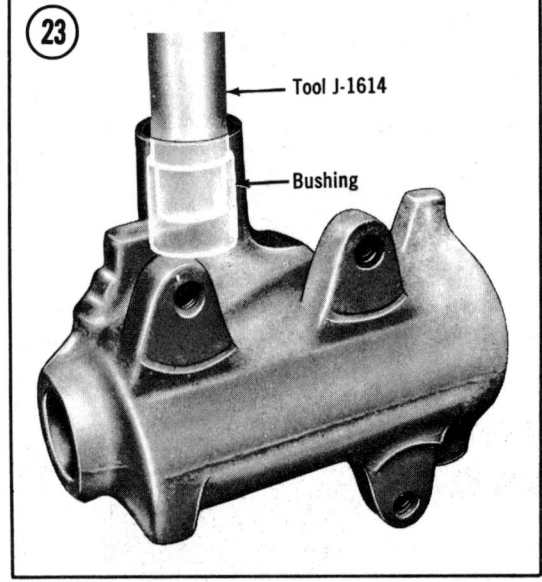

FRONT SUSPENSION AND STEERING

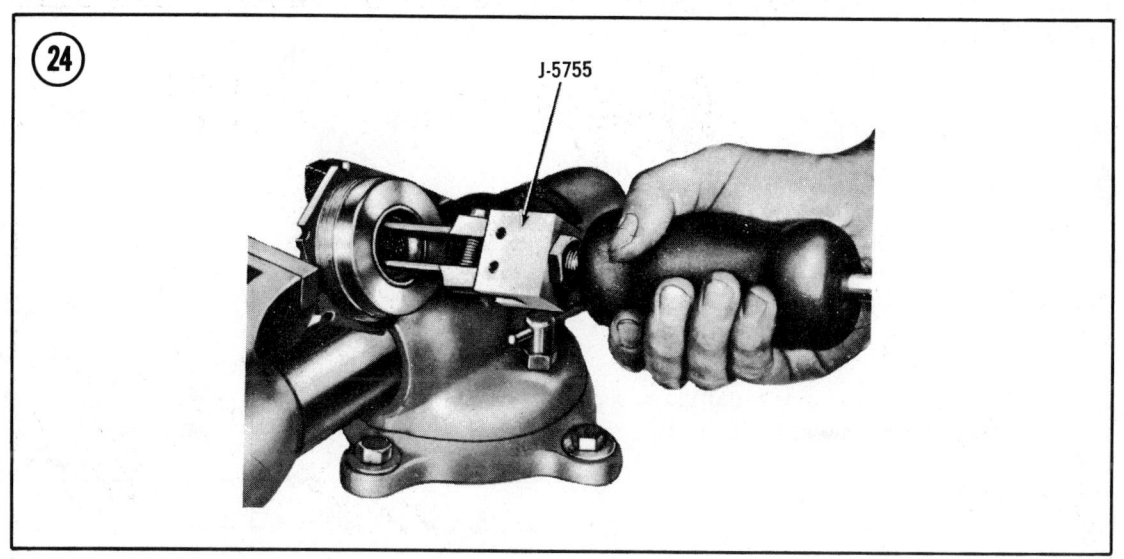

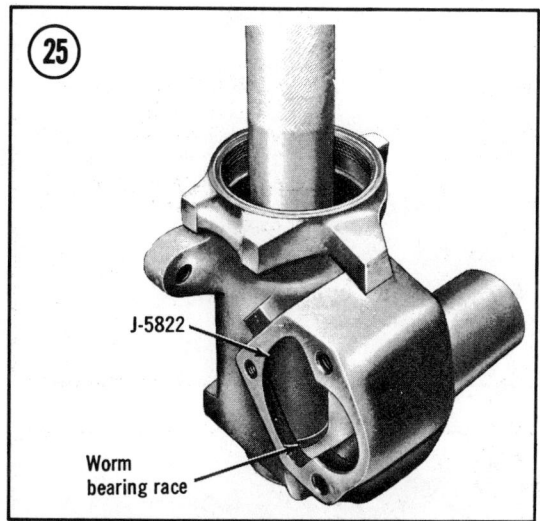

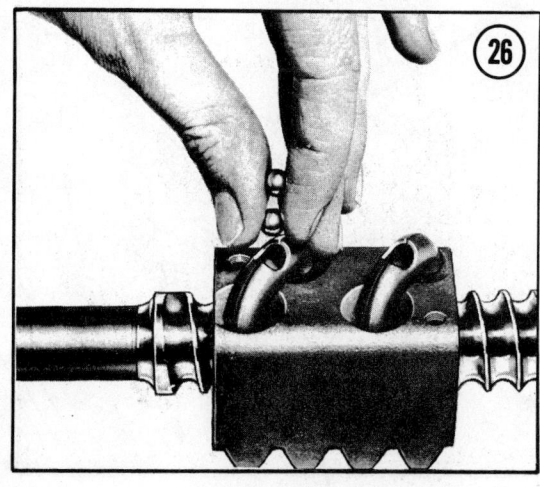

11. Lay worm shaft flat on bench. Slip ball nut over worm with ball guide holes up and the shallow end of rack teeth to the left as viewed from the steering wheel position. See **Figure 26**. Insert ball guides as shown.

12. Count 24 balls into a clean shallow container like a pie tin. Each circuit requires 24 steel balls.

13. Align grooves in the worm and nut by sighting through the ball guide holes.

14. Insert balls for one circuit into one of the guide holes one at a time while gradually turning the worm away from that hole. Continue until all 24 balls have been inserted.

15. Fill second circuit by repeating Steps 12 through 14.

16. Mount ball guide clamp on nut. Use lockwasher under clamp screw.

17. Rotate nut on worm to be sure it rotates freely. Do not bottom nut on either end of worm. If nut binds slightly, ball guide ends are probably damaged and require replacement.

18. Hold steering gear housing in vise with the worm shaft bore vertical and the side cover opening up.

19. Slip upper ball bearing over worm shaft and install worm shaft and nut assembly into housing. Feed shaft end through ball bearing race and seal.

20. Place a bearing in adjuster race and press the retainer into place with a suitable socket.

21. Install locknut and adjuster into lower end of housing until nearly all end play is removed from worm shaft. Take care while guiding end of worm shaft into bearing.

22. Assemble lash adjuster with shim in slot in end of sector shaft.

23. Check end clearance as shown in **Figure 27**. End clearance should not exceed 0.002 in. Change shim installed in Step 22 if necessary. Shims are available 0.063-0.069 in. thick only in 0.002 in. increments.

24. Install gasket on side cover and insert sector shaft into housing. Do not damage seal. Be sure to index center tooth of sector in center space of ball nut rack.

25. Align holes in side cover with holes in housing. Install upper cap screw.

26. Hold steering gear housing in vise in its approximate position when installed in car.

27. Inject steering gear lubricant into lowest side cover cap screw opening until lubricant appears in another opening. Install 2 remaining cap screws and lockwashers.

Preliminary Adjustment

Worm bearing and lash adjustments may be made before gear is installed in car.

1. Hold steering gear in vise in its approximate position when installed in car.

2. Temporarily install steering wheel on worm shaft.

3. Perform Steps 5-7, *Steering Gear Adjustment* procedure, described earlier.

STEERING LINKAGE

Tie Rod Replacement

Refer to **Figure 28**.

1. Raise car on jackstands and remove front wheels.

2. Remove cotter pins and nuts from ball studs.

3. Hold heavy hammer as backing. Tap steering arm with another hammer to loosen ball stud. See **Figure 29**.

4. Remove inner ball stud in a similar manner.

5. Loosen clamp bolts and unscrew tie rod ends.

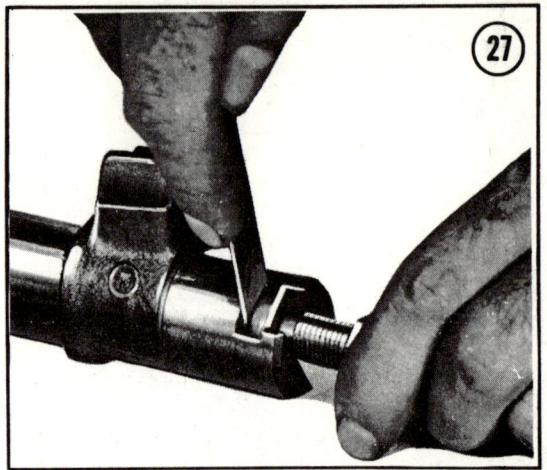

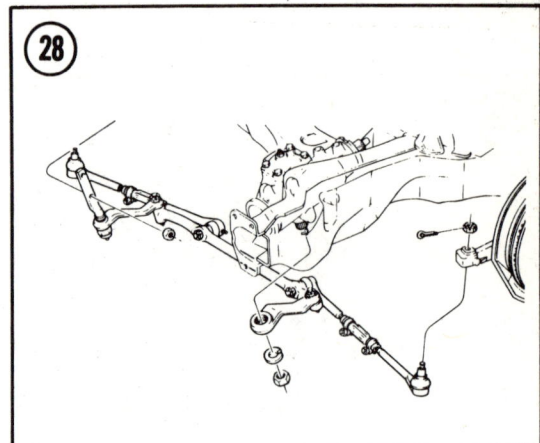

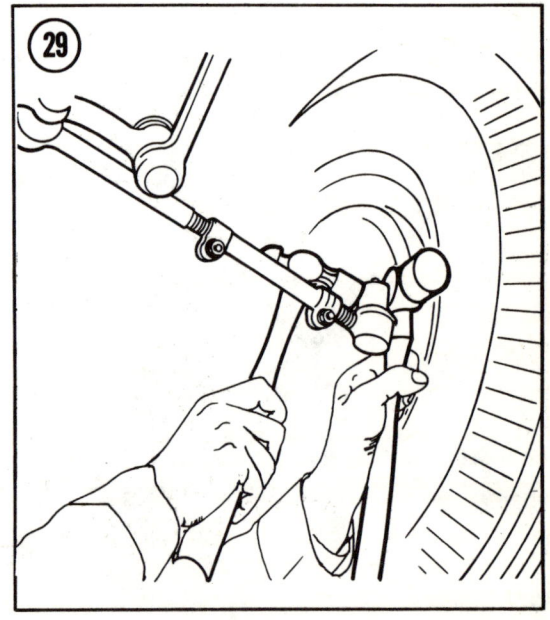

FRONT SUSPENSION AND STEERING

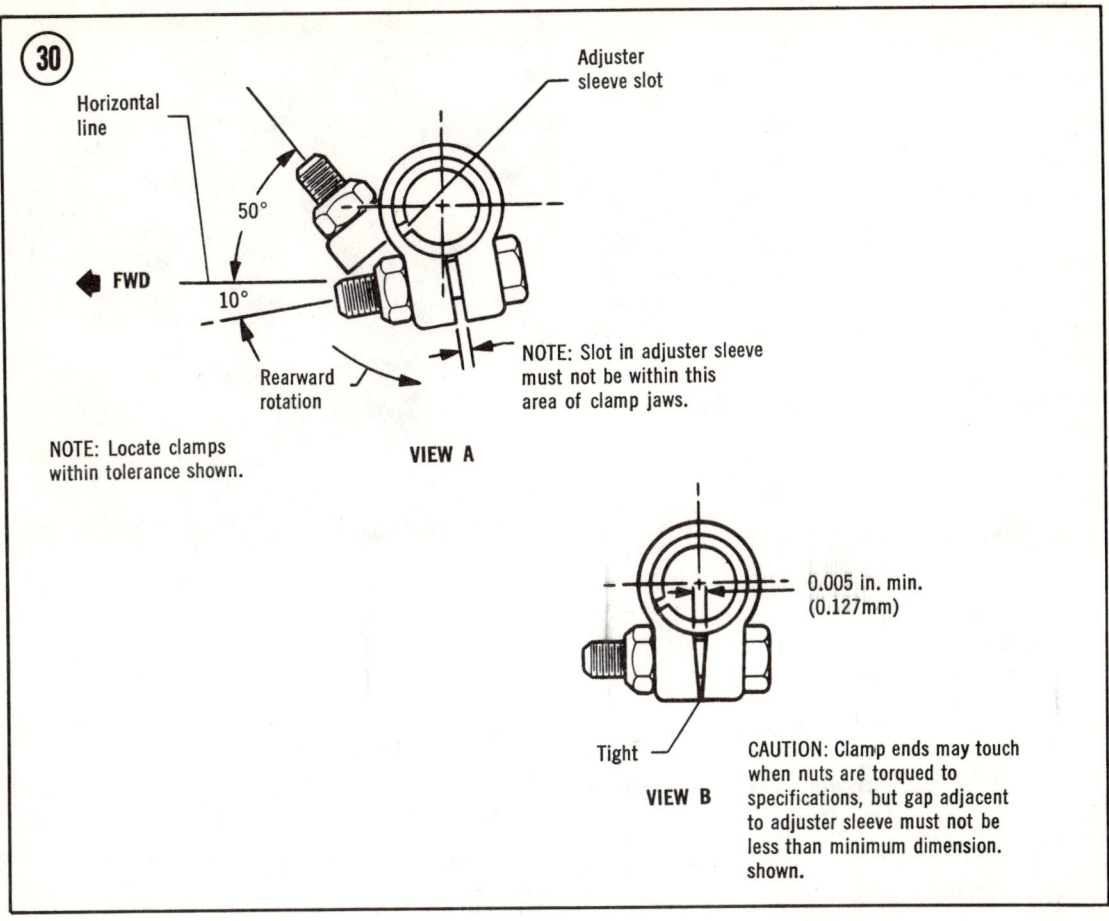

6. Install tie rod ends on tie rod. Make sure both ends are threaded on an equal amount.

7. Make sure that ball stud threads and nut threads are perfectly clean and smooth.

8. Install neoprene seals on ball studs.

9. Install ball studs in the steering arm and the relay rod.

10. Install ball stud nuts. Torque to 45 ft.-lb. Tighten additionally if necessary to align cotter pin holes; do not loosen nut.

11. Install new cotter pin.

12. Lubricate tie rod ends. See Chapter Three.

13. Adjust toe-in.

CAUTION
Before locking clamp bolts, make sure that tie rod ends are aligned with their ball studs. Otherwise, binding will result. See Figure 30 for alignment.

Steering Damper Replacement

1. Raise car on jackstands and remove front wheels.

2. Remove bolt securing damper pivot bracket at relay rod.

3. Remove nut from damper pivot at frame bracket.

4. Remove damper.

5. Installation is the reverse of these steps.

Relay Rod Replacement

See **Figure 28**.

1. Raise car on jackstands.

2. Remove steering damper.

3. Remove anchor bracket from relay rod.

4. Disconnect inner ends of tie rods from relay rod. See Steps 3 and 4, *Tie Rod Replacement*.

5. Remove cotter pin and nut on ball stud at pitman arm.

6. Detach relay rod from pitman arm. Shift linkage as required to free rod from arm.

7. Remove cotter pin and nut from idler arm.

8. Disconnect relay rod from idler arm.

9. Remove washer and seal from idler arm.

10. Clean relay rod assembly with solvent.

11. Place relay rod on idler arm stud. Make sure that stud seal and washer are in place. Install nut and tighten to 45 ft.-lb. Tighten nut additionally if necessary to align cotter pin holes.

12. Install new seal and clamp over ball at end of pitman arm.

13. Install inner spring seat and spring to relay rod.

14. Install end of rod on pitman arm.

15. Install spring seat, spring, and end plug.

16. Tighten end plug until springs are compressed and the plug bottoms out. Back off ¾ turn (slightly more if necessary) to align cotter pin holes. Install cotter pin.

17. Connect tie rod ends to relay rod.

18. Lubricate tie rod ends and pitman arm balljoint. See Chapter Three.

19. Install steering damper.

20. Adjust toe-in.

Idler Arm Removal/Installation

Refer to **Figure 28**.

1. Raise car on jackstands.

2. Remove cotter pin and nut from idler arm.

3. Disconnect relay rod from idler arm.

4. Remove idler arm from frame.

5. Remove stud seals and clean idler arm in solvent.

6. Check for damage and wear. Make sure studs turn smoothly. A grating noise indicates dirt within unit.

7. Hold idler arm in a vise and check torque required to rotate idler shaft. Torque should be from 30-55 in.-lb.

8. Attach nut to ball stud and check torque required to rotate the ball stud. The torque should be 10-35 in.-lb. Replace idler arm if torque values are not within specifications.

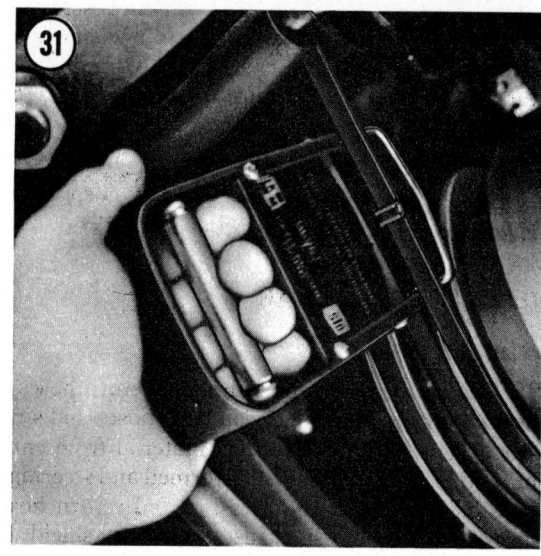

9. Mount idler arm on frame. Torque the bolts to 25-35 ft.-lb.

10. Connect relay rod to idler arm. Use new seal and tighten nut to 45 ft.-lb.

11. Install cotter pin. Tighten nut additionally if necessary to align cotter pin holes.

POWER STEERING PUMP

Belt Tension Adjustment

See **Figure 31**.

1. Loosen the power steering bracket-to-power steering pump attaching bolts.

2. Move pump, with belt in place, until belt tension is correct. This occurs when a 15 lb. force applied at the midpoint between the power steering pump pulley and the drive pulley cause ½-¾ in. belt deflection.

3. Tighten pump mounting bolts.

Removal

1. Disconnect hoses at pump. Secure ends to prevent drainage of oil. Cap or tape the ends of the hose to prevent entry of dirt.

2. Install 2 caps at pump fittings to prevent drainage of oil from pump.

3. Loosen bracket-to-pump mounting nuts.

4. Remove pump belt.

5. Remove bracket-to-pump bolts and remove pump from vehicle.

FRONT SUSPENSION AND STEERING

6. Remove drive pulley attaching nut.
7. Slide pulley from shaft.

> **CAUTION**
> *Do not hammer pulley off shaft as this will damage the pump.*

Installation

1. Slide pulley on shaft.

> **CAUTION**
> *Do not hammer pulley on shaft.*

2. Install pulley nut and torque to 35-45 ft.-lb. Use a new pulley nut.

3. Position pump assembly on vehicle and install nuts loosely.
4. Connect and tighten hose fittings.
5. Fill reservoir. Bleed pump by turning pulley backward (counterclockwise as viewed from front) until air bubbles cease to appear.
6. Install pump belt over pulley.
7. Adjust belt tension as described earlier.
8. Bleed the hydraulic system by raising the front wheels, starting the engine, and turning the steering wheel from stop to stop several times. Lower car, test operation, and repeat the procedure if the unit is not operating properly.

CHAPTER TWELVE

REAR SUSPENSION AND DRIVE SHAFT

In this rear suspension system the rear axle assembly is connected to the frame by a lower control arm and an upper control arm on each side. Individual coil springs at each wheel are located between brackets on the axle tube and spring seats in the frame. Double acting shock absorbers provide ride control.

The rubber-bushed lower control arms maintain fore and aft relationships between the axle and the chassis, while the upper control arms, which are shorter than the lower ones, control side sway and maintain the pinion nose angle. A typical rear suspension system is shown in **Figure 1**.

Two types of drive shafts have been used. The type used on your car can be recognized by inspecting the universal joints. The Cleveland drive shaft uses snap rings in the outboard ends of the yokes (see **Figure 2**) to retain the bearing caps. The Saginaw drive shaft universal joint bearing caps are retained by nylon material which is injected during manufacture (see **Figure 3**).

In this type of suspension system, the relationship of the angles of the front and rear universal joints is very important. If these angles are not almost identical, rough operation and vibration can result. The front angle is the angle between the engine-transmission center

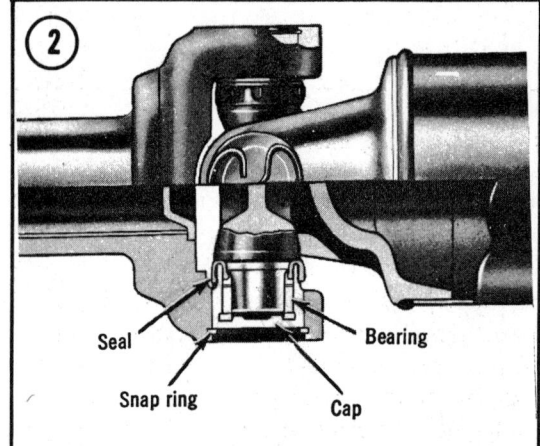

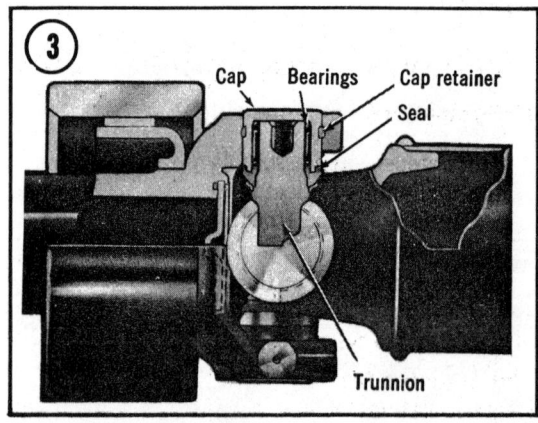

REAR SUSPENSION AND DRIVE SHAFT

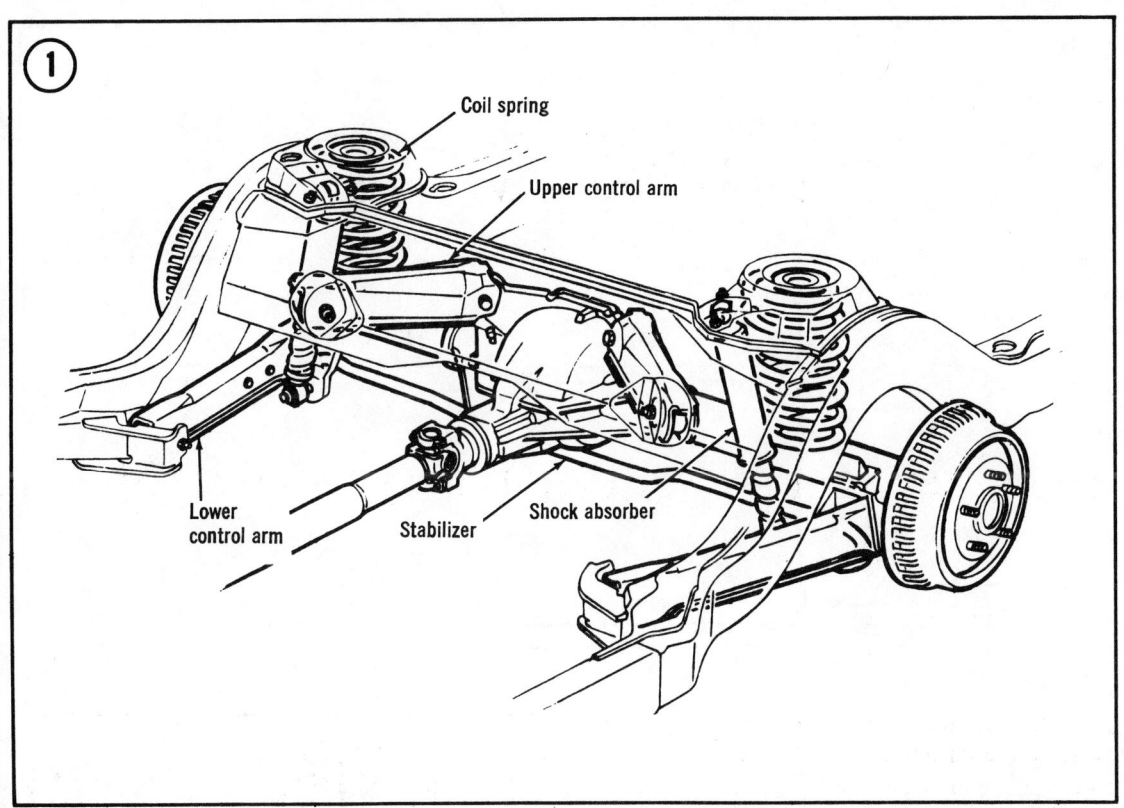

line and the drive shaft, and can be adjusted by the addition or removal of shims between the transmission rear bearing retainer and the transmission mount. The rear angle is determined by the length of the upper control arms, and the angle is adjusted by installing different length upper control arms. Special tools and skills are required to make these adjustments properly. If the procedures provided in this chapter do not clear up vibration and rough operation problems, the car should be taken to your dealer.

REAR SUSPENSION

Coil Spring Removal/Installation

NOTE: *Springs must be removed and installed one side at a time.*

1. Raise the rear of the vehicle and support at the rear lifting points (see **Figure 4**) with jackstands. Make sure the rear wheels clear the floor by several inches.

2. Support the rear axle with a jack or other adjustable lifting device.

3. Disconnect both upper control arms at the axle. If the car has a stabilizer bar, disconnect it at the side of the control arm.

4. Remove the brake hose support, if necessary, to allow for the drop when the spring is removed. It is not necessary to disconnect the brake hose.

5. Remove the shock absorber lower attachments, then slowly lower the jack under the axle until the spring can be removed. Remove the spring and insulator.

6. Install the insulator on the spring and position the spring and insulator on the axle. See **Figure 5** (typical). Upper end of coil must face as shown (within the limits listed). Left side is shown.

7. Raise the axle and reinstall the lower attachments of the shock absorber. Tighten the nut to 62-65 ft.-lb.

8. Connect the brake hose support and securely tighten the bolt.

9. Connect the stabilizer bar, if so equipped, and tighten the nut to 35 ft.-lb.

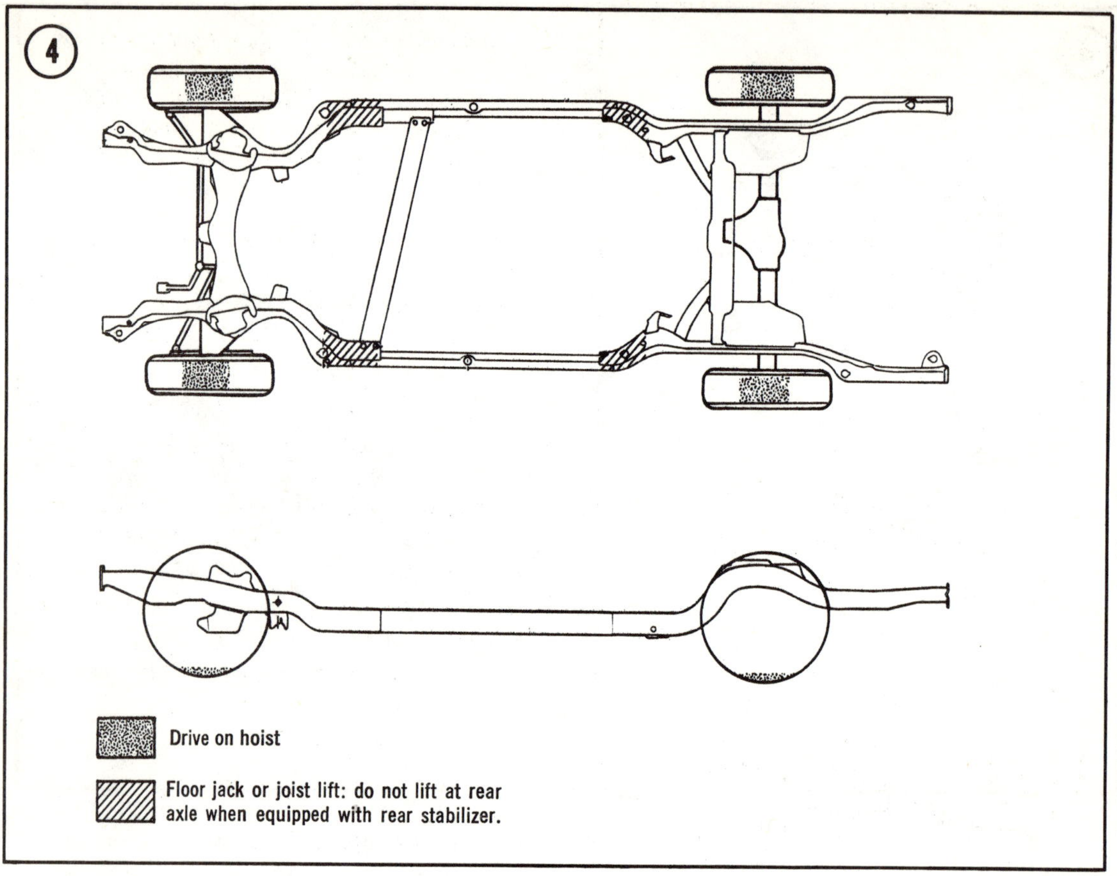

![Figure 4 - Lifting points diagram]

Drive on hoist

Floor jack or joist lift: do not lift at rear axle when equipped with rear stabilizer.

10. Connect the upper control arms to the axle and tighten the bolts to 80 ft.-lb.

11. Remove the jack from the axle and lower the rear of the vehicle.

Shock Absorber Removal/Installation

1. Raise the vehicle and place jackstands under the rear lifting points (see **Figure 4**). Make sure the rear wheels clear the floor by several inches.

2. If the vehicle is equipped with superlift shock absorbers, disconnect the air line from the shock absorber. A snap-on connector is used. See **Figure 6**.

3. Support the rear axle with a jack and remove the 2 bolts attaching the shock absorber to the upper mounting bracket. Removal of the rear wheels, while not necessary, will make this task easier.

4. Disconnect the shock absorber at the lower mounting point by holding the hex head on the end of the mounting stud and removing the nut.

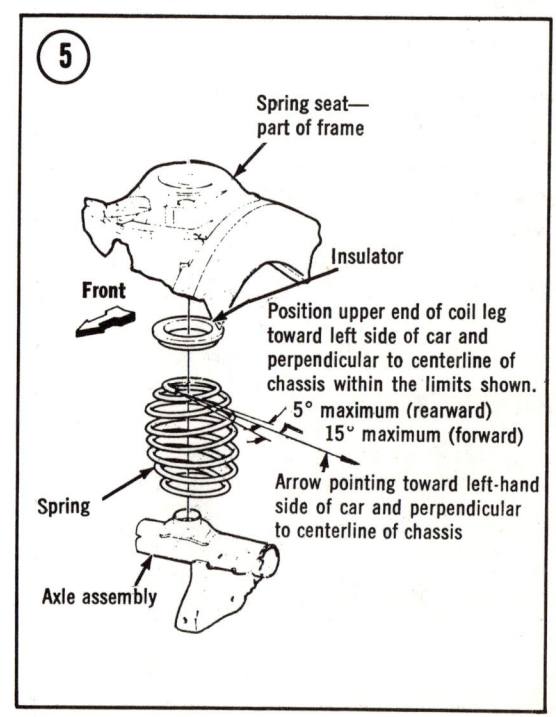

Rear Suspension and Drive Shaft

CAUTION
Failure to hold the hex head of the shock absorber lower mounting stud during removal or installation of the shock absorber can result in damage to the bond between the stud and the shock absorber bushing.

5. Loosely install the shock absorber to the upper mounting bracket with the attaching bolts (see **Figure 7**) and then position the lower mounting stud in the lower bracket and loosely install the washer and nut.

6. Tighten the upper attaching bolts to 20 ft.-lb. or upper nuts to 12 ft.-lb. (20 ft.-lb. on superlift shocks).

7. Hold the hex on the end of the lower mounting stud (see CAUTION above) and tighten the nut to 62-65 ft.-lb.

8. If superlift shocks are being installed, connect the air line to the shock absorber. Then add air to obtain a minimum of 10 psi to avoid damage to the shock absorber.

9. Lower the vehicle.

Upper or Lower Control Arm Removal/Installation

NOTE: *If both upper and lower control arms are to be removed, first remove the coil spring as described earlier in this chapter.*

1. Raise the rear of the vehicle and support it with jackstands under the rear body lifting points. See **Figure 4**.

2. Use a jack to raise the axle until tension is removed from the control arm being removed.

3. Remove the attaching bolts and remove the control arm. See **Figure 8**.

4. Installation is the reverse of these steps. Tighten nuts to 70-73 ft.-lb.

Control Arm Bushing Replacement

Removal and installation of the control arm bushings require the use of special tools and an arbor press. Remove the control arm(s) and take them to your Chevrolet dealer for bushing replacement when required.

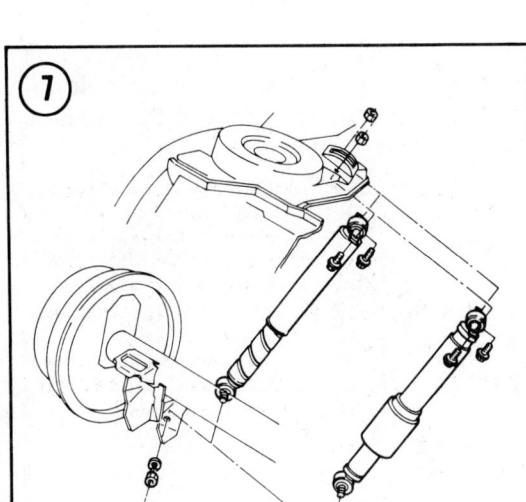

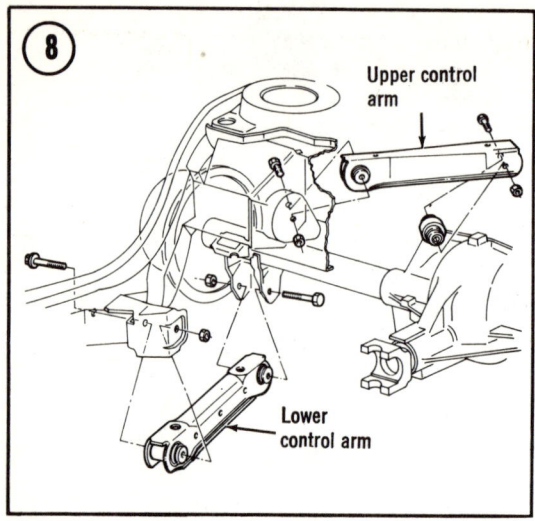

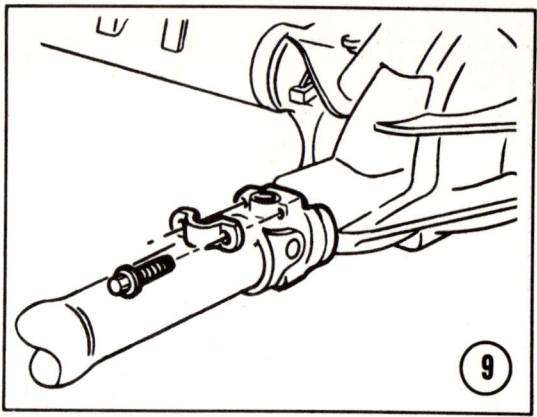

Drive Shaft Removal/Installation

1. Raise the rear of the vehicle and support it on jackstands at the rear body lifting points. See **Figure 4**.

2. Make matching marks to show the relationship of the drive shaft to the companion flange at the rear universal joint. This will permit installation of the drive shaft in its original position.

3. Remove the trunion bearing straps (see **Figure 9**) and tape the bearing cups to the trunion to prevent their loss.

4. Withdraw the drive shaft front yoke carefully from the transmission by pulling the shaft to the rear. Check for oil leakage at the transmission output shaft housing. Make sure the yoke seal is in good condition.

5. Installation is the reverse of these steps. Lubricate the yoke splines with a light coat of transmission oil before inserting the yoke. Make sure the alignment marks made during disassembly are lined up, then (after removing tape from the bearing cups) install the retaining straps and tighten the attaching bolts to 70 ft.-lb.

Cleveland Universal Joint Disassembly/Assembly

Refer to **Figure 10**.

> NOTE: *While these universal joints do not require periodic lubrication, they should be repacked with wheel bearing grease whenever they are disassembled and reassembled. Take care not to loosen or damage dust seals. The entire universal joint must be replaced if the dust seal becomes loose or damaged.*

1. Remove the bearing snap rings (6, **Figure 10**) from the trunion yoke.

> NOTE: *The following step calls for the use of an arbor press. If this tool is not available, a wide-jaw bench vise can be substituted.*

2. Support the yoke on a short piece of 1¼ in. ID pipe (see **Figure 11**) and use a suitable size socket or rod to press on the trunnion until the bearing cup is almost pressed out. Then grasp the cup and work it out of the yoke (the bearing cup cannot be fully pressed out).

3. Turn the yoke and tools over and press on the other end of the trunnion to remove the opposite bearing cup. Then remove the trunnion from the yoke.

4. Clean and inspect all parts. Check the seals (2, **Figure 10**) for looseness and damage and replace the trunnion assembly if either condition is present. Check rollers for overheating (blueness) and wear. Check trunnion for damage.

5. Pack the bearings and cups with wheel bearing grease. Also pack the reservoirs in the ends of the trunnion arms.

6. Position the trunnion in the yoke. Partially install a bearing cup into the yoke and start the trunnion into the cup. Partially install the other cup, then start the other end of the trunnion

REAR SUSPENSION AND DRIVE SHAFT

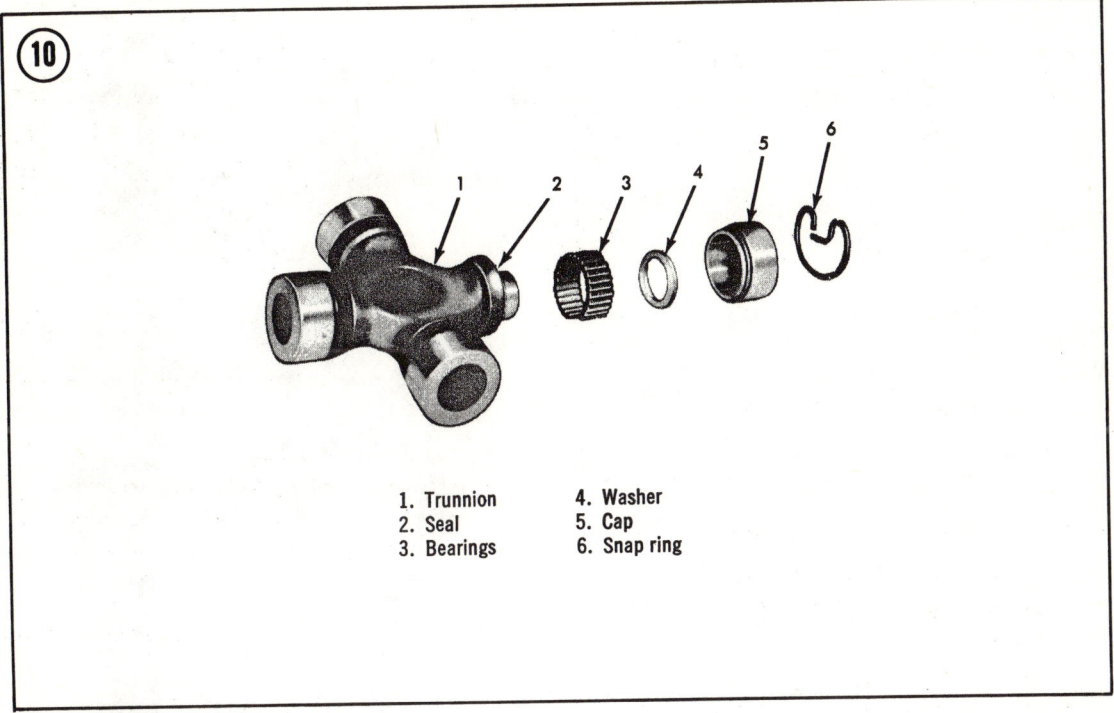

1. Trunnion
2. Seal
3. Bearings
4. Washer
5. Cap
6. Snap ring

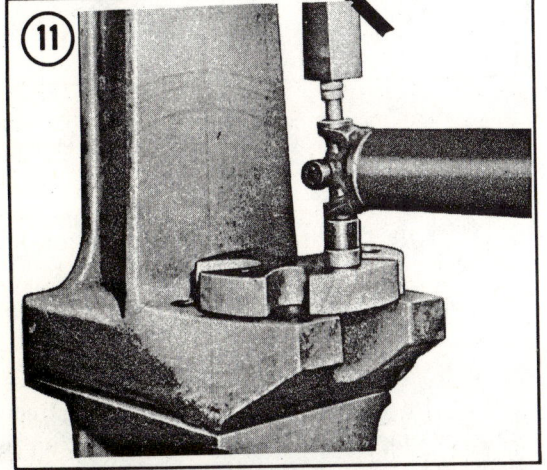

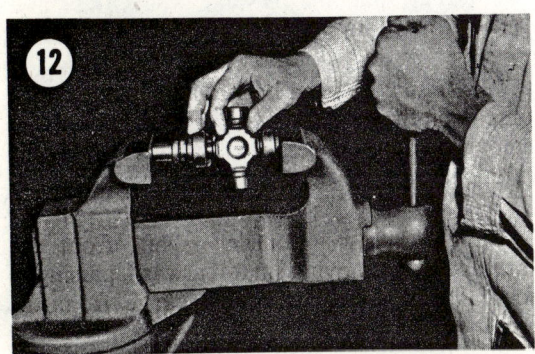

into the second cup. Press the cups into place as shown in **Figure 12**, taking care not to damage the dust seals.

7. Remove the assembly from the vise and install the snap rings.

Saginaw Universal Joint Disassembly/Assembly

Special tools are required for disassembly and assembly of the Saginaw type universal joint. Remove the drive shaft as previously described and take it to your Chevrolet dealer for rebuilding, when required.

REAR AXLE

Most rear axle maintenance and repair involves axle shafts, axle shaft bearings, seals, or the pinion shaft seal. If the correct ring and pinion gear adjustment and bearing preloads were made in production, and if there is adequate lubricant, there should be no need for differential adjustment or repair. Except for replacement of the housing or complete assembly, all rear axle repairs and adjustments can be accomplished without removal from the vehicle.

Axle Shaft, Seal or Bearing Replacement

1. Raise the vehicle by the frame, until wheels clear the ground.

> **CAUTION**
> *Use safety stands when working under the car. Do not rely solely on hydraulic or mechanical jacks.*

2. Remove the wheels and the brake drums (Chapter Ten).

3. Place a container under the differential to catch lubricant, and remove cover.

4. Unscrew the pinion shaft lockscrew, and remove the pinion shaft. See **Figure 13**.

5. Without turning differential or axles, tap each axle toward the differential, and remove C-locks. Replace the pinion shaft and lockscrew. Axle shafts can now be removed.

6. With a suitable puller, remove oil seal and bearing. See **Figure 14**. (If replacing seal only, it can be removed by inserting the lock end of the axle shaft behind the seal and prying it out of its bore. Use caution not to damage the bore with the shaft.)

7. Lubricate the new bearing and seal lip cavity with wheel bearing grease. Install, referring to **Figure 14**.

8. Now is a good time to replace any damaged lug bolts in the axle flange; this can be done with the axle installed, however. Studs can be driven out (or pressed out with tool J-6627 or J-5504). A new stud can be drawn into place by starting it with finger pressure, then threading the lug nut on backwards and tightening until bolt head is seated.

9. When reinstalling the axle shafts, use care not to damage the seals.

10. Remove the pinion shaft, insert the C-locks, and pull the shafts outward to seat the C-locks in the recessed end of the side gears before reinstalling the pinion shaft. Install cover on differential, using a new gasket, and tighten bolts to 20 ft.-lb.

Pinion Seal Replacement

1. With rear universal disconnected and rear brake drums removed, measure pinion bearing preload with an in.-lb. torque wrench on the pinion flange nut. Record the torque required to keep the pinion turning (for reference during reassembly).

2. Mark the pinion shaft-to-flange relationship for correct reassembly.

3. Holding the pinion flange, remove the self-locking nut.

4. With a suitable puller, remove pinion flange. Inspect oil seal surface, drive splines, and bearing cup surfaces. If deflector is damaged, replace it by tapping it from the flange, cleaning up the stake points and staking a new deflector into place at 3 new equally spaced locations. (Be careful not to damage the sealing surface when staking).

5. Measure the old seal location with respect to the rim of the seal bore, and pry out seal. Install the new seal to that same location after applying grease to the cavity between the seal lips. (**Figure 15**) shows the position of the pinion oil seal. Note that the seal will be seated at the inner shoulder before its outside flange reaches the rim of the differential carrier housing.)

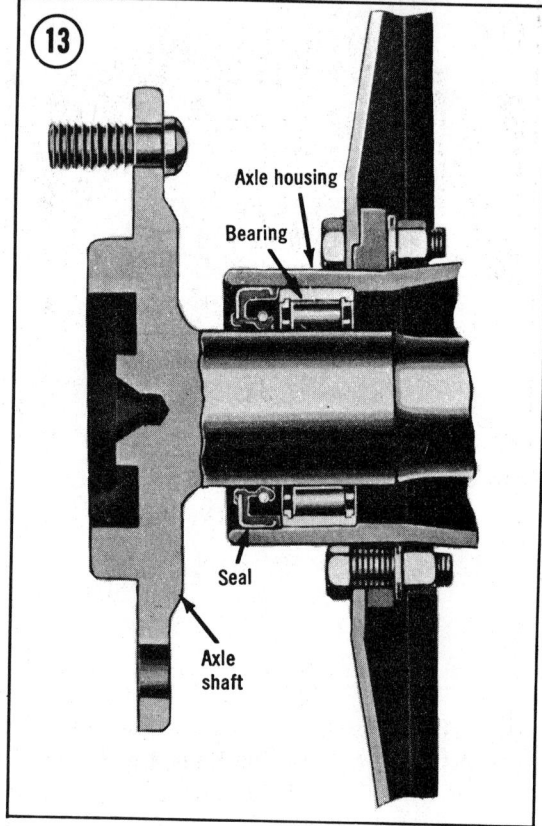

REAR SUSPENSION AND DRIVE SHAFT

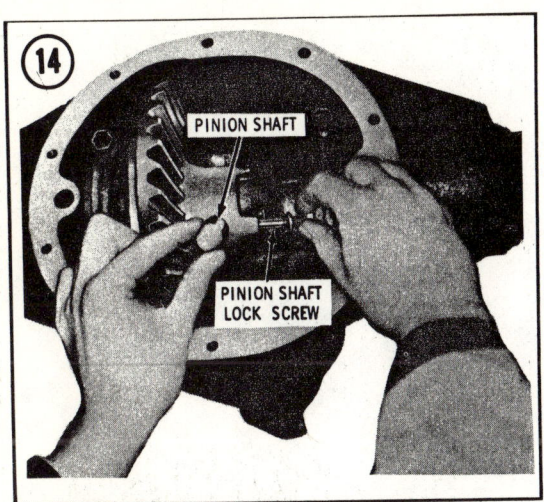

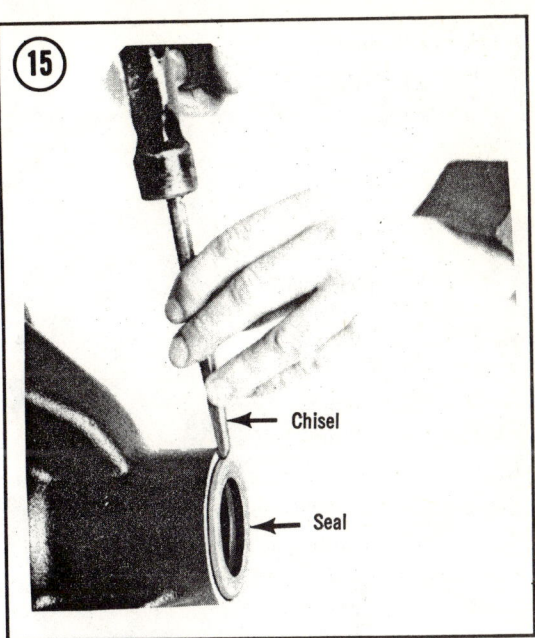

6. Start flange onto pinion shaft, located according to marks previously made, and tap lightly until 2 or 3 threads of the pinion shaft are showing.

7. Using the old nut, draw flange onto pinion shaft *only until end play is removed*. Remove the old nut, apply a non-hardening sealer between the washer and the flange, and start the new nut onto the shaft. Tighten in small increments until previously recorded (Step 1) bearing preload is reached.

Differential Removal

1. With axle shafts removed, remove the differential pinion gear shaft. Identify the differential pinions and thrust washers, also the side gears and thrust washers. Then remove, keeping gears and thrust washers together.

2. Loosen the bearing cap bolts, after marking the caps and housing, and tap each cap lightly to unseat it; do not pry caps. A pry bar may be used to help release the caps by inserting one end into the differential case web — where the spider gears are located — then prying against the axle housing.

3. Identify cups, caps, and shims, and keep them together for each side.

Cleaning and Inspection

Wash all parts and inspect them for scored, chipped, nicked, or worn bearing surfaces. Examine differential case for cracks. Inspect thrust washers for wear. Check fit of all splines.

Rear Axle Assembly/Removal/Installation

1. Raise the rear of the car and support it on jackstands placed under the rear body lifting points. See **Figure 4**.

2. Remove the drive shaft as described elsewhere in this chapter.

3. Disconnect the parking brake cable from the equalizer by removing the adjusting nut. Then remove the cable from the rear connectors and remove it from the car.

4. Disconnect the rear brake hose at the floor pan and cover the hose and pipe connectors to prevent the entrance of dirt.

5. Disconnect shock absorbers at their lower ends. See *Shock Absorber Removal/Installation*, this chapter, for instructions and CAUTION. Push shocks up out of the way.

6. Remove the springs as described in *Coil Spring Removal/Installation*, this chapter.

7. Remove the upper and lower control arms as described in *Upper or Lower Control Removal/Installation*, this chapter.

8. Roll the axle and wheel assembly out from under the car.

9. Installation is the reverse of these steps. Tighten all bolts and nuts to the specifications given in the individual procedures.

INDEX

A

Air cleaners 45-46
Air conditioning
   Condenser 218
   Discharging the system 223
   Evaporator 219
   Expansion valve 218-219
   Inspection 220
   Maintenance 219-220
   Receiver/drier 218
   Refrigerant 220-221
   System operation 214-217
   Testing 220
   Troubleshooting 221-223
Air injection reactor system 193-195
Alternator
   Delcotron, 10-SI 228-230, 238-241
   Delcotron, 1D 230-234, 235-238
Axle, rear 329-331

B

Ball-joints 311
Battery 34, 46, 224-226
Brakes
   Adjustment 293
   Bleeding 286-287
   Caliper 299-301
   Inspection 286
   Linings 287-293
   Maintenance 49-50
   Master cylinder 303-305
   Pad replacement 295-299
   Parking brake 50, 294
   Power brake unit 302-303
   Troubleshooting 28-29
   Wheel cylinder 294-295

C

Caliper, brake 299-301
Camber 309
Camshaft 120-121
Carburetor, Holley 4150 176-186
Carburetor, Rochester 1MV-1ME 142-155
Carburetor, Rochester 2GV-2GC-2GE ... 157-166
Carburetor, Rochester M2M 166-167
Carburetor, Rochester M4MC-M4MCA ... 175-176
Carburetor, Rochester 4MV 167-175
Carburetor adjustment 68-77, 203
Caster 309
Catalytic converter 188, 203-204
Charging system 9-14, 226-228
Circuit breakers 245
Chassis 49-50

Clutch
   Inspection 259
   Installation 259-260
   Linkage 257
   Pilot bearing 260
   Removal 259
   Troubleshooting 27-28
Coil spring 325-326
Combined emission control system ... 195-201
Compression test 58
Condenser 218
Connecting rod and piston 115-120
Connecting rod bearing 114-115
Control arm 327
Controlled combustion system 190-193
Coolant level 34
Cooling system
   Coolant 207
   Coolant recovery system 207
   Fan/fan clutch 206
   Maintenance 210-211
   Radiator 206, 212-213
   Radiator cap 206
   Thermostat 206
   Troubleshooting 26-27
   Water pump 207-209, 211-212
Crankshaft 121-122
Cylinder block 122
Cylinder head 97-101

D

Delcotron, 10-SI 228-230, 238-241
Delcotron, 1D 230-234, 235-238
Delcotron removal/installation 234-235
Distributor 61-65, 203, 251-252
Drive shaft 328-329
Dwell angle 65-66

E

Early fuel evaporation system 204-205
EGR 24
Electrical system
   Battery 224-226
   Charging system 226-228
   Circuit breaker 245
   Delcotron, 10-SI 228-230, 238-241
   Delcotron, 1D 230-234, 235-238
   Delcotron removal/installation .. 234-235
   Distributor 251-252
   Fuses 245-246
   Fusible links 246-249

INDEX

High energy ignition system 252-256
Lighting system. 244
Starter motor 241-244
Switches 249-250
Wiring diagrams . . . . end of book
Emission control systems
   Air injection reactor system 193-195
   Carburetor calibration 203
   Catalytic converter 188, 203-204
   Combined emission control system 195-201
   Controlled combustion system 190-193
   Distributor calibration 203
   Early fuel evaporation system 204-205
   Exhaust gas recirculation 24, 202-203
   Fuel evaporation control system 201-202
   Positive crankcase ventilation 189-190
   Troubleshooting 25-26
Engine
   Camshaft. 120-121
   Connecting rod and piston 115-120
   Connecting rod bearings 114-115
   Crankshaft. 121-122
   Cylinder block 122
   Cylinder head 97-101
   Installation 82-83
   Main bearings 112-114
   Manifolds 90-93
   Oil and filters 33-34, 43-45
   Oil pan replacement 101-102
   Oil pump (except 231 V6) 103-105
   Oil pump (231 V6) 105-109
   Overhaul 83-84
   Pistons 115-120
   Specifications 126-141
   Reassembly sequence 84-90
   Rear main oil seal 109-112
   Removal 78-82
   Rocker arm cover 93
   Troubleshooting 19-20
   Tune-up 51-67
   Valve lifters 95
   Valve mechanism 93-94
   Valve stem oil seal 95-97
Evaporator 219
Exhaust gas recirculation 202-203
Exhaust systems
   Catalytic converter 188
   Exhaust pipe 188
   Muffler 187
   Tailpipe 187
Expansion valve 218-219

F

Fan/fan clutch 206
Filter 46
Firing order 51
Fuel evaporative control 22-24
Fuel system
   Carburetor, Holley 4150 176-186
   Carburetor, Rochester 1MV-1ME 142-155
   Carburetor, Rochester 2GV-2GC-2GE . . . 157-166
   Carburetor, Rochester M2M 166-167
   Carburetor, Rochester M4MC-M4MCA . 175-176
   Carburetor, Rochester 4MV 167-175
   Filter 46
   Fuel pump 186
   Tank 186-187
Fuses 245-246
Fusible links 246-249

G

General information 1-8

H

Headlight 244
Heater 26, 213-214
High energy ignition system (HEI) 252-256
Hood latches and hinges 50

I

Ignition system
   Centrifugal and vacuum advance 253
   Electronic module 252
   Ignition coil 252, 255-256
   Maintenance 253
   Operation 253
   Pick-up assembly 253
   Spark plugs 16-17
   Troubleshooting 13-14, 253-254
   Tune-up 61-65, 66-67

L

Lighting system 25-26, 244

M

Main bearings 112-114
Maintenance (also see Tune-up)
   Air cleaners 45-46
   Air conditioning 219-220
   Battery 34, 46
   Chassis 49-50
   Coolant level 34
   Cooling system 210-211
   Engine oil and filters 33-34, 43-45
   Fuel filter 46
   Hood latches and hinges 50
   Lubricants, recommended 43
   Maintenance schedule 37-38
   Master cylinder, brake 49-50
   Parking brake 50
   Periodic maintenance 36-43
   Power steering 50
   Transmission 46-49
   Wheel bearings 49
Manifolds 90-93
Master cylinder, brake 303-305
Muffler 187

INDEX

O

Oil and filters 33-34, 43-45
Oil pan.......................... 101-102
Oil pressure light 18
Oil pump 103-109

P

Pad, brake....................... 295-299
Parking brake.................... 50, 294
Pistons........................... 115-120
Positive crankcase ventilation 21, 189-190
Power brake 302-303
Power steering 50, 322-323

R

Radiator..................... 206, 212-213
Rear main oil seal 109-112
Receiver/drier 218
Refrigerant 220-221
Ride height......................... 309
Rocker arm cover 93

S

Safety 3-4
Service hints....................... 2-3
Shift cover...................... 273, 282
Shock absorbers.............. 310, 326-327
Spark plugs 16-17, 51, 58, 59-61
Springs, front 310-311
Stabilizer bar.................... 311-312
Starter motor.................... 241-244
Starting system troubleshooting 9
Steering axis inclination 309
Suspension, rear
   Coil spring 325-326
   Control arm 327
   Shock absorber 326-327
   Troubleshooting 29
Suspension and steering, front
   Adjustments..................... 314-315
   Ball-joints...................... 311
   Maintenance..................... 50
   Power steering pump 322-323
   Shock absorbers 310
   Springs......................... 310-311
   Stabilizer bar 311-312
   Steering gear 315-320
   Steering knuckle 312
   Steering linkage 320-322
   Troubleshooting 29
   Wheel alignment 306-310
   Wheel bearings................. 312-314
   Wheel hub...................... 312
Switches......................... 249-250

T

Tail pipe......................... 187
Tank 186-187
Thermostat....................... 206
Thermostatic air cleaner 21-22
Timing 66
Tires 29-32
Toe-in............................ 309
Tools 4-8
Transmission
   Assembly, Saginaw 3-speed 270-273
   Assembly, Saginaw 4-speed 281-282
   Disassembly, Saginaw 3-speed 264-268
   Disassembly, Saginaw 4-speed 273-278
   Inspection and repair, Saginaw 3-speed .. 268-270
   Inspection and repair, Saginaw 4-speed .. 278-280
   Maintenance 46-49
   Replacement 260-261
   Shift cover.................... 273, 282
   Shift linkage adjustment 261-264
Troubleshooting
   Air conditioning................ 221-223
   Brakes.......................... 28-29
   Charging system 9-14
   Clutch 27-28
   Cooling system 26
   Electrical accessories 25-26
   Emission control systems.......... 20-24
   Engine......................... 14-18, 24-25
   Fuel system 19-20
   Starting system 9
   Steering and suspension 29
   Tires 29-32
   Tools 5-8
   Transmission 28
   Wheels 32
Tune-up
   Carburetor adjustment 67-77
   Compression test 58
   Distributor..................... 61-65
   Dwell angle 65-66
   Firing order 51
   Ignition system 61-65, 66-67
   Spark plugs 51, 58, 59-61
   Specifications 52-57
   Timing 66
   Valve clearance adjustment
     (mechanical lifters) 51

V

Valves
   Clearance adjustment (mechanical lifters)... 51
   Mechanism 93-94
   Valve stem oil seal 95-97

W

Water pump 207-209, 211-212
Wheel alignment 306-310
Wheel bearings................. 49, 312-314
Wheel cylinders 294-295
Wheel hub 312
Windshield wiper and washer 26
Wiring diagrams .... end of book

1970 ENGINE COMPARTMENT

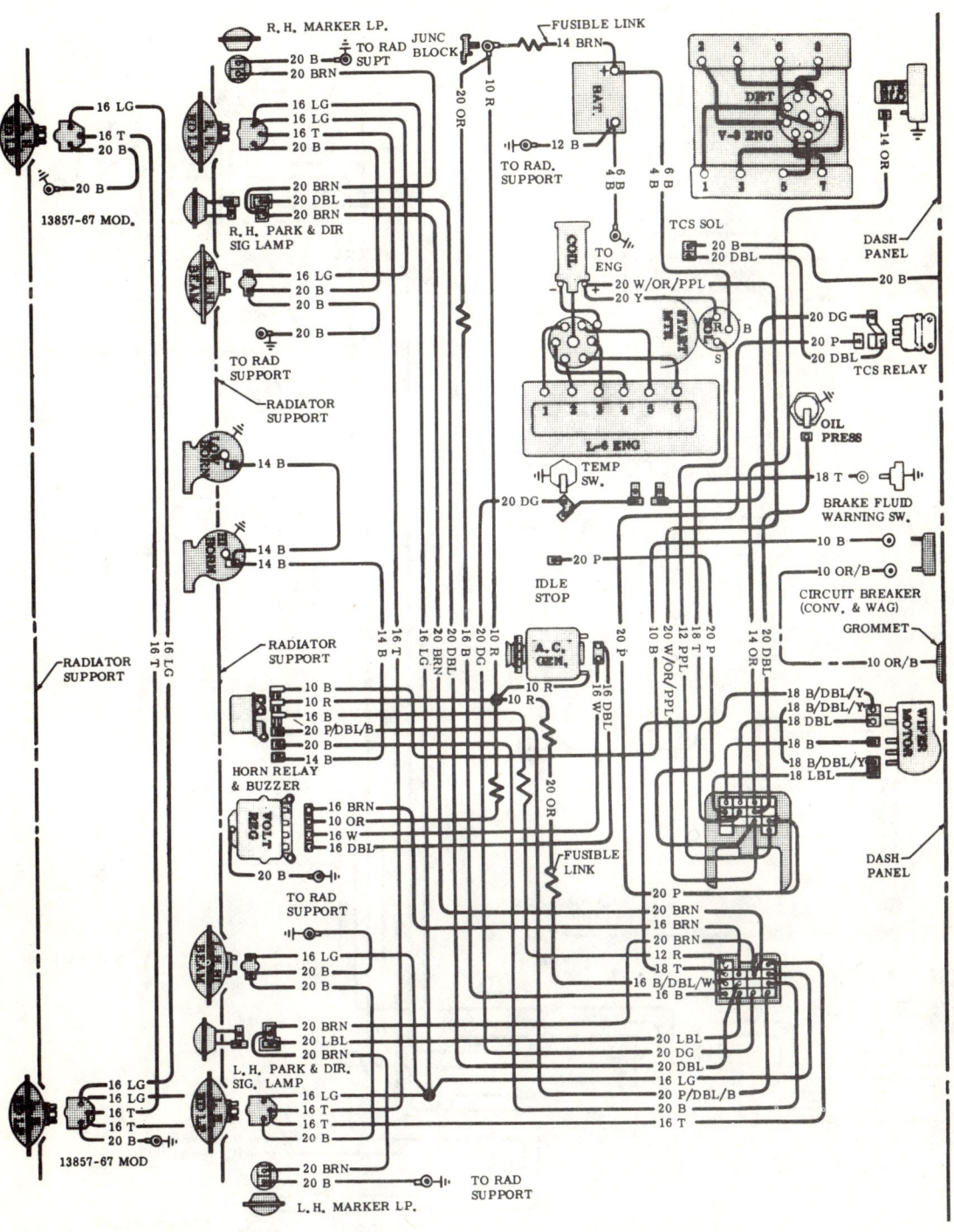

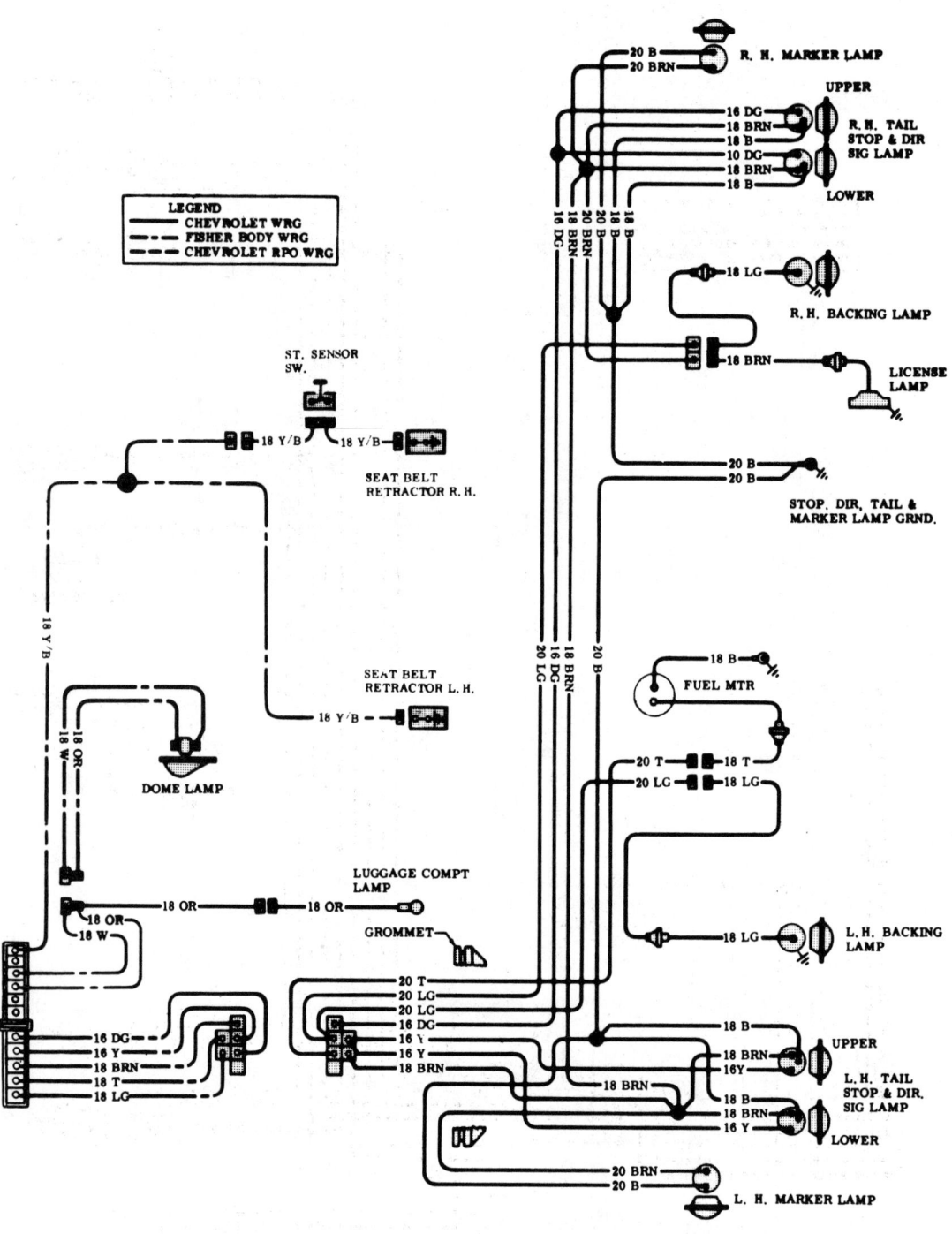

1970-1972 BODY AND REAR LIGHTING (CHEVELLE)

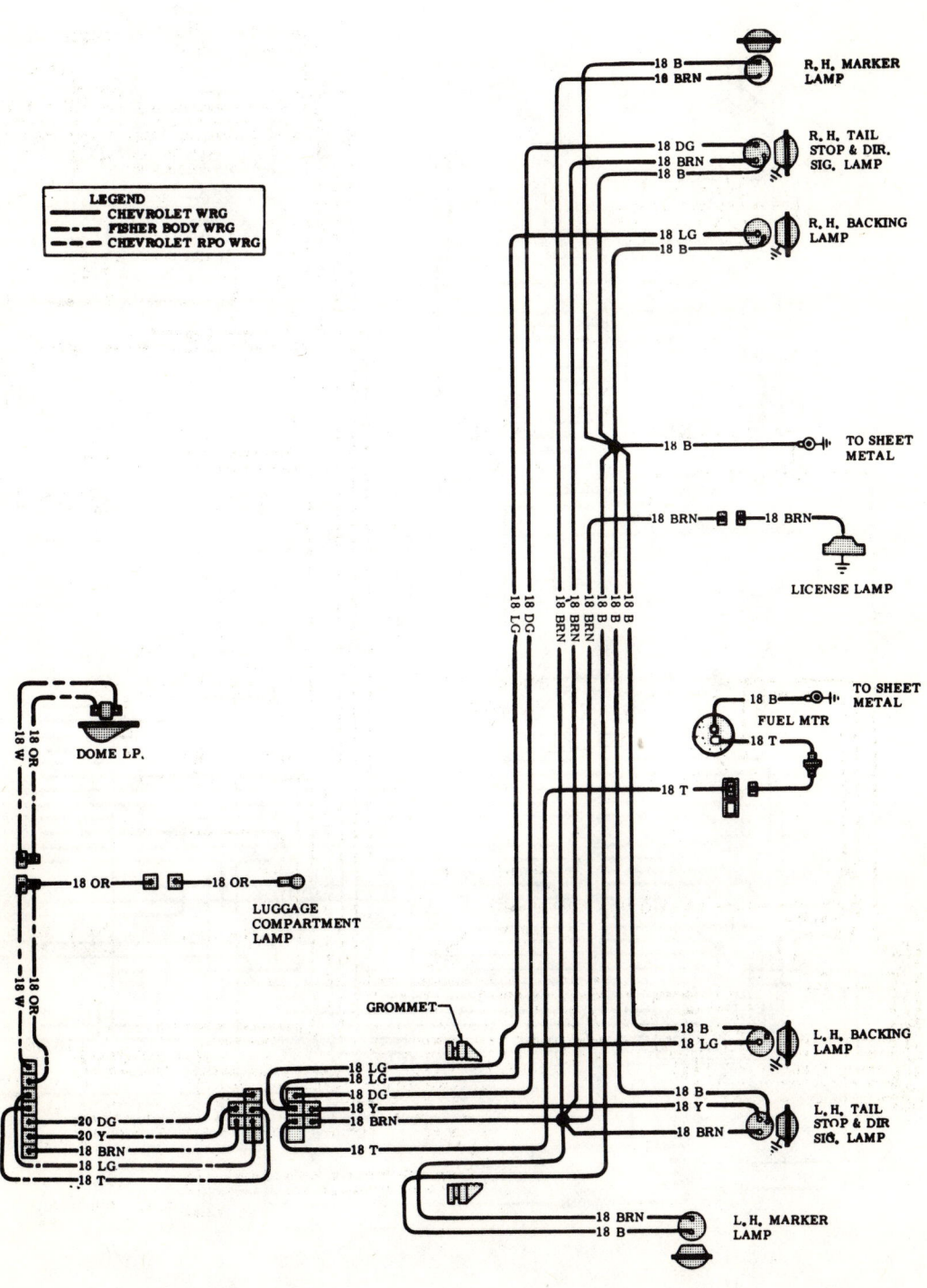

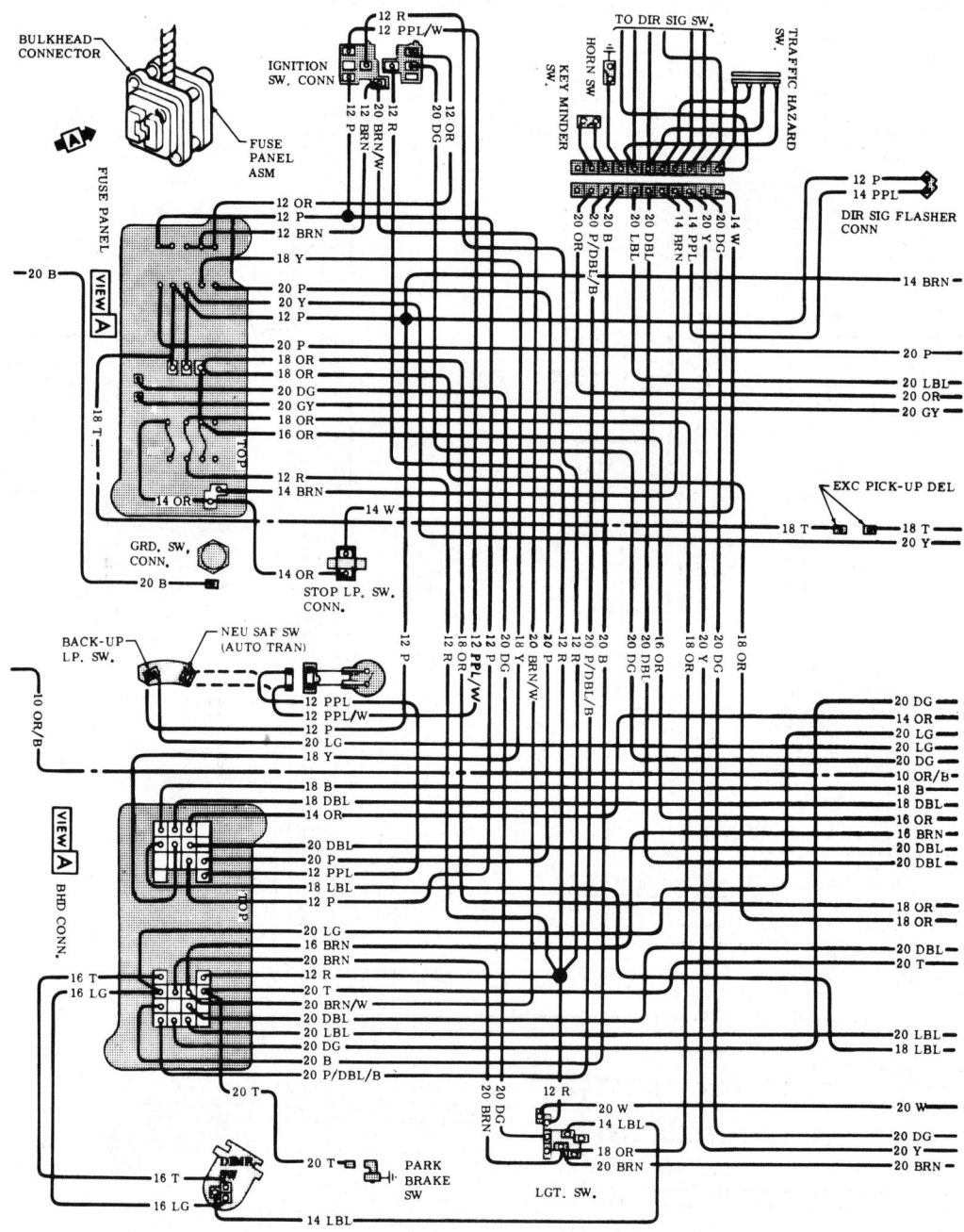

1970-1972 INSTRUMENT PANEL (EXCEPT CHEVELLE)

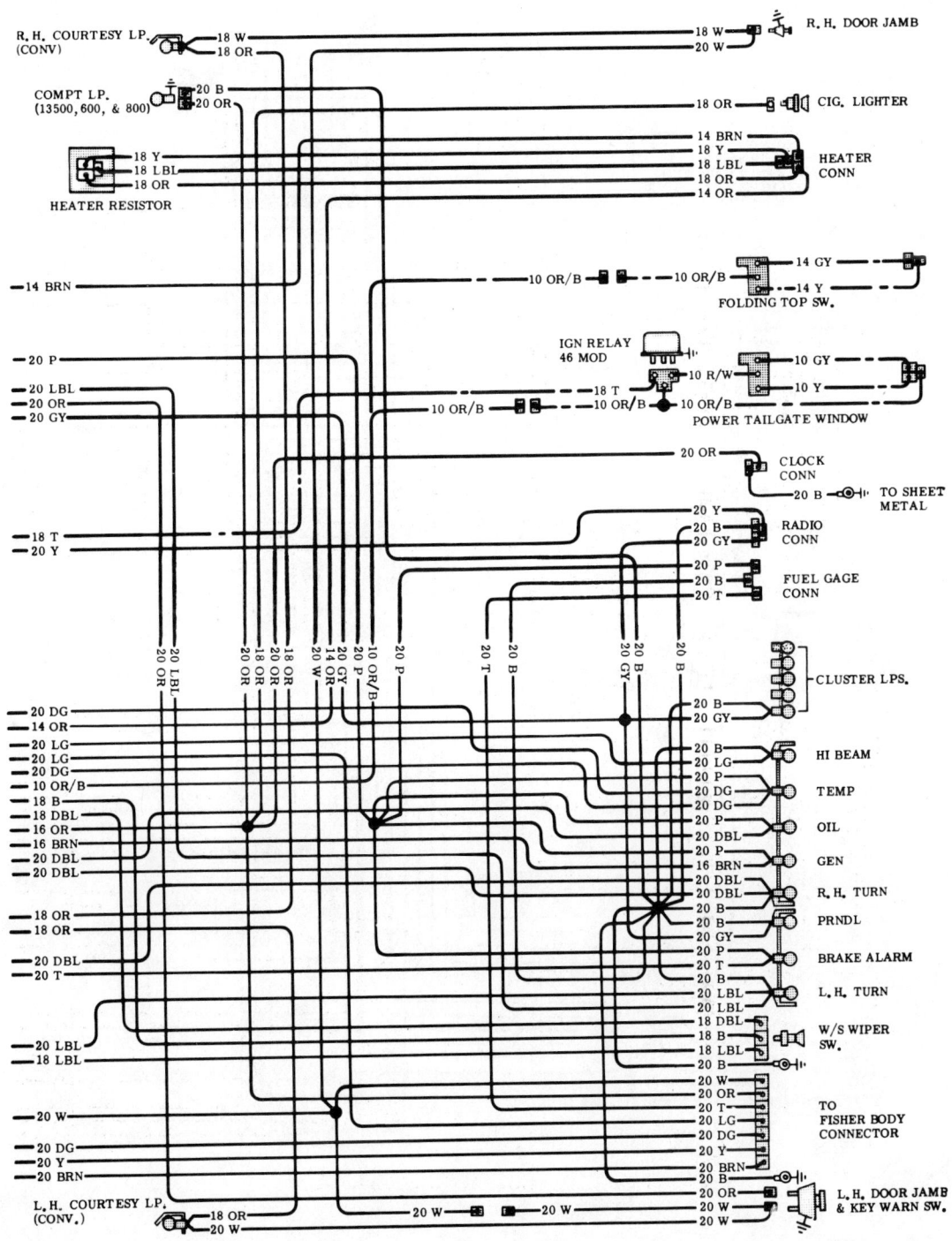

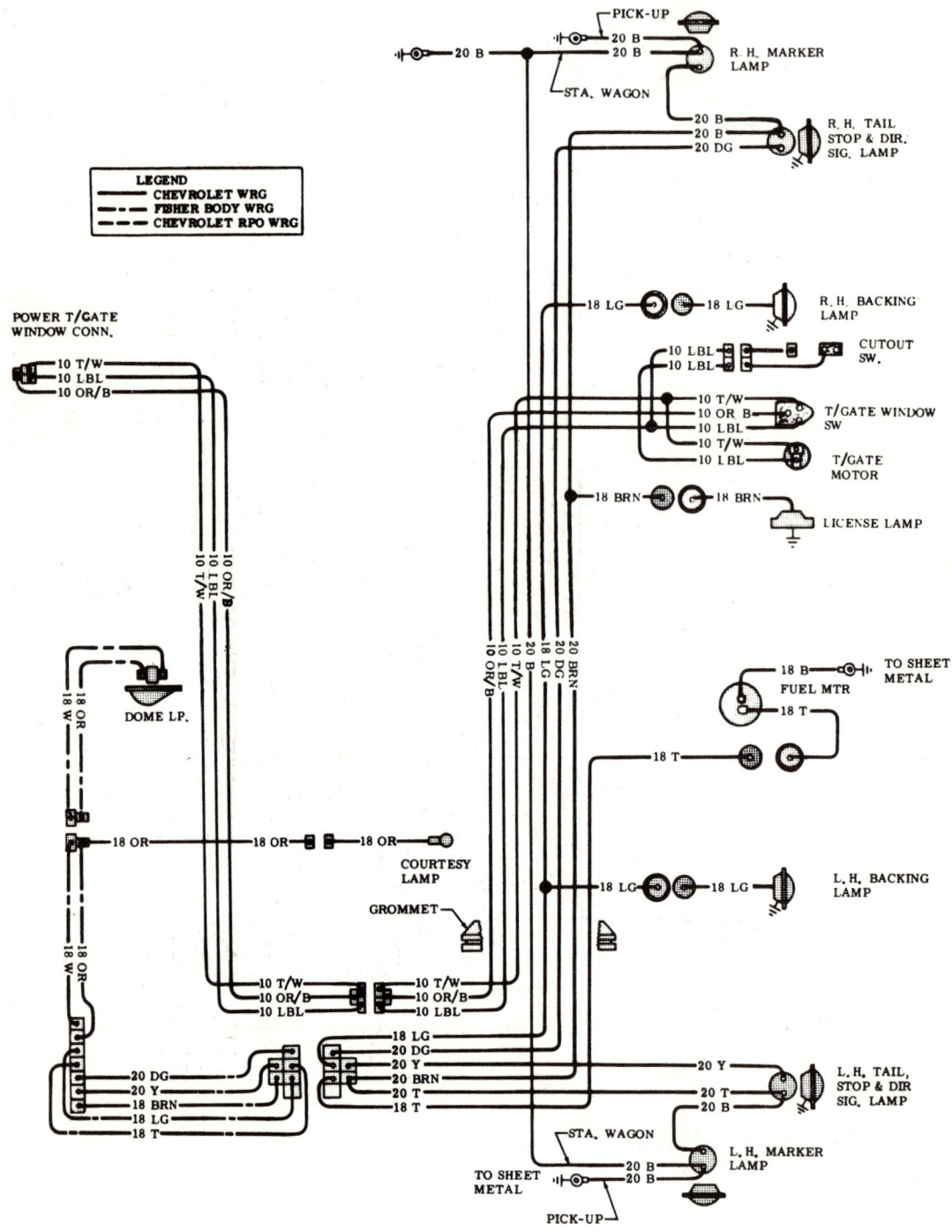

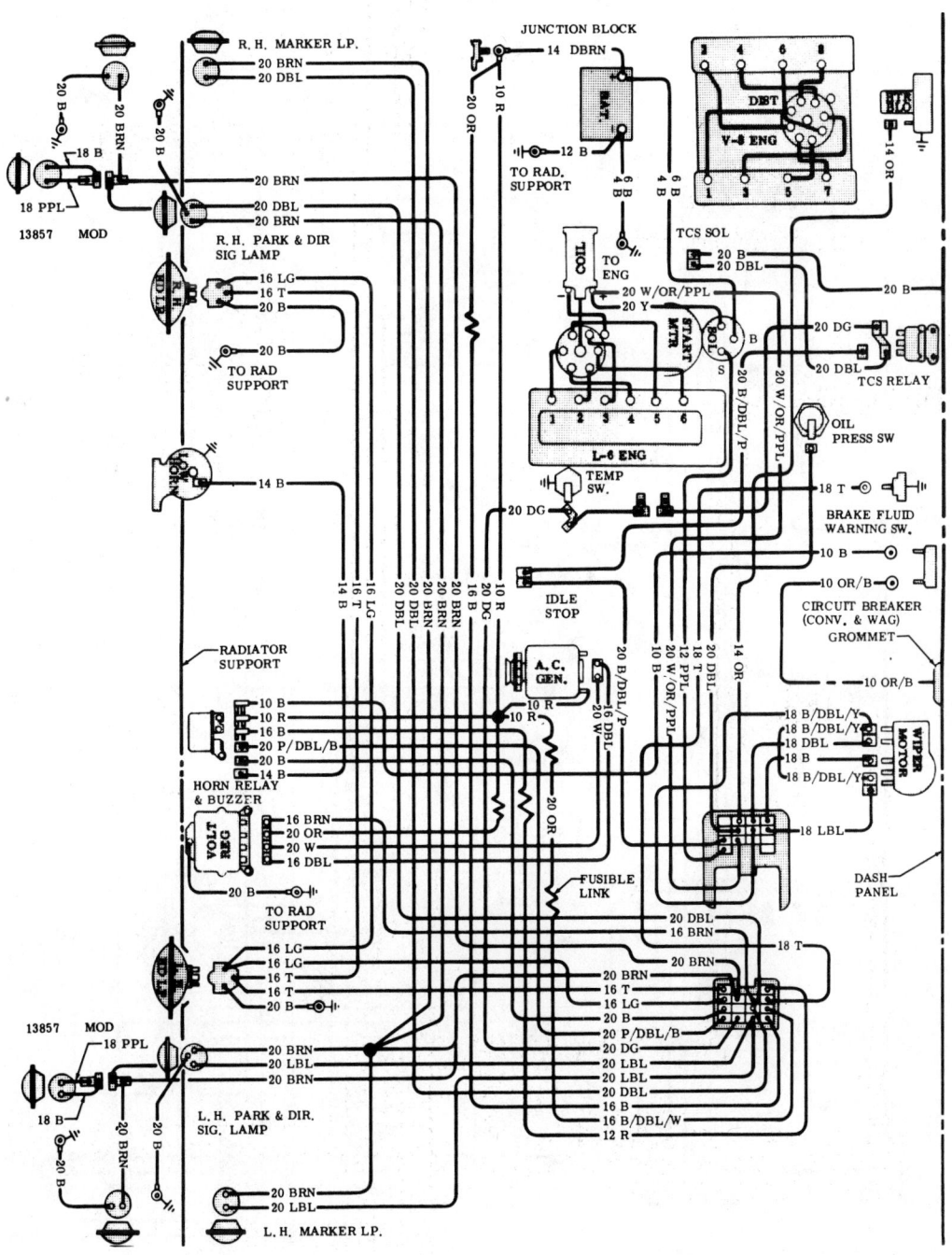
1971 ENGINE COMPARTMENT

1972 ENGINE COMPARTMENT

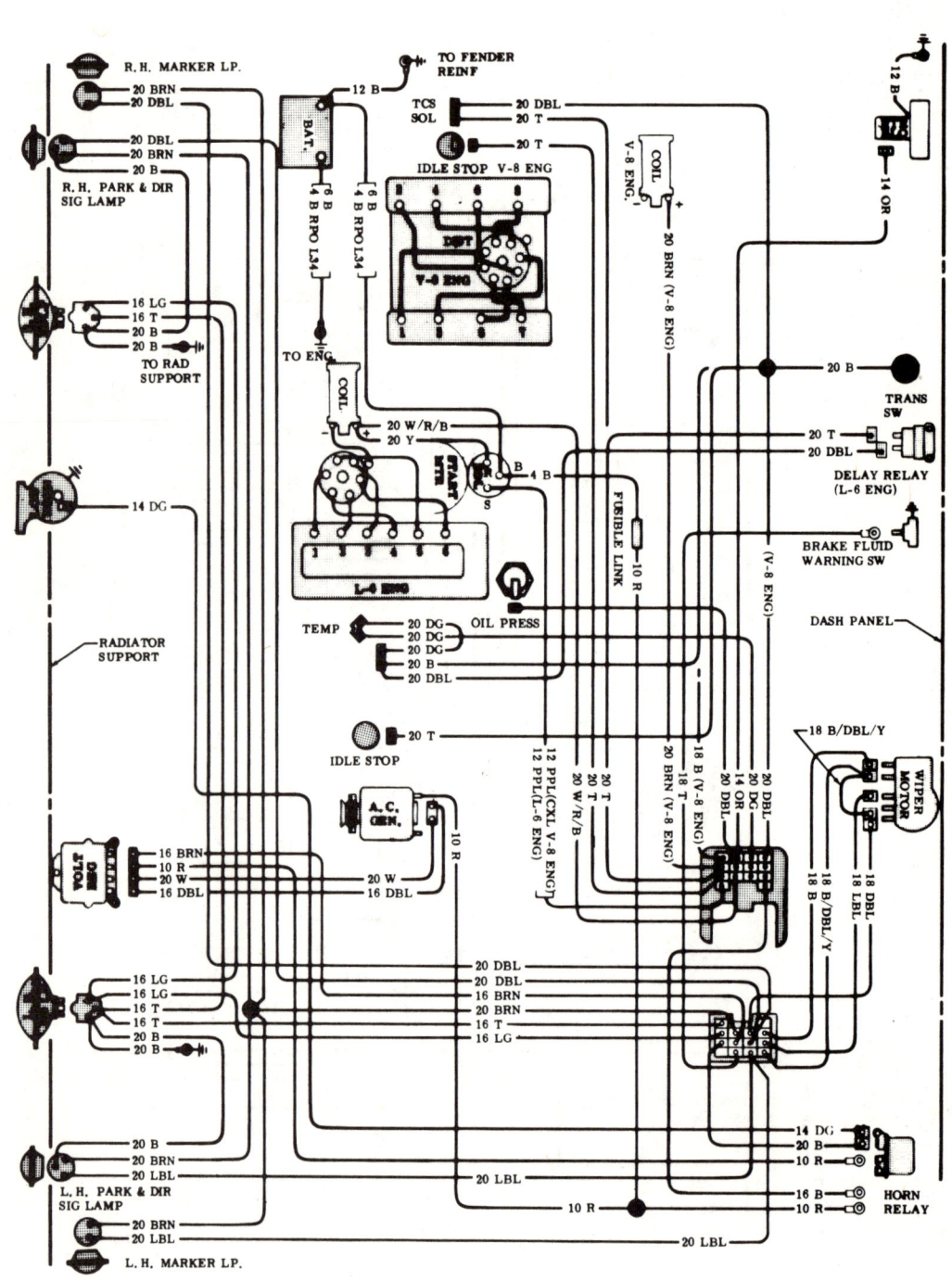

1973 ENGINE COMPARTMENT

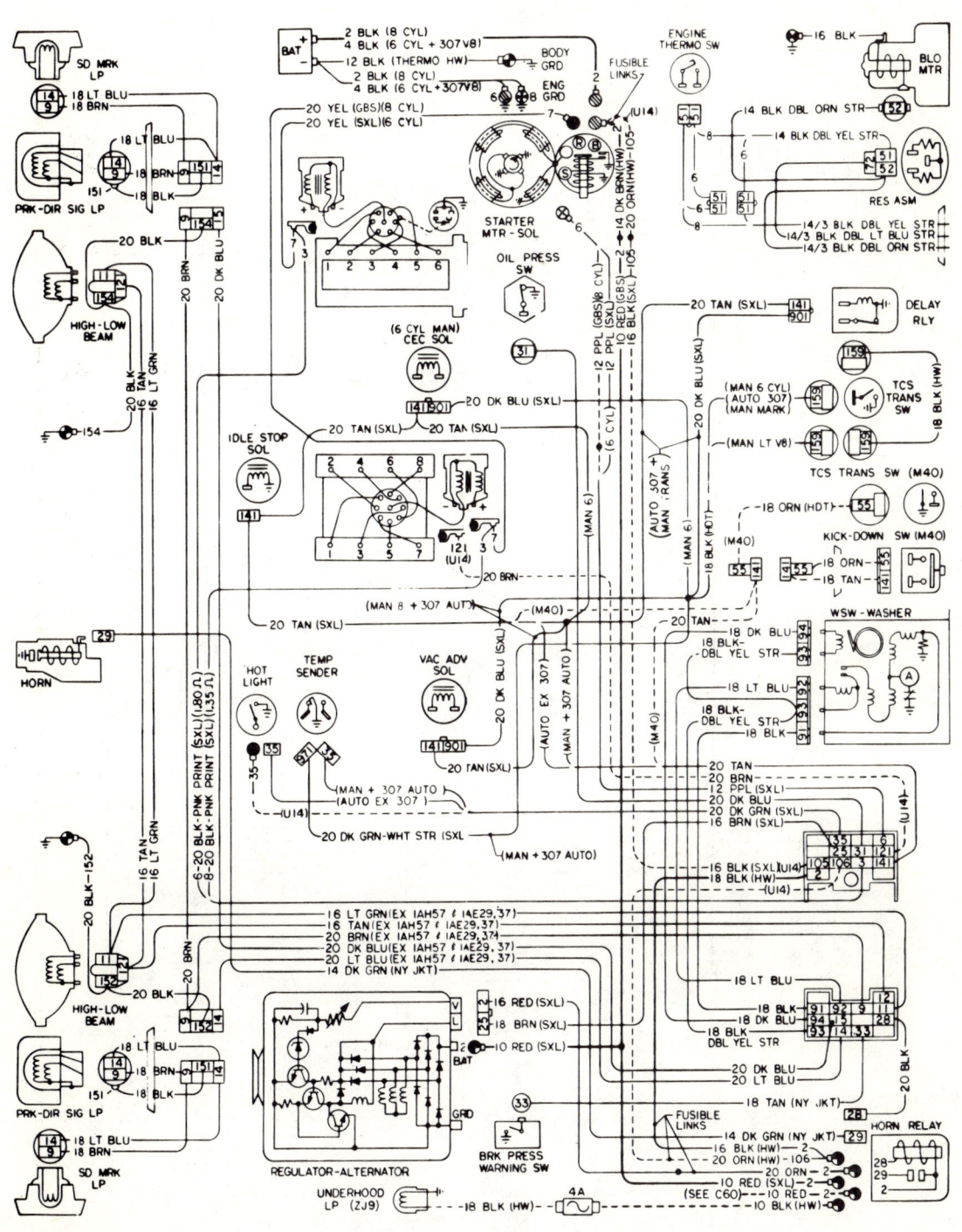

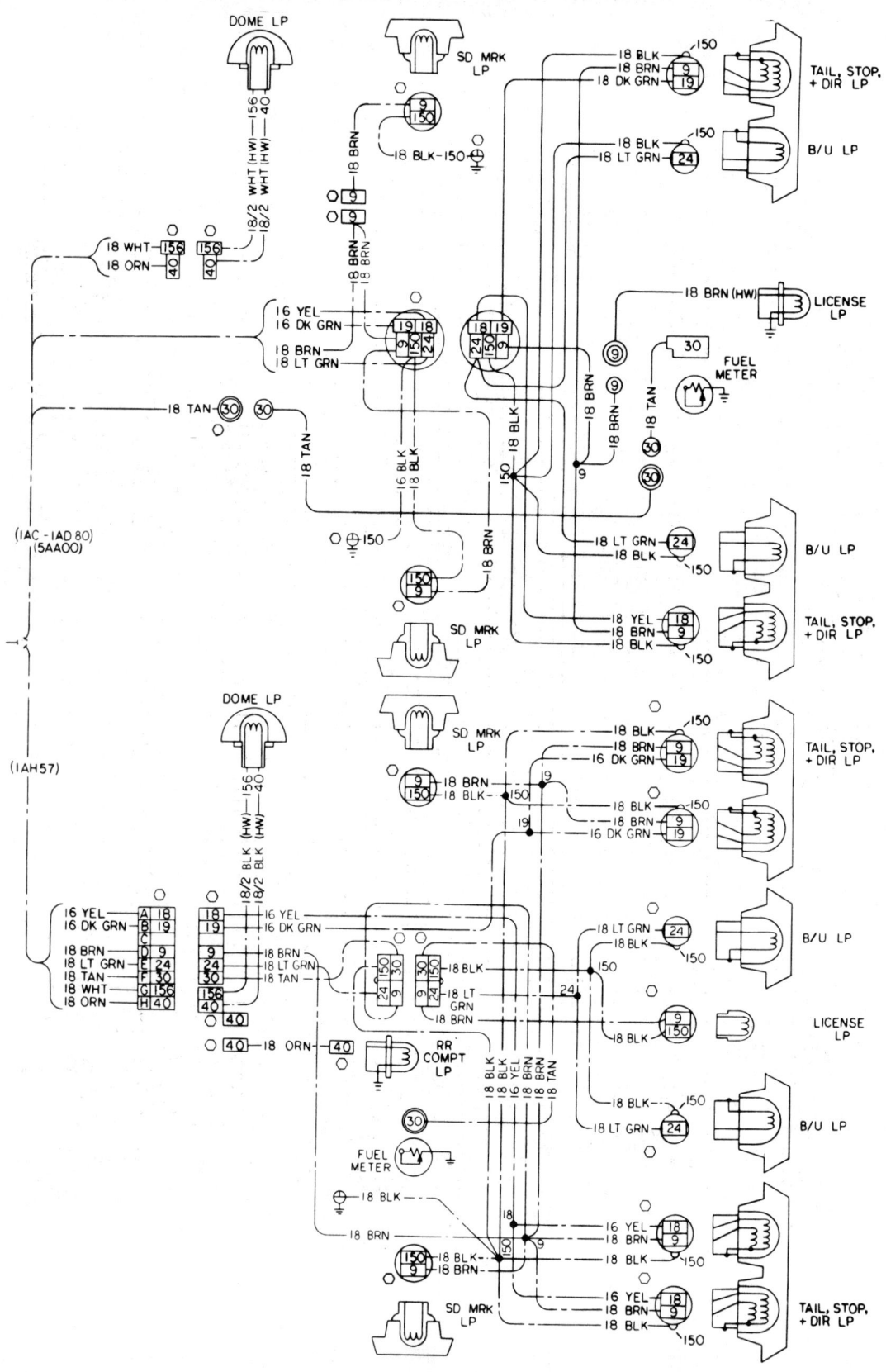

1973-1975 INSTRUMENT PANEL (EXCEPT MONTE CARLO)

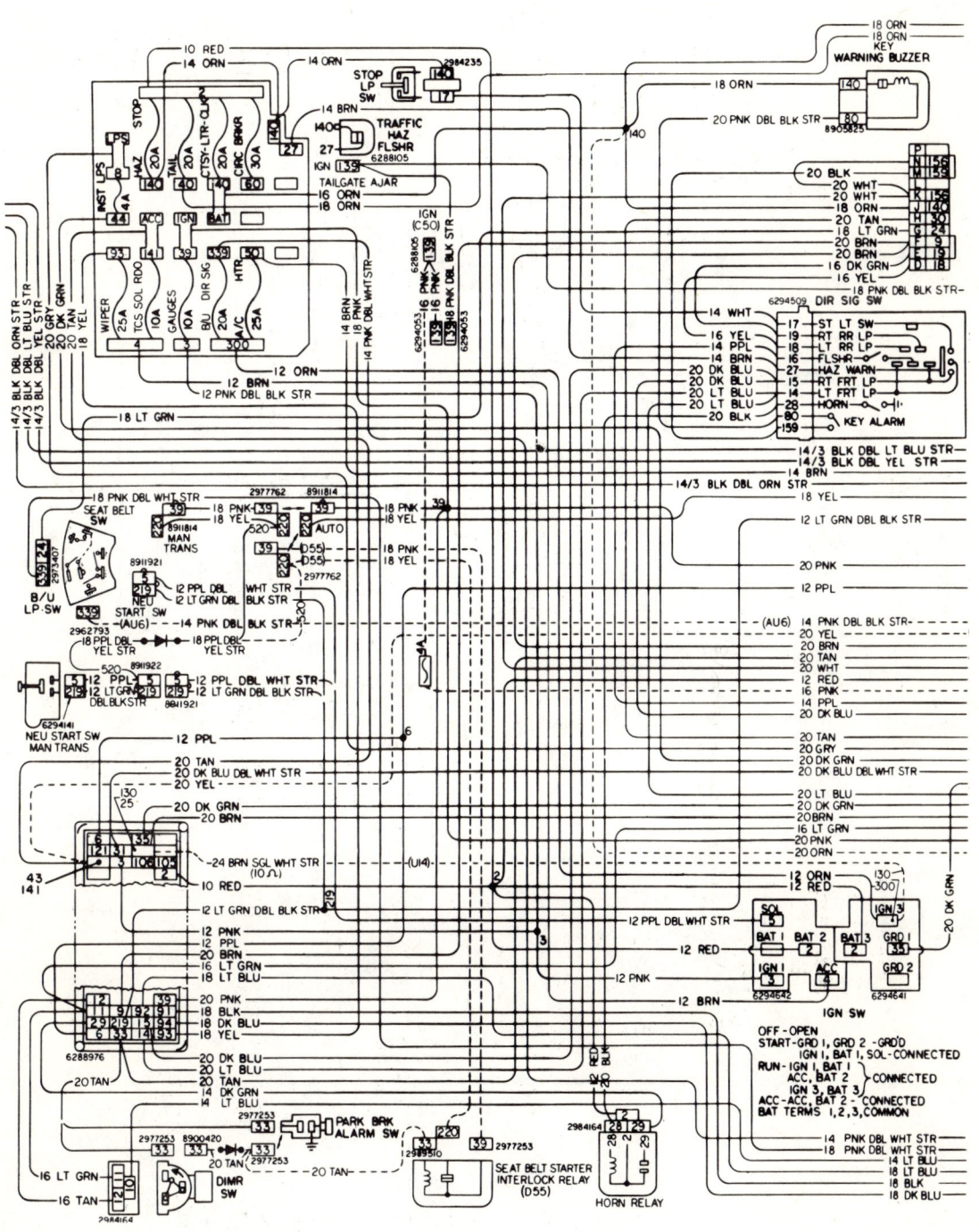

1973-1975 CRUISE CONTROL AND AIR CONDITIONING

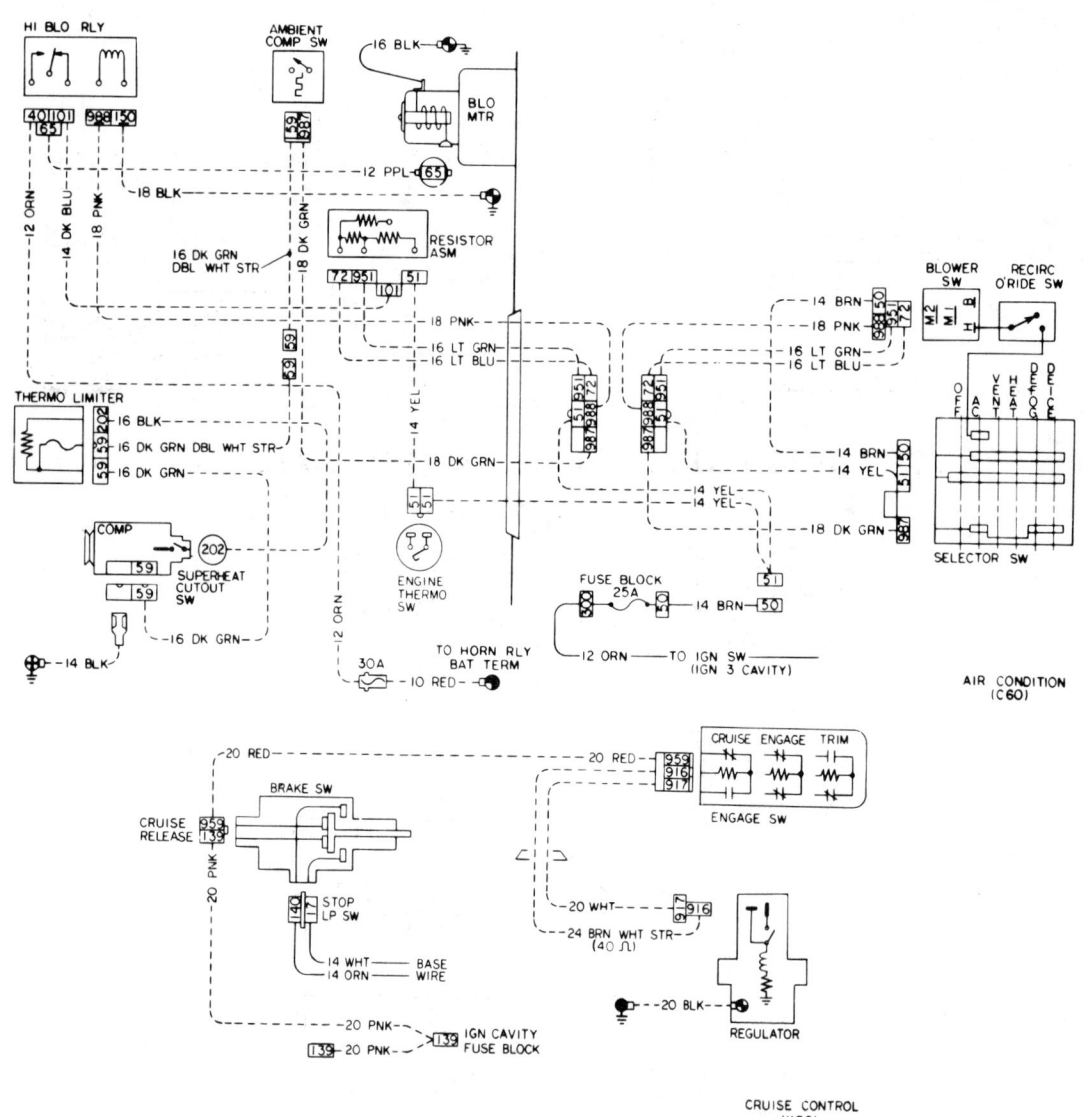

1973-1975 INSTRUMENT PANEL (MONTE CARLO)

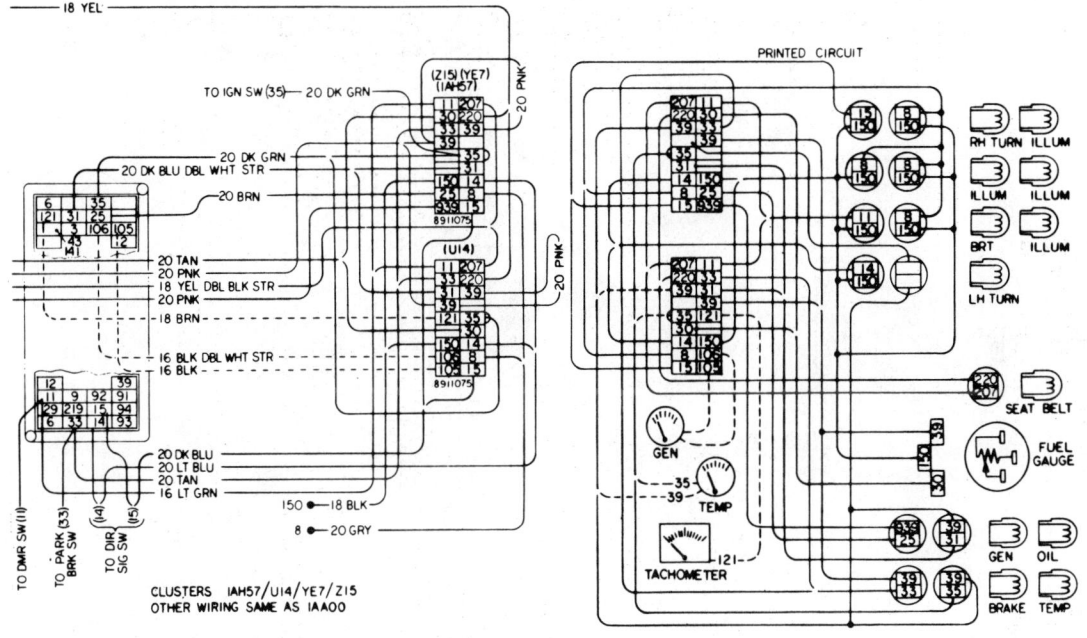

1973-1975 FRONT LIGHTING (MONTE CARLO)

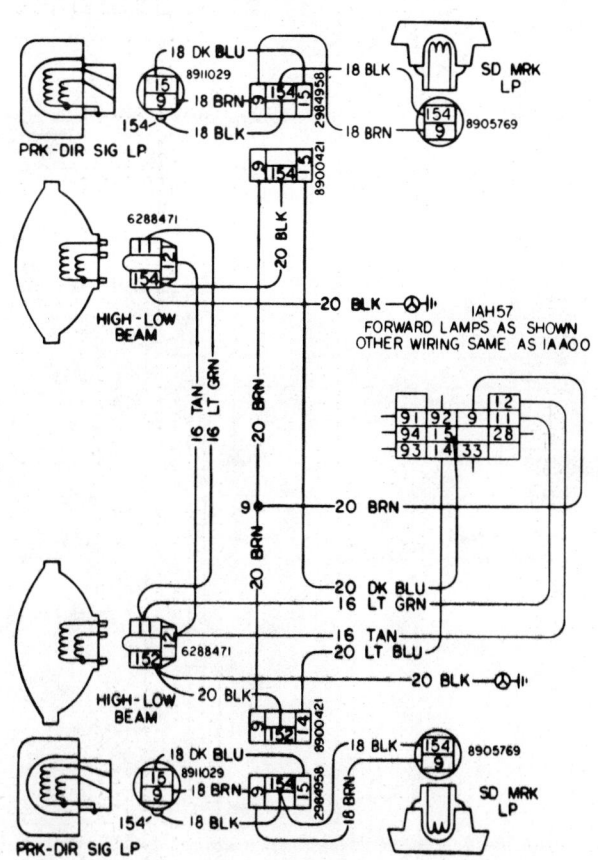

1973-1975 REAR LIGHTING (EXCEPT MONTE CARLO)

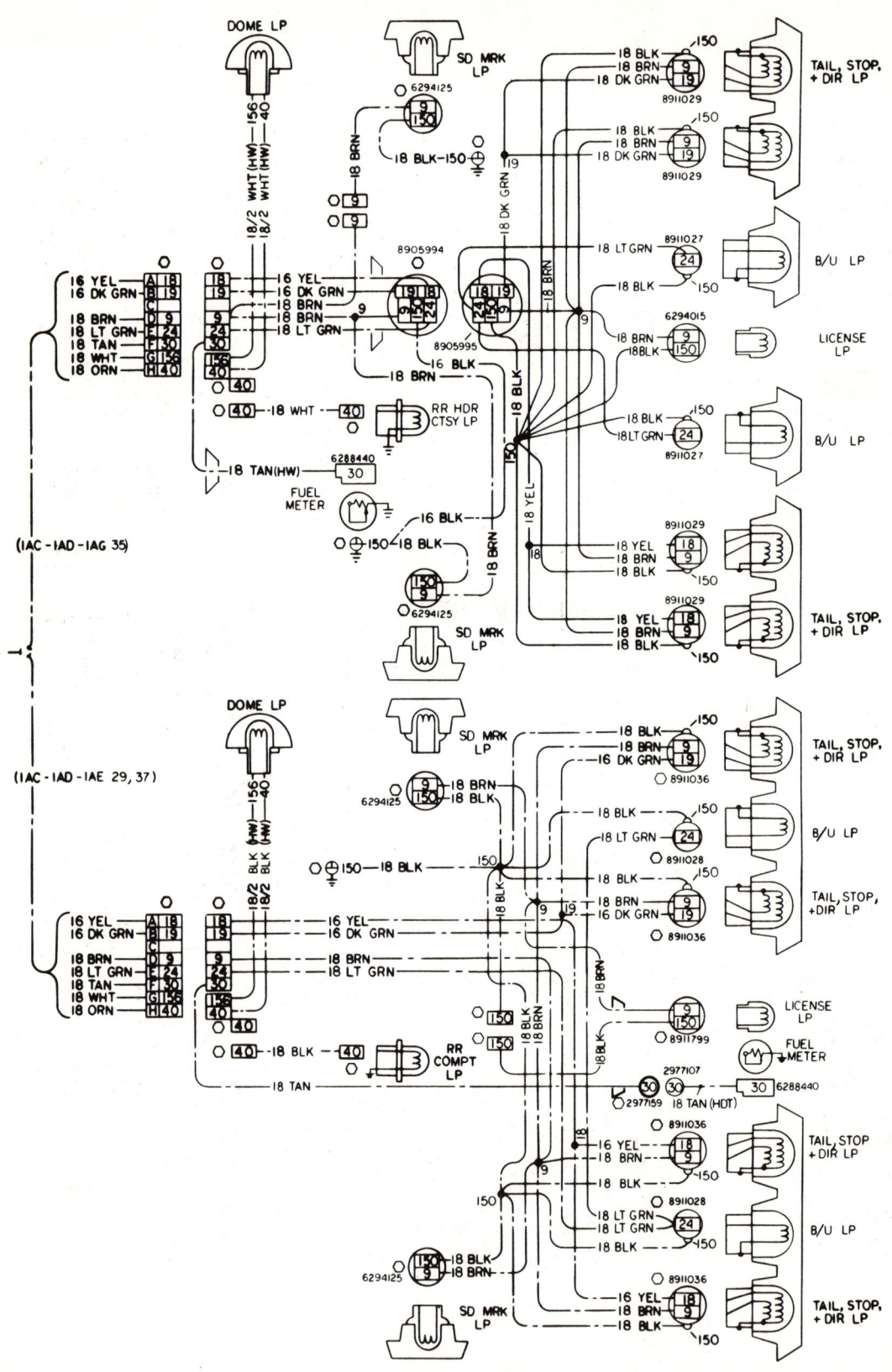

1974 ENGINE COMPARTMENT

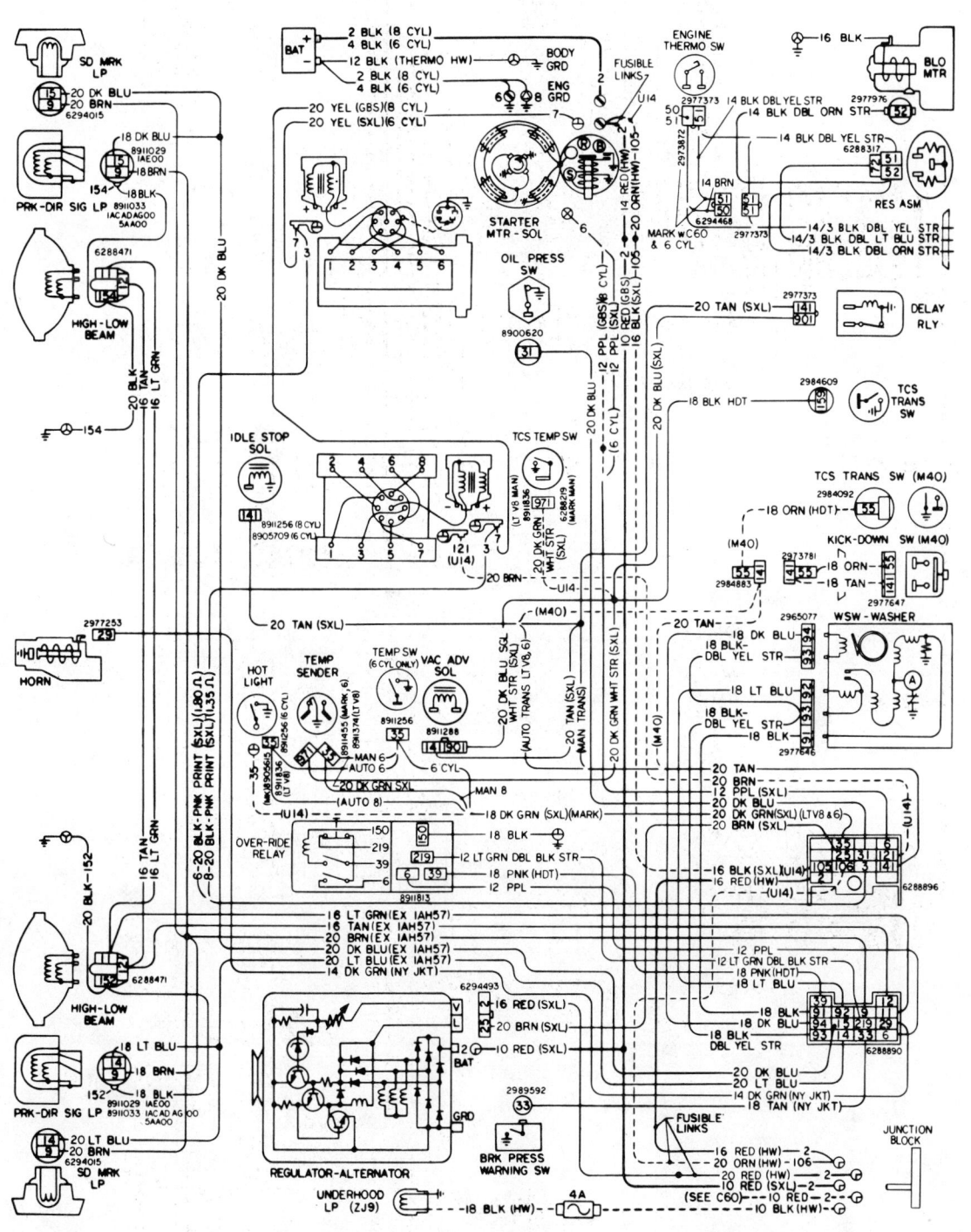

1974 INSTRUMENT PANEL (EXCEPT MONTE CARLO)

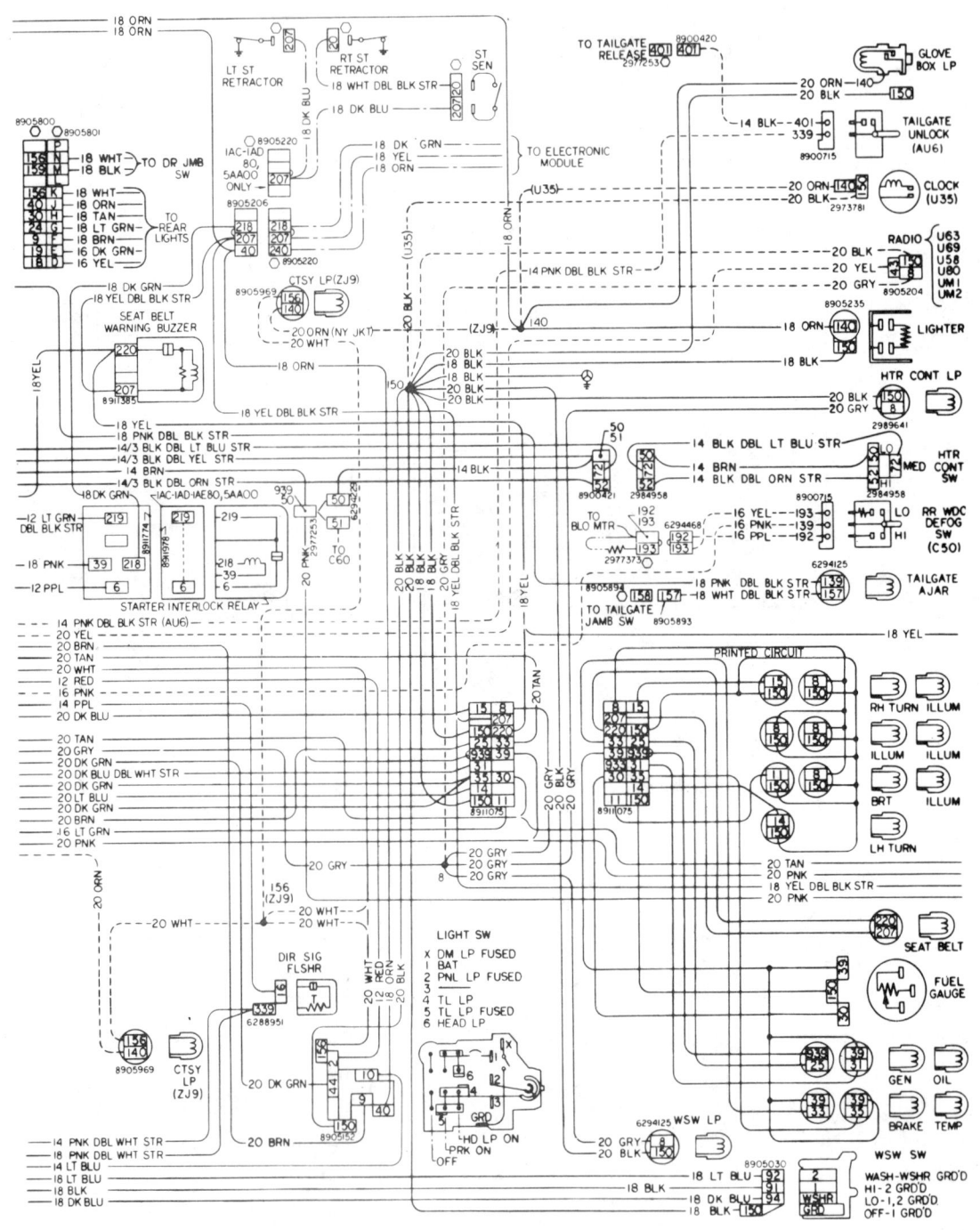

1975 ENGINE COMPARTMENT

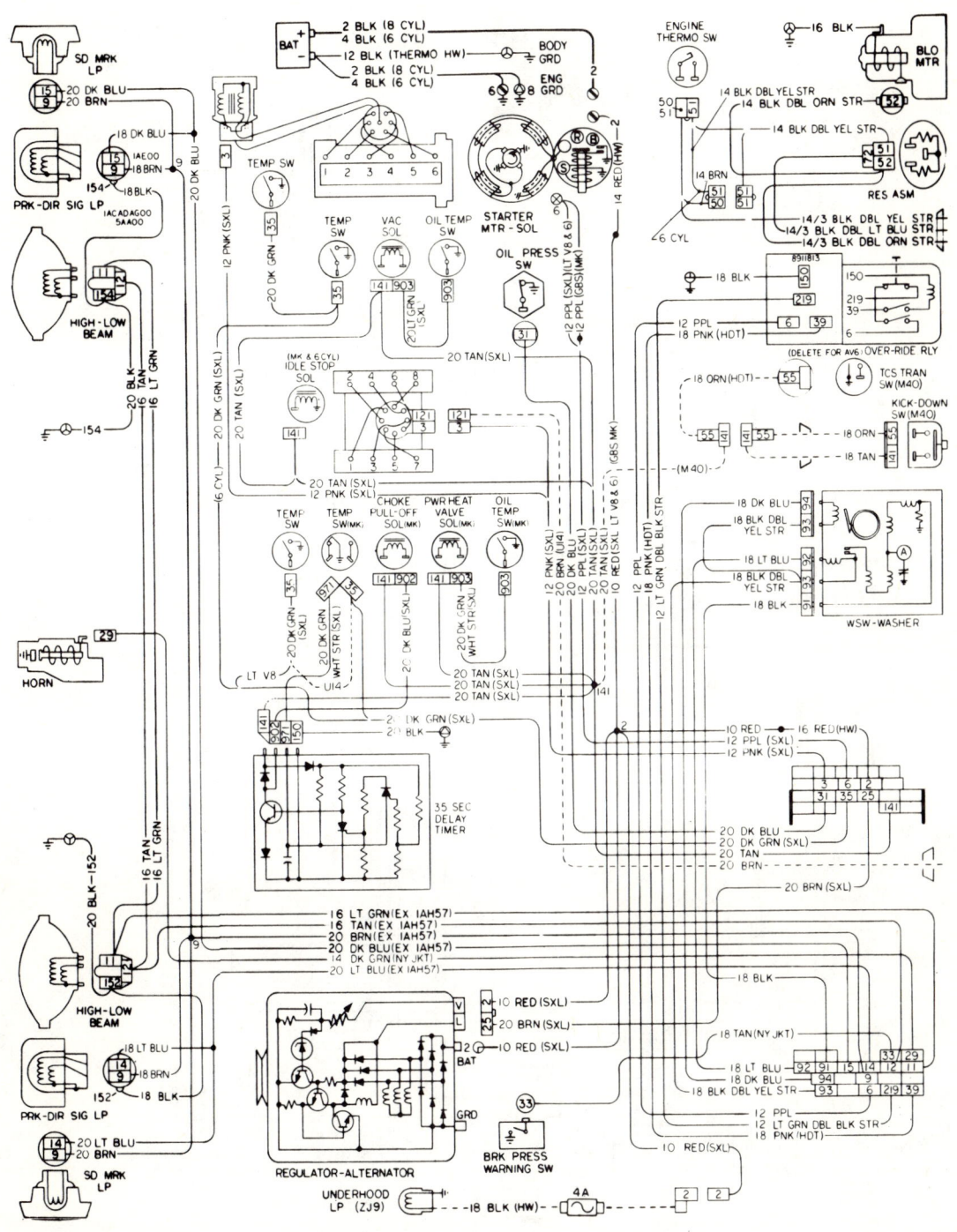

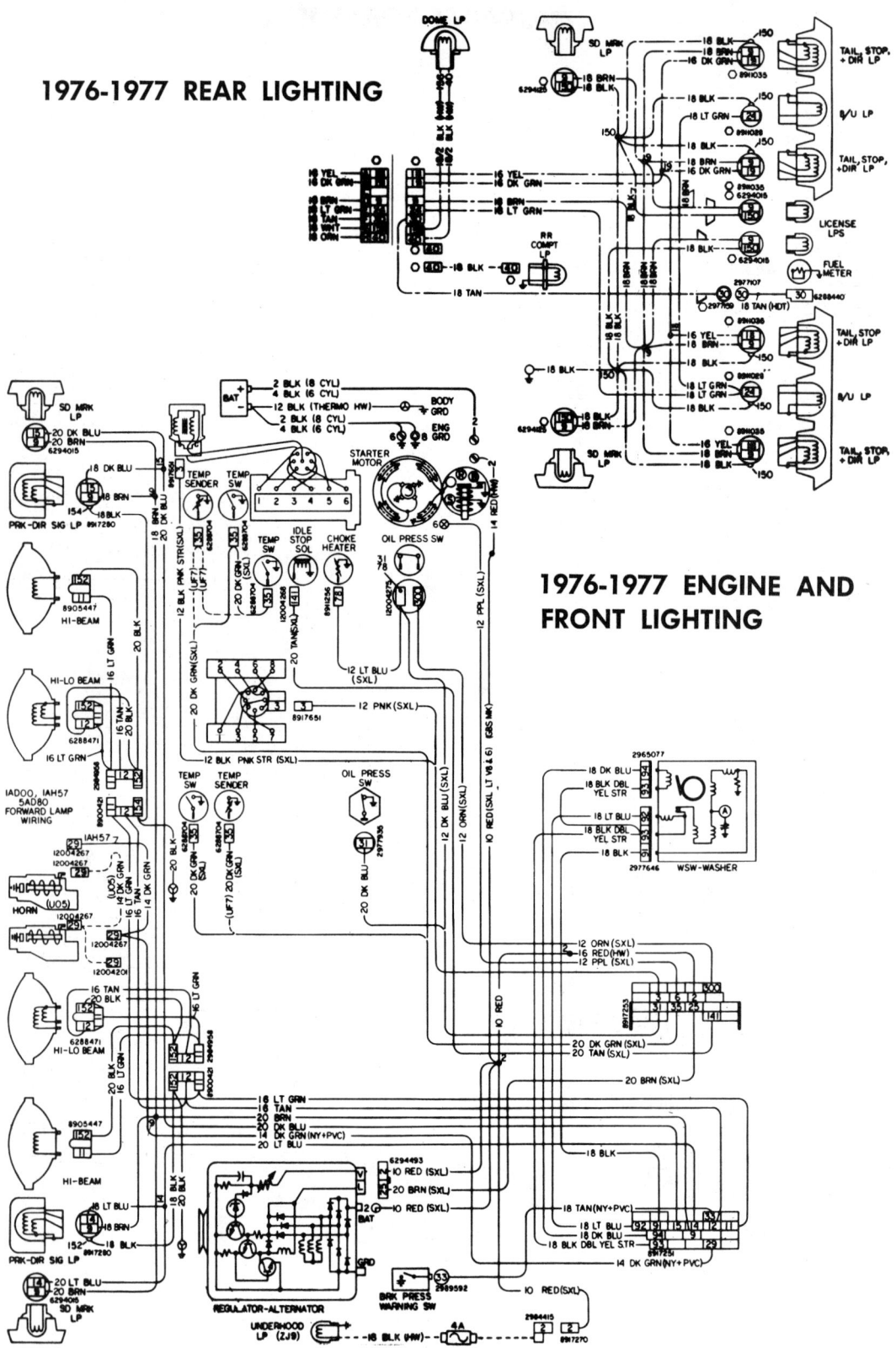

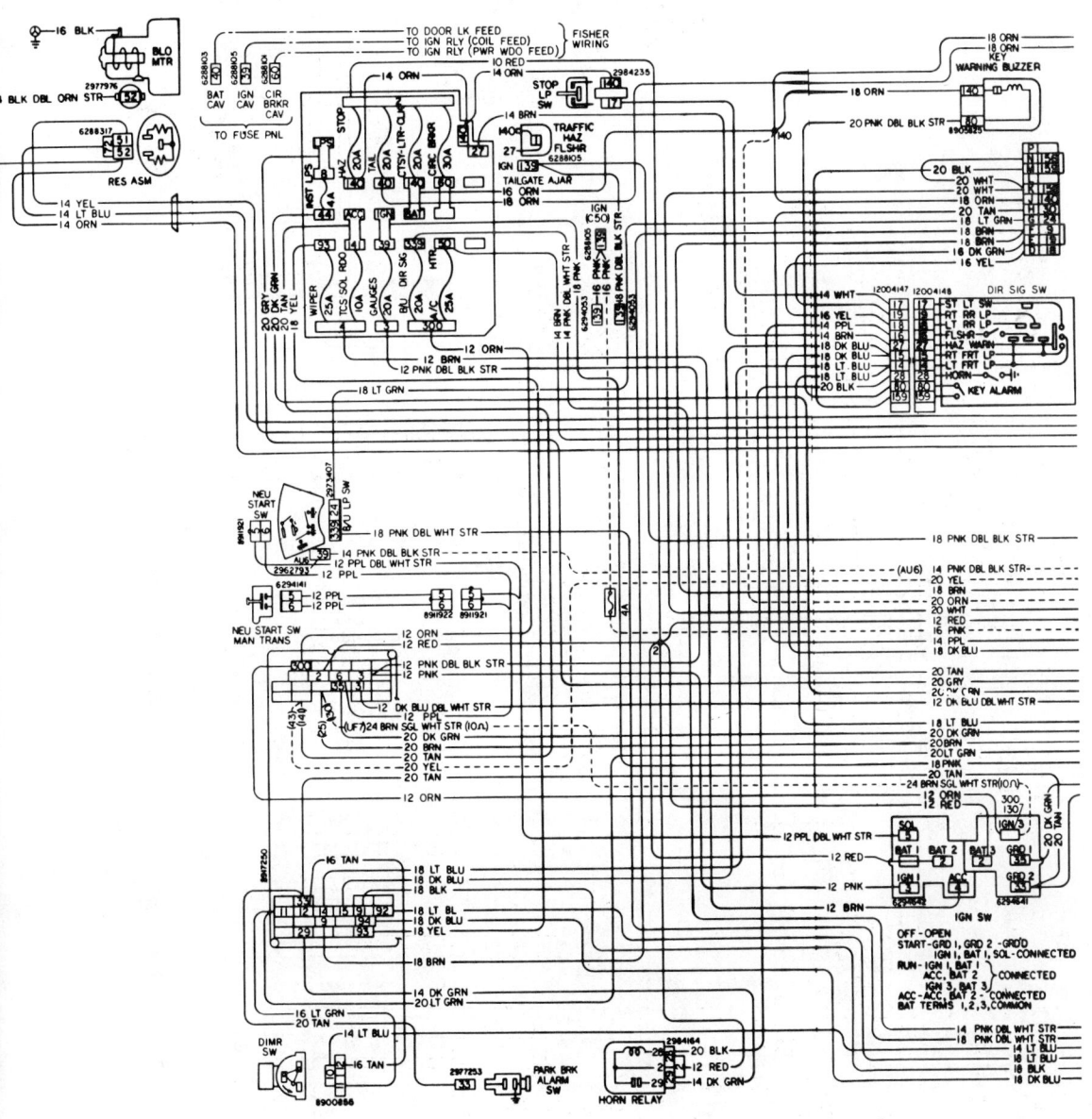

1976-1977 BODY WIRING

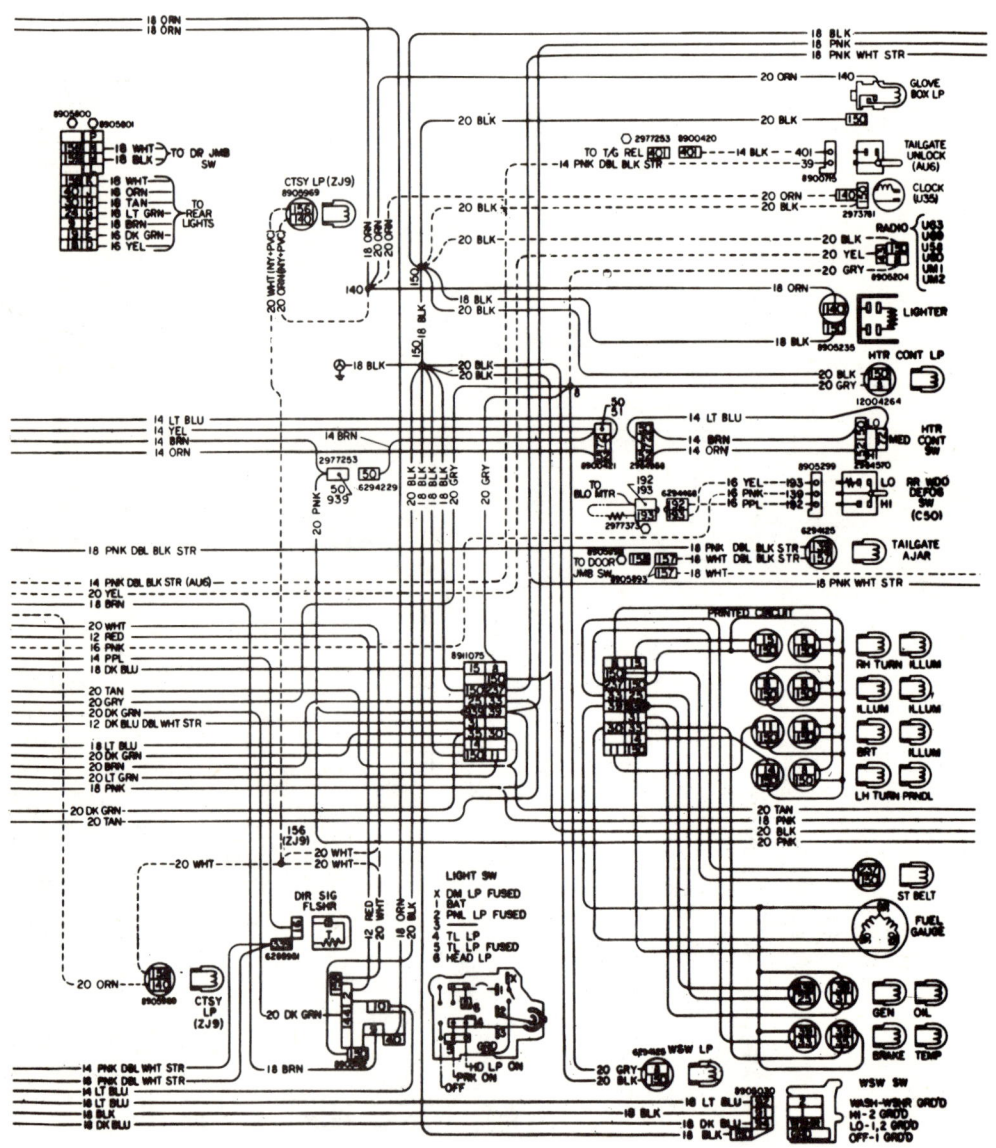

1976-1977 BODY WIRING

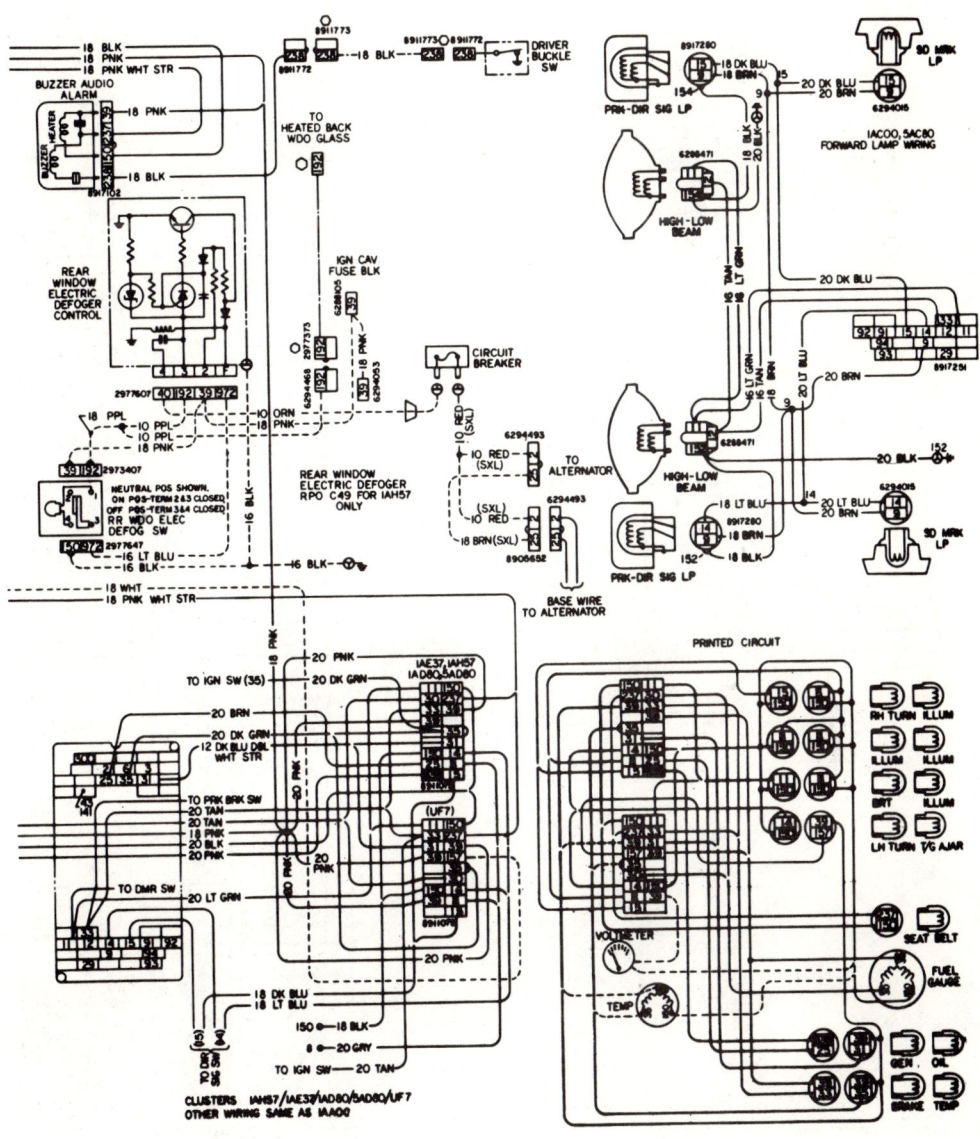

1978-1980 BODY WIRING

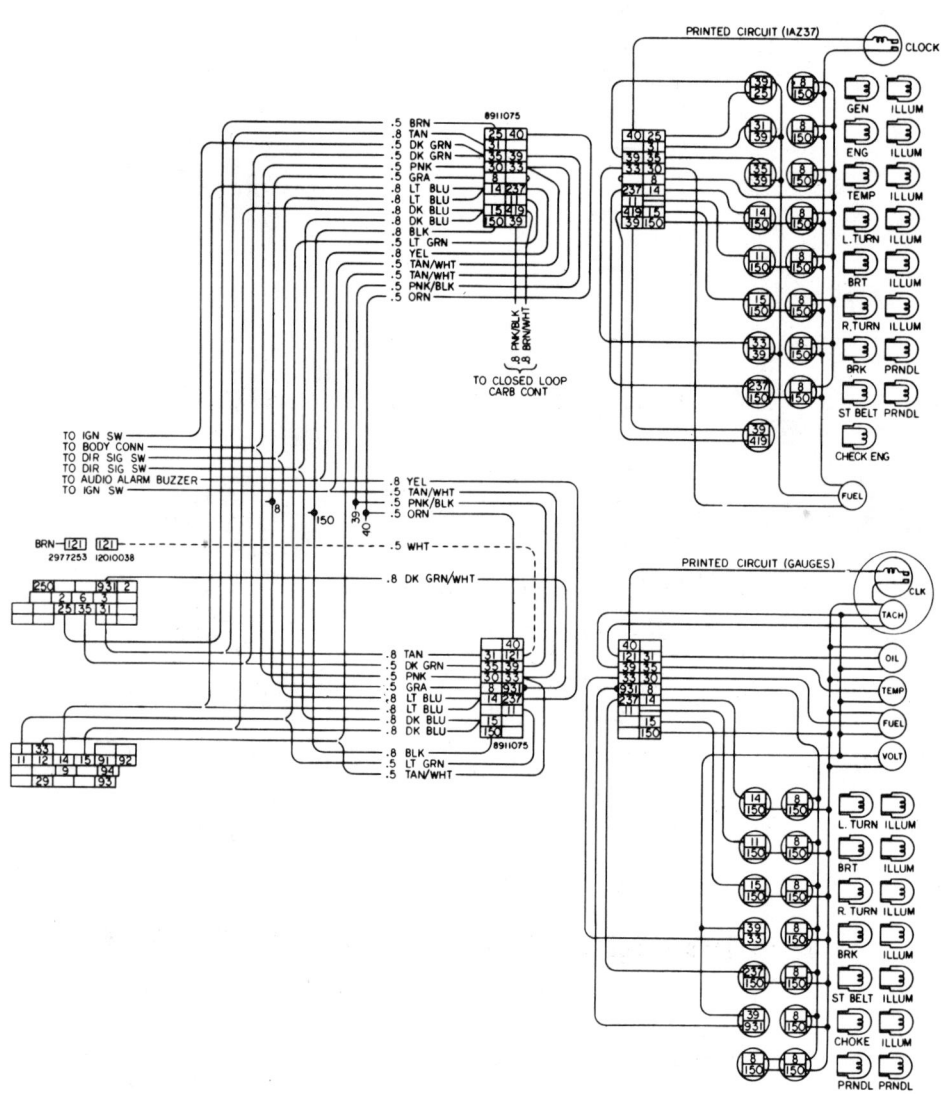

1978-1980 REAR COMPARTMENT LID RELEASE

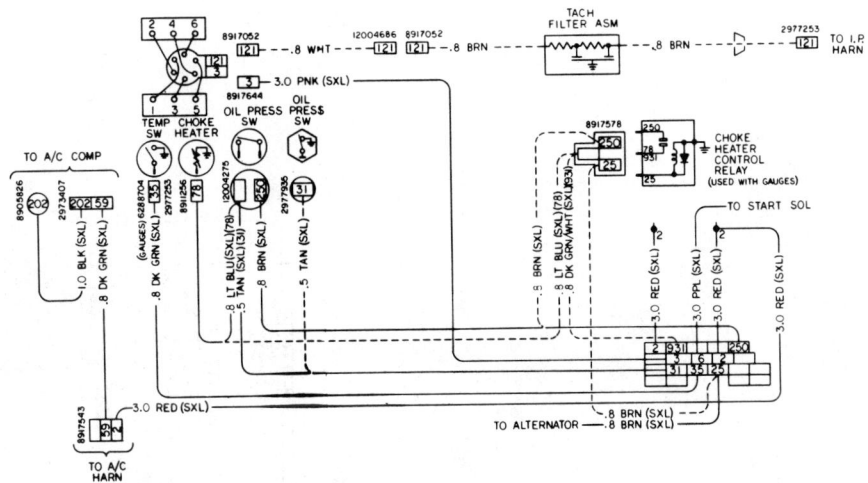

1978-1979 PULSE WIPER SYSTEM

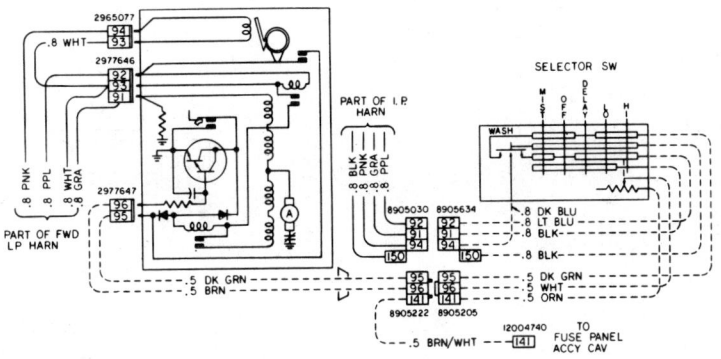

1978-1980 ELECTRIC REAR WINDOW DEFROSTER

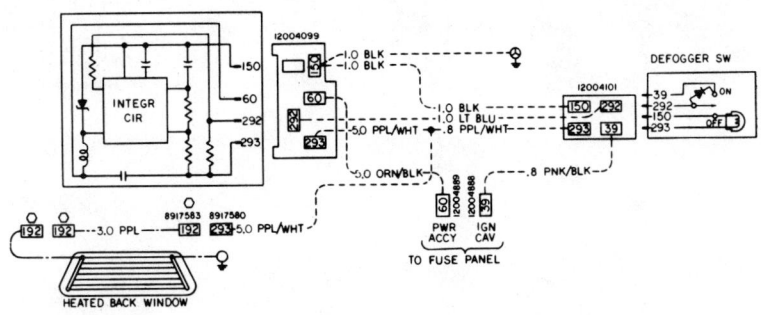

1978-1979 ENGINE AND FRONT LIGHTING

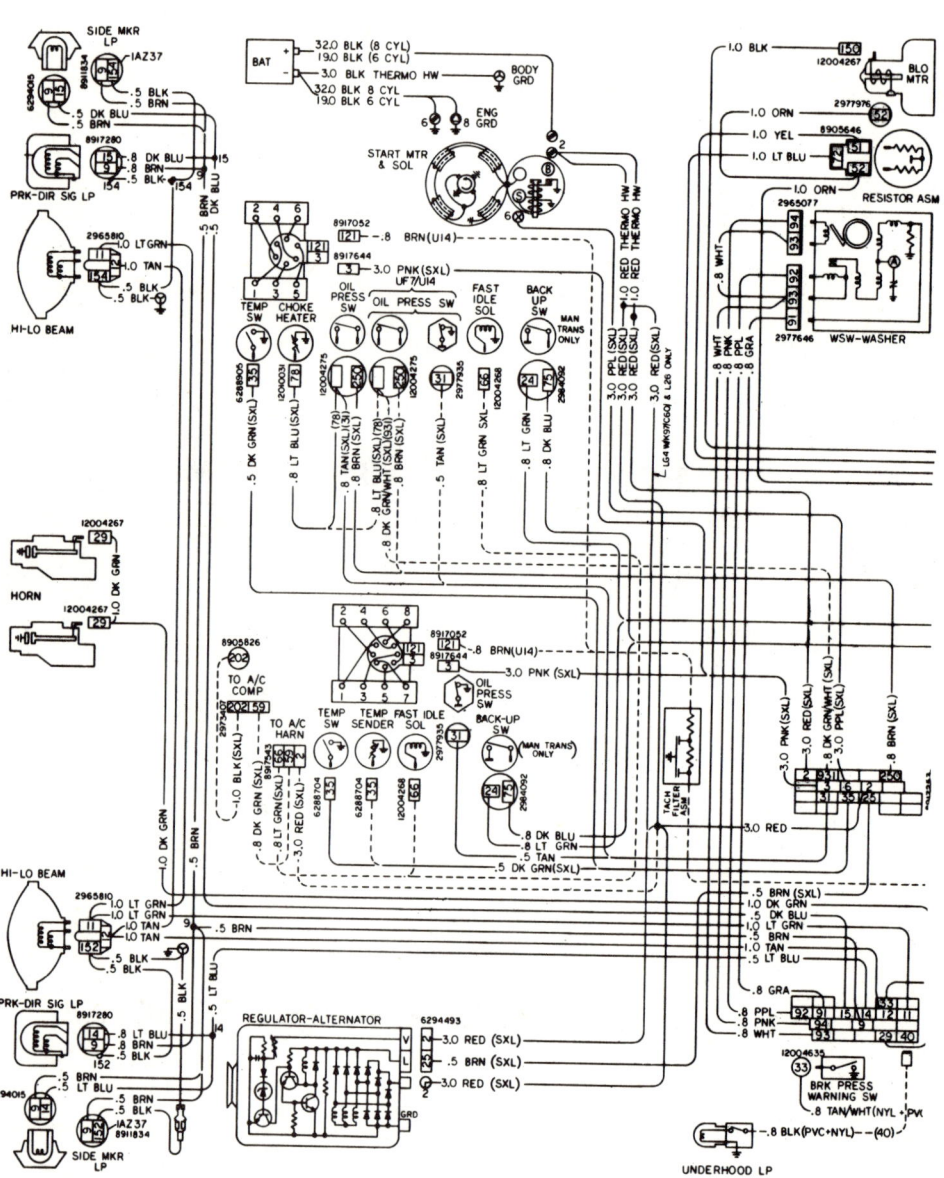

1978-1980 BODY WIRING

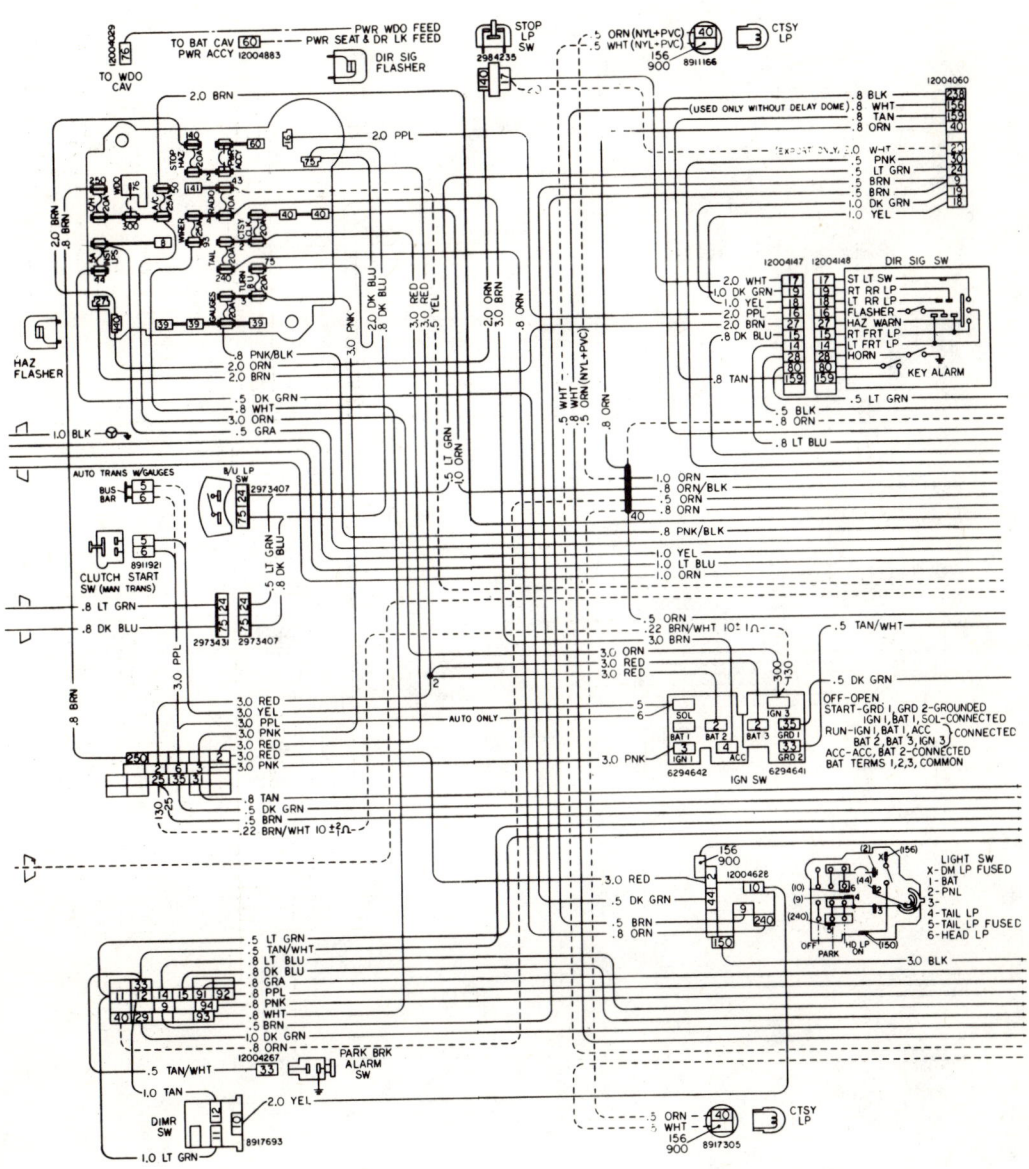

1978-1980 BODY WIRING

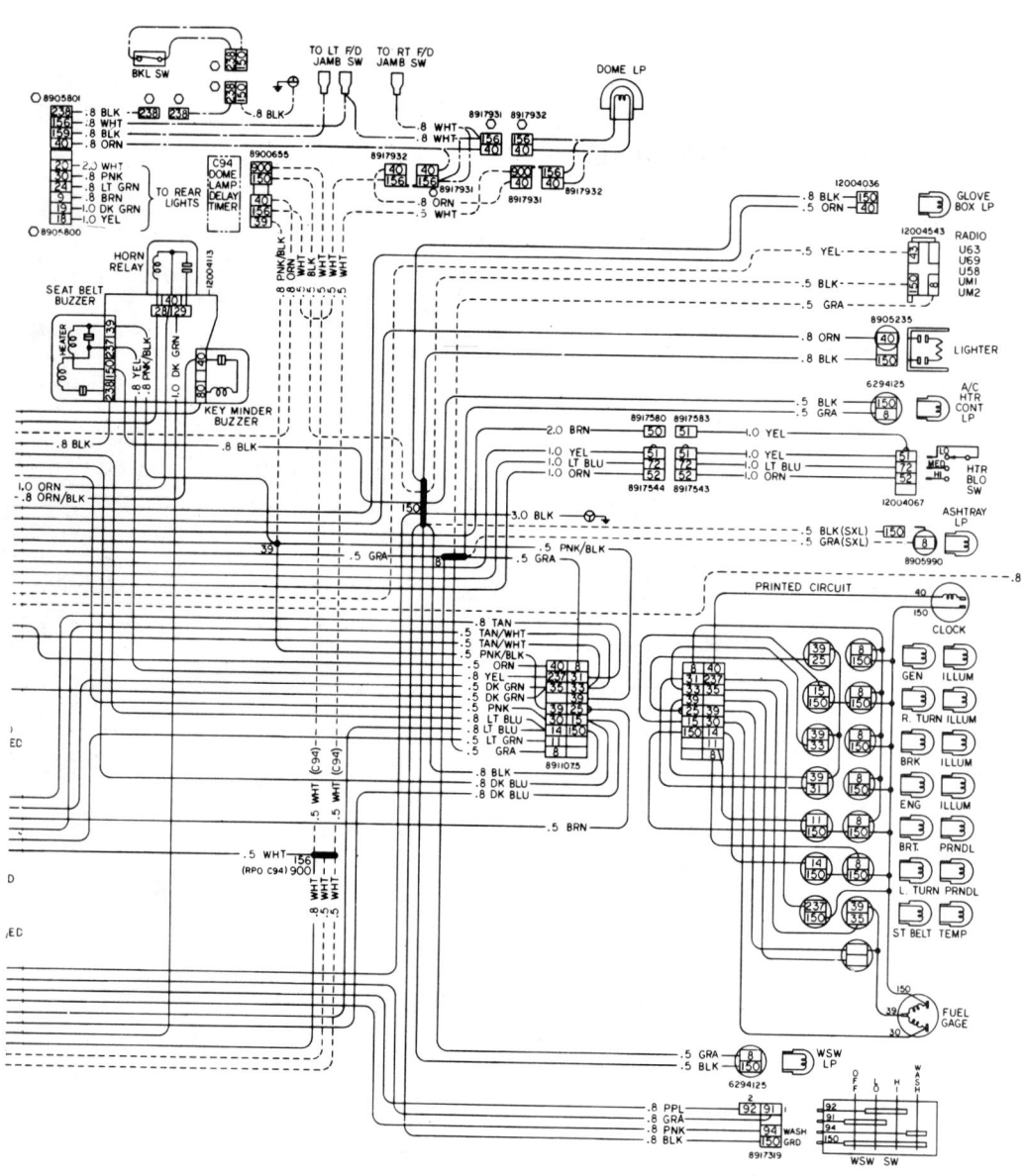

1979-1980 END GATE RELEASE

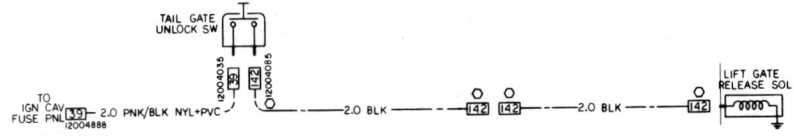

1978-1980 AIR CONDITIONING

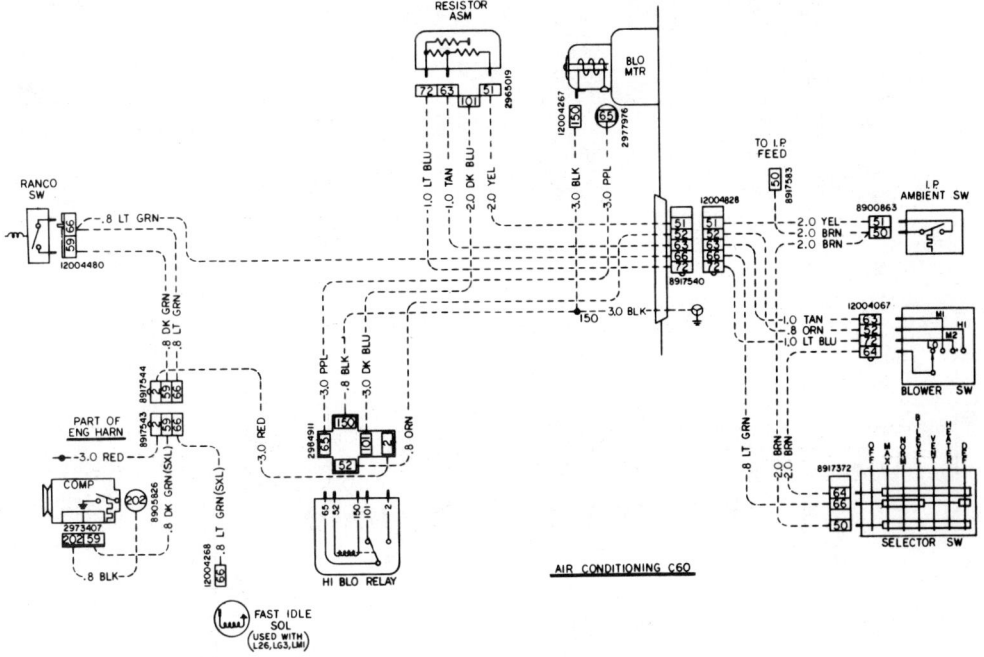

1978-1980 POWER ANTENNA

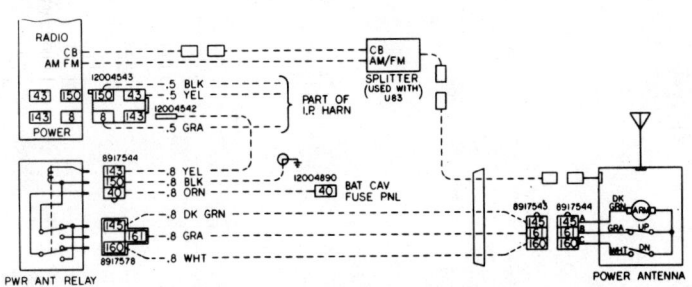

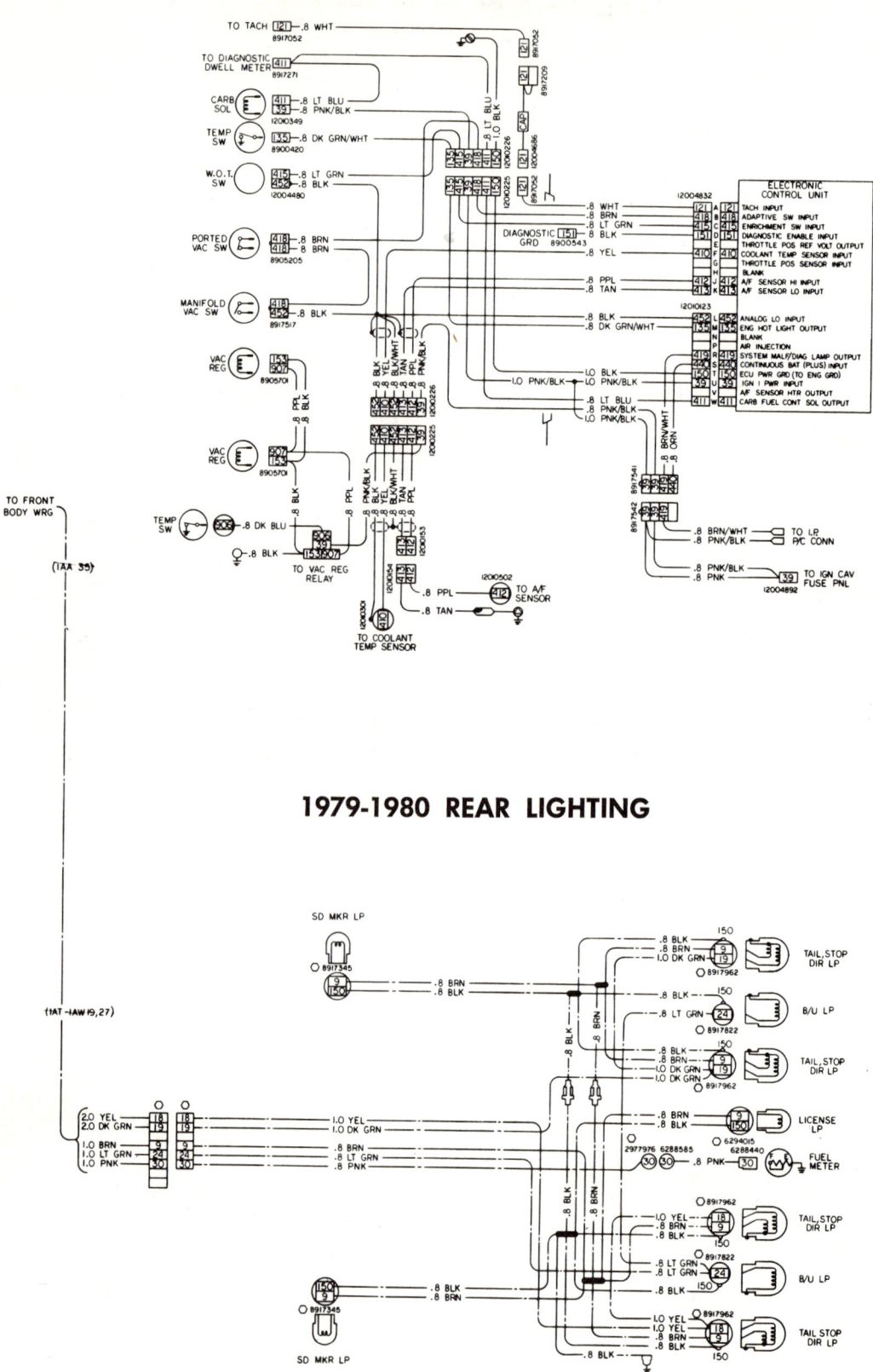

1979 CLOSED LOOP CARBURETOR CONTROL

1979-1980 REAR LIGHTING

1978-1980 CRUISE CONTROL

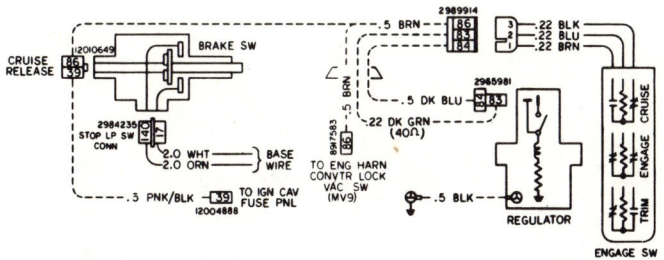

1980 CLOSED LOOP CARBURETOR CONTROL (V8)

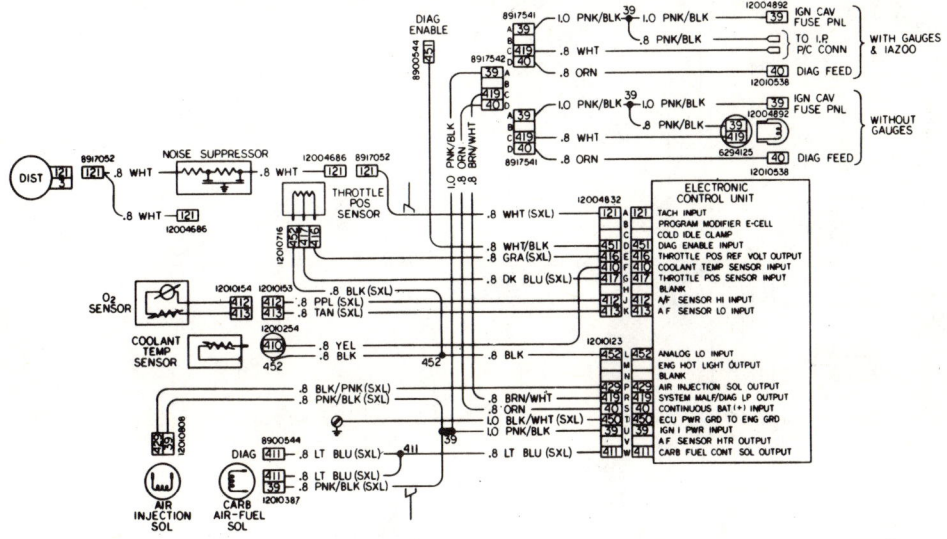

1980 PULSE WIPER SYSTEM

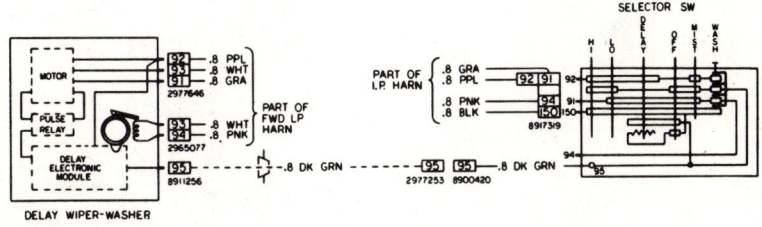

1980 ENGINE COMPARTMENT AND FRONT LIGHTING

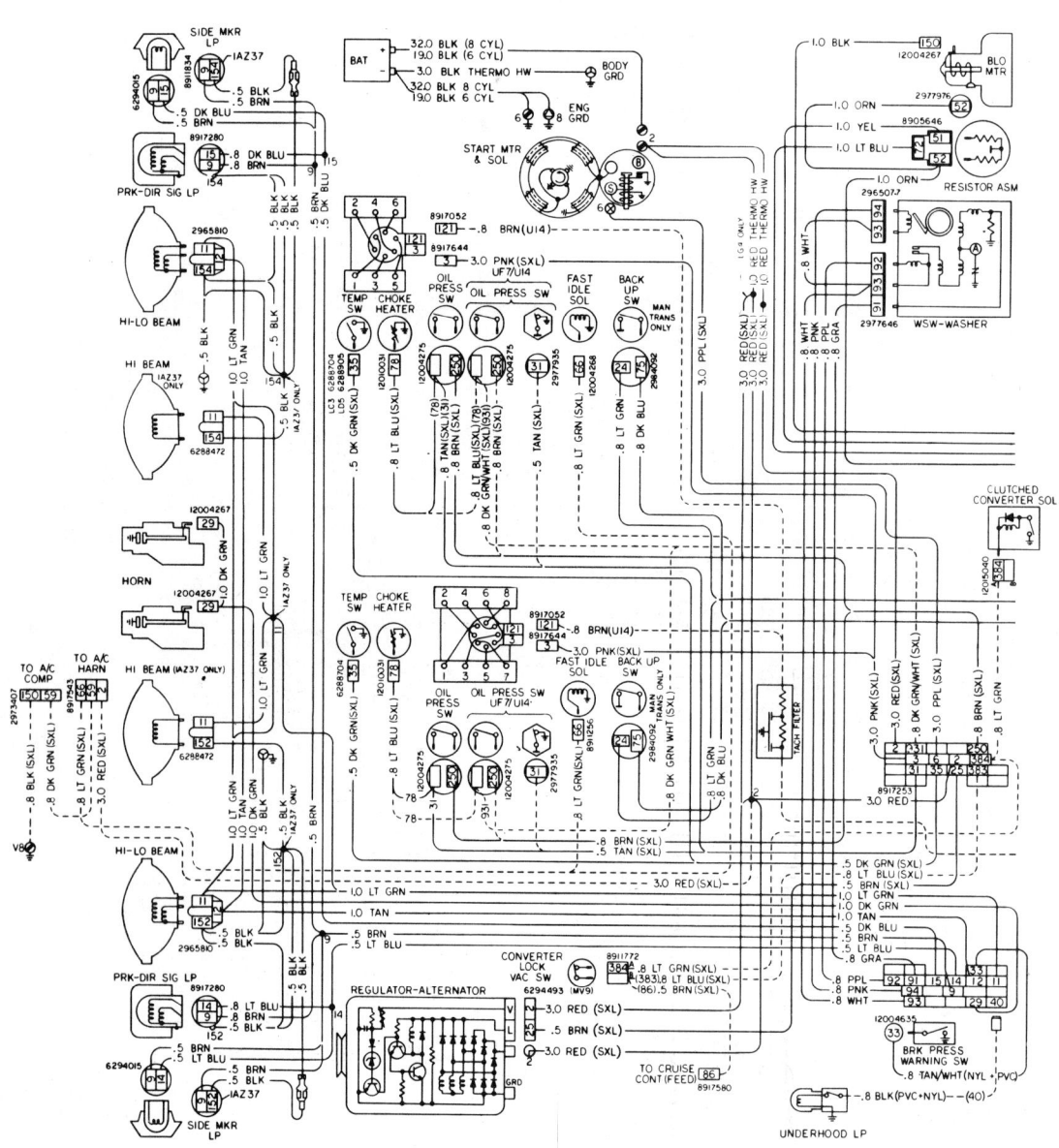

1980 CLOSED LOOP CARBURETOR CONTROL (V6)

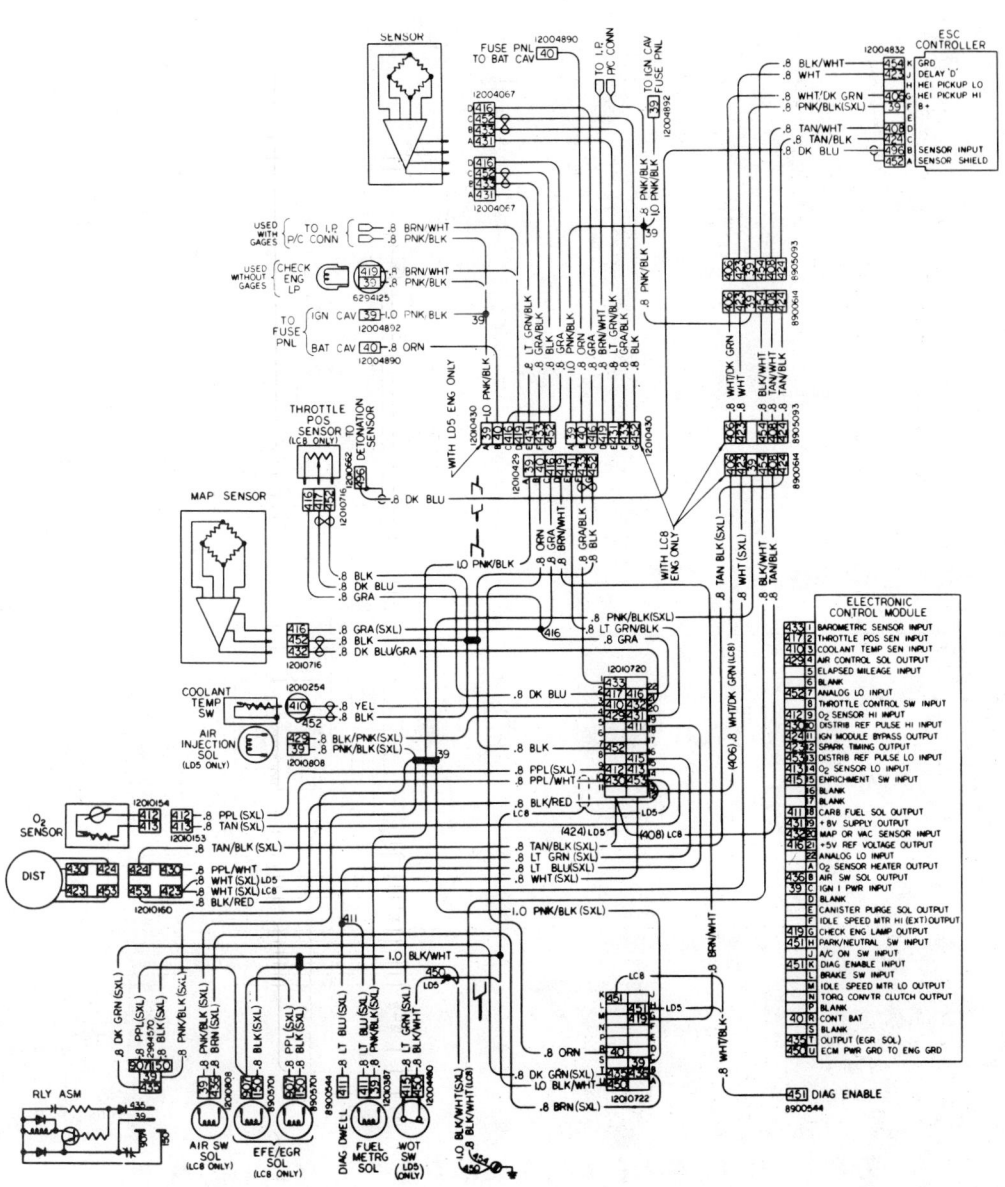

1980 ENGINE COMPARTMENT (V6 BUICK AND V6 TURBO)

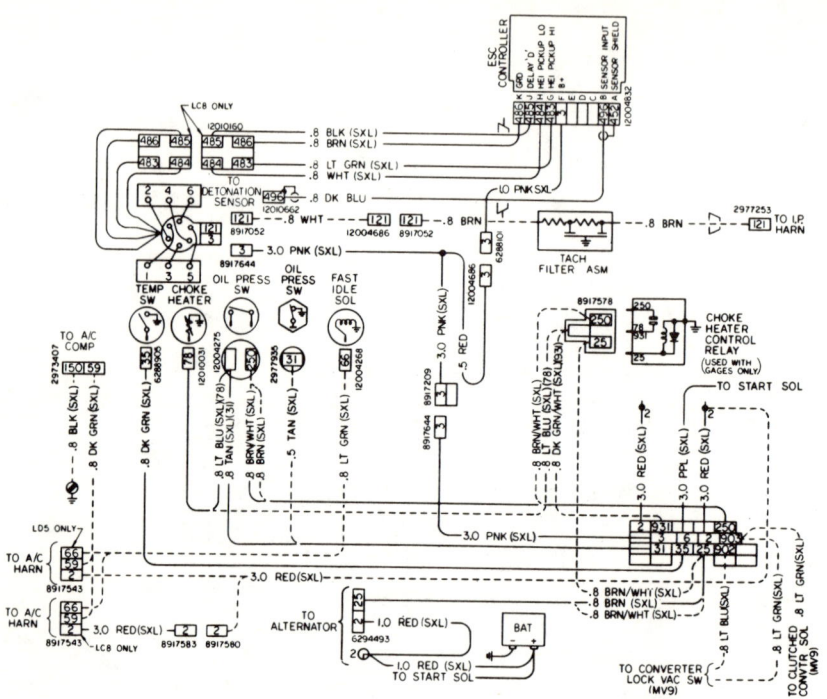

1978-1980 REAR WINDOW DEFROSTER

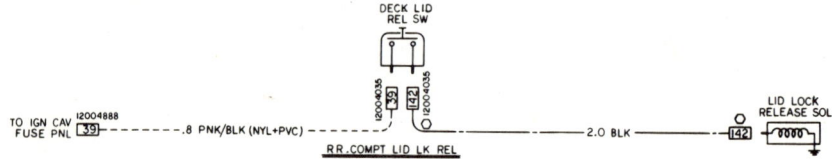

1978-1979 ENGINE COMPARTMENT (V6 BUICK)

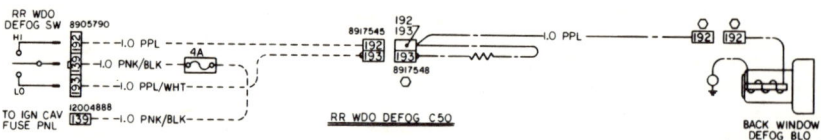